Hillslope Hydrology: Theory and Practice

Hillslope Hydrology: Theory and Practice

Edited by Zach Vaughn

MURPHY & MOORE
www.murphy-moorepublishing.com

Published by Murphy & Moore Publishing,
1 Rockefeller Plaza,
New York City, NY 10020, USA
www.murphy-moorepublishing.com

Hillslope Hydrology: Theory and Practice
Edited by Zach Vaughn

International Standard Book Number: 978-1-63987-303-6 (Hardback)

Cataloging-in-Publication Data

Hillslope hydrology : theory and practice / edited by Zach Vaughn.
p. cm.
Includes bibliographical references and index.
ISBN 978-1-63987-303-6
1. Hydrology. 2. Slopes (Physical geography). I. Vaughn, Zach.
GB665 .H55 2022
551.48--dc23

TABLE OF CONTENTS

PREFACE

Hydrology is the branch of science that is concerned with the movement and management of water on the surface of Earth, beneath the surface of Earth, as well as in the atmosphere. Hillslopes make up the flanks of valleys and the margins of eroding uplands. Hillslope hydrology is a scientific study to understand the hillslope runoff processes and the contribution of groundwater to storm runoff. Such hydrologic organization brings in distinct characteristics in soil and vegetation of hilltops, valleys and between sunny and shady slopes. The topographic gradients present on the hillslopes drive water, sediments and biogeochemical fluxes down the hillslopes. Hydrologic processes are traditionally studied at hillslope-to-catchment scales (tens of meters to kilometers). An important area of study in this field is the interplay of sediments, subsurface flow, soil erosion and overland flow. This book provides significant information of this discipline to help develop a good understanding of hillslope hydrology and its related fields. From theories to research to practical applications, case studies related to all contemporary topics of relevance to this field have been included herein. This book will help new researchers by foregrounding their knowledge in this branch.

This book is a result of research of several months to collate the most relevant data in the field.

When I was approached with the idea of this book and the proposal to edit it, I was overwhelmed. It gave me an opportunity to reach out to all those who share a common interest with me in this field. I had 3 main parameters for editing this text:

1. Accuracy - The data and information provided in this book should be up-to-date and valuable to the readers.

2. Structure - The data must be presented in a structured format for easy understanding and better grasping of the readers.

3. Universal Approach - This book not only targets students but also experts and innovators in the field, thus my aim was to present topics which are of use to all.

Thus, it took me a couple of months to finish the editing of this book.

I would like to make a special mention of my publisher who considered me worthy of this opportunity and also supported me throughout the editing process. I would also like to thank the editing team at the back-end who extended their help whenever required.

Editor

1

Dominant controls of transpiration along a hillslope transect inferred from ecohydrological measurements and thermodynamic limits

Maik Renner[1], Sibylle K. Hassler[2,6], Theresa Blume[2], Markus Weiler[3], Anke Hildebrandt[4,1], Marcus Guderle[4,1,7], Stanislaus J. Schymanski[5], and Axel Kleidon[1]

[1]Max-Planck-Institut für Biogeochemie, Jena, Germany
[2]GFZ German Research Centre for Geosciences, Section Hydrology, Potsdam, Germany
[3]Universität Freiburg, Hydrologie, Freiburg, Germany
[4]Universität Jena, Ecological Modelling Group, Jena, Germany
[5]ETH Zürich, Department of Environmental Systems Science, Zurich, Switzerland
[6]Karlsruhe Institute of Technology, Institute of Water and River Basin Management, Karlsruhe, Germany
[7]Technische Universität München, Chair for Terrestrial Ecology, Department of Ecology and Ecosystemmanagement, Munich, Germany

Correspondence to: Maik Renner (mrenner@bgc-jena.mpg.de)

Abstract. We combine ecohydrological observations of sap flow and soil moisture with thermodynamically constrained estimates of atmospheric evaporative demand to infer the dominant controls of forest transpiration in complex terrain. We hypothesize that daily variations in transpiration are dominated by variations in atmospheric demand, while site-specific controls, including limiting soil moisture, act on longer timescales.

We test these hypotheses with data of a measurement setup consisting of five sites along a valley cross section in Luxembourg. Both hillslopes are covered by forest dominated by European beech (*Fagus sylvatica* L.). Two independent measurements are used to estimate stand transpiration: (i) sap flow and (ii) diurnal variations in soil moisture, which were used to estimate the daily root water uptake. Atmospheric evaporative demand is estimated through thermodynamically constrained evaporation, which only requires absorbed solar radiation and temperature as input data without any empirical parameters. Both transpiration estimates are strongly correlated to atmospheric demand at the daily timescale. We find that neither vapor pressure deficit nor wind speed add to the explained variance, supporting the idea that they are dependent variables on land–atmosphere exchange and the surface energy budget. Estimated stand transpiration was in a similar range at the north-facing and the south-facing hillslopes despite the different aspect and the largely different stand composition. We identified an inverse relationship between sap flux density and the site-average sapwood area per tree as estimated by the site forest inventories. This suggests that tree hydraulic adaptation can compensate for heterogeneous conditions. However, during dry summer periods differences in topographic factors and stand structure can cause spatially variable transpiration rates. We conclude that absorption of solar radiation at the surface forms a dominant control for turbulent heat and mass exchange and that vegetation across the hillslope adjusts to this constraint at the tree and stand level. These findings should help to improve the description of land-surface–atmosphere exchange at regional scales.

1 Introduction

Evapotranspiration E couples water and energy balances at the land surface and is constrained by both, the supply of water and the atmospheric demand for water. Total E is composed of evaporation from intercepted water on plants and the surface, soil evaporation, and the physiological process of plant transpiration (E_T) taking water from the soil, groundwater, and possibly bedrock (Shuttleworth, 1993; Miller et al., 2010; Schwinning, 2010). Transpiration is of key importance for the (local) climate by altering the surface energy balance (Oke, 1987), and for water resources where E_T is an important loss term of the water balance (Federer, 1973; Jasechko et al., 2013). The tight coupling to photosynthesis and thus primary productivity makes E_T also central for agriculture and forestry. There is thus a need to understand the spatial and temporal variation of E_T as a result of interacting climatic and biogeophysical processes. Especially in temperate forests these biogeophysical feedbacks are poorly understood, which strongly reduces our ability to assess their role in mitigating climate change (Bonan, 2008).

Forests are often found in complex topographical settings. Thus, we have to consider that the first-order physical controls on E are strongly altered by topography: hillslope angle and aspect systematically alter the amount of absorbed solar radiation and thus atmospheric demand (Baumgartner, 1960; Lee and Baumgartner, 1966; Ivanov et al., 2008), whereas slope and the topographic setting, such as contributing area, alter the lateral redistribution of water (Famiglietti et al., 1998; Bachmair and Weiler, 2011). Apparently, forests adjust to these conditions, but little is known how vegetation and site-scale transpiration in particular respond to these topographically altered boundary conditions of water supply and demand (Tromp-van Meerveld and McDonnell, 2006). For example Holst et al. (2010) found that detailed forest hydrological models yielded different trends of simulated mean stand transpiration when comparing European beech stands on different hillslope aspects. These differences add up when simulating runoff generation from these sites and emphasizes the difficulty in the assessment of forest water balance in complex terrain.

Although detailed models of various feedbacks between plant physiology and the environmental conditions are available (Monteith, 1965; Farquhar et al., 1980; Wang and Jarvis, 1990; Whitehead, 1998; Haas et al., 2013) their applicability is generally restricted by the need for detailed physiologic, atmospheric, and soil parameters. As an alternative to improve our understanding by increasingly detailed modeling, several authors proposed to deduce and predict E through physical and physiological constraints (Calder, 1998; West et al., 1999; Raupach, 2001; Zhang et al., 2008; Wang and Bras, 2011). Applications of these fundamental constraints do not only increase our understanding of the soil–plant atmosphere continuum, they may also lead to a few but independent predictors of E. For example, Kleidon and Renner (2013b) applied thermodynamic limits of convective heat exchange to the surface–atmosphere system. They found that under the assumption of maximum convective power and no surface water limitation, atmospheric demand is only dependent on absorbed solar radiation and surface temperature, being consistent with the empirical formulations of potential evaporation of Makkink (1957) and the well-known equilibrium evaporation concept (Schmidt, 1915; Priestley and Taylor, 1972; de Bruin et al., 2016). Contrary to the classic formulation of Dalton (see Brutsaert (1982) for an overview), where vapor pressure deficit (VPD) and wind speed are used as forcing variables of potential evapotranspiration, Kleidon and Renner (2013b) argued that VPD and wind speed emerge from the land–atmosphere interaction. The practical implication of maximum convective power is that atmospheric demand can be estimated without these variables and empirical parameters.

In this study we aim to assess the dominant temporal and spatial controls of forest transpiration in complex terrain. Therefore, we use the parsimonious approach of maximum convective power to estimate potential evaporation. The approach of maximum convective power has so far only been applied for long-term annual estimates of E and large-scale geographic variations (Kleidon et al., 2014). Here we test this approach at the site level and for daily timescales in complex topographical terrain. We use data of a well-instrumented beech forest with measurement sites across a transect with a north-facing and a south-facing hillslope. The indirect assessment of E_T is based on ecohydrological measurements of sap flow and soil moisture. We present a systematic analysis to address the question of how much transpiration varies in complex terrain and what measurable site characteristics influence forest transpiration. In particular we hypothesize that potential evaporation representing the atmospheric demand for water can be estimated by thermodynamic limits of convection, which only relies on surface absorption of solar radiation and temperature. Specifically, we test how much of the daily variations of in situ transpiration observations can be explained by atmospheric demand and by how much they respond to atmospheric demand along the measurement transect. Therefore, we perform a linear regression and correlation analysis of the transpiration estimates to atmospheric demand comprising daily data of the growing season. Further, we evaluate how much additional variance can be explained by other time-varying parameters such as VPD, wind speed, and soil moisture. Differences in the strength of correlation between sites would imply that site-dependent, time-varying constraints on transpiration are relevant. In contrast, a high, consistent correlation would imply that atmospheric demand and thus energy limitation of transpiration is the dominant driver of day to day variability. The overall response of transpiration to atmospheric demand represented by the slope of the linear regression would then indicate the importance of site-dependent controls such as topographic, soil, and plant factors.

2 Methods

2.1 Site description

We analyze measurements at six different sites along a well-instrumented steep forested hillslope transect (north-facing vs. south-facing; see Fig. 1) in the Attert catchment in Luxembourg over the vegetation period of 2013. The hillslope transect is part of the Catchments As Organized Systems (CAOS) field observatory (Zehe et al., 2014) and is located in the western part of Luxembourg (5° E 48′13″, 49° N 49′34″) at about 460 m NN. The land cover of the transect is a mixed forest dominated by European beech (*Fagus sylvatica* L.). The north-facing slope has an inclination of ≈ 15° and is composed of a few dominant trees with large gaps and dense understorey mainly of young beech trees, whereas the south-facing slope is generally steeper (≈ 22°) and has no understorey and a denser canopy. Also tree species composition varies between slopes, with 97 % beech on the north-facing sites with single spruce trees and 90 % beech and 10 % oak on the south-facing sites. The valley site has 80 % beech with 10 % spruce and alder. Geologically, the site is situated in northeast–southwest-trending fold system of Schists of the Ardennes Massif. The shallow soils developed on periglacial slope deposits (Juilleret et al., 2011). The deposits are generally found at a depth of 70–90 cm.

Standard meteorological data, global radiation, air temperature, relative humidity, and wind speed were measured 2 m above ground at all sites. For the meteorological forcing we used the data from the open grassland site G1, which is located 240 m to the northwest of the forest site N1; see Fig. 1. Absorbed solar radiation $R_{sn} = (1-\alpha)R_g$ in W m^{-2} is derived from global radiation R_g and an albedo estimate of $\alpha = 0.15$, which is representative for deciduous forests (Oke, 1987).

2.2 Estimation of atmospheric demand

Our aim is to estimate the potential evaporation from first principles and with few, independent input data. Therefore we make use of the concept of thermodynamic limits of convection, which was recently established by Kleidon and Renner (2013b) and used successfully to estimate the sensitivity of the hydrologic cycle to global warming (Kleidon and Renner, 2013a; Kleidon et al., 2015) and for grid-based global-scale predictions of annual average terrestrial evaporation (Kleidon et al., 2014). Here we only illustrate how the concept is used to estimate potential evaporation; for further details the reader is referred to the mentioned publications.

Convection can be thought of as a heat engine, which converts a temperature gradient into kinetic energy (Ozawa et al., 2003). To capture the fundamental trade-off of thermodynamic limits of convective exchange, we consider a simple two-box surface–atmosphere system in steady state, which is sketched in Fig. 2. We consider the steady-state energy balance of the surface $R_{sn} = J + R_{ln}$. The surface is heated by absorption of incoming solar radiation R_{sn}. The turbulent heat fluxes J and the net longwave exchange R_{ln} both cool the surface. The turbulent heat fluxes are composed of the sensible (H) and latent heat flux (λE). Longwave radiative exchange is represented by a simplified linearized radiation $R_{ln} = k_r(T_s - T_a)$, with T_s being the temperature of the surface and T_a the temperature of the atmosphere and k_r being a constant radiative exchange coefficient. The power of the convective heat engine G is fundamentally limited by the Carnot limit:

$$G = J\frac{T_s - T_a}{T_s}. \quad (1)$$

Different from the classic Carnot engines, the convective heat engine has flexible boundary conditions, namely, the temperature gradient $T_s - T_a$ responds to the strength of the convective heat fluxes J through its coupling to the surface energy balance. Rearranging the surface energy balance

$$T_s - T_a = \frac{R_{sn} - J}{k_r} \quad (2)$$

shows that a stronger convective heat flux at a given radiation reduces the temperature gradient. This effectively yields a thermodynamic limit of the maximal power of the convective heat engine. The limit is obtained by inserting this feedback into the Carnot limit Eq. (1) and solving for the maximum with respect to J. Hypothesizing that convection actually operates at this limit (subscript opt), we obtain a closure for the strength of convective fluxes:

$$J_{opt} = \frac{R_{sn}}{2}. \quad (3)$$

Once the convective fluxes are obtained we aim to estimate the partitioning into sensible and latent heat fluxes. Using the Bowen ratio B and bulk formulations for sensible (H) and latent heat fluxes (λE), we can write

$$B = \frac{H}{\lambda E} = \frac{\rho c_p k_h}{\rho \lambda k_e}\frac{T_s - T_a}{q_s - q_a} = \frac{c_p}{\lambda}\frac{T_s - T_a}{q_s - q_a} \quad (4)$$

where $c_p = 1.2\,\mathrm{J\,K^{-1}\,kg^{-1}}$ is the heat capacity of air, $\rho = 1\,\mathrm{kg\,m^{-3}}$ the air density, $\lambda = 2.5 \times 10^6\,\mathrm{J\,kg^{-1}}$ the latent heat of vaporization, q_s and q_a being the specific humidity (g kg^{-1}) at the surface and the atmosphere, respectively. If we assume that the air at the surface and in the atmosphere is saturated with water vapor and that the slope of the saturation vapor pressure curve ($s = \partial e_{sat} / \partial T$) is invariant within this range, the vertical humidity gradient can be written as $q_s - q_a = s(T_s - T_a) \cdot 0.622 / p_s$. e_{sat} is the temperature-dependent vapor pressure at saturation, p_s the surface pressure (hPa) and 0.622 being the ratio of the gas constants for air and water vapor. Introducing the psychrometric constant with $\gamma = c_p / \lambda \cdot p_s / 0.622 \approx 65\,\mathrm{Pa\,K^{-1}}$ and assuming that the

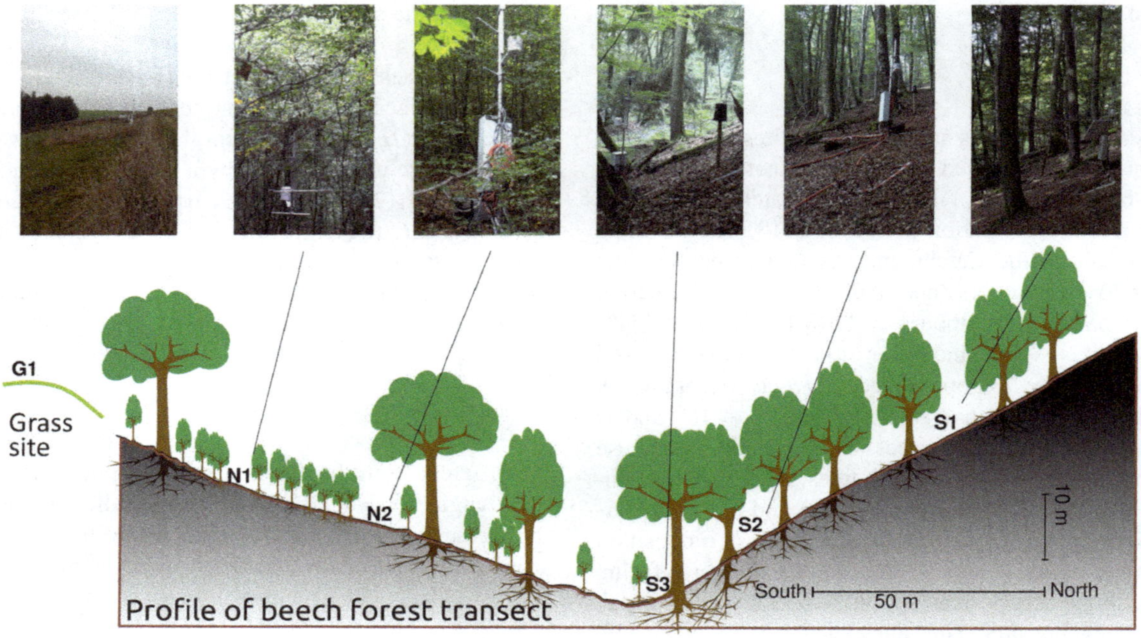

Figure 1. Measurement setup along a hillslope transect. Captital letters indicate aspect of the sites (N: north-facing, S: south-facing) and the numbering represents the position of the site on the hillslope: upslope (1), midslope (2), and downslope (3).

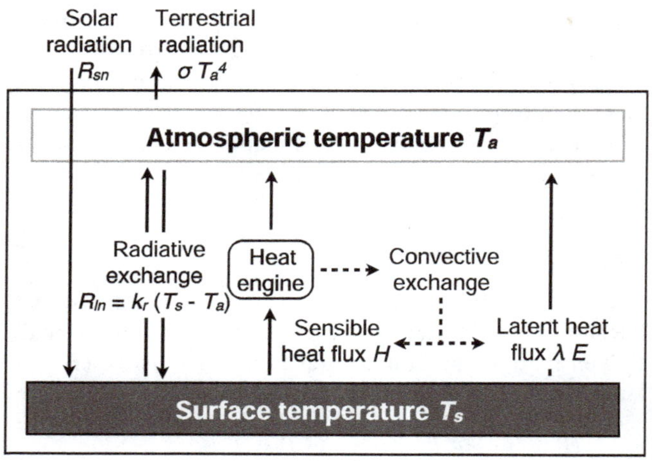

Figure 2. Land–atmosphere energy balance scheme for derivation of atmospheric demand adapted after Kleidon et al. (2014).

vertical exchange coefficients of air k_h and of vapor k_e are equal, we thus obtain the equilibrium Bowen ratio $B = \frac{\gamma}{s}$ (Schmidt, 1915; Stull, 1988). The equilibrium Bowen ratio depends only on temperature, because s is a nonlinear function of temperature and γ is approximately constant. We determine s by an empirical equation (Bohren and Albrecht, 1998): $s = s(T) = 6.11 \cdot 5417 \cdot T^{-2} \cdot e^{19.83 - 5417/T}$ with temperature T (K).

Combining the equilibrium Bowen ratio with the concept of maximum convective power we obtain an expression for the potential evaporation, herein referred to as atmospheric demand E_{opt} (Kleidon and Renner, 2013b; Kleidon et al., 2014):

$$E_{opt} = \frac{1}{\lambda} \frac{s}{s+\gamma} \frac{R_{sn}}{2}. \tag{5}$$

Hence, only absorbed solar radiation R_{sn} and temperature data are required to estimate atmospheric evaporative demand, when derived from thermodynamic limits using simplifying assumptions on longwave radiative exchange and the assumption for the equilibrium Bowen ratio.

We also compare our results with the well-established FAO Penman–Monteith grass reference evaporation equation (Allen et al., 1998):

$$E_{PM} = \frac{0.408\, s (R_n - G) + \gamma \frac{900}{T+273} u \mathrm{VPD}}{s + \gamma (1 + 0.34 u)}, \tag{6}$$

where R_n is net radiation, G the ground heat flux, T air temperature, u wind speed (both in 2 m height), and VPD is the saturation vapor pressure deficit in kilopascal (kPa). Equation (6) is a modified Penman–Monteith equation for a standardized grass surface at a daily resolution. As input variables we use the same variables as for E_{opt} (daily average air temperature, shortwave radiation) and in addition net radiation, ground heat flux, relative humidity, and wind speed. Because net radiation was not measured at our sites we used an empirical formulation for R_n (Allen et al., 1998, Eq. (39)). Therefore, data of daily minimum and maximum temperature as well as information on the day of year and latitude were required. The soil heat flux was estimated as $G = 0.1 \cdot R_n$.

2.3 Topographic effects on shortwave radiation

In complex terrain the incoming radiation is influenced by the slope and the aspect of the current position (Kondratyev and Fedorova, 1977). To account for these effects we use a topographic radiation correction method, which projects the extraterrestrial irradiance on inclined surfaces implemented in the software *r.sun* (Šúri and Hofierka, 2004), which is part of the open-source Geographic Information System (GIS) platform GRASS GIS (http://grass.osgeo.org). *r.sun* estimates potential clear-sky spatial global radiation $R_{g,pot}$ fields for each day of year. To account for cloudy conditions we use the open-field site G1 as a reference station and estimate a daily cloud factor $f_c = R_{g,G1} / R_{g,pot}$ using the closest grid point of the respective site. Then $R_{g,pot}$ is multiplied with f_c to obtain a radiation estimate for the sites in the forest $R_{g,c}$. As input for *r.sun* we use a 10 m resolution Digital Elevation Model (DEM) (with slope and aspect inputs from the DEM) and default parameters (Linke atmospheric turbidity coefficient = 2). We thus estimate global radiation on a tilted surface (aspect, slope) and also include topographic shading effects. The effect is largest when the sun angle is low (autumn–winter–spring) and decreases R_g and thus E_{opt} at the north-facing slope while increasing it on the south-facing slope; see Fig. S1 in the Supplement. These estimates are generally consistent with radiative observations in sloped terrain (Holst et al., 2005).

2.4 Biometric measurements

A forest inventory for all sites was done in March 2012. The circumference at breast height of all trees with circumference larger or equal to 4 cm was measured in a 20 m by 20 m plot for each site. Stem basal area was calculated and a total stand basal area was computed for each site.

Leaf area index (LAI) was measured with a LICOR LAI-2200 plant canopy analyzer at all forested sites in two campaigns. The summer campaign was carried out on 11 August 2012 and the winter campaign on 20 March 2014. Here we use the measurements taken with all rings of the LAI-2200. The LAI was averaged from 36 measurements points per site. The difference between summer and winter LAI should reveal the actual leaf area index without stems and topographic shading effects.

2.5 Sap flow measurements and upscaling stand transpiration

The five forested sites are instrumented with multiple sap flow sensors (four trees at each site, installed between mid-May to November). Heat pulse sensors (East 30 Sensors, Pullman, Washington 99163 USA) based on the heat ratio method (Burgess et al., 2001) have been used as they are less susceptible to natural heating gradients, and require less electrical power (Vandegehuchte and Steppe, 2013). The sensors measure the heat pulse velocity, which is corrected for wounding and sensor alignment; see Sect. A1 in the appendix. The heat pulse velocity is then converted into sap flux density, SFD ($cm^3\,cm^{-2}\,h^{-1}$) (Burgess et al., 2001):

$$\mathrm{SFD} = v_c \frac{\rho_b (c_w + m_c c_s)}{\rho_s c_s}. \quad (7)$$

Thus, to estimate the sap flux density, knowledge of xylem wood properties, namely, the basic density of wood ρ_b, the density of the sap ρ_s, the specific heat capacity of the wood matrix c_w, and of the sap c_s as well as the water content of the xylem is required. For sap we use the standard parameters for water at 20° C, with $\rho_s = 1\,\mathrm{kg\,cm^3}$, $c_s = 4182\,\mathrm{J\,kg^{-1}\,K^{-1}}$. The heat capacity of the woody matrix is generally given with $c_w = 1200\,\mathrm{J\,kg^{-1}\,K^{-1}}$ (Burgess et al., 2001). The basic density of sapwood ρ_b measured as dry weight over green volume and the moisture content of the xylem m_c are species-specific parameters. We used $\rho_b = 0.61\,\mathrm{kg\,cm^{-3}}$ and $m_c = 0.7$ for the xylem of European beech estimated by Glavac et al. (1990). They sampled about 260 trees at two different sites between 35 and 42 years old over the course of 1 year. For the thermal diffusivity of the xylem, we used a fixed value of $k_w = 2.5 \times 10^{-3}\,\mathrm{cm^2\,s^{-1}}$ (Burgess et al., 2001).

The raw measurements obtained every 30 min were quality controlled and suspect data were filtered before further analysis was performed. At three larger trees the outermost sensor readings were replaced by the second sensor depth reading because their annual mean was smaller than the inner sensors, which indicates a sensor misplacement into the bark of the tree. The sensors measure the heat pulse velocities at three different radial depths in the tree. The daily mean sap flux density per tree is obtained by an average of all readings per depth and day. Units of SFD were converted to $m^3\,m^{-2}\,d^{-1}$. An arithmetic mean of the tree-average sap flux density was used to obtain the site-average sap flux density.

Measurements of sap flux density only provide a relative measure of the velocity of the ascending xylem sap. To obtain tree water fluxes a representative xylem area per sensor depth is assumed, which is then multiplied with the corresponding sap flux density (Burgess et al., 2001). Tall trees with a diffuse-porous xylem structure such as beech are known to have a large sapwood to basal area ratio (Köstner et al., 1998). Further, the radial profile of the sap flux density varies in these trees. Therefore, deep measurements are ideally required (Gebauer et al., 2008). In the absence of these deeper measurements, we follow the reasoning of Lüttschwager and Remus (2007) to derive uncertainty ranges for the inner conducting sapwood area; see appendix.

To upscale to the site level we use the inventory data, which provides the number of trees in the stand, diameter at breast height (DBH), and species information. Daily sap flux density per sensor depth was averaged for each species and tree status (dominant vs. suppressed) for each site. Missing sap flux density data were filled by linear regression with neighboring sites. The largest data gaps were filled at sites N2 and S2. Finally, the upscaled daily stand transpiration was obtained by summing up the product of sap flux density per depth D and sap wood area per depth $A_s(D)$ for all trees and dividing by the area of the inventory A_{stand}:

$$E_{sap} = \frac{1}{A_{stand}} \sum_{tree=1}^{n} \sum_{D=1}^{3} (\mathrm{SFD}(D) A_s(D)). \tag{8}$$

2.6 Root water uptake estimation

As another means to estimate transpiration we use soil moisture measurements. Soil moisture sensors (Decagon 5TE soil moisture sensors, with an accuracy of ± 3 % volumetric water content and resolution of 0.08 %) are installed at three profiles at each site at 10–30–50 cm depth and one deeper sensor (approx. 70 cm) at one of the profiles. There is a range of methods to estimate root water uptake from soil moisture observations (Shuttleworth, 1993; Cuenca et al., 1997; Wilson et al., 2001; Schwärzel et al., 2009; Breña Naranjo et al., 2011; Guderle and Hildebrandt, 2015). Here, we employ a simplified soil water budget method to estimate daily root water uptake E_{RWU} (units mm d^{-1}). The method is based the daytime reduction of measured soil moisture integrated over the soil profile. To avoid the influence of infiltration and other fluxes, which are not caused by root water uptake E_{RWU}, we filtered the days with significant nighttime soil moisture variations as well as rainy days. A representative site value of E_{RWU} was derived by averaging the estimates of all three soil profiles. Details of the method can be found in the Appendix A3.

2.7 Statistical analyses

To test if the transpiration estimates are driven by potential atmospheric demand, we use linear ordinary least squares (OLS) regression with the transpiration estimates as response variable and E_{opt} as independent variable. The slope of the regression with sap flux density is denoted by $b_{SFD} = \frac{dv_{sap}}{dE_{opt}}$ with units [m^3 m^{-2} d^{-1}] / [mm d^{-1}]. For E_{sap} the slope is denoted $b_{sap} = \frac{dE_{sap}}{dE_{opt}}$ and for E_{RWU} the slope is denoted $b_{RWU} = \frac{dE_{RWU}}{dE_{opt}}$. The latter two regression slopes have nondimensional units [mm d^{-1}] / [mm d^{-1}]. To estimate the potential influence of other independent variables on transpiration, such as soil moisture or vapor pressure deficit, we use the residuals of the OLS regression as response variable of a further linear regression. The explanatory power of the other variable is reported by the adjusted explained variance (denoted by R^2_θ for soil moisture effects), which is a diagnostic of the OLS regression.

Note that by using time series of daily data, which are shaped by the seasonal cycle, the assumption of independence of the predictor variables in the linear regression is not often justified. The effect of serial dependence generally does not bias the regression coefficients, but reduces the statistical significance of a regression. Therefore, we estimate the standard deviation of the regression coefficient (σ_{slope}) and its reported significance with a pre-whitening procedure of Newey and West (1994) provided by function *coeftest* of the *R* packages lmtest (Zeileis and Hothorn, 2002) and sandwich (Zeileis, 2004). All data analysis was done in R (R Core Team, 2016) and the R-package data.table (Dowle et al., 2014).

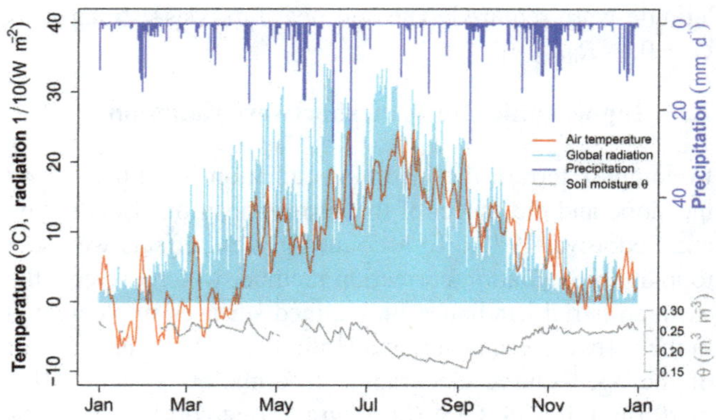

Figure 3. Meteorological forcing and average soil moisture at the grassland site (G1) for 2013. A few data gaps were filled with data from nearby grassland sites.

3 Results

3.1 Meteorological observations in 2013

Daily time series of temperature, global radiation, precipitation, and site-average soil moisture content for 2013 are shown in Fig. 3. The average annual (growing season, May–October) temperature was 7.9 (13.7) °C, with an average global radiation of 117 (169) W m^{-2}. The highest temperatures were observed in a short period in June and a longer period between mid-July and August. We observed a total annual precipitation of 866 mm (no catch correction applied), which was, however, unevenly distributed throughout the year. Most precipitation fell in June (110 mm), with low rainfall during July and August (34 and 38 mm, respectively) and more rainfall in the autumn months. Hence, the soil moisture recession lasted from July to September (bottom gray line in Fig. 3).

3.2 Site-scale transpiration estimates

Two independent site-scale transpiration estimates are evaluated in the following. The first estimate, E_{sap}, was obtained by upscaling sap flow measurements with the forest plot inventory data. The second estimate was derived by averaging E_{RWU} of the 2–3 soil profiles per site. Both transpiration

Table 1. Seasonal totals in mm of estimated stand transpiration E_{sap} and potential evaporation E_{opt} and E_{PM}. For the totals we used the period 12 May 2013–31 October 2013. For E_{sap} we present a range of estimates with minimal, linear decline and maximal estimate with respect to the depth of conducting inner sapwood (Eqn. A3, A4, and A5).

Site	E_{sap} min–lin–max	E_{opt}	E_{PM}
N1	106–128–156	240	351
N2	76–115–164	232	345
S3	127–176–244	296	387
S2	104–129–170	293	386
S1	96–122–160	295	387
G1		270	371

estimates are shown as daily time series for site N1 at the north-facing and S1 at the south-facing hillslope in Fig. 4. The north-facing sites have lower potential evaporative demand (estimated by topographically corrected incoming solar radiation) than the south-facing sites. Variations of daily soil water contents are quite similar between the sites. However, the north-facing site is on average slightly wetter ($\overline{\theta} = 0.21 \pm 0.04$) than the south-facing site ($\overline{\theta} = 0.18 \pm 0.03$). There is a large temporal variability in the atmospheric forcing represented by E_{opt}, which is also found for both transpiration estimates. The highest transpiration rates are seen in the dry and sunny period in July. Comparing the sites at the different hillslopes we find similar magnitudes of E_{sap}. However, E_{RWU} is found to be much higher than E_{sap} in spring and early summer at the north-facing sites.

Seasonal totals of E_{sap} and potential evaporation are reported in Table 1. The E_{sap} estimates for the upper north and south-facing hillslope sites are relatively similar. The highest E_{sap} is estimated at the valley bottom site S3, which amounts to a difference of about 50 mm for the seasonal total. Large uncertainty in the nominal value of E_{sap} is due to the assumed depth of the sapwood, which is estimated from data of the innermost sensor and allometric relationships of sapwood. Thereby the range between different conducting sapwood depths is largest at the valley site because there are more tall trees where this uncertainty matters.

3.3 Transpiration response to atmospheric demand

To test if the transpiration estimates are driven by potential atmospheric demand, we performed a linear regression of the site-average values with E_{opt} as independent variable. We found that E_{sap} was relatively low in May and early June after leaf out and after mid-October with leaf senescence. In order to avoid phenological effects in the statistical analysis, we restrict the following analyses to the vegetation period from 10 June to 20 October. Respective scatter plots are shown in Fig. 5 and a summary of the results is reported in Table 2.

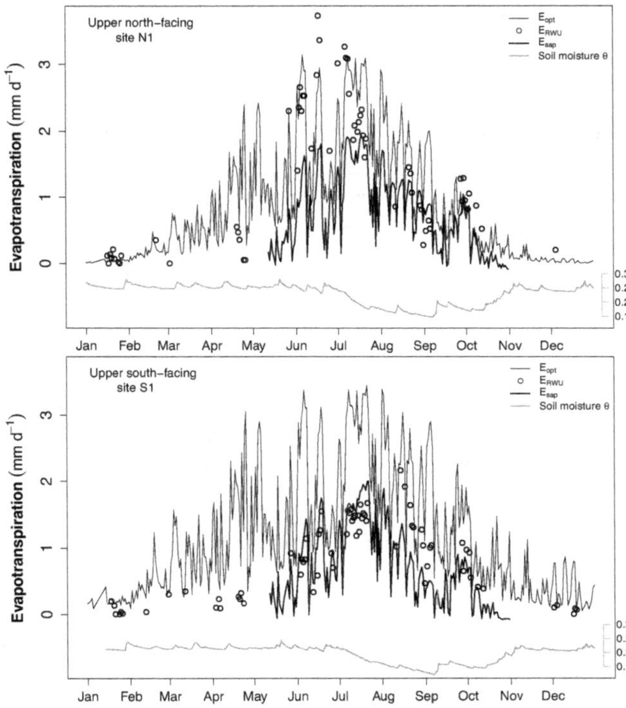

Figure 4. Time series of evapotranspiration estimates for north-facing hillslope site N1 in the left panel and south-facing site S1 in the right panel. E_{opt} is the potential evaporation derived from maximum power, E_{RWU} (blue points) and E_{sap} are transpiration estimates.

Figure 5 shows that both observations almost linearly increase with E_{opt}. For E_{sap} we find very high linear correlations with an average of $r^2_{E_{opt}} = 0.82$ (Table 2). The relatively large temporal variability in E_{sap} seen in Fig. 4 with a coefficient of variation CV = 0.68 across all forest sites is explained by the atmospheric demand E_{opt} (CV = 0.53). It is interesting to note that across all forest sites global radiation has only a slightly lower average $r^2 = 0.79$, whereas air temperature with $r^2 = 0.55$ and vapor pressure deficit with $r^2 = 0.66$ have less explanatory power. The correlation of site-average E_{RWU} to E_{opt} is much lower ($r^2 = 0.49$, averaged among all sites) than the correlation of E_{sap} to E_{opt}. All meteorological variables show lower correlations to E_{RWU} than to E_{sap}, which indicates a larger uncertainty of the soil moisture-derived transpiration estimate.

We also tested if the regression residuals show correlations with other daily observations. The reported adjusted squared correlation of a linear regression of the residual with VPD, wind speed, and soil moisture at the site level is reported in Table 2. Vapor pressure deficit does not explain residual variance of E_{sap} or E_{RWU}. Also wind speed does not show residual correlation for E_{sap} but higher positive residual correlation is found for E_{RWU} at the lower sites N2, S2, and significant at S3. Both of these variables are used as external

Table 2. Regression statistics for E_{sap} and E_{RWU} to E_{opt}, with n providing the number of observations (days). The slope and intercept are reported with the estimated standard deviation of the coefficients with $\pm\sigma$. Significance of the coefficients is indicated by stars: $p < .001$, ***; $p < .01$, **; $p < .05$, *. $r^2_{E_{opt}}$ and $r^2_{E_{PM}}$ are the linear squared correlation coefficients to E_{opt} and E_{PM}, respectively. The last three columns report the adjusted explained variance of a linear regression of the regression residuals for daily average vapor pressure deficit (R^2_{VPD}), daily average wind speed (R^2_u), and daily site-average volumetric water content (R^2_θ).

Variable	Site	n	Slope	Intercept	$r^2_{E_{opt}}$	$r^2_{E_{PM}}$	R^2_{VPD}	R^2_u	R^2_θ
E_{sap}	N1	130	0.61 ± 0.02***	−0.03 ± 0.05	0.88	0.89	0.01	0.00	0.02
	N2	129	0.57 ± 0.03***	−0.03 ± 0.05	0.88	0.89	0.01	0.01	−0.00
	S3	130	0.84 ± 0.08***	−0.36 ± 0.10***	0.83	0.87	−0.01	0.01	0.19**
	S2	109	0.62 ± 0.07***	−0.27 ± 0.09**	0.75	0.80	−0.01	0.01	0.25*
	S1	130	0.58 ± 0.07***	−0.23 ± 0.07**	0.78	0.83	−0.01	0.00	0.24**
E_{RWU}	N1	36	0.77 ± 0.18***	0.19 ± 0.27	0.52	0.50	0.04	0.03	0.64***
	N2	30	0.76 ± 0.19***	0.20 ± 0.40	0.67	0.69	−0.02	0.10	0.11
	S3	35	0.56 ± 0.11***	−0.21 ± 0.22	0.44	0.55	−0.03	0.32**	0.53***
	S2	30	0.52 ± 0.11***	0.11 ± 0.19	0.47	0.58	−0.02	0.12	0.45
	S1	39	0.44 ± 0.03***	0.13 ± 0.06*	0.63	0.60	−0.03	0.03	−0.02
	G1	28	0.75 ± 0.34*	0.32 ± 0.77	0.22	0.21	0.04	−0.04	0.63***

forcing in the Penman equation. Therefore, we calculated the squared correlation also for the FAO Penman–Monteith reference evaporation E_{PM}, Eq. (6). One can see from Table 2 that $r^2_{E_{PM}}$ is fairly similar, follows the same pattern, and is on average only slightly larger ($r^2_{E_{PM}} = 0.86$) than the correlation of E_{sap} to E_{opt}. Note that E_{PM} is on average 1.44 times larger than E_{opt} with a correlation between E_{PM} and E_{opt} of $r^2 = 0.98$.

In contrast to the meteorological variables, we found that the residuals are significantly correlated to the site-average soil moisture content at some sites. E_{sap} is significantly affected by soil moisture deficits at the south-facing sites. At these sites we find significant residual correlation of soil moisture $0.19 < R^2_\theta < 0.25$, which results in lower correlation to E_{opt} as compared to the two north-facing sites. Further, we find significant negative intercepts in the E_{sap}–E_{opt} relationship. Even more affected by soil water limitation is E_{RWU} with significant values of $R^2_\theta > 0.5$ at N1, S3, and G1. However, at S1 there is no residual correlation at the site average level, but this is probably an effect of deriving a representative site-average from E_{RWU} profile estimates, because one soil profile at S1 actually shows a significant residual correlation (Table S3 in the Supplement). The potential effect of soil moisture on transpiration is visualized by the size of the symbols in Fig. 5. Thereby, E_{RWU} tends to be above the regression line under moist and warm conditions in early summer particularly at sites N1 and G1.

Figure 5 and the regression statistics allow for a further comparison of E_{sap} and E_{RWU}. Generally there is a good agreement of both estimates with $0.5 < r^2 < 0.9$, but there are site-specific deviations between the site transpiration estimates. The upper south-facing sites S1 and S2 agree well in magnitude, while E_{sap} for site S3 appears to have an offset of about 0.5 mm d^{-1}. This offset could be influenced by the apparent site heterogeneity. The site is situated in the transition between the steep hillslope and the riparian zone of the nearby creek. The E_{sap} upscaling represents this transition zone by the 20 m by 20 m forest inventory size and assumes constant sapwood properties for all trees within this site. On the other hand, three soil moisture profiles may be too few in this heterogeneous zone, which complicates the estimation of a representative site-average E_{RWU} value. At the north-facing sites we find that E_{sap} is lower than the E_{RWU} estimates. Especially during early summer when soil moisture was still high, E_{RWU} was found to be almost twice as large as E_{sap} and of similar magnitude as E_{opt} at these sites. However, there was a better agreement in late summer, when soil moisture declined; see Fig. 4. This indicates that the root water uptake estimates tend to be higher under moist conditions, which is reflected by the positive residual correlation to soil moisture reported in Tables 2 and S3 in the Supplement.

3.3.1 Response at tree level

By using tree-averaged sap flow density measurements directly we can assess how strong single trees respond to E_{opt}. Results of a linear regression analysis of daily tree average SFD to E_{opt} are reported in Table 3 (see also Fig. S3 in the Supplement). SFD of every tree shows a strong linear relation to E_{opt}, with varying slopes and intercept terms. Because most of the variability is explained by E_{opt}, we can simply compare the slope of the regression b_{SFD}. First, we find that b_{SFD} is especially high at the tall beech trees at the north-facing sites (1.26–1.64 [m^3 m^{-2} d^{-1}] / [mm d^{-1}]), Table 3). These sampled tall trees at the north-facing sites are dominant trees without other tall trees in their vicinity. All other trees show much smaller values of b_{SFD} (0.26–0.88 [m^3 m^{-2} d^{-1}] / [mm d^{-1}]). Second, we find that b_{SFD} is on

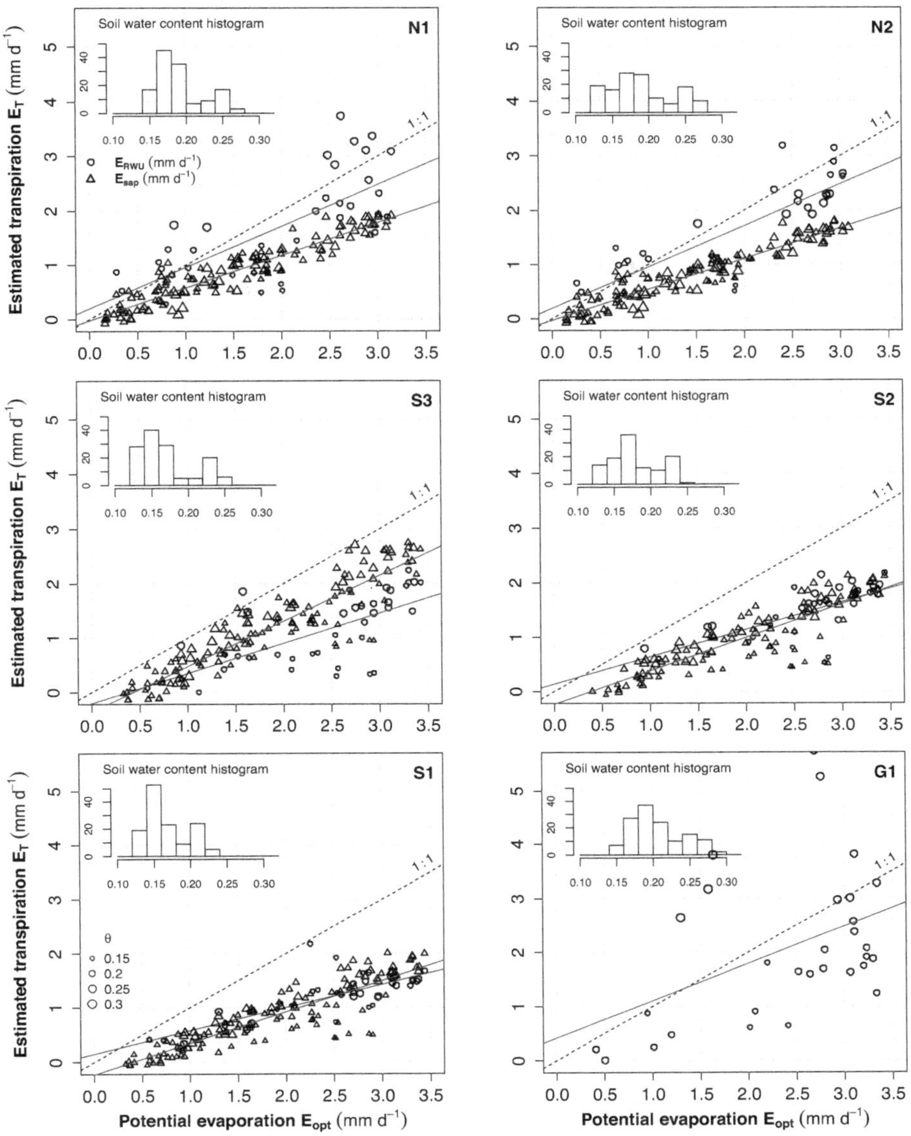

Figure 5. Site-average E_{RWU} (blue circles) and E_{sap} (red triangles) as function of E_{opt} during vegetation period (10 June–20 October) in 2013. Each panel presents data of one site (see Fig. 1 for site identifiers). The size of the symbols corresponds to the daily average soil moisture content. The histogram shows the distribution of site-average soil moisture. Solid lines depict the linear regression as tabulated in Table 2.

average larger at the north-facing than at the south-facing trees. Last, we also find that the oak tree at site S1 has the lowest b_{SFD}. For these reasons the upscaling of SFD to E_{sap} was done for each site separately and distinction was made between dominant and small trees as well as between species.

The residual regression analysis shows a similar pattern as the site-average values, with low influence of wind speed and VPD. Site-average soil moisture content (θ) shows significant effects at 9 out of 14 trees; see Table 3. Thereby all beech trees at the south-facing sites show an influence of θ on the sap flow residuals, with a maximum adjusted explained variance of $R^2_\theta = 0.4$ at the upper south-facing site S1. In addition, the 3 of 4 small beech trees at the north-facing sites show significant influences of θ.

3.3.2 Soil profile response and root water uptake profiles

Generally, the measurement setup allows one to differentiate between three different levels of aggregation of root water uptake estimates: site-average, per profile, and per depth of the sensor.

Root water uptake per profile is obtained by summing up the estimates of each sensor. We find that 15 out of 16 pro-

Table 3. Regression statistics of daily SFD as average per tree and per site as extra row. Column DBH is the diameter at breast height in centimeters. n is the number of observations (days). The slope b_{SFD} and intercept of the linear univariate regression of SFD to E_{opt} with $\pm\sigma$ reporting the estimated standard deviation of the coefficients. Significance of the coefficients is indicated by stars: $p < .001$, ***; $p < .01$, **; $p < .05$, *. $r^2_{E_{\mathrm{opt}}}$ and $r^2_{E_{\mathrm{PM}}}$ are the linear squared correlation coefficients of SFD to E_{opt} and E_{PM}, respectively. The last three columns report the adjusted explained variance of a linear regression of the regression model residuals for the variables: vapor pressure deficit (R^2_{VPD}), wind speed (R^2_u), and site-average volumetric water content (R^2_θ).

Site	Tree	Species	DBH	n	b_{SFD}	Intercept	$r^2_{E_{\mathrm{opt}}}$	$r^2_{E_{\mathrm{PM}}}$	R^2_{VPD}	R^2_u	R^2_θ
N1	1	beech	66	130	1.30 ± 0.05***	0.09 ± 0.13	0.88	0.88	0.03	0.07**	0.11**
N1	2	beech	58	130	1.26 ± 0.04***	−0.06 ± 0.08	0.92	0.84	0.00	0.03*	−0.01
N1	3	beech	9	130	0.88 ± 0.05***	−0.08 ± 0.09	0.82	0.89	0.03	−0.00	0.08*
N1	4	beech	8	130	0.63 ± 0.08***	−0.22 ± 0.08**	0.79	0.88	−0.00	0.02	0.29***
N1				130	1.02 ± 0.03***	−0.07 ± 0.07	0.90	0.91	0.01	0.01	0.00
N2	1	beech	53	97	1.31 ± 0.08***	0.23 ± 0.16	0.83	0.83	−0.01	−0.00	0.03
N2	2	beech	49	97	1.64 ± 0.09***	−0.06 ± 0.18	0.87	0.88	0.00	−0.00	−0.01
N2	3	beech	10	97	0.68 ± 0.04***	−0.11 ± 0.05*	0.89	0.82	0.01	0.00	−0.00
N2	4	beech	8	97	0.51 ± 0.08***	−0.14 ± 0.08	0.80	0.78	−0.01	0.00	0.26**
N2				97	1.04 ± 0.06***	−0.02 ± 0.11	0.87	0.88	−0.00	−0.01	0.02
S1	1	beech	43	130	0.81 ± 0.08***	−0.23 ± 0.10*	0.77	0.80	−0.01	0.01	0.34**
S1	2	oak	40	130	0.26 ± 0.01***	−0.10 ± 0.02***	0.92	0.94	0.05*	0.01	−0.01
S1	3	beech	39	97	0.39 ± 0.12**	−0.14 ± 0.12	0.64	0.81	0.01	0.01	0.40***
S1	4	beech	46	130	0.40 ± 0.05***	−0.13 ± 0.06*	0.76	0.87	−0.01	0.01	0.40***
S1				130	0.49 ± 0.05***	−0.17 ± 0.06**	0.80	0.84	−0.01	0.00	0.30**
S3	1	beech	45	124	0.52 ± 0.06***	−0.22 ± 0.07***	0.83	0.86	−0.01	0.02	0.22***
S3	2	beech	39	124	0.86 ± 0.08***	−0.35 ± 0.10***	0.84	0.90	−0.00	−0.00	0.09*
S3				124	0.69 ± 0.07***	−0.28 ± 0.08***	0.84	0.87	−0.01	0.01	0.12*

files show significant slopes b_{RWU} (Table S3 in the Supplement), which reveals a significant linear relationship with E_{opt}. The profile-based results also highlight that there is considerable within-site variability of the response of E_{RWU} to E_{opt}, with significant slopes ranging between 0.15 and 1.17. Testing the regression residuals for an influence of soil moisture content we find 10 of 16 forest soil profiles with significant residual correlation R^2_θ. At the grass site two profiles showed even larger residual correlations to θ than to E_{opt} ($R^2_\theta = 0.71, 0.72$). Testing for additional correlation of VPD and wind speed on the regression residuals of E_{RWU} to E_{opt} we find that VPD has generally low additional value with only three profiles with significant influences. Wind speed shows significant residual correlation at 3 out of 16 profiles. Root water uptake at a specific sensor level ($E_{\mathrm{RWU},d,z}$) allows one to assess at which depths plant water extraction and/or capillary rise due to soil evaporation are effective. E_{RWU} is detected at almost all soil moisture sensors during the growing season in 2013 (Fig. S4 in the Supplement). However, E_{RWU} at the deepest sensor is much lower than what is observed at the top (0–20 cm) soil layer. Comparing the grass site (G1) with the forest sites we find that the grass site has much larger E_{RWU} from the top-layer (0–20 cm) than from deeper layers. Further, we can see that the top-layer uptake decreases during the drier summer period July to September. The reduction is the strongest at the grass site, whereas the forest sites show a more evenly distributed uptake with soil depth.

3.4 Influence of topography and stand structure

The results show that most of the daily variability in both transpiration estimates is driven by atmospheric demand. The slope of the linear regression of E_{sap}, SFD, and E_{RWU} to E_{opt} provides a summary statistic of how strong a site or a tree responds to atmospheric demand. In the following we evaluate which location factors correlate with the slopes of the linear regressions. We concentrate on topographic and stand structural parameters.

The most obvious factor is the aspect of the sites. While b_{sap} does not show aspect-related differences, we find that b_{SFD} and b_{RWU} are higher at the north-facing sites and lower at the south-facing sites. This is found at the site-average level (Table 2), at the soil profile level (Table S3 in the Supplement) and at the tree level (Table 3). The differences are, however, only significant for b_{SFD} at the tree level being larger at the north-facing slope with average $b_{\mathrm{SFD}} = 1.03$

Table 4. Site topographic characteristics with average inclination of each site is given in column "slope angle". Column, $R_{g,c}$ is the average global radiation $R_{g,c}$, and E_{opt} and E_{PM} are all averaged for the vegetation period 10 June–20 October. Values of site G1 are measured directly, values of the forested sites are estimated by the topographic correction of solar radiation.

Site	Slope angle (°)	$R_{g,c}$ ($W\,m^{-2}$)	E_{opt} ($mm\,d^{-1}$)	E_{PM} ($mm\,d^{-1}$)
N1	14	153	1.46	2.19
N2	18	148	1.41	2.15
S3	30	192	1.82	2.44
S2	22	190	1.80	2.43
S1	24	191	1.81	2.44
G1	8	173	1.65	2.33

than at the south-facing slope $b_{SFD} = 0.63$, which is significant at the 0.05 level with a Student's t test.

Another topographic factor influencing the response to atmospheric demand is the inclination of the sites, where we find that the steeper the site the lower the slope of the linear regression of SFD to E_{opt} (cf. Tables 3 and 4). Both aspect and inclination affect the received solar radiation. According to the topographic correction of solar radiation, the north-facing sites received about 85 to 89 % of the solar radiation, which was observed at the grass site. In contrast the south-facing sites received about 108–110 %; see Table 4. This difference is confirmed by observations of air temperature within the forest, which reveal the same ranking across the sites as $R_{g,c}$.

Apart from the topographic data, the stand structure varies remarkably (see Table 5). At the two north-facing sites we find much understorey (> 98 % of trees smaller than DBH 15 cm) and a few tall trees, but no medium-sized trees between 15 and 50 cm DBH. The tall, dominant trees therefore have a well-lit canopy. The taller young trees form a secondary, lower but closed canopy layer. In contrast the south-facing sites S1 and S2 have about 20 medium sized trees per 20 by 20 m plot mainly between 16 and 65 cm DBH. These sites form a rather closed one-layer forest canopy. The valley site shows the largest tree size diversity with small, medium, and very tall trees. The stand composition differences of the sites result in quite different stand basal areas and thus total stand sapwood areas, which is the key factor in extrapolating SFD to site level. Although most trees are found at the north-facing sites, their total sapwood area is about half of the south-facing sites. The valley site (S3) has the largest sapwood area (Table 5). The LAI measurements also show the largest values for the valley site and slightly lower values for the south-facing site. Lower values are found at the north-facing sites. However, the differences in LAI between sites are comparably small when compared with sapwood area.

Plotting b_{SFD} as a function of the site-average sapwood area per tree ($A_s\,/\,n_{tree}$) (see Fig. 6), we find that b_{SFD} decreases with $A_s\,/\,n_{tree}$. In addition, the maximum SFD estimated as the 95 % quantile of the daily maxima show a very similar decline. Thus, both measures show higher sap flux densities in the younger and smaller north-facing stands as compared to the older south-facing stands. In contrast the sensitivity of E_{sap} is rather constant $b_{sap} \approx 0.6$ $[mm\,d^{-1}]\,/\,[mm\,d^{-1}]$ across sites at the hillslopes. Only at the valley site b_{sap} is significantly higher with ≈ 0.8 $[mm\,d^{-1}]\,/\,[mm\,d^{-1}]$. Although this site has a lower sapwood area per tree ($A_s\,/\,n_{tree}$) than the upper south-facing sites, the total sapwood area at S3 is larger, because of a few very tall trees; see Table 5.

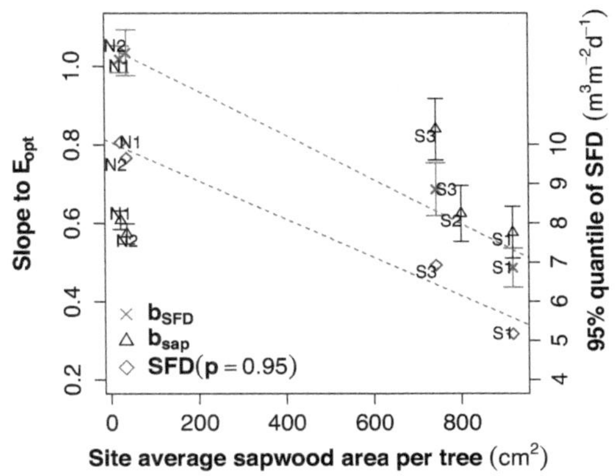

Figure 6. Sensitivity to E_{opt} as a function of the site-average sapwood area per tree computed by $A_s\,/\,n_{tree}$. On the left y axis: sensitivity of site-average sap flux density b_{SFD} (green crosses; see also Table 3) and the sensitivity of the upscaled transpiration estimate b_{sap} (red triangles). The vertical bars show the standard deviation of the sensitivity estimates tabulated in Tables 2 and 3. Right y axis: 95 % quantile of daily SFD maxima per site to estimate a robust maximum SFD. The dashed lines show linear regression model fits.

4 Discussion

4.1 Using thermodynamic limits to estimate daily potential evaporation

Generally, there are two different physical limitations of atmospheric evaporative demand, which have been used for modeling E, (i) by energy limitation, and (ii) the mass exchange limitations, which requires that the vapor is removed before it might condense again (Brutsaert, 1982). The approach of maximum convective power applied here combines both of these physical constraints and emphasizes the strong link between the energy balance and the strength of convective motion.

The maximum power-derived estimate of potential evaporation E_{opt} is quite similar to the Priestley–Taylor formulation, but has two key differences. First, the Priestley–Taylor equation contains an additional empirical factor, which has been interpreted as an effect of large-scale advection or

Table 5. Observed biometric characteristics of the five forested sites along the transect. Column names: n_{tree} ... number of trees with circumference > 4 cm in a 20 by 20 m reference plot, diameter at breast height (DBH) distribution with minimum, first quartile, mean, third quartile, and maximum (cm), total stand basal area A_b ($m^2\ ha^{-1}$), total stand sapwood area estimated from published allometric relationships A_s ($m^2\ ha^{-1}$); see Sect. 2.4. Site leaf area index (LAI) was derived from the difference of summer LAI_s and winter LAI_w.

		DBH (cm)					A ($m^2\ ha^{-1}$)		LAI ($m^2\ m^{-2}$)		
site	n_{tree}	min	Q25	mean	Q75	max	A_b	A_s	LAI_s	LAI_w	LAI
N1	346	< 1.3	2.1	3.8	4.4	65.4	25	19	7.5	2.3	5.2
N2	196	< 1.3	1.6	4.1	4.0	66.5	29	18	7.5	1.7	5.9
S3	28	1.8	12.7	30.2	40.7	80.9	75	52	8.1	1.7	6.5
S2	20	19.1	32.4	37.4	44.5	63.8	60	40	7.4	1.2	6.3
S1	17	16.4	31.8	41.1	49.8	58.9	61	39	7.1	0.7	6.4

boundary layer dynamics (Brutsaert, 1982). The second difference is the use of net shortwave radiation in our model, as opposed to net radiation in the Priestley–Taylor equation, the latter of which is not an independent forcing, as it depends on the surface temperature. Using shortwave radiation has important practical advantages because R_{sn} can be obtained from global radiation measurements and albedo information. Further topographic influences on incoming radiation can be directly computed by topographic radiation correction methods (Šúri and Hofierka, 2004). This also circumvents the necessity of collecting site-specific radiation data, which can be challenging above high vegetation like forests.

The derivation of E_{opt} is based on a representation of an atmosphere, which is in equilibrium with the underlying surface and the surface–atmosphere exchange of heat, moisture, and momentum is driven by locally absorbed solar radiation (Kleidon and Renner, 2013b). This implies that meteorological variables such as wind speed or vapor pressure deficit, which are commonly used in Penman formulations, cannot change independently, but are rather constrained by land–atmosphere interaction. In order to test if these assumptions are generally met, we tested for the effect of VPD on transpiration in our data set. The results showed that VPD and wind speed did not add consistently to the explained variance. In addition the correlation of the transpiration estimates to the FAO Penman–Monteith grass reference evaporation is on average slightly higher but follows the same patters across sites. Hence, the two additional input variables only slightly increased the predictability of atmospheric demand, which is consistent with the residual regression analysis above. Both of these results indicate that E_{opt}, which is a function of absorbed solar radiation and surface temperature only, captures the dominant drivers of daily sap velocity variations within this temperate forest without requiring further data on wind speed and VPD.

The dominance of absorbed solar radiation in explaining latent and sensible heat fluxes was also found by Best et al. (2015), who showed that simple linear regression models with solar radiation and temperature as input and thus without any information on wetness conditions had a similar performance than commonly used land-surface models at 20 diverse tower flux sites. However, an improved reproduction of sensible and latent heat fluxes was obtained by Best et al. (2015) when relative humidity was included into their regression model. It is likely that our potential evaporation estimate E_{opt} represents evaporation in a wet environment, whereas in drier environments, air humidity may become increasingly important. Interestingly, in contrast to our approach of eliminating air humidity while considering air temperature as an independent variable, Aminzadeh et al. (2016) estimated wet environment evaporation potential as a function of radiation and air humidity, while eliminating air temperature. Clearly, more investigations, encompassing larger and more diverse data sets, are needed to better understand general patterns of atmospheric control on transpiration. For example, it may well be that air humidity carries information about soil moisture and hence adds to explanatory power of transpiration models in water limited environments, while radiation represents the main control in energy-limited environments. In how far wind, air humidity, and temperature are affected by the land–atmosphere coupling and can hence be treated as internal variables, likely also depends on the scale of interest.

The derivation of E_{opt} shown in Sect. 2 was based on a range of simplifying assumptions to focus on the emerging maximum power limit in a coupled land–atmosphere system (Kleidon and Renner, 2013b; Kleidon et al., 2014). Most important features, which need to be addressed in future work are (i) changes in heat storage, (ii) a revision of the simplified scheme for longwave radiative exchange, and (iii) horizontal circulation patterns by large scale and mesoscale circulation. By assuming a steady state, heat storage effects were neglected in the derivation of the turbulent fluxes J_{opt}. However, heat storage in the surface–atmosphere system is an important mechanism to balance the seasonally and diurnally varying input of solar radiation. A comparison of $J_{opt} = R_{sn} / 2$ with the empirical net radiation estimate of Allen et al. (1998), which was used to estimate E_{PM} showed that J_{opt} is consistently lower during summer. Thus, to improve daily estimates based on thermodynamic limits, it is recommended to consider seasonal heat storage effects in the derivation. Heat storage is even more relevant at the diurnal timescale, which reduces the applicability of E_{opt} as given in Eq. (5). The relevance of considering heat storage is illustrated in Fig. 7, which shows a distinct hysteresis of both sap flow and air temperature with respect to global radiation on a sunny day. Both sap flow and temperature linearly increase with shortwave radiation in the morning hours, but remain high after midday until sunset. Thus, for the same amount of received global radiation there is a distinct difference between morning and afternoon transpiration. While moisture storage in soil and plants also affect the hysteresis, the dominant magnitude of the hysteresis is probably due to the lag of temperature to radiation leading to a VPD-radiation lag (Zhang et al., 2014). Hence, for predictions at the diurnal timescale, heat storage effects as well as boundary layer dynamics have to be accounted for.

Another limitation of the approach is the simple linearized scheme for longwave heat exchange. The description of radiative exchange affects the maximum power limit, because longwave radiation "competes" with the convective fluxes to cool the surface (Kleidon et al., 2015). Although the parameter describing the net longwave radiative exchange drops out in the maximization for the convective fluxes, the radiation scheme affects the partitioning of radiative and convective fluxes at the surface. Therefore, a more detailed description of radiation transfer processes will increase the ability to predict surface energy partitioning.

The third limitation is concerned with the spatial representativity of E_{opt}, but also other potential evaporation formulations and the role of horizontal heat and mass exchange. While the land surface is generally quite heterogeneous, the lower atmosphere is mostly well mixed (Claussen, 1991). By using the topographically corrected incoming radiation we effectively treated each site as an independent surface–atmosphere column. We thus neglected any lateral fluxes that could alter the estimated potential evaporation. The topographic correction increased the potential evaporation at the south-facing and decreased estimates for the north-facing sites. To evaluate this effect with respect to the transpiration estimates, we used the global radiation measured at the nearby grassland site as forcing for all forest sites. The correlation to the transpiration estimates was unaffected, but the topographic correction slightly amplified the aspect-related differences of the response to E_{opt} between sites (Fig. S1 in the Supplement). For a better understanding of such microclimatic effects spatial observations of the canopy surface energy balance are required to test more detailed models, which include horizontal exchange processes. Progress in this respect will be quite important for the parameterization of subgrid processes in numerical land-surface models.

4.2 Topographical and stand composition effects

Our results demonstrate a strong influence of daily variations in atmospheric demand on both transpiration estimates. This strong correlation at the daily timescale may allow us to separate the timescales of atmospheric demand and plant water limitations, which may become relevant at timescales larger than 1 day. Differences in the slope of the relationship to E_{opt} can then be compared across sites and be used to identify potential permanent site controls on transpiration.

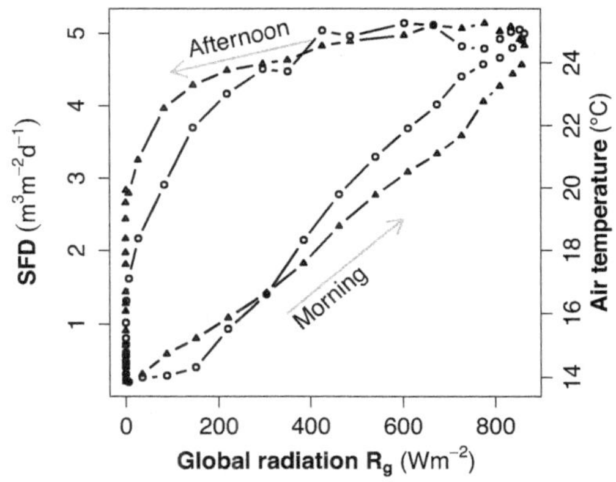

Figure 7. Hysteresis of sap flux density (left y axis) and air temperature (right y axis, red) plotted as a function of global radiation at the field site G1. Data are half-hourly values over one sunny summer day (7 July 2013). Sap flow data are taken from a beech (tree 1) at site S1.

Of the many factors that influence transpiration, stand composition and topography were the most important invariant controls during our measurement campaign. The measurement transect was placed on a valley cross section to primarily reveal the influence of hillslope aspect. Our results indicate only significant effects of aspect and hillslope angle for the sensitivity of SFD to E_{opt}, but no effects for the sensitivity of the upscaled transpiration estimate E_{sap}. E_{RWU} shows a weak relation to these topographic factors, which was, however, not significant because of the larger uncertainty of the soil moisture-based estimate.

The biometric measures of stand composition, however, also co-vary with aspect and hillslope inclination. Here we found a strongly negative linear relationship of b_{SFD} and the maximum sap flux densities to the site-average sapwood area per tree (Fig. 6). Such a decline in b_{SFD} and maximum SFD with sapwood area was found to be a universal tree hydraulic mechanism observed across different tree species (Andrade et al., 1998; Meinzer et al., 2001). Interestingly, Andrade et al. (1998) found that maximum sap velocity nonlinearly declined with sapwood area, with strongest effects at very small trees or branches with $A_s < 100\,cm^2$. Considering such an increase in b_{SFD} for the understorey vegetation at the north-facings sites in our study would increase their relevance in estimating stand transpiration.

In addition to the tree size differences, our sites show marked differences in canopy structure. The open canopy structure may explain the larger b_{SFD} of the dominant trees sampled at the north-facing sites, because tree growth could be enhanced by gaps in the canopy structure. For example Schweingruber et al. (2006) showed that annual tree ring width of beech trees increase quickly after neighboring trees have been cut. At the south-facing sites average tree size is high but there is a low number of stems. Here individual tree growth is limited by the neighboring trees with similar demands for light, water, nutrients, and space. Self-thinning or thinning by forest management has resulted in a more even-sized stand structure with old and large trees. Thus, the closed canopy structure, which is composed of well-established trees, may have lead to a more conservative water use per tree, which is reflected by low b_{SFD} and low maximum sap flux densities. These arguments are consistent with results of studies that show higher sap flux densities of thinned forest stands as compared to control stands (Morikawa et al., 1986; Bréda et al., 1995; Nahm et al., 2006).

The upscaling procedure integrates both the tree distribution within the stand and the sap flux density observations to yield the stand transpiration estimate E_{sap}. Apparently this has led to a compensation of the vastly different stand structure and sap flux densities across sites. Due to this tradeoff between SFD and the sapwood area similar transpiration rates are achieved at the hillslope sites. Therefore, forest transpiration may indeed be regarded as a conservative hydrological process (Roberts, 1983) even in this very complex terrain. Only for the valley site S3 significantly larger E_{sap} and b_{sap} was estimated. This site has comparably high maximal SFD and the largest total sapwood area, even though $A_s\,/\,n_{tree}$ is lower than observed at the upper south-facing sites. The most likely reason for the comparably dense, but tall forest is the vicinity to the nearby stream, which reduces potential soil water limitations. A similar upslope–downslope effect was recently also established by Kume et al. (2015) for a mountainous stand of Japanese cypress with the lower site having taller trees with higher SFD but same tree density.

4.3 Response to soil moisture

Due to low rainfall amounts, soil moisture decreased from July to September at all sites and all depths. Thus the soil moisture reduction may have limited transpiration. We captured this effect of soil moisture limitation of transpiration through the regression of the residuals of the transpiration to E_{opt} relationships to site-average soil moisture conditions. Results showed that soil moisture explained a significant part of the residuals of E_{sap} at the south-facing sites (Table 2), which implies that site-scale transpiration was also water limited during the dry period in August. At the north-facing sites water limitation at the site scale was only apparent during a shorter period in September. However, we found significant residual correlation to soil moisture at three of four sampled small beech trees. This suggests that small trees are more susceptible to water shortage.

We argue that soil water limitation effects on transpiration could be topographically enhanced. First, by an aspect that affects the amount of received solar radiation and thus the atmospheric demand for water. A higher atmospheric demand increases evaporation from intercepted water and from the soil, which reduces the amount of precipitation entering the soil. Second, the hillslope inclination could have enhanced lateral runoff at the steeper south-facing sites, which reduces the soil water holding capacity (Bronstert and Plate, 1997). These topographical factors of soil moisture availability are also apparent in the tree species composition of our sites. While the north-facing sites are predominantly composed of beech and a few spruce trees, the south-facing sites have about 10 % oak trees, which are known to cope well at more dry sites (Zapater et al., 2011).

4.4 Limitations of transpiration estimates

4.4.1 Limitations of sap flow observations

Generally, sap flow observations are not limited by spatial heterogeneity and complex terrain, which would limit the applicability of micrometeorological measurements (Wilson et al., 2001). Installation is relatively simple and sensors are inexpensive. Despite these advantages, we experienced the following limitations:

(a) Deriving water fluxes requires extrapolation from the point measurement at some specific place within the stem to the entire tree. However, sapwood conductivity can have radial and circumferential differences and species-specific properties (Wullschleger and King, 2000; Saveyn et al., 2008). This can easily bias sap flow estimates (Köstner et al., 1998; Shinohara et al., 2013; Vandegehuchte and Steppe, 2013). An indication of this problem is that we found different b_{SFD} at the same tree installed the year before. Comparison with measurements taken at the same trees from the previous year showed differences in b_{SFD} ranging between -0.39 and 0.23 with an average of 0.04 $[\mathrm{m^3\,m^{-2}\,d^{-1}}]\,/\,[\mathrm{mm\,d^{-1}}]$ (estimated for nine trees with more than 30 days of data each year).

(b) There is a sample bias towards larger trees as the method is more difficult with very small trees, which would require a different type of sensor, because the heat ratio method is designed only for lower sap velocities (Marshall, 1958; Burgess et al., 2001). This adds a sampling uncertainty in estimating site transpiration where much understorey exists. This is especially relevant for the north-facing sites, with a median DBH of 4 cm. This means that most trees were not sampled. The sampling rather reflected the trees that contributed most to the stand sapwood area.

(c) The inter-comparison of sap flux density measured in different trees is limited by the fact that xylem characteristics in the estimation of SFD are required (Burgess et al., 2001). Most important is the thermal diffusivity of sapwood, k_w as used in Eq. (A1). This conductivity is a function of wood density and wood water content m_c (Burgess et al., 2001; Vandegehuchte and Steppe, 2012a), both of which vary between species and trees (Gebauer et al., 2008). Xylem water content m_c, which in addition influences the apparent sap flux density through affecting the sapwood heat capacity (see Eq. 7), was shown to have seasonal changes in diffuse-porous species (Glavac et al., 1990; Hao et al., 2013). Glavac et al. (1990) found that m_c can reduce to about 25 % in sampled European beech trees during summer. Such a decline would thus reduce the apparent sap velocities. Therefore, it is recommended to use methods that take into account wood moisture content changes (Vandegehuchte and Steppe, 2012b). Here, in the absence of measurements of such wood properties we used the same parameters for all trees.

4.4.2 Limitations of root water uptake estimates from soil moisture variations

The advantage of using temporally highly resolved soil moisture readings is that it allows one to estimate root water uptake without further information on soil properties (Guderle and Hildebrandt, 2015). The accuracy of the method depends on various factors that can influence results:

(a) Data filtering: the method only applies under conditions with negligible soil water movement excluding events of infiltration, drainage, capillary rise, or hydraulic redistribution. These fluxes can have major influences on the observations of soil moisture and comprise the second term of the right-hand side of Eq. (A6). Thus, the estimates depend on the choice of suitable filter criteria. A very strict filter would reduce the number of estimates, whereas soft filter criteria may result in biased E_{RWU} estimates. Hence, seasonal or annual totals cannot be derived from this method alone. We use relatively strict filter criteria for nighttime fluxes of 0.1 Vol %, which is close to the sensor resolution. This filter criteria set the maximum accuracy per soil layer of 200 mm depth to 0.2 mm d^{-1}.

(b) Soil heterogeneities, dominant at the hillslopes, can induce large local variations in soil moisture and may lead to dissimilar / biased E_{RWU} compared to other methods (Wilson et al., 2001). Here, we found that the influence of E_{opt} and soil moisture content on E_{RWU} varied between soil profiles at a specific site (Table S3 and Fig. S5 in the Supplement). Large differences were observed at site N2, which results in a quite uncertain site-average estimate of E_{RWU}. Therefore, it is recommended to install several, representative measurement profiles when such a soil water budget method is used for transpiration estimation in heterogeneous terrain (Schwärzel et al., 2009).

(c) Deep root water uptake in response to drying topsoil may cause root water uptake below the deepest measurement depth in forest sites as observed e.g. by Teskey and Sheriff (1996); Wilson et al. (2001). Observations from the deepest sensor profile confirm root uptake at 60 cm depth (Fig. S4 in the Supplement), which also persists during the dry period. However, overall the contribution of deep root water uptake is assumed to be small, given the low observed diurnal variations.

4.4.3 Upscaling sap flow to site-scale transpiration

The estimates of site-scale transpiration based on up-scaled sap flow measurements were of similar magnitudes and correlated well with estimates derived from soil moisture variations. The seasonal estimates by E_{sap} are about 50 mm lower than other estimates for beech forests. For example Schipka et al. (2005) found 200–300 mm per year for European beech forests in Germany. Their sites, however, have been located in less steep terrain.

The comparison between E_{RWU} and E_{sap} also revealed striking differences, which could be an indication of the potential shortcomings of both methods, as discussed above. While the south-facing sites are in good agreement, E_{sap} at the north-facing sites seems to be quite low. First, this is due to the low basal area at the north-facing site. One reason could be that the assumed sapwood area of the few tall trees is much larger than that reported in the literature. Another possible reason could be that small trees (< 8 cm DBH), which were not sampled, had a significant contribution to stand transpiration. Also E_{RWU} might overestimate actual transpiration because soil evaporation would equally contribute to the diurnal signal in soil moisture. For example Bréda et al. (1993) also found consistently larger estimates of a soil water balance method than stand transpiration estimates, which was attributed to soil evaporation by the authors.

5 Summary and conclusions

We aimed to infer the dominant temporal and spatial controls on forest transpiration along a steep valley cross section through ecohydrological measurements of sap flow and soil moisture and their relation to atmospheric evaporative demand. The estimation of transpiration in space and time for forests in complex terrain is a challenge in its own right. Obtaining transpiration is only possible through indirect observations, whereby each method has its own limitations. Therefore, we used two independent observations to obtain site-scale estimates of transpiration along the hillslope transect. To estimate atmospheric demand, a formulation similar to the well-known Priestley–Taylor equilibrium evaporation concept was employed. The formulation is based on a simplified energy balance representation of the surface–atmospheric system and hypothesizing that convection operates at its upper thermodynamic limit. The formulation does not require empirical parameters and only requires data on the absorbed solar radiation and temperature. We find that at the daily timescale this approach explains most of variability in both transpiration estimates at the site and tree scale. This suggests that atmospheric demand is the dominant control on daily transpiration rates in this temperate forest. Although the well-established FAO Penman–Monteith reference evaporation yields slightly higher correlation and 20–30 % higher values, it requires additional data of net radiation, VPD and wind speed. Thereby both, VPD and wind speed did not add consistently to the explained variance and are also difficult to obtain above forests. While our results demonstrate that thermodynamic limits provide a first-order estimate for potential evaporation, we have to stress that the derivation is based on the simplest possible energy balance representation. Further refinements will probably improve the predictability of surface exchange fluxes.

Despite the prevailing topographic contrasts between the north-facing and the south-facing measurement sites, we find that up-scaled stand transpiration yields rather similar seasonal totals as well as a similar average response to atmospheric demand. This similarity is achieved through a compensation of the low sapwood area with high sap flux densities at the north facing sites, while at the south-facing sites a high sapwood area was accompanied with low sap flux densities. It appears that individual and stand average sap flux densities can vary strongly in heterogeneous terrain in order to compensate for tree size and stand structural differences through tree hydraulic mechanisms. The importance of these stand structural differences on stand transpiration thus masks the potential effects of topographical factors such as aspect and hillslope angle, which are cross correlated. However, during dry periods we find that topographic factors can enhance the response of transpiration to soil water limitation.

Despite unavoidable limitations in estimating stand transpiration and potential evaporation in complex terrain, we find that relating the employed ecohydrological observations to a thermodynamically constrained estimate of atmospheric demand enables important insights in the temporal drivers of transpiration and how they vary at the hillslope scale. First our results highlight the dominance of absorbed solar radiation as the main and independent driver of land–atmosphere exchange. Second, our results suggest an intriguing interplay of tree hydraulics and stand composition, which seemingly leads to transpiration rates close to its physical limits. We conclude that this approach should help us to better understand surface–atmosphere coupling in relation to thermodynamic constraints and how vegetation adapts to these.

Appendix A: Detailed description of transpiration estimates

A1 Measurement and correction of heat pulse velocity

We used sap flow sensors based on the heat ratio method (Burgess et al., 2001). Thereby three needles with equal distance are vertically inserted into the tree. The upper and lower needles measure the change in temperature after a short heat pulse was emitted by a heating element in the middle needle. The heat pulse velocity v_h is proportional to the logarithmic ratio of the temperature differences measured before and at t_1 usually 60 s after the heat pulse at the lower (ΔT_{dn}) and upper (ΔT_{up}) needles (Marshall, 1958; Burgess et al., 2001):

$$v_h = \frac{4k_w t_1 \log \frac{\Delta T_{up}}{\Delta T_{dn}} - x_2^2 + x_1^2}{2t_1(x_1 - x_2)}, \tag{A1}$$

whereby k_w is the thermal diffusivity of the xylem, and x_1 and x_2 are the vertical distances to the heater. The sensors have the standard distance of 0.6 cm. This distance can easily be slightly shifted during the installation into living trees, which causes a systematic bias. This sensor alignment bias can be corrected when sap flow is zero (Burgess et al., 2001). Setting $v_h = 0$ and rearranging Eq. (A1), we can estimate the distance of the upper needle x_1 while assuming $x_2 = 0.6$:

$$x_1 = \sqrt{x_2^2 - 4k_w t_1 \log \frac{\Delta T_{up}}{\Delta T_{dn}}}. \tag{A2}$$

We prefer Eq. A2 over the published correction in Burgess et al. (2001) because it allows to correct for both, positive and negative nighttime biases in heat velocities. We estimated the corrected distance x_1 by assuming zero flow during nighttime between 01:00 and 04:00 LT and the median of x_1 for the whole installation period. The installation of the sensor needles injures the surrounding xylem vessels and reduces the actual sap flow around the sensor. We applied a polynomial wounding correction $v_c = b\,V_h + c\,V_h^2 + d\,V_h^3$ with wounding correction parameters $b = 1.8558$, $c = -0.0018$, and $d = 0.003$, applicable for a sensor spacing of 0.6 cm and a drilling hole size of 2 mm, which is tabulated in Burgess et al. (2001).

A2 Upscaling of sap flow

The needles measure heat-pulse velocities at depths of 5, 18, and 30 mm within the stem. Following the user's manual we assigned for each sensor depth a representative radius of 15, 25, and 40 mm below the cambium radius r_x, which is obtained by assuming that the cambium takes 1 % of the total radius, thus $r_x = 0.99 \cdot \mathrm{DBH}\,/\,2$.

Because tall trees may have a wider active xylem depth than measured by the inserted sap flow needles (Gebauer et al., 2008), we derived three different estimates of the representative sapwood area of the innermost sensor, which provides the range and a best-guess estimate. A minimal estimate is obtained by assuming a representative annulus of 15 mm depth for the sensor and zero flow in the inner sapwood, which is calculated by $A_{\mathrm{S3,min}}$. Assuming that sap flux density remains constant at the innermost sensor level throughout the inner sapwood area provides a maximal estimate of sap flux. Most realistic is the assumption of a linear decline reaching no flow at the estimated heartwood radius $A_{\mathrm{S3,lin}}$ similar to Pausch et al. (2000):

$$A_{\mathrm{S3,min}} = \pi\left((r_x - 25\,\mathrm{mm})^2 - (r_x - 40\,\mathrm{mm})^2\right), \tag{A3}$$

$$A_{\mathrm{S3,lin}} = \pi\left((r_x - 25\,\mathrm{mm})^2 - \frac{1}{3}\left((r_x - 25\,\mathrm{mm})^2 + (r_x - 25\,\mathrm{mm})\,r_h + r_h^2\right)\right), \tag{A4}$$

$$A_{\mathrm{S3,max}} = \pi\left((r_x - 25\,\mathrm{mm})^2 - (r_x - r_h)^2\right). \tag{A5}$$

Stem sapwood area was computed using published allometric power-law relationships of the form $A_\mathrm{s} = a\ \mathrm{DBH}^b$. For beech and alder we used the relationship published in Gebauer et al. (2008), for oak we used Schmidt (2007) and Alsheimer et al. (1998) for spruce. Then the radius of the heartwood r_h is obtained from sapwood area estimates. For small trees with radius smaller than the representative annulus depth we set the area to 0.

A3 Root water uptake estimation

The soil water continuity equation at a point in the soil may be written as Cuenca et al. (1997):

$$\frac{\partial\theta}{\partial t} = -\frac{\partial q_i}{\partial x_i} + S, \tag{A6}$$

where θ is the soil moisture content, q summarizes any soil water fluxes over an orthogonal coordinate system x_i with horizontal (x, y) and vertical directions (z). The sink term S is the root water uptake, which when integrated over the soil volume V_s yields the transpiration flux E_{RWU} (Cuenca et al., 1997):

$$E_{\mathrm{RWU}} = \int_{V_\mathrm{s}} S\,dV_\mathrm{s} = \int_{V_\mathrm{s}} \frac{\partial\theta}{\partial t} dV_\mathrm{s} + \int_{V_\mathrm{s}} \frac{\partial q_i}{\partial x_i} dV_\mathrm{s}. \tag{A7}$$

Hence, to estimate E_{RWU} we need to resolve the two terms in Eq. (A7). The first term is the temporal evolution of soil moisture, which is in principle measured by the soil moisture sensors. The second term describes soil water fluxes within the soil such as downward fluxes during an infiltration event. The soil water fluxes q_i themselves depend on soil moisture, through its role in determining soil water potentials and unsaturated hydraulic conductivity. Soils along hillslopes have large spatial heterogeneity of their hydraulic properties and pose many influences on the second term of Eq. (A7), which makes continuous estimation of root water uptake difficult.

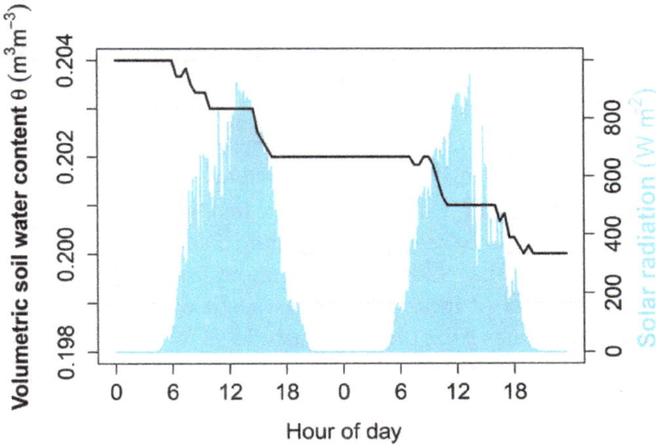

Figure A1. Observed diurnal decline in soil moisture over two sunny days in summer 2013 to illustrate the approach to estimate daily root water uptake from diurnal soil moisture variations. E_{RWU} is estimated at sensor level per day as a cumulative sum of differences from sunrise to sunset, in the absence of any nighttime variations or daytime increases in soil moisture.

There are, however, periods and locations where soil water fluxes q_i, such as drainage, capillary rise, etc., are of minor importance. Especially during dry conditions the reduction of soil moisture is dominated by root water uptake and soil evaporation during daytime, as illustrated in Fig. A1. Under these radiation-driven conditions, we observe rather constant nighttime moisture levels, which thus indicates that soil water fluxes are not active and nocturnal root–soil exchange is negligible. Thus, nighttime soil moisture dynamics are a practical filter to exclude days with dominant drainage or capillary rise fluxes. To estimate daily root water uptake we use soil moisture observations on a half-hourly basis. We first quantify the daytime change in soil moisture per sensor $\Delta\theta_{\mathrm{d,s}}$ by cumulative sum of differences of soil moisture (at 30 min intervals) from astronomical sunrise to sunset:

$$\Delta\theta_{\mathrm{d,s}} = \sum_{t=t_{\mathrm{sunrise}}}^{t=t_{\mathrm{sunset}}} (\theta_{t+1} - \theta_t). \tag{A8}$$

Assuming that the sensor observation is representative for the respective soil depth Δz of the sensor, we obtain a flux estimate $E_{\mathrm{RWU},d,z} = -\Delta\theta_{\mathrm{d,s}}\Delta z$ per soil layer. A constant soil depth of $\Delta z = 200\,\mathrm{mm}$ per sensor level was assumed for all soil profiles. We then summed up the contributions $E_{\mathrm{RWU},d,z}$ of each soil layer to obtain a daily root water uptake $E_{\mathrm{RWU},d}$ per profile. For most profiles this allowed us to estimate $E_{\mathrm{RWU},d}$ down to a depth of 600 mm, and for three profiles down to 800 mm (see Table S1 for an overview of sensor placement). During sensor installation thick roots were mostly found until a depth of ≈ 300 mm with sporadic fine roots up to 800 mm depth.

The data were filtered for (i) precipitation (daily sum < 0.1 mm and rainfall of previous day < 1 mm), (ii) only negative daytime soil moisture changes $\Delta\theta_{\mathrm{d,s}}$, and (iii) absolute cumulative nighttime changes in soil moisture $|\Delta\theta_{n,s}| < 0.1$ Vol %. Approximately 30 % of the actual data were retained, with details listed in Table S2 in the Supplement. Generally, diurnal variations of soil moisture are large in the upper soil depths compared to deeper layers. Thus, for estimating the total profile root water uptake, the upper sensors are very important. Unfortunately, we had to face sensor failure of top soil sensors at site N1, profile 2 and site S1 profiles 2 and 3; see Table S2 in the Supplement for an overview of available data. At these soil profiles we used the root water uptake estimates from the 2nd sensor level to fill the missing data. Two profiles, one at G1 and another at N2 were completely disregarded for site averaging due to sensor failures. The site-average was obtained from at least two profile-based E_{RWU} estimates. At the grass site this condition was relaxed because there were only a few days where E_{RWU} could be estimated from both profiles. At these few days the estimates of the two profiles were comparable.

Acknowledgements. This research contributes to the "Catchments As Organized Systems (CAOS)" research group (FOR 1598) funded by the German Science Foundation (DFG). AH was partially funded by grant SFB 1076 by the German Science Foundation. We thank Conrad Jackisch (KIT) for comments on an earlier draft of the paper. We thank all people involved in the field work. In particular the technical backbone Britta Kattenstroth (GFZ Potsdam), Tatiana Feskova (UFZ – Leipzig), Laurent Pfister and François Iffly (LIST, Luxembourg) and the landowners for giving access to their land. We acknowledge the encouraging and constructive comments of two reviewers, who helped to improve the manuscript.

Edited by: T. Bogaard

References

Allen, R., Pereira, L., Raes, D., and Smith, M.: Crop evapotranspiration-Guidelines for computing crop water requirements-FAO Irrigation and drainage paper 56, FAO, Rome, 300, 6541, 1998.

Alsheimer, M., Kästner, B., Falge, E., and Tenhunen, J. D.: Temporal and spatial variation in transpiration of Norway spruce stands within a forested catchment of the Fichtelgebirge, Germany, Ann. Sci. Forest Res., vol. 55, 103–123, EDP Sciences, 1998.

Aminzadeh, M., Roderick, M. L., and Or, D.: A generalized complementary relationship between actual and potential evaporation defined by a reference surface temperature, Water Resour. Res., 52, 385–406, doi:10.1002/2015WR017969, 2016.

Andrade, J. L., Meinzer, F. C., Goldstein, G., Holbrook, N. M., Cavelier, J., Jackson, P., and Silvera, K.: Regulation of water flux through trunks, branches, and leaves in trees of a lowland tropical forest, Oecologia, 115, 463–471, doi:10.1007/s004420050542, 1998.

Bachmair, S. and Weiler, M.: New Dimensions of Hillslope Hydrology, in: Forest Hydrology and Biogeochemistry, edited by: Levia, D. F., Carlyle-Moses, D., and Tanaka, T., vol. 216, 455–481, Springer Netherlands, Dordrecht, 2011.

Baumgartner, A.: Gelände und Sonnenstrahlung als Standortfaktor am Gr. Falkenstein (Bayerischer Wald), Forstwissenschaftliches Centralblatt, 79, 286–297, doi:10.1007/BF01815008, 1960.

Best, M. J., Abramowitz, G., Johnson, H. R., Pitman, A. J., Balsamo, G., Boone, A., Cuntz, M., Decharme, B., Dirmeyer, P. A., Dong, J., Ek, M., Guo, Z., Haverd, V., van den Hurk, B. J. J., Nearing, G. S., Pak, B., Peters-Lidard, C., Santanello, J. A., Stevens, L., and Vuichard, N.: The Plumbing of Land Surface Models: Benchmarking Model Performance, J. Hydrometeorol., 16, 1425–1442, doi:10.1175/JHM-D-14-0158.1, 2015.

Bohren, C. F. and Albrecht, B. A.: Atmospheric thermodynamics, Atmospheric thermodynamics, New York, Oxford, Oxford University Press, ISBN: 0195099044, 1, 1998.

Bonan, G. B.: Forests and Climate Change: Forcings, Feedbacks, and the Climate Benefits of Forests, Science, 320, 1444–1449, doi:10.1126/science.1155121, 2008.

Breña Naranjo, J. A., Weiler, M., and Stahl, K.: Sensitivity of a data-driven soil water balance model to estimate summer evapotranspiration along a forest chronosequence, Hydrol. Earth Syst. Sci., 15, 3461–3473, doi:10.5194/hess-15-3461-2011, 2011.

Bronstert, A. and Plate, E. J.: Modelling of runoff generation and soil moisture dynamics for hillslopes and micro-catchments, J. Hydrol., 198, 177–195, doi:10.1016/S0022-1694(96)03306-9, 1997.

Brutsaert, W.: Evaporation into the Atmosphere, Springer Netherlands, Dordrecht, 1982.

Bréda, N., Cochard, H., Dreyer, E., and Granier, A.: Water transfer in a mature oak stand (Quercuspetraea): seasonal evolution and effects of a severe drought, Canad. J. For. Res., 23, 1136–1143, doi:10.1139/x93-144, 1993.

Bréda, N., Granier, A., and Aussenac, G.: Effects of thinning on soil and tree water relations, transpiration and growth in an oak forest (Quercus petraea (Matt.) Liebl.), Tree Physiol., 15, 295–306, doi:10.1093/treephys/15.5.295, 1995.

Burgess, S. S. O., Adams, M. A., Turner, N. C., Beverly, C. R., Ong, C. K., Khan, A. A. H., and Bleby, T. M.: An improved heat pulse method to measure low and reverse rates of sap flow in woody plants, Tree Physiol., 21, 589–598, doi:10.1093/treephys/21.9.589, 2001.

Calder, I. R.: Water use by forests, limits and controls, Tree Physiol., 18, 625–631, doi:10.1093/treephys/18.8-9.625, 1998.

Claussen, M.: Estimation of areally-averaged surface fluxes, Bound.-Lay. Meteorol., 54, 387–410, doi:10.1007/BF00118868, 1991.

Cuenca, R. H., Stangel, D. E., and Kelly, S. F.: Soil water balance in a boreal forest, J. Geophys. Res.-Atmos., 102, 29355–29365, doi:10.1029/97JD02312, 1997.

de Bruin, H. A. R., Trigo, I. F., Bosveld, F. C., and Meirink, J. F.: A Thermodynamically Based Model for Actual Evapotranspiration of an Extensive Grass Field Close to FAO Reference, Suitable for Remote Sensing Application, J. Hydrometeorol., 17, 1373–1382, doi:10.1175/JHM-D-15-0006.1, 2016.

Dowle, M., Short, T., Lianoglou, S., and Srinivasan, A.: data. table: Extension of data. frame. R package version 1.9.4, 2014.

Famiglietti, J. S., Rudnicki, J. W., and Rodell, M.: Variability in surface moisture content along a hillslope transect: Rattlesnake Hill, Texas, J. Hydrol., 210, 259–281, doi:10.1016/S0022-1694(98)00187-5, 1998.

Farquhar, G. D., Caemmerer, S. v., and Berry, J. A.: A biochemical model of photosynthetic CO2 assimilation in leaves of C3 species, Planta, 149, 78–90, doi:10.1007/BF00386231, 1980.

Federer, C. A.: Forest Transpiration Greatly Speeds Streamflow Recession, Water Resour. Res., 9, 1599–1604, 1973.

Gebauer, T., Horna, V., and Leuschner, C.: Variability in radial sap flux density patterns and sapwood area among seven co-occurring temperate broad-leaved tree species, Tree Physiol., 28, 1821–1830, doi:10.1093/treephys/28.12.1821, 2008.

Glavac, V., Koenies, H., and Ebben, U.: Auswirkung sommerlicher Trockenheit auf die Splintholz-Wassergehalte im Stammkörper der Buche (Fagus sylvatica L.), Holz als Roh- und Werkstoff, 48, 437–441, doi:10.1007/BF02627628, 1990.

Guderle, M. and Hildebrandt, A.: Using measured soil water contents to estimate evapotranspiration and root water uptake profiles – a comparative study, Hydrol. Earth Syst. Sci., 19, 409–425, doi:10.5194/hess-19-409-2015, 2015.

Haas, E., Klatt, S., Frählich, A., Kraft, P., Werner, C., Kiese, R., Grote, R., Breuer, L., and Butterbach-Bahl, K.: LandscapeDNDC: a process model for simulation of biosphere-atmosphere-hydrosphere exchange processes at site and regional scale, Landsc. Ecol., 28, 615–636, doi:10.1007/s10980-012-9772-x, 2013.

Hao, G.-Y., Wheeler, J. K., Holbrook, N. M., and Goldstein, G.: Investigating xylem embolism formation, refilling and water storage in tree trunks using frequency domain reflectometry, J. Exp. Bot., 64, 2321–2332, doi:10.1093/jxb/ert090, 2013.

Holst, J., Grote, R., Offermann, C., Ferrio, J. P., Gessler, A., Mayer, H., and Rennenberg, H.: Water fluxes within beech stands in complex terrain, Int. J. Biometeorol., 54, 23–36, doi:10.1007/s00484-009-0248-x, 2010.

Holst, T., Rost, J., and Mayer, H.: Net radiation balance for two forested slopes on opposite sides of a valley, Int. J. Biometeorol., 49, 275–284, doi:10.1007/s00484-004-0251-1, 2005.

Ivanov, V. Y., Bras, R. L., and Vivoni, E. R.: Vegetation-hydrology dynamics in complex terrain of semiarid areas: 2. Energy-water controls of vegetation spatiotemporal dynamics and topographic niches of favorability, Water Resour. Res., 44, W03430, doi:10.1029/2006WR005595, 2008.

Jasechko, S., Sharp, Z. D., Gibson, J. J., Birks, S. J., Yi, Y., and Fawcett, P. J.: Terrestrial water fluxes dominated by transpiration, Nature, 496, 347–350, doi:10.1038/nature11983, 2013.

Juilleret, J., Iffly, J. F., Pfister, L., and Hissler, C.: Remarkable Pleistocene periglacial slope deposits in Luxembourg (Oesling): pedological implications and geosite potential, Bull. Soc. Nat. Luxembour., 112, 125–130, 2011.

Kleidon, A. and Renner, M.: A simple explanation for the sensitivity of the hydrologic cycle to surface temperature and solar radiation and its implications for global climate change, Earth Syst. Dynam., 4, 455–465, doi:10.5194/esd-4-455-2013, 2013a.

Kleidon, A. and Renner, M.: Thermodynamic limits of hydrologic cycling within the Earth system: concepts, estimates and implications, Hydrol. Earth Syst. Sci., 17, 2873–2892, doi:10.5194/hess-17-2873-2013, 2013b.

Kleidon, A., Renner, M., and Porada, P.: Estimates of the climatological land surface energy and water balance derived from maximum convective power, Hydrol. Earth Syst. Sci., 18, 2201–2218, doi:10.5194/hess-18-2201-2014, 2014.

Kleidon, A., Kravitz, B., and Renner, M.: The hydrological sensitivity to global warming and solar geoengineering derived from thermodynamic constraints, Geophys. Res. Lett., 42, 138–144, doi:10.1002/2014GL062589, 2015.

Kondratyev, K. Y. and Fedorova, M. P.: Radiation regime of inclined surfaces, Solar Energy, 1, 36–61, 1977.

Kume, T., Tsuruta, K., Komatsu, H., Shinohara, Y., Katayama, A., Ide, J., and Otsuki, K.: Differences in sap flux based stand transpiration between upper and lower slope positions in a Japanese cypress plantation watershed, Ecohydrology, doi:10.1002/eco.1709, 2015.

Köstner, B., Granier, A., and Cermák, J.: Sapflow measurements in forest stands: methods and uncertainties, Ann. For. Sci., 55, 13–27, doi:10.1051/forest:19980102, 1998.

Lee, R. and Baumgartner, A.: The Topography and Insolation Climate of a Mountainous Forest Area, For. Sci., 12, 258–267, 1966.

Lüttschwager, D. and Remus, R.: Radial distribution of sap flux density in trunks of a mature beech stand, Ann. For. Sci., 64, 431–438, doi:10.1051/forest:2007020, 2007.

Makkink, G.: Testing the Penman formula by means of lysimeters, J. Inst. Water Eng, 11, 277–288, 1957.

Marshall, D. C.: Measurement of Sap Flow in Conifers by Heat Transport. 1, Plant Physiol., 33, 385–396, 1958.

Meinzer, F. C., Goldstein, G., and Andrade, J. L.: Regulation of water flux through tropical forest canopy trees: Do universal rules apply?, Tree Physiol., 21, 19–26, doi:10.1093/treephys/21.1.19, 2001.

Miller, G. R., Chen, X., Rubin, Y., Ma, S., and Baldocchi, D. D.: Groundwater uptake by woody vegetation in a semiarid oak savanna: GROUNDWATER UPTAKE IN AN OAK SAVANNA, Water Resour. Res., 46, W10503, doi:10.1029/2009WR008902, 2010.

Monteith, J. L.: Evaporation and environment, Symposia of the Society for Experimental Biology, 19, 205–234, 1965.

Morikawa, Y., Hattori, S., and Kiyono, Y.: Transpiration of a 31-year-old Chamaecyparis obtusa Endl. stand before and after thinning, Tree Physiol., 2, 105–114, doi:10.1093/treephys/2.1-2-3.105, 1986.

Nahm, M., Holst, T., Matzarakis, A., Mayer, H., Rennenberg, H., and Geßler, A.: Soluble N compound profiles and concentrations in European beech (Fagus sylvatica L.) are influenced by local climate and thinning, Eur. J. For. Res., 125, 1–14, doi:10.1007/s10342-005-0103-5, 2006.

Newey, W. K. and West, K. D.: Automatic Lag Selection in Covariance Matrix Estimation, Rev. Econ. Stud., 61, 631–653, 1994.

Oke, T.: Boundary layer climates, Routledge, London and New York, 1987.

Ozawa, H., Ohmura, A., Lorenz, R. D., and Pujol, T.: The second law of thermodynamics and the global climate system: A review of the maximum entropy production principle, Rev. Geophys., 41, 1018, doi:10.1029/2002RG000113, 2003.

Pausch, R. C., Grote, E. E., and Dawson, T. E.: Estimating water use by sugar maple trees: considerations when using heat-pulse methods in trees with deep functional sapwood, Tree Physiol., 20, 217–227, 2000.

Priestley, C. and Taylor, R.: On the assessment of surface heat flux and evaporation using large-scale parameters, Mont. Eather Rev., 100, 81–92, 1972.

R Core Team: R: A Language and Environment for Statistical Computing, R Foundation for Statistical Computing, Vienna, Austria, http://www.R-project.org/, 2016.

Raupach, M. R.: Combination theory and equilibrium evaporation, Quarterly J. Roy. Meteorol. Soc., 127, 1149–1181, doi:10.1002/qj.49712757402, 2001.

Roberts, J.: Forest transpiration: A conservative hydrological process?, J. Hydrol., 66, 133–141, doi:10.1016/0022-1694(83)90181-6, 1983.

Saveyn, A., Steppe, K., and Lemeur, R.: Spatial variability of xylem sap flow in mature beech (Fagus sylvatica) and its diurnal dynamics in relation to microclimate, Botany, 86, 1440–1448, doi:10.1139/B08-112, 2008.

Schipka, F., Heimann, J., and Leuschner, C.: Regional variation in canopy transpiration of Central European beech forests, Oecologia, 143, 260–270, doi:10.1007/s00442-004-1798-6, 2005.

Schmidt, M.: Canopy transpiration of beech forests in Northern Bavaria - Structure and function in pure and mixed stands with oak at colline and montane sites, Doctoral thesis, Bayreuth, 2007.

Schmidt, W.: Strahlung und Verdunstung an freien Wasserflächen; ein Beitrag zum Wärmehaushalt des Weltmeers und zum Wasserhaushalt der Erde, Ann. Calender Hydrographie und Maritimen Meteorologie, 43, 111–124, 1915.

Schweingruber, F. H., Bärner, A., and Schulze, E. D.: Atlas of Woody Plant Stems: Evolution, Structure, and Environmental Modifications, Springer Science & Business Media, 2006.

Schwinning, S.: The ecohydrology of roots in rocks, Ecohydrology, 3, 238–245, doi:10.1002/eco.134, 2010.

Schwärzel, K., Menzer, A., Clausnitzer, F., Spank, U., Häntzschel, J., Grünwald, T., Kästner, B., Bernhofer, C., and Feger, K.-H.: Soil water content measurements deliver reliable estimates of water fluxes: A comparative study in a beech and a spruce stand in the Tharandt forest (Saxony, Germany), Agr. For. Meteor., 149, 1994–2006, doi:10.1016/j.agrformet.2009.07.006, 2009.

Shinohara, Y., Tsuruta, K., Ogura, A., Noto, F., Komatsu, H., Otsuki, K., and Maruyama, T.: Azimuthal and radial variations in sap flux density and effects on stand-scale transpiration estimates in a Japanese cedar forest, Tree Physiol., 33, tpt029, doi:10.1093/treephys/tpt029, 2013.

Shuttleworth, W.: Evaporation. Handbook of Hydrology, DR Maidment, Ed, McGraw-Hill, New York, NY, USA, 1993.

Stull, R. B.: An introduction to boundary layer meteorology, vol. 13, Kluwer Academic Publishers, Dordrecht, The Netherlands, 1988.

Šúri, M. and Hofierka, J.: A New GIS-based Solar Radiation Model and Its Application to Photovoltaic Assessments, T. GIS, 8, 175–190, doi:10.1111/j.1467-9671.2004.00174.x, 2004.

Teskey, R. O. and Sheriff, D. W.: Water use by Pinus radiata trees in a plantation, Tree Physiol., 16, 273–279, doi:10.1093/treephys/16.1-2.273, 1996.

Tromp-van Meerveld, H. J. and McDonnell, J. J.: On the interrelations between topography, soil depth, soil moisture, transpiration rates and species distribution at the hillslope scale, Adv. Water Resour., 29, 293–310, doi:10.1016/j.advwatres.2005.02.016, 2006.

Vandegehuchte, M. W. and Steppe, K.: Improving sap flux density measurements by correctly determining thermal diffusivity, differentiating between bound and unbound water, Tree Physiol., 32, 930–942, doi:10.1093/treephys/tps034, 2012a.

Vandegehuchte, M. W. and Steppe, K.: Sapflow+: a four-needle heat-pulse sap flow sensor enabling nonempirical sap flux density and water content measurements, New Phytol., 196, 306–317, doi:10.1111/j.1469-8137.2012.04237.x, 2012b.

Vandegehuchte, M. W. and Steppe, K.: Sap-flux density measurement methods: working principles and applicability, Funct. Plant Biol., 40, 213–223, 2013.

Wang, J. and Bras, R. L.: A model of evapotranspiration based on the theory of maximum entropy production, Water Resour. Res., 47, W03521, doi:10.1029/2010WR009392, 2011.

Wang, Y. and Jarvis, P.: Description and validation of an array model - MAESTRO, Agr. For. Meteorol., 51, 257–280, doi:10.1016/0168-1923(90)90112-J, 1990.

West, G. B., Brown, J. H., and Enquist, B. J.: A general model for the structure and allometry of plant vascular systems, Nature, 400, 664–667, doi:10.1038/23251, 1999.

Whitehead, D.: Regulation of stomatal conductance and transpiration in forest canopies, Tree Physiol., 18, 633–644, doi:10.1093/treephys/18.8-9.633, 1998.

Wilson, K. B., Hanson, P. J., Mulholland, P. J., Baldocchi, D. D., and Wullschleger, S. D.: A comparison of methods for determining forest evapotranspiration and its components: sap-flow, soil water budget, eddy covariance and catchment water balance, Agr. For. Meteorol., 106, 153–168, 2001.

Wullschleger, S. D. and King, A. W.: Radial variation in sap velocity as a function of stem diameter and sapwood thickness in yellow-poplar trees, Tree Physiol., 20, 511–518, doi:10.1093/treephys/20.8.511, 2000.

Zapater, M., Hossann, C., Bréda, N., Bréchet, C., Bonal, D., and Granier, A.: Evidence of hydraulic lift in a young beech and oak mixed forest using 18O soil water labelling, Trees, 25, 885–894, doi:10.1007/s00468-011-0563-9, 2011.

Zehe, E., Ehret, U., Pfister, L., Blume, T., Schräder, B., Westhoff, M., Jackisch, C., Schymanski, S. J., Weiler, M., Schulz, K., Allroggen, N., Tronicke, J., van Schaik, L., Dietrich, P., Scherer, U., Eccard, J., Wulfmeyer, V., and Kleidon, A.: HESS Opinions: From response units to functional units: a thermodynamic reinterpretation of the HRU concept to link spatial organization and functioning of intermediate scale catchments, Hydrol. Earth Syst. Sci., 18, 4635–4655, doi:10.5194/hess-18-4635-2014, 2014.

Zeileis, A.: Econometric computing with HC and HAC covariance matrix estimators, J. Stat. Softw., 11, 1–17, 2004.

Zeileis, A. and Hothorn, T.: Diagnostic checking in regression relationships, R News, 2, 7–10, 2002.

Zhang, L., Potter, N., Hickel, K., Zhang, Y., and Shao, Q.: Water balance modeling over variable time scales based on the Budyko framework – Model development and testing, J. Hydrol., 360, 117–131, doi:10.1016/j.jhydrol.2008.07.021, 2008.

Zhang, Q., Manzoni, S., Katul, G., Porporato, A., and Yang, D.: The hysteretic evapotranspiration-Vapor pressure deficit relation: ET-VPD hysteresis, J. Geophys. Res.-Biogeosci., 119, 125–140, doi:10.1002/2013JG002484, 2014.

2

Derivation and evaluation of landslide-triggering thresholds by a Monte Carlo approach

D. J. Peres and A. Cancelliere

Department of Civil Engineering and Architecture, University of Catania, Catania, Italy

Correspondence to: D. J. Peres (djperes@dica.unict.it)

Abstract. Assessment of landslide-triggering rainfall thresholds is useful for early warning in prone areas.

In this paper, it is shown how stochastic rainfall models and hydrological and slope stability physically based models can be advantageously combined in a Monte Carlo simulation framework to generate virtually unlimited-length synthetic rainfall and related slope stability factor of safety data, exploiting the information contained in observed rainfall records and field-measurements of soil hydraulic and geotechnical parameters. The synthetic data set, dichotomized in triggering and non-triggering rainfall events, is analyzed by receiver operating characteristics (ROC) analysis to derive stochastic-input physically based thresholds that optimize the trade-off between correct and wrong predictions. Moreover, the specific modeling framework implemented in this work, based on hourly analysis, enables one to analyze the uncertainty related to variability of rainfall intensity within events and to past rainfall (antecedent rainfall). A specific focus is dedicated to the widely used power-law rainfall intensity–duration (I–D) thresholds.

Results indicate that variability of intensity during rainfall events influences significantly rainfall intensity and duration associated with landslide triggering. Remarkably, when a time-variable rainfall-rate event is considered, the simulated triggering points may be separated with a very good approximation from the non-triggering ones by a I–D power-law equation, while a representation of rainfall as constant–intensity hyetographs globally leads to non-conservative results. This indicates that the I–D power-law equation is adequate to represent the triggering part due to transient infiltration produced by rainfall events of variable intensity and thus gives a physically based justification for this widely used threshold form, which provides results that are valid when landslide occurrence is mostly due to that part. These conditions are more likely to occur in hillslopes of low specific upslope contributing area, relatively high hydraulic conductivity and high critical wetness ratio. Otherwise, rainfall time history occurring before single rainfall events influences landslide triggering, determining whether a threshold based only on rainfall intensity and duration may be sufficient or it needs to be improved by the introduction of antecedent rainfall variables. Further analyses show that predictability of landslides decreases with soil depth, critical wetness ratio and the increase of vertical basal drainage (leakage) that occurs in the presence of a fractured bedrock.

1 Introduction

Rainfall thresholds indicating landslide triggering are useful for the development of early warning systems in prone areas (cf., e.g., Keefer et al., 1987; Fathani et al., 2009; Takara and Apip Bagiawan, 2009; Baum and Godt, 2010; Capparelli and Versace, 2011). Commonly, such thresholds are derived by the analysis of historical rainfall and landslide data, and identified by drawing a lower-bound envelope curve of the triggering event characteristics (e.g., Campbell, 1975; Caine, 1980; Cancelli and Nova, 1985; Cannon and Ellen, 1985; Aleotti, 2004; Wieczorek et al., 2000; Guzzetti et al., 2007) or by enhanced methods which consider curves associated with a given frequency of non-exceedance by triggering events (cf. Brunetti et al., 2010; Peruccacci et al., 2012). A review by Guzzetti et al. (2007) indicated the prevailing use in literature of so-called power-law rainfall intensity–duration (I–D) thresholds, which are of the form $I = a_1 D^{a_2}$, where D is rain duration to triggering and I is rain intensity

$I = W/D$, W being rainfall accumulated over duration D. The a_1 and a_2 parameters have been derived by different researchers, for specific sites, regions or the whole globe.

Many factors of uncertainty affect the reliability of empirical thresholds, such as rainfall temporal and spatial variability, uncertain knowledge of the triggering instants, simplicity of threshold equation that does not include all control variables and statistical issues as well (Peruccacci et al., 2012). Nonetheless, it can be argued that most of the uncertainty stems from the availability and quality of the data used to derive the thresholds (Glade et al., 2000; Berti et al., 2012).

In fact, adequate historical data on landslides and simultaneous rainfall are in most cases available for a relatively short period, which may not be sufficiently significant from a statistical point of view. Moreover, the identification of the triggering instant is in many cases significantly uncertain and landslide archives are seldom complete (i.e., all landslide events occurred in the historical period are not known). This has a direct consequence on threshold derivation, because critical (where critical here means corresponding to landslide triggering) duration D, assumed as the time interval from rainfall event start and the triggering instant, cannot be computed accurately. Another key factor is the criterion used for rainfall identification, and in particular how the beginning of a rainfall event is identified. Many authors either do not specify the criteria used for rainfall identification or apply qualitative criteria, and indeed only few works in literature (Aleotti, 2004; Brunetti et al., 2010; Tiranti and Rabuffetti, 2010; Berti et al., 2012) explicitly addressed this problem. This makes thresholds subjective and impairs comparisons of results obtained by different researchers, as in analyzing the data the criterion may have been modified from one rainfall event to another. Another point is that in many countries automatic rain gauge networks have been installed only quite recently, and one has to rely on analysis based on rainfall records at the daily aggregation timescale (cf. Guzzetti et al., 2007, and references therein). Since many landslides, especially the most devastating shallow rapidly moving ones, may be triggered by rainfall events of a few hours (cf., e.g., Highland and Bobrowsky, 2008), use of daily rainfall for threshold derivation in these cases is quite questionable.

Apart from data quality issues, it can be pointed out that use of characteristic variables for the representation of rainfall events, and in particular of their intensity and duration, introduces an intrinsic uncertainty factor, because these variables may not be adequate to represent all the rainfall characteristics that affect landslide triggering. In fact, rainfall events represented by the same values of duration and intensity may correspond to totally different event time histories (hyetographs) that thus may or may not result in triggering. Sirangelo and Versace (1996) proposed an empirical method based on the use of convolution between rainfall time series and a filter function, which attempts to overcome this uncertainty. Also, use of solely the duration and intensity pair (D, I) in threshold formulation implies that the effect of initial wetness on triggering rainfall is neglected. Regarding this issue, several authors have added to D and I antecedent rainfall as a control parameter, though the empirical analyses have not yet provided unequivocal indications on the role of antecedent rainfall and different researchers used diverse temporal horizons for its computation (Guzzetti et al., 2007).

Another important point is that many thresholds have been derived by analyzing triggering events only and thus neglecting the non-triggering ones. This may lead to an underestimation of the triggering conditions, i.e., to thresholds that implemented in a early warning system tend to produce an unacceptable degree of false alarms, causing populations to no longer rely on early warnings (so-called cry-wolf effect; cf., e.g., Barnes et al., 2007). In fact, thresholds always should be provided with a measure of their reliability. To this end, Berti et al. (2012) proposed Bayesian probabilistic analysis to evaluate landslide-triggering thresholds in the presence of uncertainty. Receiver operating characteristics (ROC) analysis (cf., e.g., Wilks, 2011), based on the analysis of correct and wrong predictions, may be advantageously applied as well (cf., e.g., Staley et al., 2013).

Alternatively to empirical models, physically based models that couple hydrological and slope stability analysis (Montgomery and Dietrich, 1994; Wu and Sidle, 1995; Iverson, 2000; Baum et al., 2002; D'Odorico et al., 2005; Rosso et al., 2006; Baum et al., 2008) have been proposed to assess landslide triggering by rainfall, with the advantage that they take explicitly into account the meteorological, hydrological and geomechanical processes and variables that determine landslide triggering. From such models physically based thresholds may be derived (cf., e.g., Rosso et al., 2006; Salciarini et al., 2008). Such thresholds generally deviate from a straight line in the $\log(D)$–$\log(I)$ plane; this casts some doubts on the use of the power-law as a proper functional form for deriving rainfall thresholds. In other words, because such thresholds were derived from a physically based model, this may be interpreted as an evidence that the use of the power-law form is not supported from a physically based standpoint. Nevertheless, in such studies the meteorological aspects were analyzed in a simplistic way, because the thresholds do not consider variability of rainfall intensity during events and the initial conditions are not computed as a function of rainfall time history preceding the current event.

In spite of the limitations that we have put into evidence above, I–D rainfall thresholds are widely applied for landslide early warning systems. Perhaps their success is due to the fact that simple forms of the threshold are more easily understood by stakeholders and decision makers than the more complex, albeit more accurate, physically based models.

In this paper, a Monte Carlo-based methodology to derive and evaluate rainfall landslide-triggering thresholds is proposed, which makes use of an existing body of stochastic and physically based models. The approach combines research findings in the fields of rainfall and landslide hydrological modeling to provide an output that can be easily implemented in a early warning system, i.e., a landslide-triggering threshold, based on rainfall monitoring, of the same type that is commonly derived by direct empirical analysis of observed rainfall and landslide data. In particular, from the Monte Carlo simulations synthetic rainfall series are generated by a stochastic model and corresponding triggering/non-triggering conditions are identified by an hydrological and slope stability model. The generated data set is then analyzed to derive and evaluate I–D thresholds that take into account the variability of both rainfall intensity within events and initial conditions determined by past rainfall, as well as triggering/non-triggering events to measure uncertainty by ROC analysis. Furthermore, the derived stochastic-input physically based thresholds are compared with the constant-intensity physically based thresholds, which result from the simplistic assumption mentioned above (uniform hyetographs and prefixed initial conditions) in order to assess the effect on landslide triggering of rain intensity variability during events and variable initial conditions, computed as dependent by past rainfall time history. This analysis is related to the one by D'Odorico et al. (2005), in which the effect of rain intensity variability within events is studied by considering beta-shaped hyetographs inputs to the model of Iverson (2000) for derivation of hillslope response. Nevertheless, in their work, the variability of initial conditions as dependent from antecedent rainfall has not been considered because the steady-state asymptotic solution of Montgomery and Dietrich (1994) is utilized for computation of initial conditions. From their study they conclude that beta-shaped non-uniform hyetographs have a stronger destabilizing effect than uniform hyetographs of the same volume, since the associated return period of slope instability resulted higher in this last case. In this study we instead use hyetographs generated by a Neyman–Scott rectangular pulses (NSRP) stochastic model (Neyman and Scott, 1958; Kavvas and Delleur, 1975; Cox and Isham, 1980; Rodriguez-Iturbe et al., 1987a) and account for variability of initial conditions using a water table recession model to derive the initial water table height from the response to rainfall events preceding the current one, based on a linear reservoir mass-conservation equation with similar assumptions adopted by Rosso et al. (2006). The transient response to rainfall events is computed by a model based on the transient rainfall infiltration and grid-based regional slope-stability (TRIGRS) program (Baum et al., 2008). An application of the proposed methodology is carried out to the highly landslide-prone area of the Peloritani mountains, north-eastern Sicily, Italy. A sensitivity analysis on some of the most important control variables is carried out to analyze their effect on landslide-triggering thresholds and the associated uncertainty.

2 Monte Carlo synthetic data generation

The Monte Carlo simulation procedure for synthetic rainfall–landslide data generation consists of the following steps:

1. A stochastic rainfall model, calibrated on observations at a selected site, is used to generate a 1000-year long hourly rainfall time series. In particular we use a NSRP model (see Appendix A)

2. The synthetic rainfall time series is pre-processed in order to identify rainfall events and their inter-arrival durations. In particular, when two wet spells are separated by a dry time interval less than Δt_{min}, these are considered to belong to the same rainfall event; otherwise two separate rainfall events are considered. Details on the choice of the Δt_{min} simulation parameter are given at the end of this section

3. Some of the generated rainfall events are removed from the analysis because, according to the hydrological model, they will produce no significant variation of pressure head distribution, as their instantaneous (hourly) intensity is too low. In particular the events having maximum intensity less than i_{min} are removed from the analysis. We assume i_{min} equal to the leakage flux limit, given by $c_\text{d} K_\text{S}(1-\cos^2\delta)$, c_d being the vertical leakage ratio, K_S the saturated hydraulic conductivity and δ the slope of the hillslope (see Appendix B)

4. Application of previous steps leads to the generation of N_{RE} individual rainfall events

5. An initial value of the water table height is fixed to start simulations of the hydrological response for the whole rainfall time series. For the analyzed case-study area and many similar cases, it may be assumed that at the beginning of each hydrological year the water table is at the basal boundary, because an almost totally dry season had come prior to it (this may be a slightly conservative assumption, since pressure head at the soil–bedrock interface may assume negative values after a long dry season). As this is valid also for the first year, simulation for first event is conducted considering the water table at the soil–bedrock interface

6. The response to the generated rainfall events is simulated by the models described in Appendix B and the following procedure to be applied for $i = 1, 2, \ldots, N_{\text{RE}}$:

   a. Response in terms of pressure head ψ within rainfall events is computed using the TRIGRS model (Baum et al., 2008, 2010) (see Sect. B2). As pressure head rise may continue after the end of rainfall, the TRIGRS transient response simulation interval is prolonged $\Delta t_\text{a} = \Delta t_{\text{min}} - 1$ h after the ending time $t_{\text{end},i}$ of rainfall events.

b. The instant $t_{f,i} = \max(t_{end,i}, t_{max,i})$ is looked for, where t_{max} being the time instant at which maximum transient pressure head occurs. It follows that the final response to rainfall event i, in terms of water table height, is $\psi(d_{LZ}, t_{f,i})/\beta$, where $\beta = \cos^2\delta$ (slope parallel flow is assumed), and d_{LZ} is the soil depth. Moreover, the time interval $\Delta t_{i+1} = t_{i+1}^{(in)} - t_{f,i}$ is computed, with $t_{i+1}^{(in)}$ being the instant at which rainfall event $i+1$ begins.

c. The water table height at the beginning of rainfall event $i+1$ is computed by a sub-horizontal drainage model (see Sect. B1) which uses $\psi(d_{LZ}, t_{f,i})/\beta$ and Δt_{i+1}.

7. The result is a series of maximum pressure head, or minimum factor of safety FS responses (computed by infinite-slope stability analysis; see Sect. B3), corresponding to the N_{RE} rainfall events. The rainfall and the FS series are together analyzed to derive and evaluate landslide-triggering thresholds via a ROC-based approach (see Sect. 3).

Regarding the choice of the inter-event time Δt_{min} – an issue that is the focus of some works in literature (e.g., Restrepo-Posada and Eagleson, 1982; Bontà and Rao, 1988) – we have followed an approach analogous to that used by Balistrocchi et al. (2009) and Balistrocchi and Bacchi (2011) – for which the inter-event time may be assumed as the minimum time needed to avoid overlapping of the response produced by two subsequent rainfall events. To this end we considered that the temporal peak of pressure head due to an individual rainfall event may be reached, as mentioned above, at an instant significantly after rainfall ceases. Hence, a criterion for selecting the inter-event time has been that of choosing a value that approximates the dry time interval that contains the peak pressure head response relatively to all the N_{RE} simulated rainfall events. In our case, from preliminary simulations a $\Delta t_{min} = 24$ h appeared suitable for the hydraulic and geotechnical soil properties which are considered in this work (see Sect. 3).

Figure 1 summarizes the main steps of the described Monte Carlo methodology.

3 Threshold derivation and evaluation

3.1 Triggering and non-triggering rainfall identification

For a hillslope of given properties, Monte Carlo simulations lead to a series of time instants at which the factor of safety drops below the value of 1 (a FS = 1 down crossing).

A triggering rainfall may be associated with each down crossing, though it is noteworthy to point out that some uncertainty is present in the link between the actual failure of the slope and its theoretical instability. Nevertheless, following several works in literature (e.g., Iverson, 2000; Rosso et al., 2006; Baum et al., 2010) this uncertainty has been not taken into account here, though it may affect at a certain degree the way that rainfall events are classified as triggering and non-triggering and the subsequent ROC-based analysis (Sect. 3.2).

We investigate thresholds that are based on rainfall intensity I and duration D. Various procedures have been used to identify and compute I and D, as discussed in the "Introduction" section of this paper and by Berti et al. (2012). From a general standpoint, this procedure may be disconnected from the way event separation has been performed to compute the triggering instants with the methods described in Sect. 2. Nevertheless, for consistency with the event separation criterion that is considered in the Monte Carlo simulation methodology, it is preferable to base the procedure for identification of triggering and non-triggering events on the same inter-event time Δt_{min} used in Monte Carlo simulations.

Based on the considerations above, we adopt the following procedure for triggering and non-triggering rainfall identification. First, rainfall events are separated when their dry inter-arrival is longer than Δt_{min}. Rainfall events then have a total duration D_{tot} and mean intensity $I_{tot} = W_{tot}/D_{tot}$, where W_{tot} is the total event cumulative rainfall. For a triggering event, triggering may occur before or after the end of the rainfall event. In the first case, the critical duration D_{CR} is the time interval that starts at the beginning of the rainfall event and finishes at the triggering instant, and critical intensity is given by $I_{CR} = W_{CR}/D_{CR}$, where W_{CR} is rainfall accumulated over duration D_{CR}. In the second case it is instead characterized with D_{tot} and I_{tot}. Moreover, the P_0 events that have at their beginning a water table height $h_i \geq d_{LZ}\zeta_{CR}$, ζ_{CR} being the critical wetness ratio (corresponding to FS ≤ 1; see Eq. B6), are removed from the analysis, as the triggering is due to the preceding events, which have already been included in the set of triggering points. Non-triggering events are represented by D_{tot} and I_{tot}.

In our case (see Sect. 2) $\Delta t_{min} = 24$ h. It is worthwhile to write that with this choice the procedure of triggering rainfall identification happens to be equal to the one that was applied by Brunetti et al. (2010) in analyzing empirical data (observed landslides instead of simulated).

3.2 Uncertainty and ROC-based evaluation and optimization of thresholds

The analysis of the Monte Carlo simulations produces two sets: the set of positives P, i.e., of triggering events, and the set of negatives N, i.e., of non-triggering events. These sets may be represented as scatterplots in a double-logarithmic (D, I) plot, and in general there is a region where both sets are present – lets say, an intersection region $P \cap N$. In our framework this is due to two separate factors:

– To a given (D, I) pair there may be corresponding diverse variable NSRP-simulated hyetographs, because I

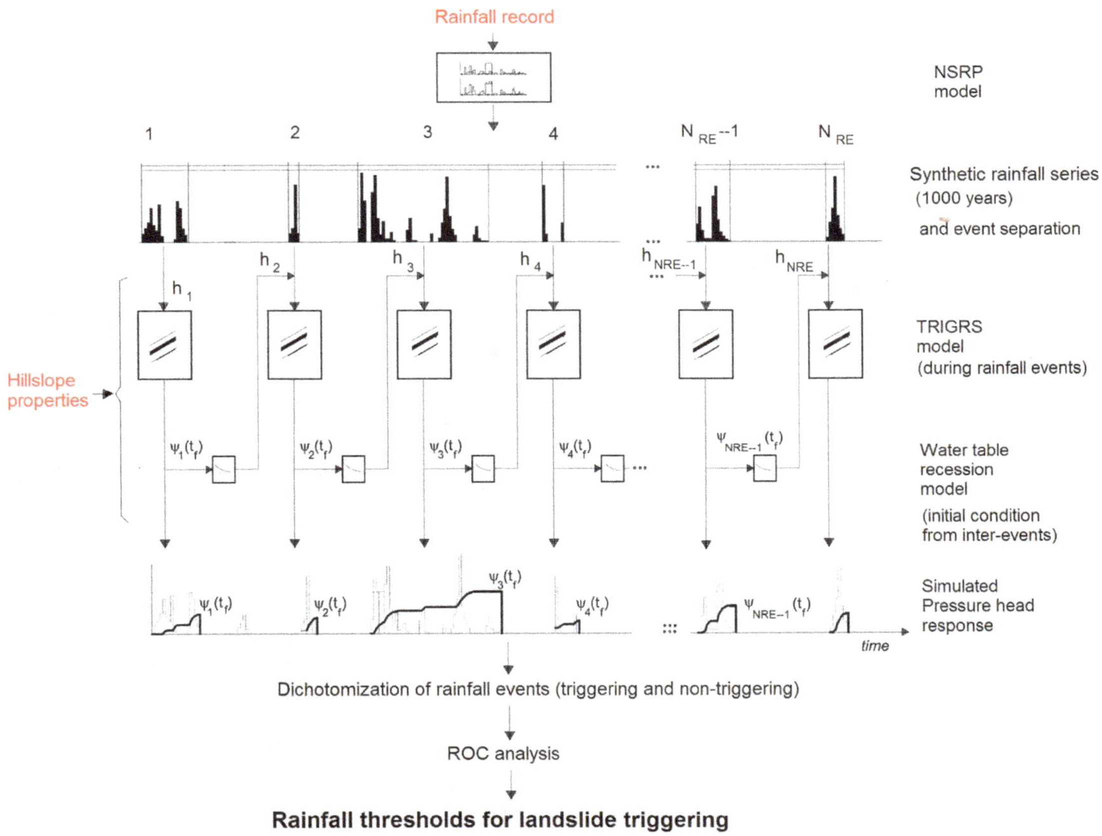

Figure 1. Scheme of the Monte Carlo methodology for derivation of landslide-triggering thresholds. Text in red indicates briefly the input required, while text in black indicates models or modeling phases.

is the mean intensity $I = W/D$ (rain-intensity variability within events)

- To a given (D, I) pair there may be corresponding diverse initial conditions (variability of initial conditions, due to variability of rainfall before the current event).

The first uncertainty factor is analyzed by letting the initial water table height $h_i = 0$ for $i = 1, \ldots N_{\mathrm{RE}}$ in performing the Monte Carlo experiments (indicated in the ensuing text as $\psi_0 = 0$ case; see also Appendix B). To investigate the second uncertainty factor, those experiments are compared with the complete ones, where the effect of initial conditions depending on past rainfall time history is taken into account by Eq. (B2), and considering different levels of memory, by varying the parameters that appear in the water table recession constant τ_M (again see Eq. B2).

Moreover, we compare the results with thresholds derived from the model by assuming uniform hyetographs as input and a prefixed initial condition (constant-intensity physically based thresholds) (cf., e.g., Rosso et al., 2006; Salciarini et al., 2008; Tarolli et al., 2011). In this case a univocal triggering threshold exists $I = f(D)$, for given hillslope properties, and the two factors of uncertainty illustrated above are not taken into account. Due to the analytical complexity of the TRIGRS (see Sect. B2) unsaturated model it is possible to determine these thresholds only numerically (not in closed form). Hence, we have derived these thresholds by simulation of infiltration and slope stability using constant-intensity hyetographs in the (D, I) domain discretized at a sufficient level, and searching the triggering curve by interpolation of the results. In doing this we have assumed an initial water table height at the soil–bedrock interface in order to properly compare results with the stochastic-input physically based thresholds of the $\psi_0 = 0$ case.

As a consequence of the presence of the region $P \cap N$, when a triggering rainfall threshold is fixed – e.g., a power-law one $I = a_1 D^{a_2}$ – the four cases of true positives, true negatives (correct predictions), false positives and false negatives (wrong predictions) can occur, as illustrated in Table 1. In general, to each pair of parameters a_1 and a_2 corresponds a prediction performance that may be measured by indices based on the number of occurrences in the four cases, denoted respectively as TP (true positive), FN (false negative), TN (true negative) and FP (false positive) (or ROC-based indices). In order to derive optimal thresholds one may maximize an objective function based on these quantities. Several

Table 1. Confusion matrix for the possible success and failure cases of a warning process based on a landslide-triggering threshold $I = f(D)$.

		Actual	
		Landslide (P)	No landslide (N)
Predicted	Landslide: $I \geq f(D)$	true positive, TP	false positive, FP
	No landslide: $I < f(D)$	false negative, FN	true negative, TN

indexes do exist and their advantages and drawbacks have been discussed by different researchers (cf. Murphy, 1996; Stephenson, 2000; Frattini et al., 2010).

Among the various possibilities, we consider for threshold evaluation the use of the true skill statistic (TSS) (also known as Hanssen–Kuipers discriminant; Hanssen and Kuipers, 1965), which was introduced by Peirce (1884), and is given by the difference between the true positive rate $\mathrm{TPR} = \frac{\mathrm{TP}}{P} = \frac{\mathrm{TP}}{\mathrm{TP+FN}}$ (also known as sensitivity or hit rate, or recall or probability of detection) and the false positive rate $\mathrm{FPR} = \frac{\mathrm{FP}}{N} = \frac{\mathrm{FP}}{\mathrm{TN+FP}}$ (also known as probability of false detection or 1 – specificity):

$$\mathrm{TSS} = \mathrm{TPR} - \mathrm{FPR}. \tag{1}$$

It is TSS = 0 for TPR = FPR (random guess) and TSS = 1 for a perfect prediction (TPR = 1 and FPR = 0). In fact this index TSS is bounded in the interval $[-1, 1]$, but negative values are fictitious as an inversion of the triggering threshold use brings TSS to its absolute value, which is always in the interval $[0, 1]$ (i.e., saying that values below the threshold trigger landslides and, vice versa, values above the threshold do not trigger landslides). Different weights may be given to the TP, TN, FP and FN, as pointed out by Peirce (1884), in order to account for the fact that a FN is more harmful than a FP (see also Peres and Cancelliere, 2012, 2013). Since data on the possible weights to assume are usually scarce, in this paper we prefer to proceed in a more simple and standard manner, where this different weighting is not considered.

We estimate the best performing power-law threshold $I = a_1 D^{a_2}$ as the one that gives the maximum value of $\mathrm{TSS} = \mathrm{TSS}(a_1, a_2)$.

At the same time the simulation-optimization methodology enables one to evaluate the use of I–D power-law thresholds, as the value of the objective function is a measure of the maximum performances that can be expected from the adopted functional form for the threshold, and thus a measure of its validity.

It is noteworthy to highlight that the real uncertainty associated with this threshold generally yields different – likely worse – performances of that assessed here, since uncertainty factors are more than the ones related to the stochastic nature of rainfall listed at the beginning of this section.

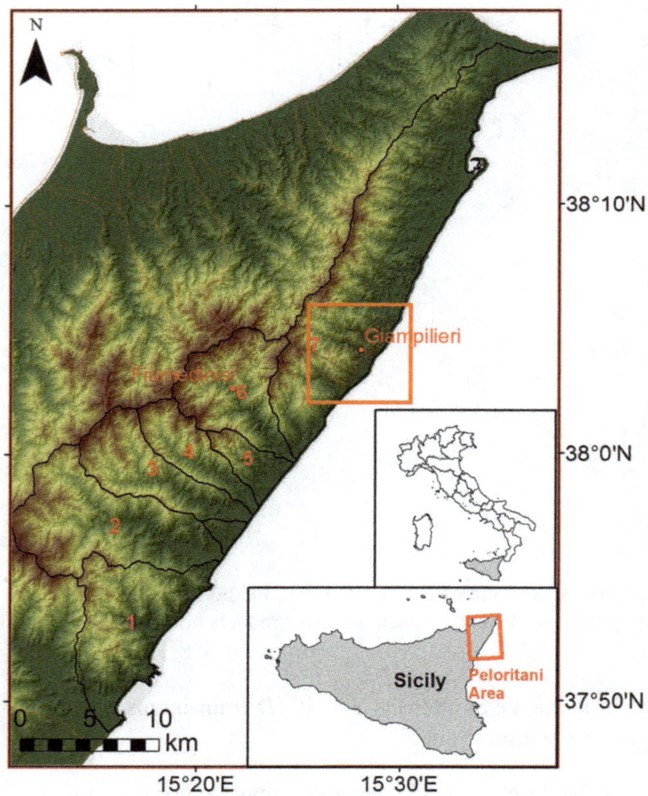

Figure 2. Map showing the location of landslide-prone study area of Peloritani mountains, Italy. The area may be subdivided into seven catchments as highlighted in the map. Relief is shown by a 5 m×5 m resolution digital terrain model based on lidar measurements in the year 2007. Location of Fiumedinisi raingauge is shown as well. The inner box contains the area surrounding the Giampilieri municipality, where a widespread rapidly moving landslide event killed 37 people on 1 October 2009 (see Fig. 3).

4 Investigated area and data

4.1 Geological setting and soil properties

An application of the described methodology is carried out to the Peloritani mountains near the Ionian coastal area, in north-eastern Sicily, Italy (Fig. 2). The mountain ridge extends longitudinally for about 50 km, with a SW–NE orientation, resulting in peaks higher than 1200 m. This area can be subdivided into seven basins: (1) minor basins between Alcantara and Agrò (70.0 km^2), (2) Agrò basin (81.8 km^2), (3) Savoca basin (44.2 km^2), (4) Pagliara basin (27.1 km^2), (5) minor basins between Pagliara and Fiumedinisi (27.1 km^2), (6) Fiumedinisi basin (49 km^2) and (7) minor basins between Fiumedinisi and Cape Peloro (172.9 km^2).

The study area belongs geologically to the Calabrian arc and represents the inner chain of the Apennin–Maghrebian mountain belt; moreover, since the middle Pleistocene, the entire Calabrian arc has undergone strong tectonic uplift. Present-day activity is testified by the strong historical seismicity (De Guidi and Scudero, 2013, and references therein). In the area outcropping lithologies consist of (Lentini et al., 2000) (1) phyllites of the Mandanici units; (2) paragneiss and mica schists of the Mela units; and (3) gneiss and paragneiss metamorphic rocks (Aspromonte units). Further information on the geological setting, is given in Lentini et al. (2000), Goswami et al. (2011) and De Guidi and Scudero (2013).

In the last decade, this area has been hit by highly damaging diffused shallow landslides. Precisely, widespread landslide events occurred in this area on (I) 15 September 2006 (areas 4, 5, 6 and 7), (II) 25 October 2007 (area 7), (III) 24 September 2009 (areas 1, 2, 3, 4, 5) and (IV) 1 October 2009 (area 7). The associated areas indicated in parentheses have been derived from newspapers archives (cf., e.g., http://gazzettadelsud.virtualnewspaper.it/gdsstorico/), which also present further information on the events. Among these events, the one that occurred on 1 October 2009 was the most severe. On that date, landslides were triggered by a rainfall event of more that 220 mm in less than 7 h, causing 37 deaths and innumerable injured people, most of them in the municipality of Giampilieri (Ardizzone et al., 2012).

A map of the landslides which occurred on 1 October 2009, derived from interpretation of orthophotos, is available from website http://www.regione.sicilia.it/presidenza/protezionecivile/documenti/rischi/r_idrogeologico/documenti/20091001_CartaDissesti.jpg. This map presents in red the slide/erosion areas and in orange the propagation/deposition areas. Figure 3 shows the slides derived from red areas of that map. The analysis of the specific upslope contributing area A/B (ratio between the upslope draining area A and the contour length B; see Appendix B) and the slope δ within the slide areas, based on a pre-event DTM at a 5 m resolution, shows that the most populated class of A/B is centered on the value of 10 m, while the mean slope within the range of theoretical potentially unstable slopes $29° \leq \delta \leq 47°$ is slightly lower than 40°. Moreover, the values of $A/B = 10$ m and $\delta = 40°$ correspond to a portion of the Peloritani mountains for which it starts to be worthy of issuing landslide early warnings. Hence these values may be adopted for the successive derivation of a threshold for the area (see Sect. 5).

Core samples collected in the area indicate the presence of surficial debris material $d_{LZ} = 2$ m deep, covering a fractured bedrock strata. The debris cover consists of a sandy loam with a significant proportion of gravel (up to 50 %) which

Table 2. Parameters of the NSRP rainfall model resulting from calibration on Fiumedinisi rainfall data, for the four homogeneous rainy seasons (the Weibull shape parameter has been fixed to $b = 0.6$).

Parameter	Jan, Feb, Mar	Sep, Oct	Nov	Dec
λ [h^{-1}]	0.002295	0.021195	0.001485	0.003185
ν	44.28	1.57	42.41	42.61
β [h^{-1}]	0.010161	2.1179	0.0059551	0.0098760
η [h^{-1}]	0.72113	0.83999	0.94053	0.67735
ξ [h^b mm^{-b}]	1.13441	0.46260	0.69261	1.03521

corresponds to a gravelly sandy loam according to USDA soil classification. The assumption of a leaking basal boundary, characterized by a given c_d ratio (see Sect. B2) is realistic, given the fractured bedrock strata. We assume that the hydraulic and geotechnical properties of Table 3 may represent the natural heterogeneity within the study area.

Spatial variability of each of the soil properties could be included in our model simulations. Nonetheless, detailed information on how the properties are distributed spatially is unavailable. Hence we preferred to carry out a sensitivity analysis, by varying the hydraulic conductivity K_S, the leakage ratio c_d and the soil depth d_{LZ} according to Table 4, and the critical wetness ratio in the range $0 < \zeta_{CR} < 1$. To proceed in this way enables one to better analyze the way model results are influenced by these variables rather than assuming that they are distributed spatially with interpolating laws of difficult validation. Since slope mainly affects slope stability (Eq. B5) rather than the infiltration process, variation of slope is indirectly taken into account by variation of ζ_{CR}. It is noteworthy to mention that an alternative approach may be to consider model parameters generated according to a probability distribution, as proposed by the TRIGRS-P modification of the TRIGRS code, developed by Raia et al. (2014).

4.2 Rainfall data and NSRP model calibration

Climate in the Peloritani area is Mediterranean with with hot and dry summers, and precipitation – mainly convective – falling mostly in the period from October to January.

For calibration of the NSRP model, the rainfall series measured at the Fiumedinisi rain gauge from 21 February 2002 to 9 February 2011 (almost 9 years) has been used (see Fig. 2). Based on a preliminary analysis of monthly statistics, six homogeneous rainfall seasons have been identified: (1) September and October, (2) November, (3) December, (4) January–March, (5) April and (6) May–August (see Fig. 4 and its caption). Separate sets of parameters of the NSRP model have been determined for each one of the four rainy seasons (in total $5 \cdot 4 = 20$ parameter values), while the last two seasons have been considered to have negligible rainfall. The Weibull shape parameter b has been fixed to 0.6 for all seasons, based on different trials. Parameters obtained from calibration are shown in Table 2.

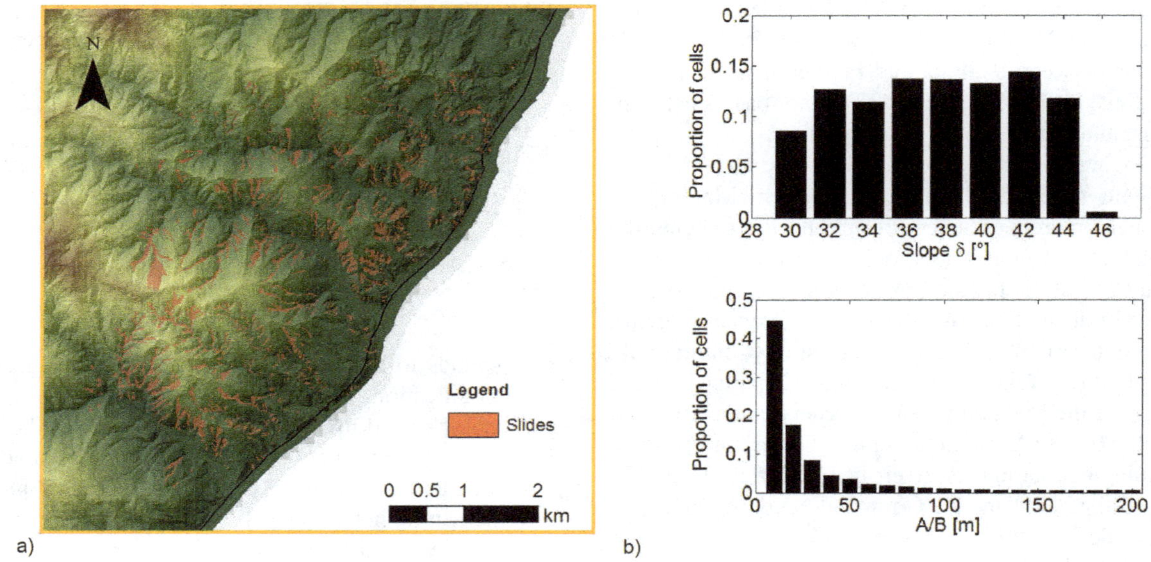

Figure 3. **(a)** Enlargement of inner box of the map of Fig. 2 showing slides occurred on 1 October 2009. Polygons were derived from slide (red) areas reported in map of Civil Protection Department of the Sicilian Region (source: http://www.regione.sicilia.it/presidenza/protezionecivile/documenti/rischi/r_idrogeologico/documenti/20091001_CartaDissesti.jpg). In all, 643 slides, occupying a total area of $2.30\,\mathrm{km}^2$, are present in the map. **(b)** Statistical distribution, within the slide areas, of slope δ and of upslope specific contributing area A/B.

Table 3. Soil properties considered for application of the model to the Peloritani mountains case-study area.

ϕ' [°]	c' [kPa]	γ_s [$\mathrm{N\,m^{-3}}$]	θ_s [–]	θ_r [–]	K_s [$\mathrm{m\,s^{-1}}$]	D_0 [$\mathrm{m^2\,s^{-1}}$]	α [$\mathrm{m^{-1}}$]	d_{LZ} [m]	δ [°]	ζ_{CR} [–]	c_d [–]
37	5.7	19 000	0.35	0.045	2×10^{-5}	5×10^{-5}	3.5	2	40	0.4645	0.1

From the assumed inter-event time $\Delta t_{min} = 24\,\mathrm{h}$ and soil properties of Table 3 the resulting number of rainfall events is $N_{RE} = 19\,826$ (in average 19.83 events per year). This number derives from the initial 28 751 events from which the events with hourly intensities below $i_{min} = c_d K_s (1 - \cos^2\delta) = 2.975\,\mathrm{mm\,h^{-1}}$ were cut. These values are statistically comparable to the ones on the observed series (19.18 events/year from 28.91 events/year before the cut of under-leakage events).

5 Results and discussion

5.1 Derivation and evaluation of rainfall thresholds

In Fig. 5 the scatterplots of triggering and non-triggering events in the $\log(D)$–$\log(I)$ plane, derived from analysis of Monte Carlo simulations, are shown for the $\psi_0 = 0$ case and for specific upslope contributing areas $A/B = 10, 20\,\mathrm{m}$ ($\tau_M = 2.75, 5.49$ days). Related results are also shown in Table 5. In the figure, red points represent triggering rainfall events, or the set of positives P, while green points represent the non-triggering ones, or the set of negatives N.

Optimal thresholds have been derived by maximization of the TSS index (see Eq. 1), preliminarily by considering both the power-law coefficient a_1 and exponent a_2 as variable parameters. Inspection of the results revealed minimal changes of the exponent a_2 with changing ratio A/B, and so a second optimization has been carried out only with reference to the a_1 parameter, fixing the exponent a_2 to its mean value of $a_2 = -0.8$. Fixing the exponents forces the different thresholds corresponding to different A/B ratios to be parallel and therefore to not intersect each other; this is somehow consistent with the fact that as the A/B increases, landslides are generally more likely to occur for less severe rainfall events because of increased past-rainfall memory.

From the case of $\psi_0 = 0$, i.e., of an initial water table at the soil–bedrock interface for all events (Fig. 5a), where simulated uncertainty of triggering is due only to the variation of rain intensity within events, it is seen that in this case the region in which triggering and non-triggering events coexist is quite narrow; moreover, a power-law relation between I and D dichotomizes well between triggering and non-triggering conditions. In fact, the optimal power-law threshold in this case has a reliability of TSS $= 0.991$, practically

Table 4. Varied soil properties considered for sensitivity analysis.

K_S $[\mathrm{m\,s^{-1}}]$	D_0 $[\mathrm{m^2\,s^{-1}}]$	c_d $[-]$	d_{LZ} [m]	$\tau_M(A/B = 10\,\mathrm{m})$ [days]
1×10^{-5} (36 mm h^{-1})	2.5×10^{-5}	0.1	1, 1.5, 2	5.5
2×10^{-5} (72 mm h^{-1})	5×10^{-5}	0.05, 0.1, 0.2	1, 1.5, 2	2.7
3×10^{-5} (108 mm h^{-1})	7.5×10^{-5}	0.1	1, 1.5, 2	1.8

Table 5. ROC-based indices for the derived best power-law stochastic-input physically based thresholds (S) and comparison with constant-intensity input physically based ones (C).

$\frac{A}{B}$ [m]		a_1 [mm h^{-1}]	TP	TN	FN	FP	P_0	TPR	FPR	TSS
0 ($\psi_0 = 0$)	S	101.49	81	19 558	0	187	0	1.000	0.009	0.991
	C	–	52	19 744	29	1	0	0.642	0.000	0.642
10	S	71.52	104	19 037	11	672	2	0.904	0.034	0.870
	C	–	52	19 708	63	1	2	0.452	0.000	0.452
20	S	42.95	164	17 131	26	2488	17	0.863	0.127	0.736
	C	–	52	19 618	138	1	17	0.274	0.000	0.274

equal to the ideal value of 1. Additional insights on the effect of variability of rainfall intensity within events may be derived comparing the scatterplots for this $\psi_0 = 0$ case with the constant-intensity physically based threshold (also determined considering $\psi_0 = 0$) represented in the plots of Fig. 5 as a dashed black line. Figure 5a reveals that the deterministic threshold approximates the lower envelope curve of critical events (red dots) for short durations – that, for the analyzed data, corresponds to durations less than about 12 h. For higher durations, this is no longer true and variable–intensity hyetographs start to have a higher destabilizing effect than the constant-intensity ones of same rainfall volume. The variability of rainfall intensity within events leads to a deviation from the deterministic line of the triggering NSRP rainfall-event points, making the scatter of triggering points more similar to a straight line than to the curved deterministic threshold. This behavior is essentially due to the presence of the leakage term $q_l = \min\{c_d K_s(1-\cos^2\delta), q(d_u, t)\}$, whose effect is stronger for uniform hyetographs than for variable ones, since in the former there are no peaks of intensity. In particular, a uniform hyetograph produces no water table rise if intensity is below a rate slightly greater than $c_d K_s(1-\cos^2\delta)$, because all infiltrating water, after percolating through the unsaturated-zone, goes to basal loss. The same does not generally occur for a variable intensity hyetograph of the same volume, because instantaneous intensity may be significantly higher than the event mean intensity W_{tot}/D_{tot}, and consequently a water table rise is produced. The opposite behavior for short durations is due to the fact that in this case variable hyetographs may have peaks of intensity higher than infiltration capacity, and thus not all rainfall infiltrates into the soil. Due to these reasons, the model deterministic threshold results in a poor erformance (TSS = 0.642).

The above results lead to conclude that it is important to account for variability of intensity during events and that landslide occurrence related to the transient part of the response to rainfall events can be represented with good approximation by a I–D power-law equation. This provides a physically based justification for such a widely used threshold form, which turns out to be valid when landslide occurrence is mostly due to the transient part of the hillslope response to rainfall.

For the $A/B = 10$ m case (Fig. 5b), which may represent prevalent conditions for the Peloritani mountains area (see Sect. 4.1), scattering of the red dots increases due to the introduced variability of initial conditions. Consequently, performances of predictions based only on intensity and duration of rainfall events become worse. Simulations for larger values of the specific catchment area (e.g., $A/B = 20$ m, Fig. 5c) confirm this conclusion.

Based on these results, it may be stated that, for a given climatic input, performances of thresholds which do not account for past rainfall time history (antecedent rainfall) are expected to decrease as the water table recession time constant τ_M increases. Rainfall time history occurring before single rainfall events generally influences landslide triggering, determining whether a threshold based only on rainfall intensity and duration may be sufficient or the I–D threshold needs to be improved by the introduction of antecedent rainfall variables.

Finally, the threshold

$$I = 71.52 D^{-0.8} \tag{2}$$

may be a reasonable choice for the Peloritani mountains area since performances are still high, since TSS = 0.862.

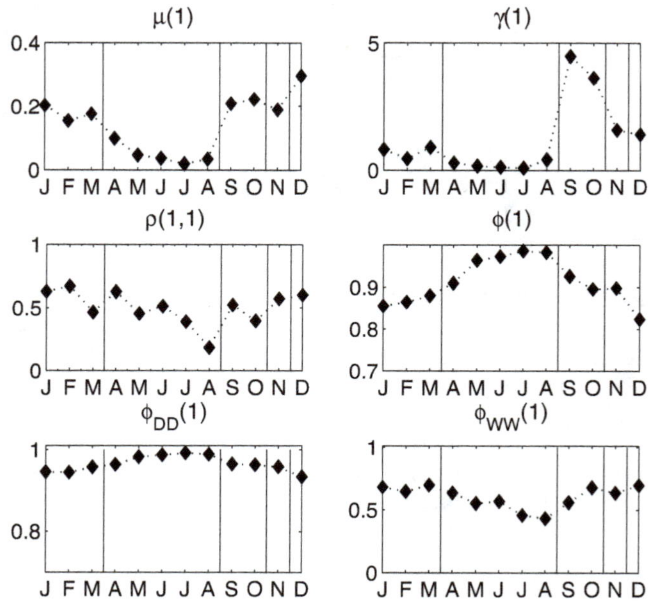

Figure 4. Moments for each month for Fiumedinisi SIAS hourly rainfall data. In particular μ denotes the mean, γ the variance, $\rho(1,1)$ the linear autocorrelation coefficient at lag = 1, ϕ the probability of a dry interval, ϕ_{DD} the probability that a given interval is dry after another dry one, ϕ_{WW} the probability that a given interval is wet after another wet one. These moments have been used in calibration of the NSRP model via the method of moments. It can be seen that there are low differences of most of the moments within the following groups of months: Sept–Oct, Nov, Dec, Jan–Mar, Apr, May–Aug. A separate set of NSRP model parameters was calibrated for each of the first four of these seasons, while the period April–August has been neglected from the successive analyses because precipitation rates are so low that it is very unlikely that a triggering event occurs in such period.

5.2 Validation of the threshold using observed data and comparison with other thresholds

The Monte Carlo simulation technique provides a framework that is useful for exploiting the information contained in the observed rainfall series and the physics of the modeled phenomenon. Nonetheless, it remains important to validate the results against observed data, to check if the models are capable of reproducing characteristics of interest which are not directly taken into account in model calibration and development.

Here we perform a global validation by comparing the derived threshold (Eq. 2) with the triggering and non-triggering observed rainfall events.

In particular we have derived from the series the rainfall events with the same criterion adopted in Monte Carlo simulations. Yet the events in the months neglected there (April–August) and the events with intensities below the leakage flow $c_d K_S(1-\cos^2(\delta))$ were not removed here in the observed record, for the test to be unbiased to this preprocessing

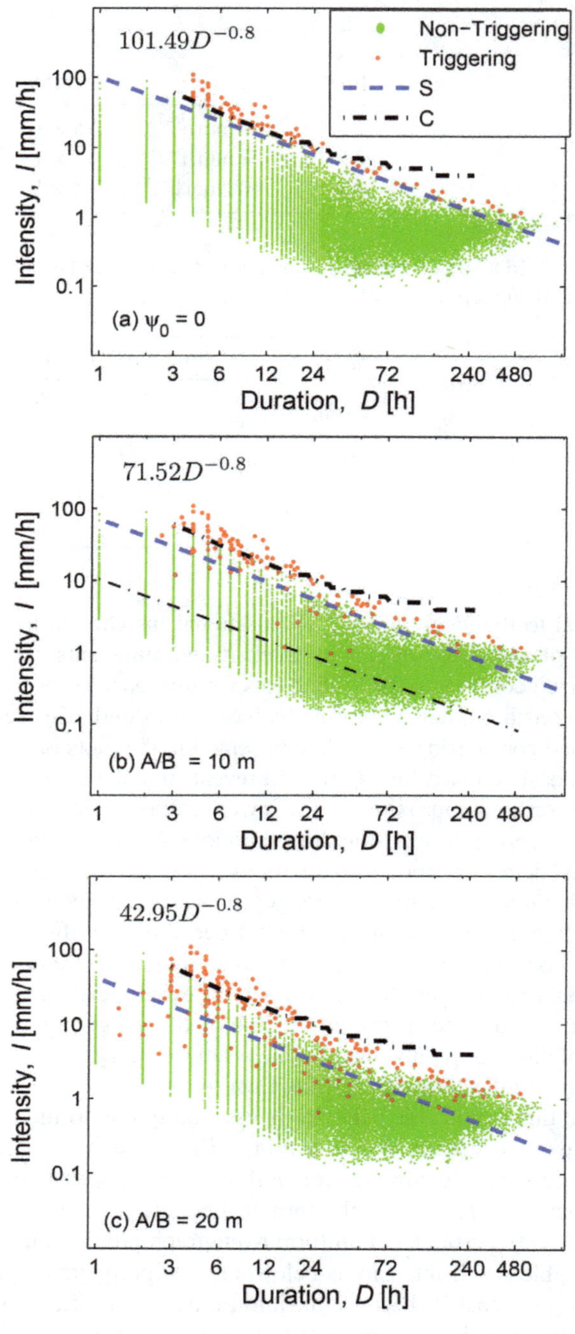

Figure 5. Derivation of thresholds from ROC optimization of Monte Carlo simulations. Red points represent triggering simulated rainfall, while green ones represent the non-triggering. The best power-law stochastic-input physically based thresholds (S) are derived by maximization of the TSS ROC-based index. The constant-intensity input physically based threshold (C) is determined considering the response to uniform hyetographs and water table initially at the basal boundary. **(a)** zero memory case $\psi_0 = 0$, **(b)** $A/B = 10$ m (threshold derived by Gariano et al. (2013) for Sicily is shown as well, thin dotted line, for comparison with the one derived) and **(c)** $A/B = 20$ m.

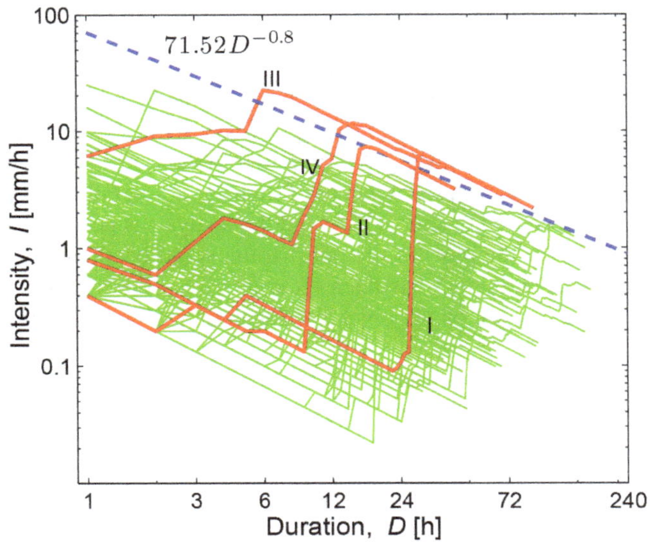

Figure 6. Validation of threshold-derivation procedure with observed rainfall events. Red lines indicate mean intensity $I(D) = W(D)/D$ time histories that exceed the derived threshold. Green lines represent observed events that do not exceed the threshold. Threshold is exceeded for all and only the four dates in which landslides occurred in the Peloritani area: (I) 15 September 2006, (II) 25 October 2007, (III) 24 September 2009 and (IV) 1 October 2009.

of data. This resulted in 190 events, whose temporal evolution of accumulated intensities $I(D) = W(D)/D$ has been compared with the derived threshold, as shown in Fig. 6. The figure indicates positive validation of the methodology, as the events in the I–D plane that exceed the threshold are all and only the four events that have triggered landslides in the considered period (red-line time histories). This is the best result one can obtain from this test, but it is perhaps noteworthy to clarify that it is expected that in the long period the same test will not perform without errors, consistently with the Monte Carlo simulations and the way that the threshold was derived.

Comparison with other thresholds may also help in understanding how reliable the performed analysis is. Gariano et al. (2013) proposed for Sicily the threshold $E = 10.4D^{0.22}$, where $E = I \times D$ is cumulative event rainfall, and hence threshold is equivalent to $I = 10.4D^{-0.78}$. This threshold has been derived considering only observed triggering events and it is corresponding to an exceedance frequency of 1%. It is firstly interesting to notice that the exponent is practically equal to the one that results from our analyses ($a_2 = -0.8$). Furthermore, as can be seen from Fig. 5b this threshold exceeds one triggering event of the Monte Carlo simulated data, which constitutes the 1% of the triggering-rainfall data set (see Table 5: $0.01 \times (\mathrm{TP}+\mathrm{FN}) = 0.01 \times (104+11) = 1.15$). This result is a further support to the validity of the performed Monte Carlo analysis and highlights the importance to take into account non-triggering rainfall in assessing threshold performance.

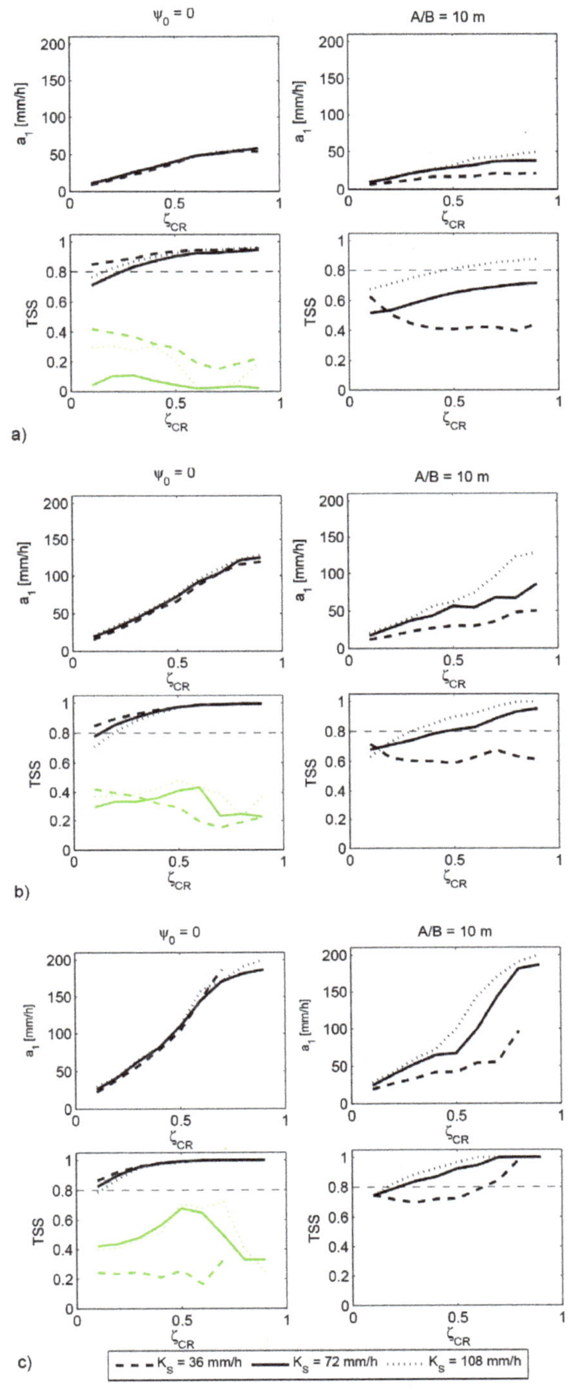

Figure 7. Sensitivity of triggering thresholds and of the relative performances to hydraulic conductivity K_S (cf. Table 4). Plots show, for the two cases of zero memory ($\psi_0 = 0$) and $A/B = 10$ m, how the a_1 power-law coefficient of optimized stochastic-input physically based thresholds and relative TSS performance index vary with the critical wetness ratio ζ_{CR}. Performances of the constant-intensity physically based thresholds are shown as well (in green). Different soil depths are considered: **(a)** $d_{LZ} = 1$ m, **(b)** $d_{LZ} = 1.5$ m and **(c)** $d_{LZ} = 2$ m.

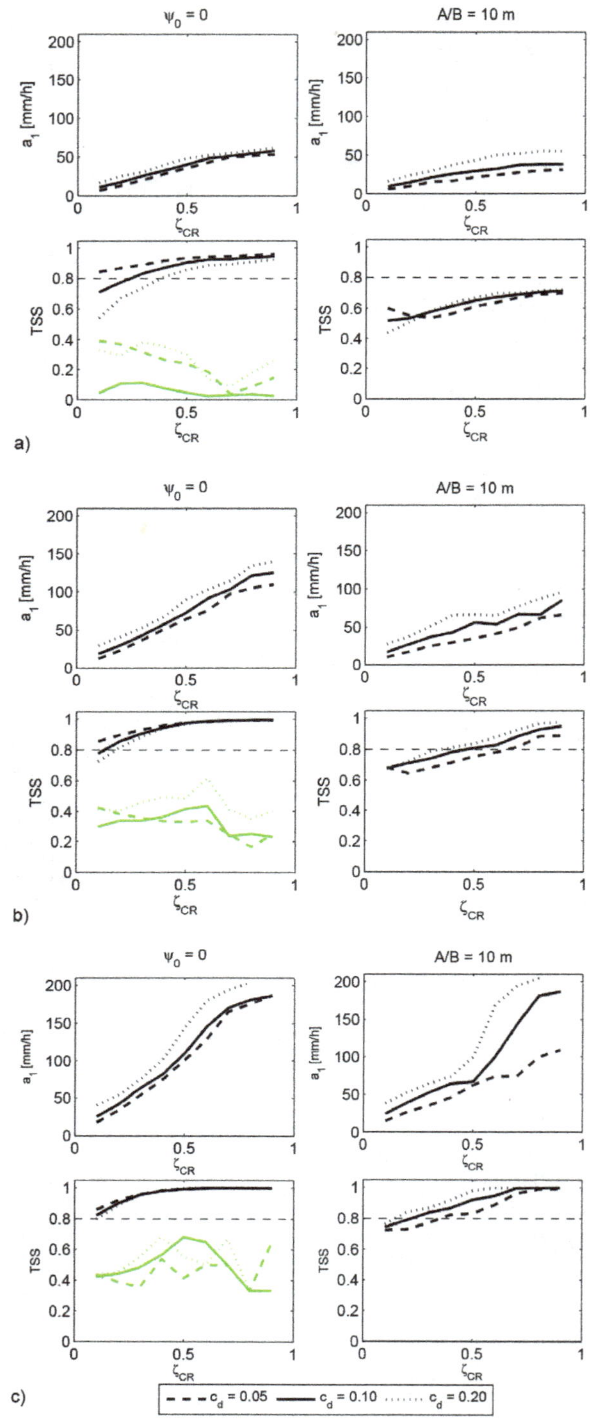

Figure 8. Sensitivity of triggering thresholds and of the relative performances to the leakage ratio c_d (fractured bedrock) (cf. Table 4). Plots show, for the two cases of zero memory ($\psi_0 = 0$) and $A/B = 10\,\text{m}$, how the a_1 power-law coefficient of optimized stochastic-input physically based thresholds and relative TSS performance index vary with the critical wetness ratio ζ_{CR} . Performances of the constant-intensity physically based thresholds are shown as well (in green). Different soil depths are considered: **(a)** $d_{LZ} = 1\,\text{m}$, **(b)** $d_{LZ} = 1.5\,\text{m}$ and **(c)** $d_{LZ} = 2\,\text{m}$.

5.3 Sensitivity analysis

A sensitivity analysis has been conducted with respect to the following variables (Sect. 4 and Table 4): the hydraulic conductivity K_S, the leakage ratio c_d, the soil depth d_{LZ} and the critical wetness ratio ζ_{CR}. Plots similar to those of Fig. 5 can be derived for each set of values of such variables. For brevity those plots are not shown here and the analysis is performed considering the plots of the optimal threshold coefficient a_1 (again the exponent has been fixed to $a_2 = -0.80$) and the maximum value of the objective function TSS as functions of ζ_{CR}.

The results of this analysis are shown in Figs. 7–8, which can be commented on as follows:

- Sensitivity to hydraulic conductivity: in the $\psi_0 = 0$ case the variation of K_S induces relevant changes neither in the threshold nor in the performance TSS. It can however be hypothesized that considering more low values of K_S, infiltrating rates more strongly depend on how rainfall is distributed within the event and thus uncertainty increases. Conversely, to an higher K_S variation neither of the threshold nor of the performance may be observed, since infiltration capacity will always be higher than rainfall intensity, which then infiltrates totally. In the $\frac{A}{B} = 10\,\text{m}$ case the variations of a_1 and TSS are relevant and due to increased memory with decreasing K_S; the threshold decreases with K_S and the associated uncertainty increases (lower TSS).

- Sensitivity to leakage ratio: both in the $\psi_0 = 0$ and the $\frac{A}{B} = 10\,\text{m}$ cases, the increase of the c_d ratio induces an increase of the threshold and in the relative uncertainty. Such a variation is of comparable magnitude in the two cases. This happens because the variation of c_d affects only pressure head response during rainfall events, but does not affect significantly memory due to antecedent rainfall. Sensitivity to c_d increases with soil depth, because the increased water absorption in the unsaturated strata and the consequent increased damping and smoothing effect induces an increase of the portion of infiltrating water that goes to leakage. Indeed this affects more the threshold (a_1) than the relative performance (TSS).

- Variation of ζ_{CR} and of soil depth d_{LZ}: with increasing ζ_{CR} the threshold increases, while the associated uncertainty decreases. The threshold and relative performances decrease with soil depth. This indicates that landslides become less predictable as soil depth d_{LZ} and ζ_{CR} diminish.

Generally, antecedent rainfall has to be taken into account to improve performance of landslide-triggering thresholds based on rainfall. Nonetheless, the use of only the I and D variables may still lead to good performing thresholds when memory is relatively low, soil thickness is not too shallow and hillslope is naturally not close to instability (ζ_{CR} is relatively high). In fact I–D power-law thresholds resulted in good performance (TSS > 0.8) when $\tau_M \leq 3$ days, $\zeta_{\mathrm{CR}} > 0.5$ and $d_{\mathrm{LZ}} \geq 1.5$ m.

The constant-rainfall physically based thresholds always perform poorly. This confirms that variability of intensity during rainfall events influences significantly rainfall intensity and duration associated with landslide triggering.

6 Conclusions

In this work it has been shown how stochastic rainfall models and hydrological and slope stability physically based models can be advantageously combined in a Monte Carlo simulation framework to derive and evaluate landslide-triggering thresholds. The approach synthesizes research findings in the fields of rainfall and landslide hydrological modeling to provide an output that is easily implemented in a early warning system, i.e., a landslide-triggering threshold, based on rainfall monitoring, of the same type that is commonly derived by direct empirical analysis of observed rainfall and landslide data. The advantages of the approach consist in a better exploitation of the information contained in observed rainfall series and measurements of hydraulic, geotechnical soil properties and geomorphological analysis. Because both triggering and non-triggering rainfall events are taken into account, the approach enables a more correct derivation and evaluation of thresholds, for which well-known prediction-skill receiver operating characteristic (ROC) analysis may be advantageously used to reduce subjectivity in the identification of thresholds and to estimate the convenience of the use of the threshold within a landslide early warning system.

Furthermore, the specific modeling framework implemented in this work enabled to analyze some general issues on landslide-triggering phenomena regarding its controlling factors and uncertainty related to variability of rainfall intensity within events and past rainfall (antecedent rainfall), with a particular focus on the widely used power-law rainfall intensity–duration threshold form. In particular, from the application to the Peloritani mountains area in north-eastern Sicily (Italy) and the conducted sensitivity analysis on various controlling parameters, the following conclusions can be drawn: (1) variability of intensity during rainfall events significantly influences rainfall intensity and duration associated with landslide triggering. In particular constant-intensity input thresholds perform conservatively only for low rainfall durations, while the opposite occurs for events of longer duration. On the other hand, when a time-variable rainfall-rate event is considered, the simulated triggering points may be separated with a very good approximation (i.e., true skill statistic is close to 1) from the non-triggering ones by a I–D power-law equation. This indicates that this widely used model is adequate to represent the triggering part due to transient infiltration produced by rainfall events. Thus, this gives a physically based justification for such a widely used threshold form, which turns out to be valid when landslide occurrence is mostly due to that part. This depends, for a given rainfall climate, mostly on the timing of recession of the saturated zone occurring during dry inter-event intervals (in our model represented by the constant τ_{M}), but also on the other soil hydraulic and geotechnical parameters, and in particular on soil depth d_{LZ}, which must not be too shallow, and critical wetness ratio ζ_{CR}, that must be not too low. For instance, for the case-study area, the I–D power-law threshold performs with a TSS > 0.80 when it is $\tau_M \leq 3$ days and $d_{\mathrm{LZ}} \geq 1.5$ m and $\zeta_{\mathrm{CR}} > 0.50$. (2) In general, rainfall time history occurring before single rainfall events influences landslide triggering, determining whether a threshold based only on rainfall intensity and duration may be sufficient or the I–D threshold needs to be improved by the introduction of antecedent rainfall variables. (3) Sensitivity analysis indicated that in general threshold performance is affected by the depth of the basal boundary and the critical wetness ratio that represents the natural degree of stability of the hillslope. In particular, uncertainty of landslide-triggering prediction increases as the soil depth or the critical wetness decrease. Hence, it is more difficult to predict landslides the more an hillslope is shallow and the more it is naturally close to instability. A decrease of performance is obtained as basal drainage (leakage) increases as well. Hence the I–D power-law may not be performing adequately in the case that the bedrock is significantly fractured and the soil cover is very shallow. Results also indicate that hydraulic conductivity mainly influences memory of past rainfall and only slightly affects the uncertainty related to variability of rainfall intensity within events.

Further ongoing research is oriented to introduce additional information in the derivation of the thresholds, such as antecedent precipitation as well as indexes representative of the shape of the hyetograph.

Appendix A: Stochastic rainfall model

Stochastic rainfall point models are aimed at the generation realistic synthetic time series of (virtually) unlimited length, by calibration based on a observed rainfall series. The uninitiated reader is invited to read Neyman and Scott (1958), Kavvas and Delleur (1975), Waymire and Gupta (1981a, b, c), Rodriguez-Iturbe et al. (1987a, b), Salas (1993) and Cowpertwait et al. (1996). Here we give some specific details on the NSRP model, for it being the one used in this work to model rainfall at a site.

NSRP process belongs to the so-called class of cluster models (cf., e.g., Salas, 1993). The NSRP cluster model consists in a two-level mechanism process. This process is related to a conceptualization of meteorological processes that originate rainfall events (Foufoula-Georgiou and Guttorp, 1989), and it is obtained by the following steps (see Fig. A1):

- First, clusters – also known as storm-generating systems, or simply storms – arrive governed by a Poisson process of parameter λt (this is the first-level mechanism)
- For each cluster origin, rectangular pulses (rain cells) are generated (this is the second-level mechanism). The number of pulses C associated with each storm is extracted from another separate Poisson distribution. In order to have realizations of C not less than one, it is assumed that $C' = C - 1$, with $c' = 0, 1, 2, \ldots$ (which implies $c = 1, 2, 3, \ldots$), is Poisson distributed with mean $\nu - 1$
- Each cell has origin at time $\tau_{i,j}$ with $j = 1, 2, \ldots, c_i$ measured from t_i, according to an exponential random variable of parameter β
- A rectangular pulse of duration $d_{i,j}$ and intensity $x_{i,j}$ is associated with each rain cell. Pulses have duration exponentially distributed with parameter η while intensities X are extracted from a Weibull distribution (cf. Cowpertwait et al., 1996), which has cumulative distribution function $F(x; \xi, b) = 1 - \exp(-\xi x^b)$
- Finally, the total intensity at any point in time is given by the sum of the intensities of all active cells at that point.

We have calibrated the NSRP model by the method of moments, i.e., using the properties of the aggregated NSRP process $Y(\tau)$ at different timescales of aggregation τ (cf., e.g., Rodriguez-Iturbe et al., 1987a, b; Cowpertwait et al., 1996; Calenda and Napolitano, 1999). According to this method, model parameters, i.e., λ, ν, β, η and ξ (b is typically fixed, in the range $0.6 \leq b \leq 0.9$; see Cowpertwait et al., 1996) are estimated using at least as many moments as the parameters of the model, considering different statistics (moments) at various time aggregations, and solving the related equation system, where the theoretical expressions, containing the parameters, are equated to the sample moments. Theoretical moments of $Y(\tau)$, such as the mean $\mu(\tau)$, variance $\gamma(\tau)$ and autocorrelation at lag k, $\rho(\tau, k)$, are given by formulas derived by Rodriguez-Iturbe et al. (1987a). Transition probabilities were derived as well, by Cowpertwait (1991), and have been included in the calibration process. We have solved the nonlinear equation system by numerical minimization of an objective function $S(\lambda, \nu, \beta, \eta, \xi)$, that measures the global relative error between theoretical and sample moments (cf. Cowpertwait et al., 1996).

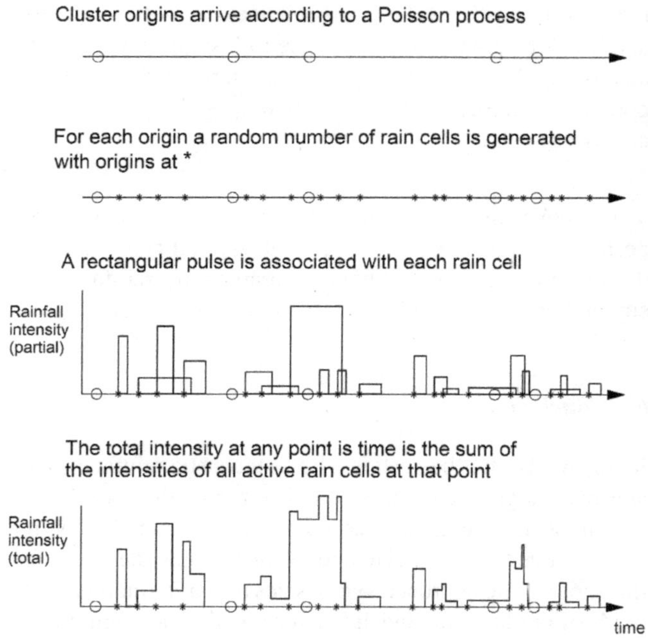

Figure A1. Representation of the NSRP stochastic process for at-site rainfall modeling (adapted from Cowpertwait et al., 1996).

Though calibration is conducted taking into account seasonality, by calibrating the model separately for the various homogeneous seasons within the year (Sect. 4.2), it is noteworthy to point out that the generated series is globally stationary, and consequently eventual annual non-stationarity due to climate change is not taken into account. In other words, the generated series represents possible realizations of the rainfall events under current climate conditions, the final aim being of deriving thresholds suitable under the present climate and not to assess how climate change may affect them.

Appendix B: Hillslope hydrological and stability model

The total pressure head response ψ of an hillslope soil to a rainfall event is given by the sum of a transient part ψ_1 and an initial part ψ_0 (cf. Iverson, 2000; D'Odorico et al., 2005).

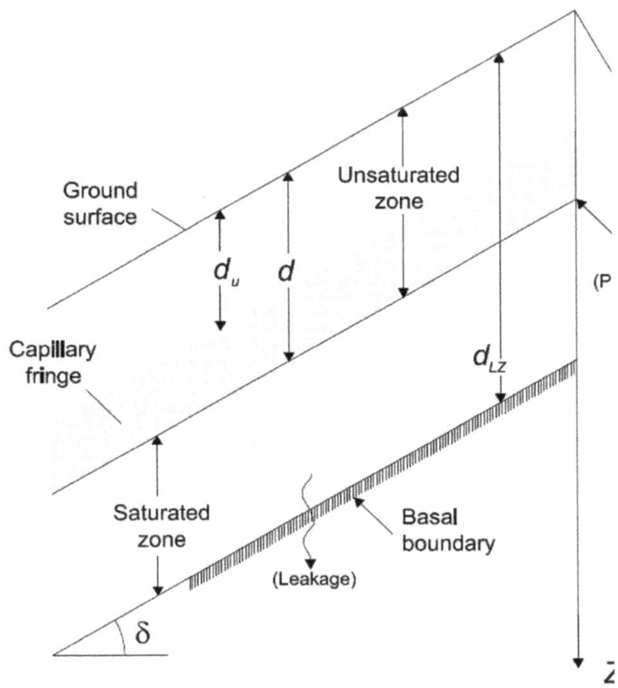

Figure B1. Soil 1-D vertical scheme used to model infiltration and slope stability based on the TRIGRS unsaturated model (adapted from Baum et al., 2008).

The transient part is due to infiltration of event rainfall, while the initial part depends on rainfall time history before the current rainfall event. As pointed out by Iverson (2000), for soils that are relatively shallow, i.e., when the ratio between soil depth and the square root of the upslope contributing area is small, $\varepsilon = d_{LZ}/\sqrt{A} \ll 1$, the prevailing process that determines ψ_1 is 1-D vertical infiltration, while in the dry periods in between events, the prevailing process is sub-horizontal drainage.

Based on these considerations, we use a vertical infiltration model for computing ψ_1, the TRIGRS unsaturated model (Baum et al., 2008, 2010), and a linear reservoir sub-horizontal drainage model to compute the initial conditions from the water table height at the end of the rainfall event preceding the current one (which in turn depends on past rainfall time history). This latter model is derived from a mass-conservation equation of soil water coupled with the Darcy's law used to describe seepage flow, where for simplicity the soil volumetric strain is neglected (the variation of porosity with pressure head is assumed null). A similar conceptualization is the basis of the model proposed by Rosso et al. (2006).

In order to understand the controlling factors of landslide triggering, it is useful to separate the analysis of the response to rainfall in terms of the transient part only. This may be done by performing the simulations assuming $\psi_0 = 0$ at the beginning of events.

From the pressure head response, the factor of safety FS for slope stability is computed, using a infinite slope model.

The description of these model components follows.

B1 Initial conditions model

The initial condition to rainfall event i is computed from the response at the end of rainfall event $i-1$, using a water table height h recession model between storms based on the following mass-conservation equation (Rosso et al., 2006):

$$B\ h\ K_s \sin\delta = -A\ (\theta_s - \theta_r)\frac{dh}{dt}, \quad \text{(B1)}$$

where A is the contributing area draining across the contour length B of the lower boundary of the hillslope, δ is the inclination of the hillslope, K_s is the saturated hydraulic conductivity, and $\theta_s - \theta_r$ is soil porosity, θ_s and θ_r being the saturated and residual soil water contents, respectively. The ratio A/B, which can be computed based on a digital terrain model (DTM), is the well-known specific upslope contributing area, an important variable on which the topographic control on shallow landslide triggering depends (Montgomery and Dietrich, 1994). For instance, it is $A/B = B\ N_d$, where N_d is the number of cells draining into the local one, if one determines flow paths via the non-dispersive single direction (D8) method (O'Callaghan and Mark, 1984). Other methods consider multiple flow directions (cf. Holmgren, 1994).

B3 Slope stability model

For analysis of hillslope stability we assume an infinite slope scheme, and compute minimum factor of safety FS with the following formula (Taylor, 1948):

$$\mathrm{FS}(d_{\mathrm{LZ}},t)=\frac{\tan\phi'}{\tan\delta}+\frac{c'-\psi(d_{\mathrm{LZ}},t)\gamma_{\mathrm{w}}\tan\phi'}{\gamma_{\mathrm{s}}d_{\mathrm{LZ}}\sin\delta\cos\delta}, \tag{B5}$$

where c' is soil cohesion for effective stress, ϕ' is the soil friction angle for effective stress, γ_{w} is the unit weight of groundwater, γ_{s} is the soil unit weight and δ is the slope angle. In this scheme the failure occurs at the basal boundary $Z=d_{\mathrm{LZ}}$, because the pressure head results maximum at that depth.

It is useful to consider the critical wetness ratio, derived from Eq. (B5) letting $\mathrm{FS}=1$, which is a parameter that for a given hillslope (given slope δ and soil depth d_{LZ}) depends only on the geotechnical characteristics of the soil:

$$\zeta_{\mathrm{CR}}=\frac{\gamma_{\mathrm{s}}}{\gamma_{\mathrm{w}}}\left[\left(\frac{c'}{\gamma_{\mathrm{s}}d_{\mathrm{LZ}}\sin\delta\cos\delta}-1\right)\frac{\tan\delta}{\tan\phi'}+1\right]. \tag{B6}$$

The ζ_{CR} varies from 0 to 1, respectively, for an unconditionally unstable and a unconditionally stable hillslope (Montgomery and Dietrich, 1994), and hence it indicates the natural degree of stability of the hillslope.

$K_{\mathrm{s}}\exp\{\alpha(\psi-\psi_{\mathrm{cf}})\}$:

$$\frac{\alpha_1(\theta_{\mathrm{s}}-\theta_{\mathrm{r}})}{K_{\mathrm{S}}}\frac{\partial K}{\partial t}=\frac{\partial^2 K}{\partial Z^2}-\alpha_1\frac{\partial K}{\partial Z}, \tag{B4}$$

where K_{s} is the saturated hydraulic conductivity, α is the SWCC parameter, $\psi_{\mathrm{cf}}=-1/\alpha$ is the pressure head at the top of the capillary fringe, θ_{r} is the residual water content, θ_{s} is the water content at saturation and $\alpha_1=\alpha\cos^2\delta$.

A closed-form solution to this equation for $\delta=0$ has been provided by Srivastava and Yeh (1991) and extended to a sloping surface by Savage et al. (2004), and used in the TRIGRS unsaturated model (Baum et al., 2008, 2010).

The solution to Richards' equation provides the pore pressure profile in the unsaturated zone, and a flux to the saturated zone $q(d_{\mathrm{u}},t)$. Because of the partial absorption of water within the unsaturated zone, this flux turns out to be damped and smoothed with respect to the infiltrating flux at the ground surface (cf. Fig. 8 of Baum et al., 2008). The TRIGRS model then computes water table rise using $q(d_{\mathrm{u}},t)$ and subtracting from it a leakage flow rate given by $q_{\mathrm{l}}=\min\{c_{\mathrm{d}}K_{\mathrm{s}}(1-\cos^2\delta),q(d_{\mathrm{u}},t)\}$ (vertical drainage at the basal boundary, which is not assumed perfectly impervious), where c_{d} represents the ratio between saturated hydraulic conductivities of the basal boundary layer and of the regolith surficial layer. In the case that no specific information on c_{d} ratio is available, a reasonable value may be $c_{\mathrm{d}}=0.1$ (cf. Baum et al., 2010), which means that hydraulic conductivity of the layer below depth $Z=d_{\mathrm{LZ}}$ is of one order of magnitude less than the regolith surficial layer.

The resulting water table rise is computed by comparing this excess flux accumulating at the top of the capillary fringe to the available pore space directly above it.

Pressure head rise is assumed transient in the saturated zone as well, and computed by formulas adapted from analogous heat-flow problems (see Baum et al., 2008).

The solution to Eq. (B1) is used to compute the water table height at the beginning of rainfall event i:

$$\begin{aligned}h_i&=\frac{\psi(d_{\mathrm{LZ}},t_{\mathrm{f},i-1})}{\cos^2\delta}\exp\left(-\frac{K_{\mathrm{s}}\sin\delta}{\frac{A}{B}(\theta_{\mathrm{s}}-\theta_{\mathrm{r}})}\Delta t_i\right)\\&=\frac{\psi(d_{\mathrm{LZ}},t_{\mathrm{f},i-1})}{\cos^2\delta}\exp\left(-\frac{\Delta t_i}{\tau_M}\right),\end{aligned} \tag{B2}$$

where $\psi(d_{LZ},t_{\mathrm{f},i-1})$ and the inter-arrival time Δt_i are defined in Sect. 2. The time constant τ_M regulates the pressure head memory from one event to another.

The initial pressure head distribution above the water table is computed accordingly with assumptions of the transient vertical infiltration model (see next Sect. B2), letting the steady (initial) surface flux $I_{\mathrm{ZLT}}=0$, which yields the following equation (see Baum et al., 2010; Baum and Godt, 2013):

$$\psi(Z,t=0)=-(d_u-Z)\cos\delta-\frac{1}{\alpha}, \tag{B3}$$

for depths $Z\le d_u$, d_u being the depth to the top of the capillary fringe and α the parameter of Gardner's (1958) exponential soil–water characteristic curve (cf. Fig. B1).

B2 Transient infiltration model

Reference scheme for a simulated hillslope is shown in Fig. B1. Infiltration in the unsaturated zone is modeled through Richards' (1931) vertical-infiltration equation for a sloping surface particularized for the Gardner's (1958) exponential soil–water characteristic curve $K(\psi)=$

Acknowledgements. This research was partially funded by the Italian Education, University and Research Ministry (MIUR), PON project no. 01_01503 Integrated Systems for Hydrogeological Risk Monitoring, Early Warning and Mitigation Along the Main Lifelines, CUP B31H11000370005. Rex L. Baum of United States Geological Survey and Jorge A. Ramirez of Colorado State University are acknowledged for their useful suggestions. The authors would like to thank the four anonymous reviewers and S. L. Gariano for their useful comments.

Edited by: T. Bogaard

References

Aleotti, P.: A warning system for rainfall-induced shallow failures, Eng. Geol., 73, 247–265, 2004.

Ardizzone, F., Basile, G., Cardinali, M., Casagli, N., Del Conte, S., Del Ventisette, C., Fiorucci, F., Garfagnoli, F., Gigli, G., Guzzetti, F., Iovine, G., Mondini, A., Moretti, S., Panebianco, M., Raspini, F., Reichenbach, P., Rossi, M., Tanteri, L., and Terranova, O.: Landslide inventory map for the Briga and the Giampilieri catchments, NE Sicily, Italy, J. Maps, 8, 176–180, 2012.

Balistrocchi, M. and Bacchi, B.: Modelling the statistical dependence of rainfall event variables through copula functions, Hydrol. Earth Syst. Sci., 15, 1959–1977, doi:10.5194/hess-15-1959-2011, 2011.

Balistrocchi, M., Grossi, G., and Bacchi, B.: An analytical probabilistic model of the quality efficiency of a sewer tank, Water Resour. Res., 45, W12420, doi:10.1029/2009WR007822, 2009.

Barnes, L., Gruntfest, E., Hayden, M., Schultz, D., and Benight, C.: False alarms and close calls: A conceptual model of warning accuracy, Weather Forecast., 22, 1140–1147, 2007.

Baum, R. L. and Godt, J. W.: Early warning of rainfall-induced shallow landslides and debris flows in the USA, Landslides, 7, 259–272, 2010.

Baum, R. L. and Godt, J. W.: Correction to "Estimating the timing and location of shallow rainfall-induced landslides using a model for transient, unsaturated infiltration", J. Geophys. Res.-Earth Surf., 118, 1999–1999, doi:10.1002/jgrf.20100, 2013.

Baum, R. L., Savage, W. Z., and Godt, J. W.: TRIGRS – A FORTRAN Program for Transient Rainfall Infiltration and Grid-Based Regional Slope-Stability Analysis, US Geological Survey Open-File Report 02-0424, Reston, Virginia, 2002.

Baum, R. L., Savage, W. Z., and Godt, J. W.: TRIGRS – A FORTRAN program for transient rainfall infiltration and grid-based regional slope-stability analysis, version 2.0., US Geological Survey Open-File Report 2008-1159, Reston, Virginia, 2008.

Baum, R. L., Godt, J. W., and Savage, W. Z.: Estimating the timing and location of shallow rainfall-induced landslides using a model for transient, unsaturated infiltration, J. Geophys. Res., 115, F03013, doi:10.1029/2009JF001321, 2010.

Berti, M., Martina, M., Franceschini, S., Pignone, S., Simoni, A., and Pizziolo, M.: Probabilistic rainfall thresholds for landslide occurrence using a Bayesian approach, J. Geophys. Res.-Earth Surf., 117, F04006, doi:10.1029/2012JF002367, 2012.

Bonta, J. V. and Rao, A. R.: Factors affecting the identification of independent storm events, J. Hydrol., 98, 275–293, doi:10.1016/0022-1694(88)90018-2, 1988.

Brunetti, M. T., Peruccacci, S., Rossi, M., Luciani, S., Valigi, D., and Guzzetti, F.: Rainfall thresholds for the possible occurrence of landslides in Italy, Nat. Hazards Earth Syst. Sci., 10, 447–458, doi:10.5194/nhess-10-447-2010, 2010.

Caine, N.: The Rainfall Intensity: Duration Control of Shallow Landslides and Debris Flows, Geograf. Annal. Series A, 62, 23–27, 1980.

Calenda, G. and Napolitano, F.: Parameter estimation of Neyman-Scott processes for temporal point rainfall simulation, J. Hydrol., 225, 45–66, 1999.

Campbell, R. H.: Soil Slips, Debris Flows, and Rainstorms in the Santa Monica Mountains and Vicinity, Southern California, Professional Paper 851, US Govt. Print. Off., Washington, D.C., 1975.

Cancelli, A. and Nova, R.: Landslides in soil debris cover triggered by rainstorms in Valtellina (central Alps – Italy), paper presented at the 4th International Conference on Landslides, Tokyo, August 1985, 267–272, Jpn. Geol. Soc., Tokyo, 1985.

Cannon, S. H. and Ellen, S. D.: Rainfall conditions for abundant debris avalanches in the San Francisco Bay Region, California, Calif. Geol., 38, 267–272, 1985.

Capparelli, G. and Versace, P.: FLaIR and SUSHI: two mathematical models for early warning of landslides induced by rainfall, Landslides, 8, 67–79, 2011.

Cowpertwait, P.: Further developments of the Neyman-Scott clustered point process for modeling rainfall, Water Resour. Res., 27, 1431–1438, 1991.

Cowpertwait, P. S. P., O'Connell, P. E., Metcalfe, A. V., and Mawdsley, J.: Stochastic point process modelling of rainfall. I. Single-site fitting and validation, J. Hydrol., 175, 17–46, 1996.

Cox, D. R. and Isham, V.: Point Processes, CRC Press, Boca Raton, FL, 1980.

De Guidi, G. and Scudero, S.: Landslide susceptibility assessment in the Peloritani Mts. (Sicily, Italy) and clues for tectonic control of relief processes, Nat. Hazards Earth Syst. Sci., 13, 949–963, doi:10.5194/nhess-13-949-2013, 2013.

D'Odorico, P., Fagherazzi, S., and Rigon, R.: Potential for landsliding: Dependence on hyetograph characteristics, J. Geophys. Res.-Earth Surf., 110, F01007, doi:10.1029/2004JF000127, 2005.

Fathani, T. F., Karnawati, D., Sassa, K., Fukuoka, H., and Honda, K.: Development of landslide monitoring and early warning system in Indonesia, in: Proceedings of The First World Landslide Forum, edited by: Casagli, N., Fanti, R., and Tofani, V., 195–198, 18–21 November 2008, United Nations University, Tokyo, Japan, 2009.

Foufoula-Georgiou, E. and Guttorp, P.: Assessment of a Class of Neyman-Scott Models for Temporal Rainfall, J. Geophys. Res., 92, 9679–9682, 1989.

Frattini, P., Crosta, G., and Carrara, A.: Techniques for evaluating the performance of landslide susceptibility models, Eng. Geol., 111, 62–72, 2010.

Gardner, W. R.: Some steady-state solutions of the unsaturated moisture flow equation with evaporation from a water table, Soil Sci., 85, 228–232, 1958.

Gariano, S., Iovine, G., Antronico, L., Brunetti, M., Melillo, M., Peruccacci, S., and Terranova, O.: Rainfall thresholds for the initiation of shallow landslides in Sicily, Rendiconti Online Societa Geologica Italiana, 24, 155–157, 2013.

Glade, T., Crozier, M., and Smith, P.: Applying Probability Determination to Refine Landslide-triggering Rainfall Thresholds Using an Empirical "Antecedent Daily Rainfall Model", Pure Appl. Geophys., 157, 1059–1079, 2000.

Goswami, R., Mitchell, N., and Brocklehurst, S.: Distribution and causes of landslides in the eastern Peloritani of NE Sicily and western Aspromonte of SW Calabria, Italy, Geomorphology, 132, 111–122, 2011.

Guzzetti, F., Peruccacci, S., Rossi, M., and Stark, C. P.: Rainfall thresholds for the initiation of landslides in central and southern Europe, Meteorol. Atmos. Phys., 98, 239–267, 2007.

Hanssen, A. W. and Kuipers, W. J. A.: On the Relationship between the Frequency of Rain and Various Meteorological Parameters: (with Reference to the Problem of Objective Forecasting), Vol. 81, Mededelingen en verhandelingen, 1965.

Highland, L. M. and Bobrowsky, P.: The landslide handbook – A guide to understanding landslides, Tech. Rep. 1325, p. 129, US Geological Survey, Reston, Virginia, 2008.

Holmgren, P.: Multiple flow direction algorithms for runoff modelling in grid based elevation models: an empirical evaluation, Hydrol. Process., 8, 327–334, 1994.

Iverson, R. M.: Landslide triggering by rain infiltration, Water Resour. Res., 36, 1897–1910, 2000.

Kavvas, M., L. and Delleur, J., W.: The stochastic and cronologic structure of rainfall sequence: Application to Indiana, Tech. Rep. 1325, p. 129, Tech. Rep. 57, Water Resour. Res. Cent., Purdue University, West Lafayette, Ind., 1975.

Keefer, D. K., Wilson, R. C., Mark, R. K., Brabb, E. E., Brown, W. M., Ellen, S. D., Harp, E. L., Wieczorek, G. F., Alger, C. S., and Zatkin, R. S.: Real-Time Landslide Warning During Heavy Rainfall, Science, 238, 921–925, 1987.

Lentini, F., Catalano, S., and Carbone, S.: Note illustrative della Carta Geologica della Provincia di Messina, scala 1 : 50.000, Tech. rep., Provincia Regionale di Messina, Assessorato Servizio Territorio – Servizio Geologico, 2000.

Montgomery, D. R. and Dietrich, W. E.: A physically based model for the topographic control on shallow landsliding, Water Resour. Res., 30, 1153–1171, 1994.

Murphy, A.: The Finley affair: A signal event in the history of forecast verification, Weather Forecast., 11, 3–20, 1996.

Neyman, J. and Scott, E. L.: Statistical Approach to Problems of Cosmology, J. Roy. Stat. Soc. B (Methodological), 20, 1–43, 1958.

O'Callaghan, J. F. and Mark, D. M.: The extraction of drainage networks from digital elevation data, Comput. Vision, Graph. Image Process., 28, 323–344, 1984.

Peirce, C. S.: The numerical measure of the success of predictions, Science, 4, 453–454, 1884.

Peres, D. and Cancelliere, A.: Development of rainfall thresholds for landslide early warning in Sicily: A method based on the use of rainfall annual maxima series, Rendiconti Online Societa Geologica Italiana, 21, 580–582, 2012.

Peres, D. J. and Cancelliere, A.: Defining Rainfall Thresholds for Early Warning of Rainfall-Triggered Landslides: The Case of North-East Sicily, in: Landslide Science and Practice, edited by: Margottini, C., Canuti, P., and Sassa, K., 257–263, Springer Berlin Heidelberg, doi:10.1007/978-3-642-31337-0_33, 2013.

Peruccacci, S., Brunetti, M. T., Luciani, S., Vennari, C., and Guzzetti, F.: Lithological and seasonal control on rainfall thresholds for the possible initiation of landslides in central Italy, Geomorphology, 139–140, 79–90, 2012.

Raia, S., Alvioli, M., Rossi, M., Baum, R. L., Godt, J. W., and Guzzetti, F.: Improving predictive power of physically based rainfall-induced shallow landslide models: a probabilistic approach, Geosci. Model Dev., 7, 495–514, doi:10.5194/gmd-7-495-2014, 2014.

Restrepo-Posada, P. J. and Eagleson, P. S.: Identification of independent rainstorms, J. Hydrol., 55, 303–319, doi:10.1016/0022-1694(82)90136-6, 1982.

Richards, L.: Capillary conduction of liquids through porous mediums, Physics, 1, 219–233, 1931.

Rodriguez-Iturbe, I., Cox, D. R., and Isham, V.: Some Models for Rainfall Based on Stochastic Point Processes, Proc. Roy. Soc. London A, 410, 269–288, 1987a.

Rodriguez-Iturbe, I., De Power, B. F., and Valdes, J. B.: Rectangular Pulses Point Process Models for Rainfall: Analysis of Empirical Data, J. Geophys. Res., 92, 9645–9656, 1987b.

Rosso, R., Rulli, M. C., and Vannucchi, G.: A physically based model for the hydrologic control on shallow landsliding, Water Resour. Res., 42, 1–16, 2006.

Salas, J. D.: Analysis and Modeling of Hydrologic Time Series, Chap. 19, Handbook of Hydrology, McGraw-Hill, 1993.

Salciarini, D., Godt, J. W., Savage, W. Z., Baum, R. L., and Conversini, P.: Modeling landslide recurrence in Seattle, Washington, USA, Eng. Geol., 102, 227–237, 2008.

Savage, W. Z., Godt, J. W., and Baum, R. L.: Landslides: Evaluation and Stabilization, Proceedings of the 9th International Symposium on Landslides, Vol. 1, Chap. Modeling time dependent aerial slope stability, 23–36, Millpress, Balkema, London, 2004.

Sirangelo, B. and Versace, P.: A real time forecasting model for landslides triggered by rainfall, Meccanica, 31, 73–85, 1996.

Srivastava, R. and Yeh, T. C. J.: Analytical solutions for one dimensional, transient infiltration toward the water table in homogeneous and layered soils, Water Resour. Res., 27, 753–762, 1991.

Staley, D. M., Kean, J. W., Cannon, S. H., Schmidt, K. M., and Laber, J. L.: Objective definition of rainfall intensity – duration thresholds for the initiation of post-fire debris flows in southern California, Landslides, 10, 547–562, doi:10.1007/s10346-012-0341-9, 2013.

Stephenson, D.: Use of the 'odds ratio' for diagnosing forecast skill, Weather Forecas., 15, 221–232, 2000.

Takara, K. and Apip Bagiawan, A.: Study on early warning system for debris flow and landslide in the Citarum River Basin, Indonesia, in: Proceedings of The First World Landslide Forum, edited by: Casagli, N., Fanti, R., and Tofani, V., 573–576, 18–21 November 2008 United Nations University, Tokyo, Japan, 2009.

Tarolli, P., Borga, M., Chang, K.-T., and Chiang, S.-H.: Modeling shallow landsliding susceptibility by incorporating heavy rainfall statistical properties, Geomorphology, 133, 199–211, 2011.

Taylor, D.: Fundamentals of Soil Mechanics, John Wiley, New York, 1948.

Tiranti, D. and Rabuffetti, D.: Estimation of rainfall thresholds triggering shallow landslides for an operational warning system implementation, Landslides, 7, 471–481, 2010.

Waymire, E. and Gupta, V. K.: The mathematical structure of rainfall representations: 1. A review of the stochastic rainfall models, Water Resour. Res., 17, 1261–1272, 1981a.

Waymire, E. and Gupta, V. K.: The mathematical structure of rainfall representations: 2. A review of the theory of point processes, Water Resour. Res., 17, 1273–1285, 1981b.

Waymire, E. and Gupta, V. K.: The mathematical structure of rainfall representations: 3. Some applications of the point process theory to rainfall processes, Water Resour. Res., 17, 1287–1294, 1981c.

Wieczorek, G. F., Morgan, B. A., and Campbell, R. H.: Debris-flow hazards in the Blue Ridge of central Virginia, Environ. Eng. Geol., 6, 3–23, 2000.

Wilks, D.: Statistical Methods in the Atmospheric Sciences, 3rd Edn., Mededelingen en verhandelingen, 2011.

Wu, W. and Sidle, R. C.: A Distributed Slope Stability Model for Steep Forested Basins, Water Resour. Res., 31, 2097–2110, 1995.

3

Monitoring hillslope moisture dynamics with surface ERT for enhancing spatial significance of hydrometric point measurements

R. Hübner[1], **K. Heller**[1], **T. Günther**[2], **and A. Kleber**[1]

[1]Institute of Geography, Dresden University of Technology, Helmholtzstr. 10, 01069 Dresden, Germany
[2]Leibniz Institute for Applied Geophysics (LIAG), Stilleweg 2, 30655 Hanover, Germany

Correspondence to: R. Hübner (rico.huebner@tu-dresden.de)

Abstract. Besides floodplains, hillslopes are basic units that mainly control water movement and flow pathways within catchments of subdued mountain ranges. The structure of their shallow subsurface affects water balance, e.g. infiltration, retention, and runoff. Nevertheless, there is still a gap in the knowledge of the hydrological dynamics on hillslopes, notably due to the lack of generalization and transferability. This study presents a robust multi-method framework of electrical resistivity tomography (ERT) in addition to hydrometric point measurements, transferring hydrometric data into higher spatial scales to obtain additional patterns of distribution and dynamics of soil moisture on a hillslope. A geoelectrical monitoring in a small catchment in the eastern Ore Mountains was carried out at weekly intervals from May to December 2008 to image seasonal moisture dynamics on the hillslope scale. To link water content and electrical resistivity, the parameters of Archie's law were determined using different core samples. To optimize inversion parameters and methods, the derived spatial and temporal water content distribution was compared to tensiometer data. The results from ERT measurements show a strong correlation with the hydrometric data. The response is congruent to the soil tension data. Water content calculated from the ERT profile shows similar variations as that of water content from soil moisture sensors. Consequently, soil moisture dynamics on the hillslope scale may be determined not only by expensive invasive punctual hydrometric measurements, but also by minimally invasive time-lapse ERT, provided that pedo-/petrophysical relationships are known. Since ERT integrates larger spatial scales, a combination with hydrometric point measurements improves the understanding of the ongoing hydrological processes and better suits identification of heterogeneities.

1 Introduction

The knowledge of system-internal water flow pathways and the response to precipitation on different spatial and temporal scales is essential for the prediction of hydrological and hydrochemical dynamics within catchments (Uhlenbrook et al., 2008; Wenninger et al., 2004). Understanding the processes involved is of particular importance for improving precipitation-runoff and pollutant-transport models (Di Baldassarre and Uhlenbrook, 2012).

Hillslopes are important links between the atmosphere and the water input into catchments. They mainly control different runoff components and residence times (Uhlenbrook et al., 2008). Several studies have addressed hillslope hydrology (Anderson and Burt, 1990; Kirkby, 1980; Kleber and Schellenberger, 1998; McDonnell et al., 2001; Tromp-van Meerveld, 2004; Uchida et al., 2006). A major problem is that the spatial and temporal variability of the hydrological response due to different natural settings – e.g. geomorphological, pedological, lithological characteristics and the spatial heterogeneity – make it difficult to generalize and to transfer results to ungauged basins (McDonnell et al., 2007).

In catchments of Central European subdued mountain ranges, the shallow subsurface of hillslopes is mostly covered by Pleistocene periglacial slope deposits (Kleber and Terhorst, 2013). These slope deposits have developed in different layers. In the literature normally three layers are classified (upper layer – LH, intermediate layer – LM, basal layer – LB: classification according to AD-hoc AG-Boden, 2005; Kleber and Terhorst, 2013). Sometimes locally a fourth layer (*Oberlage* AD-hoc AG-Boden, 2005) could be found. The occurrence of these layers can vary spatially and has different regional and local characteristics. Due to the sedimentological and substrate-specific properties, e.g. grain-size distribution, clast content and texture, they remarkably influence near-surface water balance (e.g. infiltration, percolation) and are of particular importance for near-surface runoff, e.g. interflow (Chifflard et al., 2008; Kleber, 2004; Kleber and Schellenberger, 1998; Sauer et al., 2001; Scholten, 1999; Völkel et al., 2002a, b; Heller, 2012; Moldenhauer et al., 2013).

Most of the prior studies were based on invasive and extensive hydrometric point measurements or on tracer investigations. Punctual hydrometric measurements may modify flow pathways and are not sufficient in the case of significant spatial heterogeneities in the subsurface. Tracer experiments, e.g. using isotopes, integrate much larger scales up to entire catchments but provide less direct insights into ongoing processes. Internal hydrological processes may be complex and due to the spatio-temporal interlinking of near-surface processes and groundwater dynamics, there is still a lack of knowledge regarding runoff generation in watersheds (McDonnell, 2003; Tilch et al., 2006; Uhlenbrook, 2005). For an efficient and accurate modelling of the hydrological behaviour at the crucial hillslope scale, additional methods are needed especially to improve the understanding of these complex processes in order to enhance the model hypotheses. Hydrogeophysical methods are capable of closing the gap between large-scale depth-limited remote-sensing methods and invasive punctual hydrometric arrays (Robinson et al., 2008a, b; Lesmes and Friedman, 2006; Uhlenbrook et al., 2008).

Many studies show the potential of electrical resistivity tomography (ERT) for hydrological investigation by means of synthetic case studies for aquifer transport characterization (Kemna et al., 2004; Vanderborght et al., 2005), imaging water flow on soil cores (Bechtold et al., 2012; Binley et al., 1996a, b; Garré et al., 2010, 2011; Koestel et al., 2008, 2009a, b), cross-borehole imaging of tracers (Daily et al., 1992; Oldenborger et al., 2007; Ramirez et al., 1993; Singha and Gorelick, 2005; Slater et al., 2000), or imaging of tracer injection or irrigation with surface ERT (Cassiani et al., 2006; De Morais et al., 2008; Descloitres et al., 2008a; Michot et al., 2003; Perri et al., 2012). However, some research has been conducted under natural conditions to characterize water content change, infiltration or discharge by use of cross-borehole ERT (French and Binley, 2004), surface ERT (Brunet et al., 2010; Benderitter and Schott, 1999; Descloitres et al., 2008b; Massuel et al., 2006; Miller et al., 2008) or a combined surface cross-borehole ERT array (Beff et al., 2013; Zhou et al., 2001).

Besides hydrogeophysical methods such as electromagnetics (EM) (Popp et al., 2013; Robinson et al., 2012; Tromp-van Meerveld and McDonnell, 2009), time-lapse ERT have been frequently applied to hillslope investigation in the runoff and interflow (Uhlenbrook et al., 2008; Cassiani et al., 2009) or preferential flow context (Leslie and Heinse, 2013). However, the use of ERT for monitoring hydrological dynamics on hillslopes with layered structure is still rare.

The objective of this paper is to show the potential of minimally invasive surface time-lapse ERT as a robust methodological framework for monitoring long-term changes in soil moisture and to improve the spatial resolution of punctual hydrometric measurements (e.g. tensiometer and soil moisture sensors) on a hillslope with periglacial cover beds. Furthermore, we want to show the ability of ERT for mapping spatially heterogeneous structures and water content distributions of the shallow subsurface. With a multi-method approach, we attempt to demonstrate the possibility to adequately transfer hydrometric data to higher spatial scales and to obtain additional patterns of soil water dynamics on a hillslope. These scales are fundamental for achieving a better understanding of the influence of the layered subsurface on water fluxes (e.g. infiltration, percolation or interflow) and the response to different amounts of precipitation on hillslopes.

2 Material and methods

2.1 Study site

The study area covers 6 ha of a forested spring catchment in the Eastern Ore Mountains, eastern Germany, which is located in the Freiberger Mulde catchment (Fig. 1).

Annual precipitation averages 930 mm, mean annual temperature is 6.6 °C. The altitude ranges from 521 to 575 m a.s.l. with a predominant land cover of spruce forest (*Picea abies*, approx. 30 years). The slope angle of the catchment ranges from 0.05 to 22.5° with an average of 7°. Bedrock is gneiss overlain by periglacial cover beds with up to three layers (LH, LM, LB, with no occurrence of the *Oberlage*, see Heller, 2012). The upper layer (LH) with a thickness of 0.3–0.65 m consists of silty–loamy material with a bulk density of 1.2 g m^{-3} and many roots (see Table 1). In the central part of the catchment, a silty–loamy intermediate layer (LM) follows with higher bulk density and a thickness of up to 0.55 m. The ubiquitous sandy–loamy basal layer (LB) is characterized by even higher bulk density and longitudinal axes of coarse clasts oriented parallel to the slope. Downslope it may reach a thickness of at least 3 m (see Fig. 1).

2.2 Laboratory work

Quality, amount and distribution of pore water exert a huge influence on resistivity [1] and form the link between electrical and hydrological properties. The empirical relationship of Archie's law (Archie, 1942) describes the connection between electrical resistivity and saturation in porous media.

[1] In this context the term "resistivity" always refers to " specific electrical resistivity".

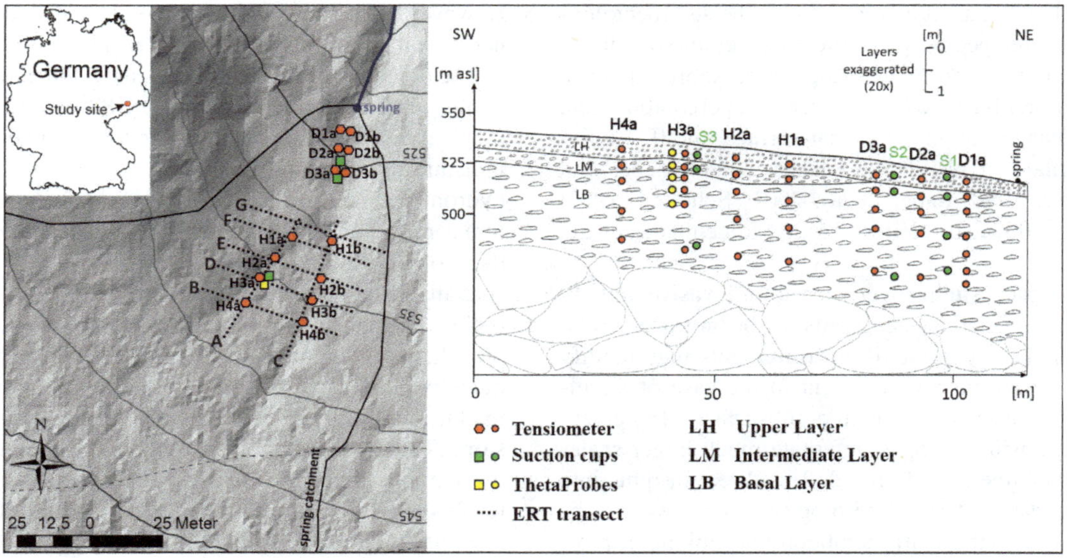

Figure 1. Study site with locations of ERT profiles and hydrometric stations (left panel; data source: ATKIS®-DGM2, Landesvermessungsamt Sachsen, 2008) and profile section with installation depths of tensiometers, soil moisture sensors and suction cups (right panel).

Table 1. Properties of cover beds from the study site ($n \geq 15$ per layer) – adapted from Moldenhauer et al. (2013).

Layer	Soil horizon	Colour (moist)	Soil texture: Clay (%)	Silt (%)	Sand (%)	Clasts (%)	Bulk density ($\mathrm{g\,m^{-3}}$)	Porosity	Hydraulic conductivity ($\mathrm{cm\,d^{-1}}$)*
LH	A/Bw	10YR/5/8	14	52	34	36	1.2	0.55	27
LM	2Bg	10YR/5/4	12	53	35	43	1.5	0.43	9
LB	3CBg	10YR/5/3	7	22	71	56	1.7	0.36	52

* Field-saturated hydraulic conductivity measured using the compact constant head permeameter (CCHP) method (Amoozegar, 1989).

Instead of saturation we use the volumetric water content θ with

$$\rho_{\mathrm{eff}} = F_\theta \rho_{\mathrm{w}} \theta^{-n_\theta}, \tag{1}$$

where ρ_{eff} is the bulk resistivity of the soil probe and ρ_{w} is the resistivity of the pore fluid. The formation factor F_θ describes the increase of resistivity due to an insulating solid matrix and constitutes an intrinsic measure of material microgeometry (Schön, 2004; Lesmes and Friedman, 2006). The exponent n_θ is an empirical constant, which depends on the distribution of water within the pore space (Schön, 2004).

This model disregards the surface conductivity, which may occur due to interactions between pore water and soil matrix, especially with a high percentage of small grain sizes. In our study the curve fitting could be carried out very well without accounting for surface conductivity.

To investigate the pedo-/petrophysical relationship between resistivity and water content, 14 undisturbed soil core specimens (diameter = 36 mm, length = 40 mm) taken at different depths (0.3–1.4 m) were analysed. After dehydration in a drying chamber, the samples were saturated. The saturation was done successively by stepwise injection in the middle of the soil core to achieve a better moisture distribution within the sample. Using a four-point array, electrical resistivity was measured for different saturation conditions during the saturation process. A calibrating solution with known resistivity was used to determine the geometric factor. Particle sizes were determined by sieving and the pipette method, using $Na_4P_2O_7$ as a dispersant (Klute, 1986, p. 393, 399–404, but with the sand–silt boundary at 0.063 mm).

Brunet et al. (2010) described remarkable conductivity increases of low mineralized water due to contact with the soil matrix. This may cause variation of resistivity with time. To minimize this effect we used spring water with high conductivity (approx. $\sigma_{\mathrm{w25}} = 150\,\mu\mathrm{S\,cm^{-1}}$ / $\rho_{\mathrm{w25}} = 66\,\Omega\,\mathrm{m}$ for $T = 25\,°\mathrm{C}$). This corresponds to the mean conductivity of soil water in the study area, which is influenced by long-term contact with the subsoil.

Aside from the invariant parameters F_θ and n_θ, the resistivity of the pore water must be known to calculate the water content from resistivity values. Because it was not possible to extract pore water under dry conditions in summer, only

Table 2. Median pore water conductivity $\tilde{\sigma}_W$, resistivity $\tilde{\rho}_W$ and mean resistivity $\bar{\rho}_W$ with standard deviation (SD_W) per depth ($n \geq 11$ per sampling depth).

Depth [m]	0.3	0.6	0.85	1.05	1.65	2.3
$\tilde{\sigma}_W$ [$\mu S\,cm^{-1}$]	72.4	107.8	111.6	114.7	135	156.7
$\tilde{\rho}_W$ [$\Omega\,m$]	138.1	92.8	89.6	87.2	74.1	63.8
$\bar{\rho}_W$ [$\Omega\,m$]	135.7	92.3	88.9	86.8	75.1	63.7
SD_W [$\Omega\,m$]	16.9	5.0	4.5	4.8	5.4	7.3

a few measurements of pore water conductivity could be carried out in late spring and early autumn. To calculate water content from resistivity obtained by field surveys, the median value over the entire time period of ρ_W for each depth was used (see Table 2). Interim values between the extraction depths were linearly interpolated.

After reforming Eq. (1) it is possible, with known parameters F_θ and n_θ and measured variables ρ_{eff} and ρ_W, to calculate volumetric water content:

$$\frac{\rho_{eff}}{F_\theta \rho_W}^{\frac{1}{-n_\theta}} = \theta. \quad (2)$$

As water saturation (S) is defined as the ratio between water content and porosity (Φ), it is also possible to calculate the degree of saturation using

$$\frac{\rho_{eff}}{F_\theta \rho_W}^{\frac{1}{-n_\theta}} \frac{1}{\Phi} = S. \quad (3)$$

The porosity (Φ) was calculated with

$$\Phi = 1 - \frac{\rho_{bulk}}{\rho_{particle}}. \quad (4)$$

The bulk density (ρ_{bulk}) was determined using undisturbed core samples. The particle density ($\rho_{particle}$) was measured with a capillary-stoppered pycnometer. The maximum sample depth for undisturbed soil cores was < 2 m. Below, the porosity had to be transferred according to grain size distribution, clasts and compaction from percussion drilling.

2.3 Field work

2.3.1 ERT mapping

In addition to conventional percussion drilling, at the end of October 2008 we measured seven ERT profiles to survey the subsurface resistivity distribution (A–G in Fig. 1). A and C are parallel to the slope inclination of approx. 9°, connecting inflection points of contour lines. B, D, E, F and G are perpendicular to these profiles (∡ A102.5°, ∡ C90°). This arrangement allows identifying potential 3-D effects, which may cause inaccurate interpretation of the subsurface resistivity distribution. To improve the mapping results aided by hydrometric data, the profiles were located close to the tensiometer stations (distance < 2 m). For all resistivity measurements, the instrument "4 Point light hp" from "LGM – Lippmann Geophysical Equipment" with 50 electrodes was used. Because of the expected interferences (e.g. by roots or clasts) and the multiple-layered stratification of periglacial cover beds, a Wenner array was found to be the most suitable configuration for the study area. This is characterized by low geometric factors (K), a high vertical resolution for laterally bedded subsurface structures, and a good signal-to-noise ratio (Dahlin and Zhou, 2004). To improve the spatial resolution, a Wenner-β array was measured additionally. With an electrode spacing of 1 m, this results in a combined data set with 784 data points for each pseudo-section with a maximum depth of investigation of 9.36 m (Wenner-β: depth of investigation for radial dipole in homogeneous ground $0.195L$ with L the maximum electrode separation in metres, according to Roy and Apparao, 1971; Apparao, 1991; Barker, 1989).

Horizontal resolution of a multi-electrode array is for shallow parts of the subsurface of the order of electrode distances. However, vertical resolution is far better, as the depth-of-investigation curves indicate (Roy and Apparao, 1971; Barker, 1989). This is further improved by measuring two electrode arrays (Wenner and Wenner-β) with different sensitivity curves so that we can expect a vertical resolution of the order of about 0.2 m in the case of excellent data quality.

2.3.2 Joint hydrometric and ERT monitoring

Since November 2007 soil water tension has been measured using 76 recording tensiometers (T8, UMS) arranged in 14 survey points along the slope at 5–7 different depths (see Fig. 1). Additionally, at the survey point H3a five soil moisture sensors (ThetaProbe, ML2x, Delta-T) were installed to measure volumetric water content. A V-notch weir with a pressure meter was used to quantify spring discharge. Rainfall was recorded by four precipitation gauges with tipping bucket (R. M. Young Co., 200 cm^2, resolution: 0.1 mm with max. 7 mm min^{-1}). For determination of pore water conductivity and resistivity, soil water was extracted with suction cups (VS-pro, UMS) at four depths at three locations (S1, S2, S3; Fig. 1) and cumulated as a weekly mixed sample.

Time lapse ERT measurements were performed with the same equipment, electrode array and spacing used for the mapping. The two time lapse profiles are congruent with profiles A and B (see Fig. 1). From May to December 2008, twenty seven time lapse measurements were carried out within almost weekly intervals. Contact resistance was checked before each measurement and was within the range of 0.2 kΩ to max. 1 kΩ over the whole measuring period. This range is very favourable and does not influence the measurements as numerical studies show (Rücker and Günther, 2011).

To compare time lapse measurements and to apply sophisticated inversion routines, the location of electrodes needs to remain constant. For current injection we used stainless steel electrodes (diameter 6 mm, length 150 mm), completely plunged into the ground, thus avoiding shifting of electrodes, except for natural soil creep. In the numerical computations, electrodes are considered points, which is not the case for the present ratio of length to distance. However, numerical computations with real electrode lengths show that the deviations are negligible, particularly if the points are placed at about half the electrode length (Rücker and Günther, 2011).

Subsoil temperature, especially in the upper layers, is characterized by distinct annual and daily variations. Therefore, the temperature dependence of resistivity must be considered when comparing different time steps. The installed tensiometers are able to measure soil temperature simultaneously. These data have been used to correct resistivity measurements to a standard temperature. Comparing several existing models for the correction of soil electrical conductivity measurements, Ma et al. (2011) conclude that the model (Eq. 5) proposed by Keller and Frischknecht (1966) is practicable within the temperature range of environmental monitoring:

$$\rho_{25} = \rho_t \left(1 + \delta \left(T - 25\,°\mathrm{C}\right)\right). \tag{5}$$

With this equation the inverted resistivity (ρ_t) at the temperature (T) was corrected to a resistivity at a soil temperature of 25 °C (ρ_{25}). The empirical parameter δ is the temperature slope compensation, with $\delta = 0.025\,°\mathrm{C}^{-1}$ being commonly used for geophysical applications (Keller and Frischknecht, 1966; Hayashi, 2004; Ma et al., 2011).

2.3.3 ERT data inversion

For inversion of the ERT data, we used the BERT Code (Günther et al., 2006). In order to account for the present topography, we used an unstructured triangular discretization of the subsurface and applied finite element forward calculations. For static inversion, a smoothness-constraint objective function is minimized that consists of the error-weighted misfit between measured data **d** and model response **f(m)**, and a model roughness:

$$\Phi = \|\mathbf{D}(\mathbf{d} - \mathbf{f}(\mathbf{m}))\|_2^2 + \lambda \|\mathbf{C}\mathbf{m}\|_2^2 \rightarrow \min. \tag{6}$$

The regularization parameter λ defines the strength of regularization imposed by the smoothness matrix **C** and needs to be chosen such that the data are fitted within expected accuracy, which is incorporated in the data weighting matrix **D**. In our case, values of $\lambda = 30$ provided sufficient data fit. See Günther et al. (2006) for details of the minimization procedure, and Beff et al. (2013) or Bechtold et al. (2012) for specific modifications in hydrological applications.

For time lapse inversion, i.e. calculating the temporal changes in resistivity, there are three different methodical approaches: (i) inverting the models for each point in time separately, (ii) using the initial model as reference model for the time step, (iii) or inverting the differences of the two data sets (Miller et al., 2008). With our data, each method generates insufficient results with unsubstantiated artifacts. An increase on the surface was always followed by a decrease below and vice versa. These systematic changes cannot be explained or related to any natural process. Descloitres et al. (2003, 2008b) showed with synthetic data that time lapse inversion may produce artifacts due to the smoothness constraints especially with changes caused by shallow infiltration (decrease of resistivity), as mostly expected in our case.

As smoothness constraints are the main reason of these problems, we avoid the smoothness operator in the time lapse inversion and minimize a different objective function for the subsequent time steps:

$$\Phi = \|\mathbf{D}(\mathbf{d}^n - \mathbf{f}(\mathbf{m}^n))\|_2^2 + \lambda \|\mathbf{m}^n - \mathbf{m}^{n-1}\|_2^2 \rightarrow \min. \tag{7}$$

Beginning from the static inversion, the subsequent models are found by reference model inversion. Only the total difference between the models of subsequent time steps $n-1$ and n is used for regularization (minimum-length constraints). A higher regularization parameter of $\lambda = 100$ proved optimal for time lapse inversion concerning both data fit and in comparison to the hydrometric results.

In order to find representative resistivity values as a function of depth, which are independent on small-scale heterogeneities, we subdivide the model down to a depth of 3 m into seven layers according to the boundaries of the described layering (see Table 1) and installation depth of hydrometric devices (see Fig. 1) (0–0.2, 0.2–0.4, 0.4–0.9, 0.9–1.2, 1.2–1.5, 1.5–2.0 and 2.0–3.0 m). The representative values are median resistivities in the layers from the stations H1a–H4a and H4b–H4a for profiles A and B, respectively.

3 Results

3.1 Laboratory

Within the separately analysed samples, non-linear curve fitting was carried out. Using the method of least squares, the data could be fitted using a power function in the form of Archie's law (Eq. 1, $0.973 < r < 0.999$).

The exponent n_θ shows a positive correlation to small grain sizes, primarily medium silt (6.3–20 µm, $r = 0.909$), but in the same case a negative correlation to grain sizes > 630 µm including clast content ($r = -0.852$) (see Fig. 2).

The amount of silt as well as the clast content are important distinctive attributes to differentiate the basal layer from the overlying intermediate or upper layer (Table 1). Two different "electrical" layers may be identified. This is due to the fact that the exponent is strongly influenced by grain size, which shows a remarkable change at the upper boundary of LB. On the other hand, grain size distribution and clast content are very similar between LH and LM, so that these may not be differentiated using ERT. Figure 3 shows the aggregation of the 14 single samples into two regions with different depth ranges.

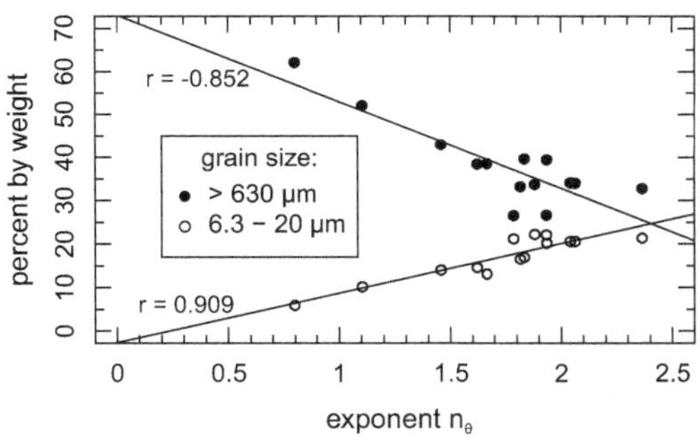

Figure 2. Exponent n_θ in dependence of different grain sizes.

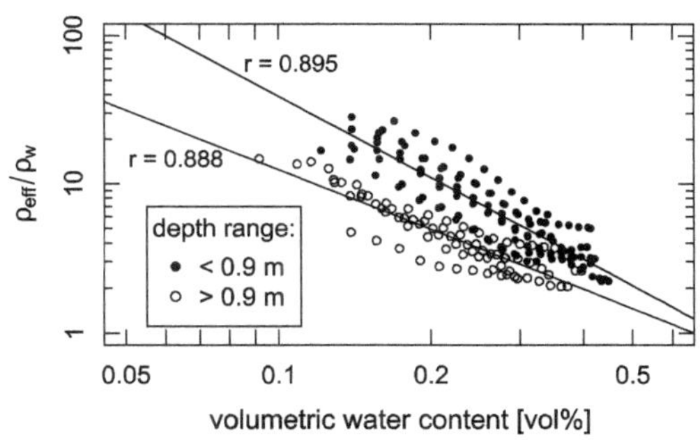

Figure 3. Volumetric water content in dependence of resistivity ratio ($\rho_{\mathrm{eff}}/\rho_{\mathrm{W}}$) for two different depth ranges.

Table 3. Fitted water content formation factor (F_θ), water content exponent (n_θ) and mean squared error for $\rho_{\mathrm{eff}}/\rho_{\mathrm{W}}$ (MSE) for Eqs. (1)–(3) of the two different depth ranges.

Depth range	F_θ	n_θ	r^*	MSE
< 0.9 m	0.577	1.83	0.895	2.8
≥ 0.9 m	0.587	1.34	0.888	1.4

* $p < 0.01$.

The first depth range comprises the upper and the intermediate layer. These two periglacial layers are characterized by a high amount of silt (mostly medium silt) and comparatively low clast content. The exponent n_θ ranges from 1.8 to 2.3. The second depth range is represented by the basal layer. This is characterized by a higher amount of coarse material at the expense of fine grain sizes. In this depth range n_θ ranges from 0.7 to 1.8. Within each of these two depth ranges, we assume, analogously to the properties of the substrate, similar electrical properties with a threshold at 0.9 m. The threshold depth of 0.9 m is not developed as an exact, continuous boundary. Rather it is a short transition zone, because the samples right from this depth may have properties of the shallow or the deeper region, similar to the geomorphological differentiation between the basal and intermediate layer, whose boundary varies between depths of 0.8 to 1 m. By combining samples from different depths into two regions, it was possible to derive the parameter for Eqs. (1)–(3) for each region (Table 3).

This relationship between water content and resistivity, shown in Fig. 3 and Table 3, is only a mean value for each depth range. In the first depth range (0–0.9 m), especially close to the surface, the differences in soil or electrical properties between the samples even at the same depth may vary. This higher variation may be explained by intense biotic activity near the surface, enhancing small-scale heterogeneity compared to deeper parts of the soil.

The fitted curves of both regions are quite similar, except for n_θ. The adapted values for F_θ are almost identical (0.577 vs. 0.587, Table 3). With high saturation, the difference of resistivity between the depth ranges is small and primarily influenced by the conductivity of the pore fluid, but increases with decreasing water content. As a result of the higher exponent, LH and LM react more sensitively to water content changes than LB, especially at low presaturations. Related to this, small water content changes cause larger changes in resistivity than in the deeper region.

3.2 ERT mapping

At our study site the resistivity of the subsoil ranges from nearly 100 Ω m up to more than 4000 Ω m. The distribution may be divided in two main areas, the "inner" area between the depression lines and the "outer" area at the hillsides which differ in their depth profiles. (see Fig. 4)

At the intersection between the longitudinal and diagonal profiles, a good match of the calculated resistivity models can be found at shallow depth. With increasing depth, the differences become more notable – e.g. A × B: depth < 1 m, average deviation 8% ($\sigma = 5.4\,\%$); depth 1–7 m, average deviation 20% ($\sigma = 10\,\%$); and depth > 7 m, average deviation 43% ($\sigma = 6.6\,\%$). To exclude potential errors (e.g. electrode positioning errors), the data quality may be evaluated by comparing normal and reciprocal measurements, i.e. interchanging potential and current electrodes (LaBrecque et al., 1996; Zhou and Dahlin, 2003). For profiles A and B repeated measurements with reciprocal electrode configuration were conducted. Thereby, no large errors (max ±1.2 %) could be found between normal and reciprocal measurements. Because of the absence of large potential errors, the increasing deviation with depth may be only explained by the inversion process, decreasing sensitivity, less spatial resolution or potential 3-D effects.

The resistivity distribution of the subsurface is characterized by large-scale and small-scale heterogeneities, but also distinct patterns may be identified. At shallow depth up to 0.9 m, the study area is characterized by high resistivity. This comprises the upper and the intermediate layer.

Since the laboratory results indicate similar electrical properties, remarkable differences between upper and intermediate layers only occur if water content deviates. There

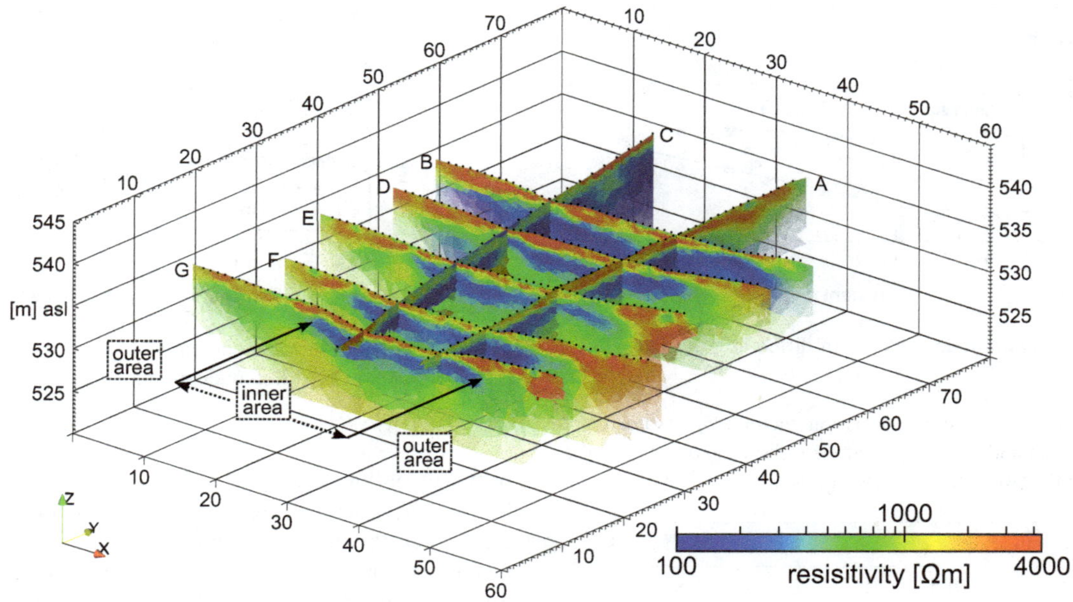

Figure 4. Resistivity results from ERT mapping (October 2008) of the study area: pseudo 3-D view of the profiles A to G.

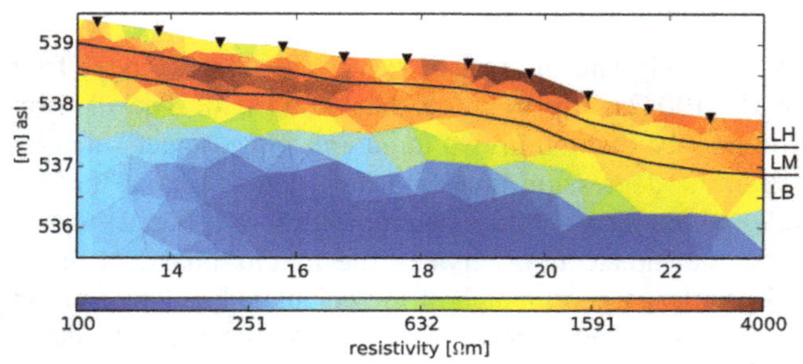

Figure 5. ERT section of profile A with plotted layer boundaries (date: 21 October 2008).

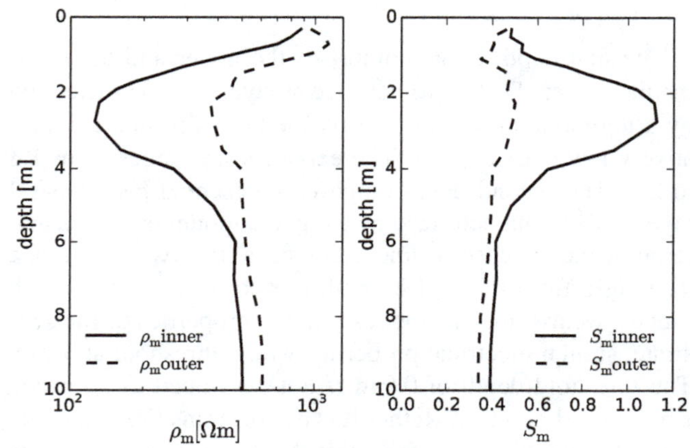

Figure 6. Median resistivity (left panel) and median water saturation (right panel) per depth for the inner region (between the depression lines) and outer region (hillslopes) (date: 21 October 2008).

are areas where the intermediate layer has higher resistivity, suggesting lower water content (see Fig. 5).

The hydrometric data show the driest conditions in 0.55–0.65 m (see Fig. 8) which is consistent with the high median resistivity of the intermediate layer at the time of data acquisition (see Fig. 6).

Resistivity decreases in greater depths (starting at 1 m). Thus, the basal layer is characterized by lower resistivity compared to the overlying layers. However, this is not constant in the lateral direction. Two different patterns are found. In the "inner" area between the two depression lines (approx. between profiles A and C), the resistivity of the basal layer is lower than in the "outer" area (the hillsides) (see Fig. 6). Between the depression lines LB is characterized as a connected zone of low resistivity. A calculation of saturation using Eq. (3) and the porosity from Table 1 indicates that this may be interpreted as a connected saturated zone (Fig. 6).

Due to the slope gradient, water from the hillsides and upper parts of the catchment flows into the direction of the depression lines, where it concentrates and forms a local slope groundwater reservoir. This results in a maximum decrease of resistivity in this zone as observed in all measured profiles at depths of 1.5–4.5 m (see Figs. 4 and 6). Percussion drilling confirmed that the thickness of LB exceeds 3.5 m downslope. Therefore, we assume that the entire saturated zone is located within the basal layer and since it is connected to the spring, it is also the source of the base flow. According to this, the shape of the surface may be partially transferred to the subsurface to identify regions of different hydrogeological conditions. Convex areas indicate dryer conditions in

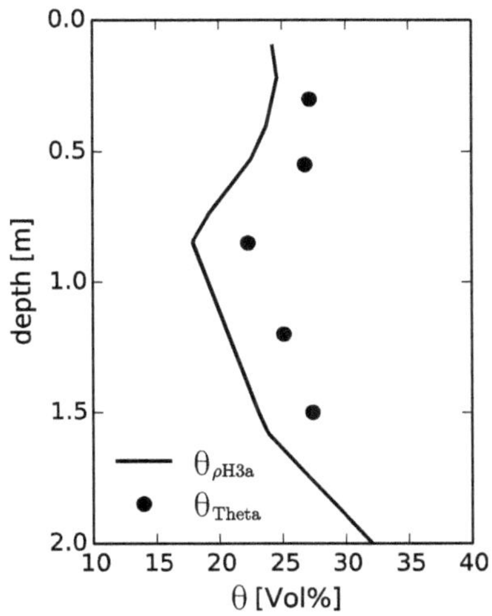

Figure 7. Volumetric water content calculated from resistivity data close to H3a in comparison with soil moisture sensors (date: 21 October 2008).

the basal layer in comparison to concave or elongated parts of the hillslope, which may act as local aquifers.

It is not feasible to relate a resistivity to the underlying gneiss or its regolith. Percussion drilling was only realized down to 4 m depth where bedrock could not be reached. If the maximum thickness of the basal layer is equal to the saturated zone, as obtained by resistivity data, the change from basal layer to underlying gneiss may be set at a depth around 4.5 m.

ERT mapping of the spatial distribution of periglacial cover beds is associated with several restrictions. In our study area, stratification is concealed by the influence of pore water, the main factor driving resistivity. On the other hand this fact may be used to improve the understanding of the moisture conditions of the subsurface.

To check the equations obtained in the lab and also to compare directly with hydrometric data, we used the water contents from the soil moisture sensors at H3a. Figure 7 compares water content, calculated with temperature-corrected resistivity ($\theta_{\rho H3a}$ profile A close to H3a), with water content from the ThetaProbes (θ_{Theta}) at time of mapping.

The values of $\theta_{\rho H3a}$ and θ_{Theta} show depth profiles of similar shape, but the values differ slightly. The resistivity depth profile shows a shift of −4.5 Vol% in comparison to the ThetaProbes. The different positions of the two probe locations could be one reason for this mismatch. Other reasons could be the inversion process of the resistivity data or differing pore water resistivity from the used median value (Table 2).

Because the data of the resistivity measurements and also the ThetaProbes may contain biased errors (e.g. caused by clast content or by the installation procedure), it is difficult to draw reliable conclusion as to which depth profile is more accurate.

3.3 Joint hydrometric and ERT monitoring

During the period May to December 2008 the spring discharge varied between 0.07 and 1.67 L s^{-1}. Median soil water tension of the study area, related to depth and time (see Fig. 8), indicates the impact of soil moisture on spring discharge. During summer increasing evapotranspiration causes the drying-out of soil. The spring showed only a slight reaction to precipitation events. Rainfall could only balance the soil water deficit and caused no runoff. Therefore, there is almost no runoff generation in the summer season. Primarily base flow dominates and decreasing discharge is mainly caused by saturation excess overland flow from the area surrounding the spring.

In contrast, during winter season (starting in November) at all depths lower tensions (< 90 hPa) were measured. Less evapotranspiration results in a replenishment of the storage water reservoirs in the subsurface. Due to the moist conditions, high presaturations predominate and cause a rapid runoff response with rain and the high discharges within the winter season.

Furthermore, there is an influence of the layered subsurface on soil moisture and runoff response. Until the beginning of May and again from December the low tensions of the upper parts of LB indicate saturated conditions, in contrast to the deeper LB with higher tensions (see Fig. 8). Due to the anisotropic hydraulic properties (low vertical compared to horizontal hydraulic conductivity) the percolation into deeper parts of LB decreases. The seepage water is concentrated as backwater in the LM and the upper parts of LB. Because of the high lateral hydraulic conductivity this saturated depth range is mainly involved in runoff and causes strong interflow.

As the hydrometric data show, the first period from May to October was mainly characterized by drying of the subsurface. After that, humid conditions began to dominate (see Fig. 8). Major changes occur at shallow depth and proceed to depth, though remarkably attenuated. Each depth has its own characteristics, its own variation in time and shows different hydrological and electrical response. To better distinguish the results and to deal with the subsurface layered structure, a depth- or layer-based analysis is appropriate.

Figure 9 shows the trend of median resistivity for each depth range for the entire time series of profile A between H1a and H4a and profile B between H4b and H4a, in comparison with daily accumulated precipitation.

The resistivity of profile A clearly correlates with profile B (Table 4). This correlation is more pronounced at shallow depths. The absolute values are similar, except for the near-surface part of LH (0–0.2 m) and parts of LB (1.5–2.0 m). These two depth ranges have higher resistivity values at profile B than A at all points in time, due to the different positions. Profile A is completely situated in one of the depression lines, in which higher soil water contents can be expected in general.

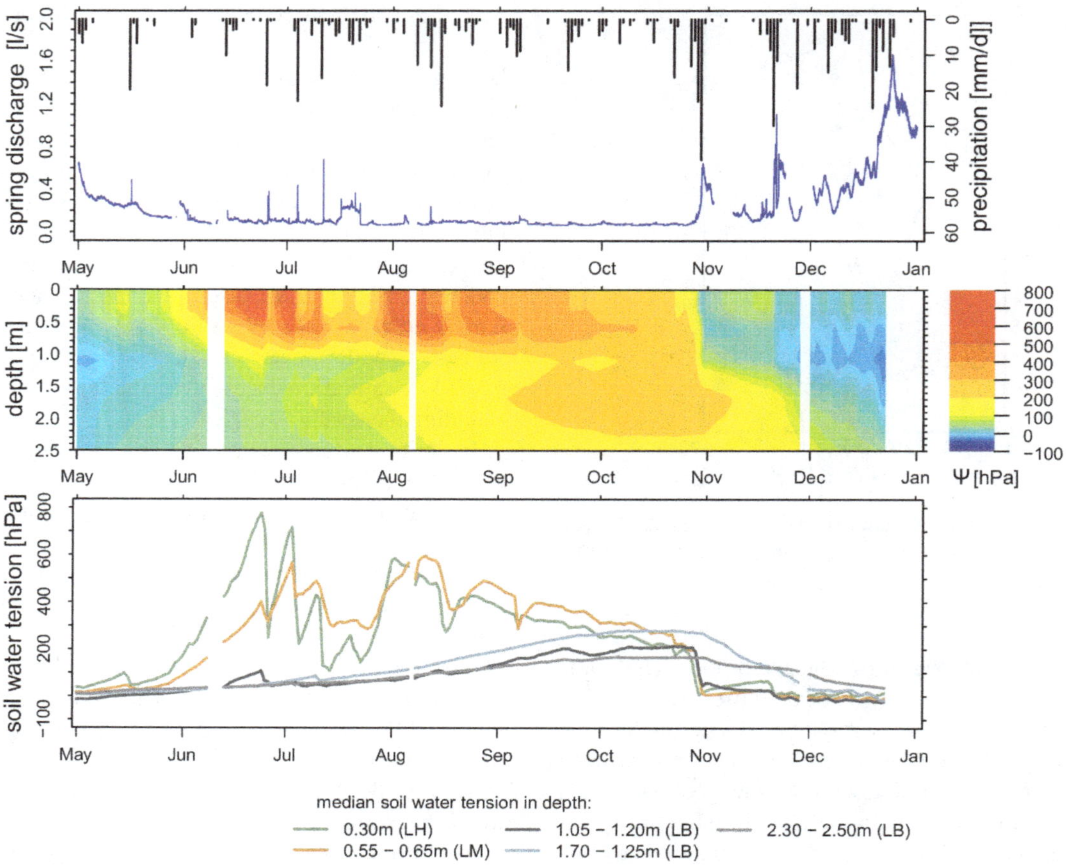

Figure 8. Spring discharge in comparison with daily precipitation (top panel), image of median soil water tension of the shallow subsurface (middle panel) and median soil water tension for different depths (bottom panel) – adapted from Heller (2012).

Table 4. Correlation between median resistivity of profiles A ($\rho_{\text{profile A}}$) and B ($\rho_{\text{profile B}}$) and between subsequent resistivity ratio of profile A ($\rho_{\text{timestep}}/\rho_{\text{initial}}$) and cumulative precipitation during the time step (ppt).

Depth [m]	$r_{(\rho_{\text{profile A}},\rho_{\text{profile B}})}$ [a]	$r_{\left(\frac{\rho_{\text{timestep}}}{\rho_{\text{initial}}},\text{ppt}\right)}$
0–0.2	0.977	−0.773[a]
0.2–0.4	0.988	−0.770[a]
0.4–0.9	0.987	−0.804[a]
0.9–1.2	0.987	−0.586[a]
1.2–1.5	0.852	−0.378[b]
1.5–2.0	0.831	−0.078[b]
2.0–3.0	0.878	0.173[b]

[a] $p < 0.01$; [b] $p > 0.01$.

During the measuring period, the upper layer (0–0.2 and 0.2–0.4 m) reacts with similar resistivity variations as the intermediate layer (0.4–0.9 m). Resistivity of the intermediate layer may temporarily exceed the upper layer (e.g. profile A, October–December).

The temporal changes in resistivity decrease with depth. Short time variations are limited down to 2 m. Below, the differences are marginal with only a continuous slight increase during the investigated period.

The variation of resistivity is significantly influenced by rainfall. As shown in Table 4, the upper and intermediate layers (< 0.9 m) show a strong negative correlation with the cumulated amount of precipitation (ppt). This correlation decreases with depth. Upper parts of the basal layer (0.9–1.5 m) respond slightly and with a delay to intense rain events or enduring dry periods. Depths > 1.5 m show no direct correlation with rainfall. Water cannot infiltrate straight to greater depths because of decreasing hydraulic conductivity, evaporation, storage, or consumption of water by roots.

One problem is the temporal resolution. Because of the time intervals (usually ≥ 1 week), we are not able to resolve the entire temporal heterogeneity of the subsurface, which may lead to misinterpretation. For example, during the period from 3 to 16 September, the amount of 33 mm rain seems not to affect the resistivity of profile A. However, 32 of these 33 mm had already been fallen by 7 September. At profile B with an additional measurement on 9 September, resistivity

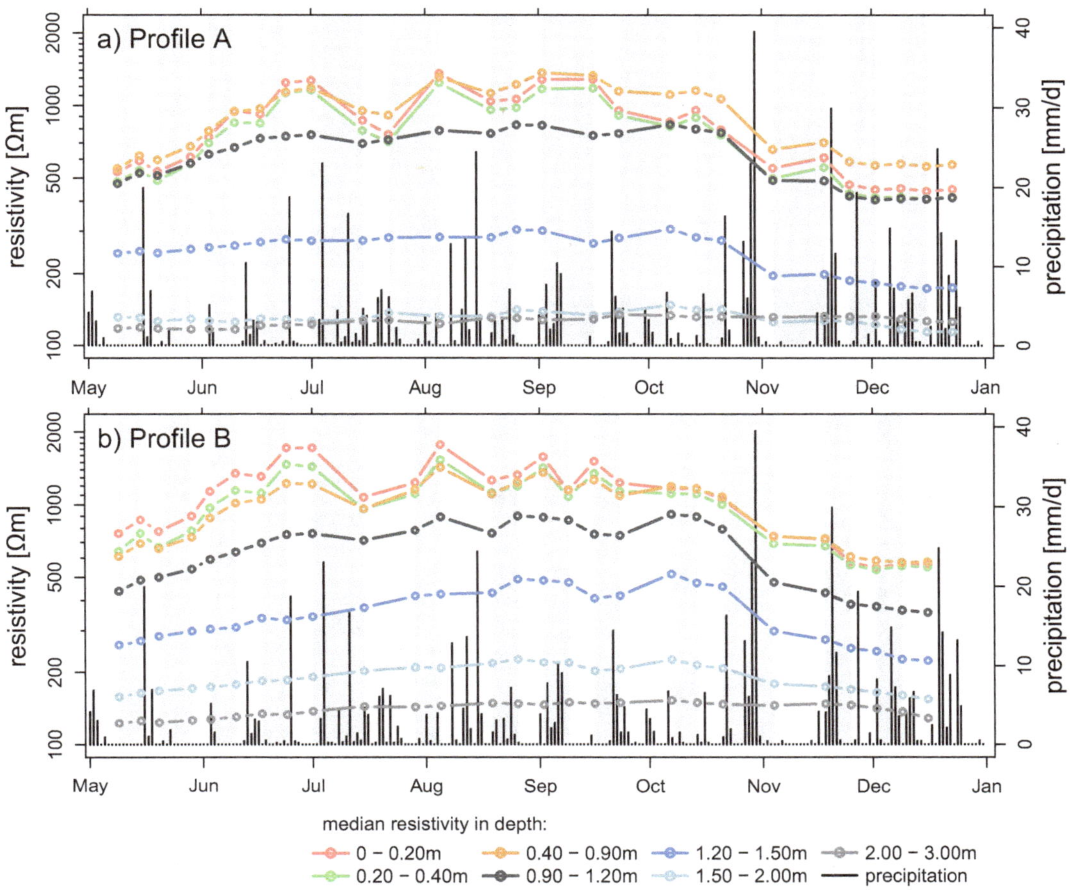

Figure 9. Trend of median resistivity for different depth ranges for **(a)** profile A and **(b)** profile B in comparison with daily precipitation (grey and white shaded regions for visualization of ERT time intervals).

at shallow depth decreases first and after that increases back to the initial level of 3 September (see Fig. 9b). Due to the missing time step, this alteration is not traced in profile A (see Fig. 9a).

This issue is also evident when comparing the resistivity with the soil suction data. With the higher temporal resolution of the tensiometer it is possible to resolve short time events, e.g. single rain events (Fig. 8), which cannot be rendered with the resistivity survey (see Fig. 9).

During the investigation period, different trends could be identified. The initial conditions in April and early May are characterized by a highly saturated subsurface. This is indicated by low soil water tension, high spring discharge and high water content. Due to the humid conditions at the beginning of the measurements, the conductivity of the shallow subsurface is high and the observed resistivity is low relative to the seasonal variations.

The first period between May and October is mainly characterized by increasing resistivity. The accumulated precipitation from 9 May to 21 October is only 337 mm. In combination with increasing evapotranspiration, this causes a drying of the subsurface (see Fig. 8). As a result of drying, at shallow depths (< 0.9 m) resistivity quickly increases until July. Below, the increase proceeds slightly, but continuously until October.

As mentioned above, resistivity, especially of LH and LM (up to 0.9 m), shows a high short time variability and is strongly associated with the amount of precipitation (ppt) (Table 4). During the investigated period three different response types could be identified that are exemplarily illustrated in Fig. 10 and compared to soil water tension.

1. A small amount of precipitation (see 23 September–7 October, ppt = 23 mm) causes a short deferment of increasing resistivity of LH and LM during the summer period. The values of initial state and time step are of the same order of magnitude. Within the temporal resolution, only a slight decrease could be recorded. Deeper parts are not affected and dry continuously. Constant discharge indicates that there is no runoff generation during this period. This amount of rain is only able to balance the deficit caused by evaporation at shallow

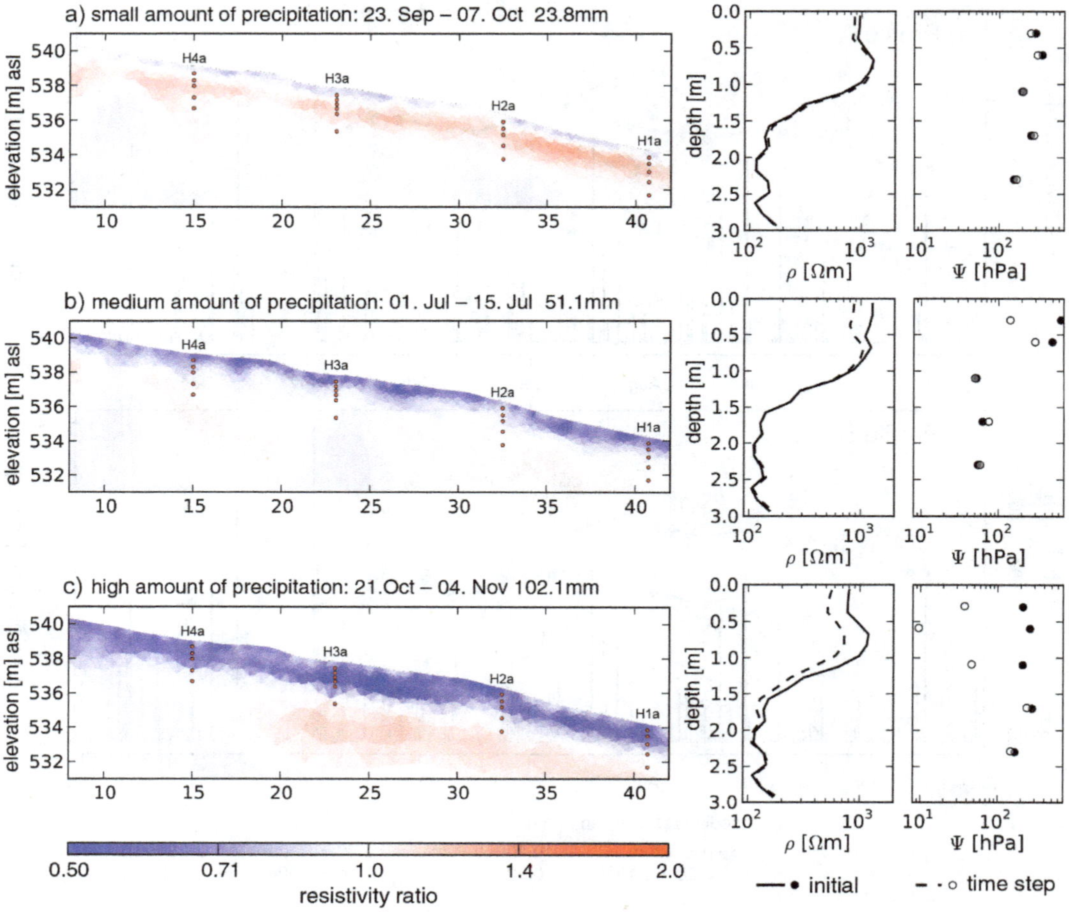

Figure 10. Ratio between subsequent resistivity (first column), median resistivity as a function of depth (second column) and median soil water tension (third column) for three exemplary precipitation responses: **(a)** small amount, **(b)** medium amount and **(c)** high amount.

depths, at least within the temporal resolution of measurements.

2. A medium amount of precipitation (see 1 July–15 July, ppt = 51.1 mm) causes a distinctive reaction at shallow depth. Resistivity at these depths shows a sharp decrease by comparatively the same ratio (~ 0.7). However, the signal is not traced into the deeper ground (> 1.2 m), which remains completely unaffected. So vertical seepage dominates in LH and LM, which leads to recharge of soil water. The water is predominantly fixed by capillary force; hence it does not percolate into deeper layers. The short rise of discharge is caused by saturation overland flow in the spring bog.

3. A high amount of precipitation (see 22 October–4 November, ppt = 102.1 mm) results in a strong response down to 2.0 m and affects LH, LM as well as parts of LB. Such a heavy rain period does not induce larger resistivity changes in LH and LM than the medium rain period, but influences deeper regions at the same order of magnitude as above. The water infiltrates to the upper, but does not reach the deeper parts of the basal layer (2–3 m). The vertical seepage is limited and therefore the increasing spring discharge may only be caused by lateral subsurface flow, such as interflow in the unsaturated subsoil.

After the major rain event at the end of October, resistivity values remain constant until the next time step. Due to precipitation of 102.1 mm during the period from 19 November to 16 December, resistivity drops below the initial state and shows highly saturated conditions.

A comparison of water content obtained by soil moisture sensors (θ_{Theta}) and water content calculated from resistivity data for different depths over time at profile A close to the hydrometric station H3a ($\theta_{\rho\text{H3a}}$) using Eq. (2) is shown in Fig. 11

At shallow depth (≤ 0.85 m), $\theta_{\rho\text{H3a}}$ correlates closely with θ_{Theta} (Table 5). However, there is a shift of the curves during the whole period. The volumetric water content from resistivity data is consequently smaller than from the soil moisture

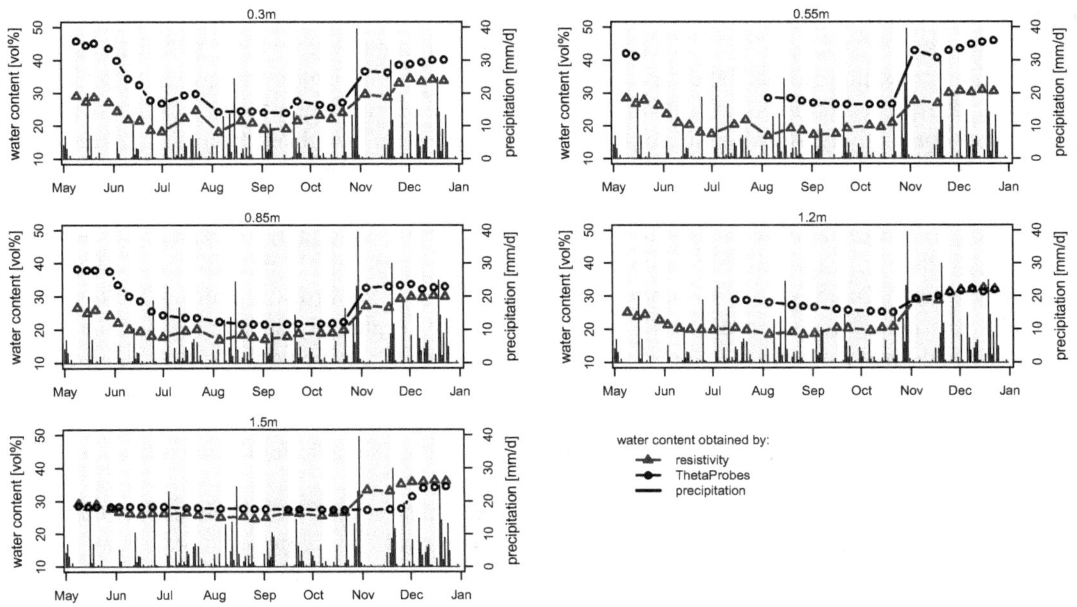

Figure 11. Trend of volumetric water content, obtained by resistivity data and soil moisture sensors (ThetaProbes) with daily precipitation for different depths.

Table 5. Correlation of volumetric water content calculated from resistivity values ($\theta_{\rho H3a}$) and water content from soil moisture sensors (θ_{Theta}) and correlation of resistivity at H3a (ρ_{H3a}) and soil suction at H3a (Ψ_{H3a}).

Depth [m]	$r_{(\theta_{\rho H3a}, \theta_{Theta})}$[a]	$r_{(\rho_{H3a}, \Psi_{H3a})}$
0.30	0.863	0.993[a]
0.55	0.957	0.904[a]
0.85	0.885	0.905[a]
1.20	0.136	0.120[b]
1.50	0.619	0.566[a]

[a] $p < 0.01$; [b] $p > 0.01$.

sensors. In dry periods (e.g. July–October), the difference is less than under humid conditions (e.g. May).

In deeper parts the variations are attenuated. At a depth of 1.2 m there is almost no response over the year, until the heavy rain period at the end of October.

At 1.5 m depth the response of the soil moisture sensor is marginal until December, but thereafter shows an increase. In contrast, $\theta_{\rho H3a}$ shows already in late October a reaction to the heavy rain event, which is not reproducible with the ThetaProbe.

The same holds true for the correlation between resistivity (ρ_{H3a}) and soil suction at H3a (Ψ_{H3a}) (see Table 5). The resistivity of LH and LM fits well to the tensiometer data at the same depth, but in deeper parts it deviates.

These deviations between resistivity data and hydrometric measurements may have different causes. Both methods contain measuring errors, just as the laboratory and other hydrometric (e.g. soil–water resistivity) measurements. Furthermore, the soil moisture sensors and tensiometers measure punctual values. Heller (2012) demonstrated with dye infiltration experiments that preferential flow is an important process in our study area. Hence, hydrometric point measurements may over- or underestimate soil moisture, depending on whether they are inside or outside a preferential pathway. Therefore the data are very limited, with restricted validity for the entire depth range or layer. In contrast, ERT has the advantage of integrating over a larger measuring volume, which makes it more suitable for extensive depth-related interpretations.

4 Conclusions

In drainage basins, hillslopes link precipitation to river runoff. Runoff components, different flow pathways and residence times are mainly influenced by the properties of the hillslope, especially the shallow subsurface. The knowledge of these properties is one of the keys to characterize the runoff dynamics in catchments. According to this, we used ERT for mapping the spatial heterogeneity of the subsurface structure on a hillslope with particular focus on mid-latitude slope deposits (cover beds).

ERT makes it possible to differentiate between LH and LM as one unit and LB as another. Like the intrinsic properties (e.g. sedimentological), LH and LM have very similar electrical characteristics. Therefore, they may only be distinguished by ERT if water contents are different or change differently with time.

In contrast, the sediments within LB have their own electrical characteristics. The pedo-/petrophysical relationship, with neglecting surface conductivity, shows equal formation factors to LH and LM, but different exponents. With the lower exponent, LB is characterized by lower resistivity at the same water content. Therefore, the resistivity of LB is lower in the entire study area, which is further reinforced by the increasing mineralization of pore water with depth.

From the results of field measurements and pedo-/petrophysical parameter determination in the laboratory we have been able to monitor seasonal changes in subsurface resistivity and its relationship to precipitation and soil moisture on the hillslope scale with a minimally invasive method directly. In combination with commonly used hydrometric approaches, we improved our understanding of the allocation, distribution, and movement of water in the subsurface. Different amounts of precipitation affect the subsurface moisture conditions differently and accordingly different depths take part in runoff generation.

Because pore water (saturation and conductivity) is the main driver for resistivity, we have arrived at some comprehensive interpretations of the subsurface moisture conditions. The high resistivities of LH and LM indicate low water contents, whereas LB is divided into two different moisture zones. On the hillsides water saturation of LB is less than between the depression lines, where low resistivity shows high water saturation and implies a local slope groundwater reservoir.

During the investigation period, temperature-corrected resistivity showed distinct seasonal variations due to changes in moisture conditions, primarily influenced by precipitation and evapotranspiration. Close to the surface, these variations are very evident and decline with increasing depth, mainly limited to a depth of 2 m. This primarily affects LH, LM, and the upper parts of LB, since it may be assumed that deeper parts are already saturated and changes are only possible due to changes in water conductivity.

In summer the subsurface continuously dries, starting at the surface and proceeding to depth. This drying is temporarily interrupted by precipitation. Penetration depth and intensity of the response strongly depend on the amount of precipitation. During periods with a small amount of precipitation, infiltration is limited to LH. There is no runoff generation, and greater depths remain unaffected which leads after repeated occurrence to drier conditions within LM compared to LH. In contrast to this, a response caused by a medium amount of precipitation includes LM and a small increase in spring discharge. The main source of this runoff is saturated overland flow from the surface surrounding the spring. With a high amount of precipitation, changes in resistivity point to vertical seepage down to 2 m. Due to lateral subsurface flow within LH, LM and the upper parts of LB, the discharge of the spring strongly increases.

The results from ERT measurements show a strong correlation to the hydrometric data. The average resistivity response is congruent to the average soil tension data. Water content obtained with soil moisture sensors shows similar variations as calculated from the closest ERT profile. Consequently, soil moisture on the hillslope scale may be determined not only by punctual hydrometric measurements, but also by minimally invasive ERT monitoring, provided pedo-/petrophysical relationships are known. By the use of ERT, expansive invasive hydrometric measurements may be reduced or partially substituted without losing information – but rather enhancing the spatial significance of these conventional point measurements. A combination improves the spatial understanding of the ongoing hydrological processes and is more suitable for the identification of heterogeneities.

Cassiani et al. (2009) pointed out that a combination of geophysical and hydrometric data may be used for quantitative estimation of hillslope moisture conditions. Our study has shown that this may also be applied to mid-latitude hillslopes covered by periglacial slope deposits. Nevertheless, there are some restrictions requiring further improvements.

One shortcoming is the temporal resolution. Some hydrological responses especially at hillslopes may proceed very quickly. The major goal for further research should be to increase the temporal resolution of ERT measurements to at least trace single rain events. This could be realized with automated data acquisitions as described in Kuras et al. (2009).

Another aim should be to improve the spatial resolution. A high-resolution ERT in combination with additional cross-borehole measurements would be more suitable to deal with small-scale heterogeneities and to overcome the problem of decreasing sensitivity with depth.

Acknowledgements. We acknowledge support by the German Research Foundation and the Open Access Publication Funds of the TU Dresden. We would also like to thank M. Leopold, J. Boaga and two anonymous reviewer for their helpful comments and suggestions that improved the quality of the paper.

Edited by: N. Romano

References

AD-hoc AG-Boden: Bodenkundliche Kartieranleitung, 5th Edn., Bundesanst. für Geowiss. und Rohstoffe in Zusammenarb. mit den Staatl. Geol. Diensten, Hannover, 2005.

Amoozegar, A.: A compact constant-head permeameter for measuring saturated hydraulic conductivity of the Vadose Zone, Soil Sci. Soc. Am. J., 53, 1356–1361, 1989.

Anderson, M. G. and Burt, T. P. (Eds.): Process studies in hillslope hydrology, Wiley, Chichester, West Sussex, England, New York, 1990.

Apparao, A.: Geoelectric profiling, Geoexploration, 27, 351–389, 1991.

Archie, G.: The electrical resistivity log as an aid in determining some reservoir characteristics, T. Am. I. Min. Met. Eng., 146, 54–61, 1942.

ATKIS®-DGM2, Landesvermessungsamt Sachsen – Amtliches Topographisch-Kartographische Informationssystem – Digital elevation model, resolution: 2m, 2008

Barker, R.: Depth of investigation of collinear symmetrical four-electrode arrays, Geophysics, 54, 1031–1037, 1989.

Bechtold, M., Vanderborght, J., Weihermueller, L., Herbst, M., Günther, T., Ippisch, O., Kasteel, R., and Vereecken, H.: Upward transport in a three-dimensional heterogeneous laboratory soil under evaporation conditions, Vadose Zone J., 11, doi:10.2136/vzj2011.0066, 2012.

Beff, L., Günther, T., Vandoorne, B., Couvreur, V., and Javaux, M.: Three-dimensional monitoring of soil water content in a maize field using Electrical Resistivity Tomography, Hydrol. Earth Syst. Sci., 17, 595–609, doi:10.5194/hess-17-595-2013, 2013.

Benderitter, Y. and Schott, J. J.: Short time variation of the resistivity in an unsaturated soil: the relationship with rainfall, Eur. J. Environ. Eng. Geophys., 4, 37–49, 1999.

Binley, A., Henry-Poulter, S., and Shaw, B.: Examination of solute transport in an undisturbed soil column using electrical resistance tomography, Water Resour. Res., 32, 763–769, 1996a.

Binley, A., Shaw, B., and Henry-Poulter, S.: Flow pathways in porous media: electrical resistance tomography and dye staining image verification, Meas. Sci. Technol., 7, 384–390, 1996b.

Brunet, P., Clément, R., and Bouvier, C.: Monitoring soil water content and deficit using Electrical Resistivity Tomography (ERT) a case study in the Cevennes area, France, J. Hydrol., 48, 146–153, 2010.

Cassiani, G., Bruno, V., Villa, A., Fusi, N., and Binley, A.: A saline trace test monitored via time-lapse surface electrical resistivity tomography, J. Appl. Geophys., 59, 244–259, 2006.

Cassiani, G., Godio, A., Stocco, S., Villa, A., Deiana, R., Frattini, P., and Rossi, M.: Monitoring the hydrologic behaviour of a mountain slope via time-lapse electrical resistivity tomography, Near Surf. Geophys., 7, 475–486, 2009.

Chifflard, P., Didszun, J., and Zepp, H.: Skalenübergreifende Prozess-Studien zur Abflussbildung in Gebieten mit periglazialen Deckschichten (Sauerland, Deutschland), Grundwasser, 13, 27–41, 2008.

Dahlin, T. and Zhou, B.: A numerical comparison of 2-D resistivity imaging with 10 electrode arrays, Geophys. Prospect., 52, 379–398, 2004.

Daily, W., Ramirez, A., LaBrecque, D. J., and Nitao, J.: Electrical resistivity tomography of vadose water movement, Water Resour. Res., 28, 1429–1442, 1992.

De Morais, F., De Almeida Prado Bacellar, L., and Aranha, P. R. A.: Study of flow in vadose zone from electrical resistivity surveys, Rev. Bras. Geofis., 26, 115–122, 2008.

Descloitres, M., Ribolzi, O., and Le Troquer, Y.: Study of infiltration in a Sahelian gully erosion area using time-lapse resistivity mapping, Catena, 53, 229–253, 2003.

Descloitres, M., Ribolzi, O., Troquer, Y. L., and Thiébaux, J. P.: Study of water tension differences in heterogeneous sandy soils using surface ERT, J. Appl. Geophys., 64, 83–98, 2008a.

Descloitres, M., Ruiz, L., Sekhar, M., Legchenko, A., Braun, J., Mohan Kumar, M. S., and Subramanian, S.: Characterization of seasonal local recharge using electrical resistivity tomography and magnetic resonance sounding, Hydrol. Process., 22, 384–394, 2008b.

Di Baldassarre, G. and Uhlenbrook, S.: Is the current flood of data enough? A treatise on research needs for the improvement of flood modelling, Hydrol. Process., 26, 153–158, 2012.

French, H. and Binley, A.: Snowmelt infiltration: monitoring temporal and spatial variability using time-lapse electrical resistivity, J. Hydrol., 297, 174–186, 2004.

Garré, S., Koestel, J., Günther, T., Javaux, M., Vanderborght, J., and Vereecken, H.: Comparison of heterogeneous transport processes observed with electrical resistivity tomography in two soils, Vadose Zone J., 9, 336–349, 2010.

Garré, S., Javaux, M., Vanderborght, J., Pages, L., and Vereecken, H.: Three-dimensional electrical resistivity tomography to monitor root zone water dynamics, Vadose Zone J., 10, 412–424, 2011.

Günther, T., Rücker, C., and Spitzer, K.: Three-dimensional modelling and inversion of dc resistivity data incorporating topography – II. Inversion, Geophys. J. Int., 166, 506–517, 2006.

Hayashi, M.: Temperature-electrical conductivity relation of water for environmental monitoring and geophysical data inversion, Environ. Monit. Assess., 96, 119–128, 2004.

Heller, K.: Einfluss periglazialer Deckschichten auf die oberflächennahen Fließwege am Hang – eine Prozessstudie im Osterzgebirge, Sachsen, Ph.D. thesis, Faculty of Environmental Sciences, TU Dresden, available at: http://nbn-resolving.de/urn:nbn:de:bsz:14-qucosa-98437 (last access: 30 April 2014), 2012.

Keller, G. V. and Frischknecht, F. C.: Electrical Methods in Geophysical Prospecting, Pergamon Press, Oxford, 1966.

Kemna, A., Vanderborght, J., Hardelauf, H., and Vereecken, H.: Quantitative imaging of 3-D solute transport using 2-D time-lapse ERT: a synthetic feasibility study, in: 17th EEGS Symposium on the Application of Geophysics to Engineering and Environmental Problems, Colorado, USA, 342–353, 2004.

Kirkby, M. J. (Ed.): Hillslope Hydrology, Wiley, Chichester, 1980.

Kleber, A.: Lateraler Wasserfluss in Hangsedimenten unter Wald, in: Stoff- und Wasserhaushalt in Einzugsgebieten, Beiträge zur EU-Wasserrahmenrichtlinie, edited by: Lorz, C. and Haase, D., Springer Verlag, Heidelberg, 7–22, 2004.

Kleber, A. and Schellenberger, A.: Slope hydrology triggered by cover-beds, With an example from the Frankenwald Mountains, northeastern Bavaria, Z. Geomorphol., 42, 469–482, 1998.

Kleber, A. and Terhorst, B. (Eds.): Mid-Latitude Slope Deposits (Cover Beds), vol. 66 of Developments in sedimentology, 1st Edn., Elsevier, Amsterdam, Boston, Heidelberg, London, New York, Oxford, Paris, San Diego, San Francisco, Singapore, Sydney, Tokyo, 2013.

Klute, A. (Ed.): Methods of Soil Analysis, Part 1: Physical and Mineralogical Methods, 2nd Edn., ASA, SSA, Madison, Wisconsin, 1986.

Koestel, J., Kemna, A., Javaux, M., Binley, A., and Vereecken, H.: Quantitative imaging of solute transport in an unsaturated and undisturbed soil monolith with 3-D ERT and TDR, Water Resour. Res., 44, W12411, doi:10.1029/2007WR006755, 2008.

Koestel, J., Vanderborght, J., Javaux, M., Kemna, A., Binley, A., and Vereecken, H.: Noninvasive 3-D transport characterization in a sandy soil using ERT: 1. Investigating the validity of ERT-derived transport parameters, Vadose Zone J., 8, 711–722, 2009a.

Koestel, J., Vanderborght, J., Javaux, M., Kemna, A., Binley, A., and Vereecken, H.: Noninvasive 3-D transport characterization in a sandy soil using ERT: 2. Transport process inference, Vadose Zone J., 8, 723–734, 2009b.

Kuras, O., Pritchard, J. D., Meldrum, P. I., Chambers, J. E., Wilkinson, P. B., Ogilvy, R. D., and Wealthall, G. P.: Monitoring hydraulic processes with automated time-lapse electrical resistivity tomography (ALERT), CR Geosci., 341, 868–885, 2009.

LaBrecque, D. J., Miletto, M., Daily, W., Ramirez, A., and Owen, E.: The effects of noise on Occam's inversion of resistivity tomography data, Geophysics, 61, 538–548, 1996.

Leslie, I. N. and Heinse, R.: Characterizing soil-pipe networks with pseudo-three-dimensional resistivity tomography on forested hillslopes with restrictive horizons, Vadose Zone J., 12, doi:10.2136/vzj2012.0200, 2013.

Lesmes, D. P. and Friedman, S. P.: Relationships between the electrical and hydrogeological properties of rocks and soils, in: Hydrogeophysics, edited by: Rubin, Y. and Hubbard, S. S., Springer, Dordrecht, 87–128, 2006.

Ma, R., McBratney, A., Whelan, B., Minasny, B., and Short, M.: Comparing temperature correction models for soil electrical conductivity measurement, Precis. Agric., 12, 55–66, 2011.

Massuel, S., Favreau, G., Descloitres, M., Le Troquer, Y., Albouy, Y., and Cappelaere, B.: Deep infiltration through a sandy alluvial fan in semiarid Niger inferred from electrical conductivity survey, vadose zone chemistry and hydrological modelling, Catena, 67, 105–118, 2006.

McDonnell, J. J.: Where does water go when it rains? Moving beyond the variable source area concept of rainfall–runoff response, Hydrol. Process., 17, 1869–1875, 2003.

McDonnell, J. J., Tanaka, T., Mitchell, M. J., and Ohte, N.: Hydrology and biogeochemistry of forested catchments, Hydrol. Process., 15, 1673–1674, 2001.

McDonnell, J. J., Sivapalan, M., Vaché, K., Dunn, S., Grant, G., Haggerty, R., Hinz, C., Hooper, R., Kirchner, J., Roderick, M. L., Selker, J., and Weiler, M.: Moving beyond heterogeneity and process complexity: a new vision for watershed hydrology, Water Resour. Res., 43, W07301, doi:10.1029/2006WR005467, 2007.

Michot, D., Benderitter, Y., Dorigny, A., Nicoullaud, B., King, D., and Tabbagh, A.: Spatial and temporal monitoring of soil water content with an irrigated corn crop cover using surface electrical resistivity tomography, Water Resour. Res., 39, SBH 14-1–SBH 14-20, 2003.

Miller, C. R., Routh, P. S., Brosten, T. R., and McNamara, J. P.: Application of time-lapse ERT imaging to watershed characterization, Geophysics, 73, G7–G17, 2008.

Moldenhauer, K.-M., Heller, K., Chifflard, P., Hübner, R., and Kleber, A.: Influence of cover beds on slope hydrology, in: Mid-Latitude Slope Deposits (Cover Beds), edited by: Kleber, A. and Terhorst, B., vol. 66 of Developments in Sedimentology, Elsevier, Amsterdam etc., 127–152, 2013.

Oldenborger, G. A., Knoll, M. D., Routh, P. S., and LaBrecque, D. J.: Time-lapse ERT monitoring of an injection/withdrawal experiment in a shallow unconfined aquifer, Geophysics, 72, F177–F188, 2007.

Perri, M. T., Cassiani, G., Gervasio, I., Deiana, R., and Binley, A.: A saline tracer test monitored via both surface and cross-borehole electrical resistivity tomography: comparison of time-lapse results, J. Appl. Geophys., 79, 6–16, 2012.

Popp, S., Altdorff, D., and Dietrich, P.: Assessment of shallow subsurface characterisation with non-invasive geophysical methods at the intermediate hill-slope scale, Hydrol. Earth Syst. Sci., 17, 1297–1307, doi:10.5194/hess-17-1297-2013, 2013.

Ramirez, A., Daily, W., LaBrecque, D. J., Owen, E., and Chesnut, D.: Monitoring an underground steam injection process using electrical resistance tomography, Water Resour. Res., 29, 73–87, 1993.

Robinson, D. A., Binley, A., Crook, N., Day-Lewis, F. D., Ferré, T. P. A., Grauch, V. J. S., Knight, R., Knoll, M. D., Lakshmi, V., Miller, R., Nyquist, J., Pellerin, L., Singha, K., and Slater, L.: Advancing process-based watershed hydrological research using near-surface geophysics: a vision for, and review of, electrical and magnetic geophysical methods, Hydrol. Process., 22, 3604–3635, 2008a.

Robinson, D. A., Campbell, C. S., Hopmans, J. W., Hornbuckle, B. K., Jones, S. B., Knight, R., Ogden, F., Selker, J., and Wendroth, O.: Soil moisture measurement for ecological and hydrological watershed-scale observatories: a review, Vadose Zone J., 7, 358–389, 2008b.

Robinson, D. A., Abdu, H., Lebron, I., and Jones, S. B.: Imaging of hill-slope soil moisture wetting patterns in a semi-arid oak savanna catchment using time-lapse electromagnetic induction, J. Hydrol., 416, 39–49, 2012.

Roy, A. and Apparao, A.: Depth of investigation in direct current methods, Geophysics, 36, 943–959, 1971.

Rücker, C. and Günther, T.: The simulation of Finite ERT electrodes using the complete electrode model, Geophysics, 76(4), F227–238, 2011.

Sauer, D., Scholten, T., and Felix-Henningsen, P.: Verbreitung und Eigenschaften periglaziärer Lagen im östlichen Westerwald in Abhängigkeit von Gestein, Exposition und Relief, Mitt. Dtsch. Bodenkdl. Ges., 96, 551–552, 2001.

Scholten, T.: Periglaziäre Lagen in Mittelgebirgslandschaften – Verbreitungssystematik, Eigenschaften und Bedeutung für den Landschaftswasser- und stoffhaushalt, in: Tagungsbeiträge IFZ-Workshop Ressourcensicherung in der Kulturlandschaft, Selbstverlag, Justus-Liebig-Universität, Gießen, 11–15, 1999.

Schön, J. H.: Physical Properties of Rocks: Fundamentals and Principles of Petrophysics, vol. 18 of Handbook of Geophysical Exploration: Seismic Exploration, Elsevier, Oxford, 2004.

Singha, K. and Gorelick, S. M.: Saline tracer visualized with three-dimensional electrical resistivity tomography: field-scale spatial moment analysis, Water Resour. Res., 41, W05023, doi:10.1029/2004WR003460, 2005.

Slater, L., Binley, A., Daily, W., and Johnson, R.: Cross-hole electrical imaging of a controlled saline tracer injection, J. Appl. Geophys., 44, 85–102, 2000.

Tilch, N., Uhlenbrook, S., Didszun, J., Wenninger, J., Kirnbauer, R., Zillgens, B., and Leibundgut, C.: Hydrologische Prozessforschung zur Hochwasserentstehung im Löhnersbach-Einzugsgebiet (Kitzbüheler Alpen, Österreich), Hydrol. Wasserbewirts., 50, 67–78, 2006.

Tromp-van Meerveld, H. J.: Hillslope hydrology: from patterns to processes, Ph.D. thesis, Oregon State University, Corvallis, 2004.

Tromp-van Meerveld, H. J. and McDonnell, J. J.: Assessment of multi-frequency electromagnetic induction for determining soil moisture patterns at the hillslope scale, J. Hydrol., 368, 56–67, 2009.

Uchida, T., McDonnell, J. J., and Asano, Y.: Functional intercomparison of hillslopes and small catchments by examining water source, flowpath and mean residence time, J. Hydrol., 327, 627–642, 2006.

Uhlenbrook, S.: Von der Abflussbildungsforschung zur prozessorientierten Modellierung – ein Review, Hydrol. Wasserbewirts., 49, 13–24, 2005.

Uhlenbrook, S., Didszun, J., and Wenninger, J.: Source areas and mixing of runoff components at the hillslope scale – a multi-technical approach, Hydrolog. Sci. J., 53, 741–753, 2008.

Vanderborght, J., Kemna, A., Hardelauf, H., and Vereecken, H.: Potential of electrical resistivity tomography to infer aquifer transport characteristics from tracer studies: a synthetic case study, Water Resour. Res., 41, W06013, doi:10.1029/2004WR003774, 2005.

Völkel, J., Leopold, M., Mahr, A., and Raab, T.: Zur Bedeutung kaltzeitlicher Hangsedimente in zentraleuropäischen Mittelgebirgslandschaften und zu Fragen ihrer Terminologie, Petermann. Geogr. Mitt., 146, 50–59, 2002a.

Völkel, J., Zepp, H., and Kleber, A.: Periglaziale Deckschichten in Mittelgebirgen – ein offenes Forschungsfeld, Ber. deuts. Landesk., 76, 101–114, 2002b.

Wenninger, J., Uhlenbrook, S., Tilch, N., and Leibundgut, C.: Experimental evidence of fast groundwater responses in a hillslope/floodplain area in the Black Forest Mountains, Germany, Hydrol. Process., 18, 3305–3322, 2004.

Zhou, B. and Dahlin, T.: Properties and effects of measurement errors on 2-D resistivity imaging surveying, Near Surf. Geophys., 1, 105–117, 2003.

Zhou, Q. Y., Shimada, J., and Sato, A.: Three-dimensional spatial and temporal monitoring of soil water content using electrical resistivity tomography, Water Resour. Res., 37, 273–285, 2001.

4

Coevolution of volcanic catchments in Japan

Takeo Yoshida[1,2] and Peter A. Troch[1]

[1]University of Arizona, Tucson, AZ, USA
[2]National Agricultural and Food Research Organization, Tsukuba, Ibaraki, Japan

Correspondence to: Takeo Yoshida (takeoys@affrc.go.jp)

Abstract. Present-day landscapes have evolved over time through interactions between the prevailing climates and geological settings. Understanding the linkage between spatial patterns of landforms, soils, and vegetation in landscapes and their hydrological response is critical to make quantitative predictions in ungaged basins. Catchment coevolution is a theoretical framework that seeks to formulate hypotheses about the mechanisms and conditions that determine the historical development of catchments and how such evolution affects their hydrological response. In this study, we selected 14 volcanic catchments of different ages (from 0.225 to 82.2 Ma) in Japan. We derived indices of landscape properties (drainage density and slope–area relationship) as well as hydrological response (annual water balance, baseflow index, and flow–duration curves) and examined their relation with catchment age and climate (through the aridity index). We found a significant correlation between drainage density and baseflow index with age, but not with climate. The intra-annual flow variability was also significantly related to catchments age. Younger catchments tended to have lower peak flows and higher low flows, while older catchments exhibited more flashy runoff. The decrease in baseflow with catchment age is consistent with the existing hypothesis that in volcanic landscapes the major flow pathways change over time from deep groundwater flow to shallow subsurface flow. The drainage density of our catchments decreased with age, contrary to previous findings in a set of similar, but younger volcanic catchments in the Oregon Cascades, in which drainage density increased with age. In that case, older catchments were thought to show more landscape incision due to increasing near-surface lateral flow paths. Our results suggests two competing hypotheses on the evolution of drainage density in mature catchments. One is that as catchments continue to age, the hydrologically active channels retreat because less recharge leads to lower average aquifer levels and less baseflow. The other hypothesis is that the active channels do not undergo much surface dissection after the catchments reach maturity.

1 Introduction

The hydrological functions that partition incoming water, energy, and carbon at the land surface vary from place to place, reflecting spatial variability in land surface characteristics (landforms, soils, vegetation). Because of this spatial variability, hydrological predictions in ungaged basins are highly uncertain (Wagener et al., 2007; Sivapalan, 2003). Improving these predictions will require a better understanding of the different interactions between landscape features and hydrological response, especially in a changing environment (McDonnell et al., 2007; Wagener et al., 2013).

Present-day land surface characteristics in pristine landscapes have co-evolved over time through various interactions between the prevailing climates and geological settings (bedrock type, tectonic uplift) (Blöschl et al., 2013; Troch et al., 2015). Hydrologists have coined the term catchment coevolution as a theoretical framework that seeks to formulate explanatory hypotheses about spatial patterns of landscape characteristics and the corresponding hydrological response, based on the historical development of catchments (Sivapalan et al., 2012; Troch et al., 2013; Harman and Troch, 2014). The results of historical development are captured in the soil and hillslope catena profiles, the geometry of channel networks and the spatial distribution of plant functional types. This landscape organization determines the hydrological functions that partition incoming water, energy, and carbon. A detailed mechanistic reconstruction of how catchments have evolved to their present state is a daunting task, given the many processes involved and the unknown initial conditions during landscape formation.

Therefore, formulating and testing explanatory hypotheses of catchment coevolution and hydrological response requires a different approach from the classic mechanistic Newtonian approach. Recently, an alternative approach, called Darwinian hydrology, was proposed that seeks to document patterns of variation in populations of hydrological systems, and explain these patterns in terms of the processes and conditions of their historical development (Wagener et al., 2013; Harman and Troch, 2014). Harman and Troch (2014) summarized what the Darwinian approach might look like in watershed science as follows: (1) collect extensive and detailed observations of the pattern to be explained, (2) conceive a hypothesis that accounts for as many of the observations as possible, (3) derive from this hypothesis a subsequent set of circumstances that also must hold if the hypothesis is true, and (4) critically test whether these consequences do hold. In this study, we build upon this framework and explore empirically the links between landscape properties, hydrological response, and time since formation (age) of catchments of similar bedrock geology (volcanic rock deposits of different geologic age).

It may be argued that climate exerts a first-order control on hydrologic partitioning and that physical, chemical, and biological catchment properties only have a second-order impact (e.g., Berghuijs et al., 2014; Sivapalan et al., 2011). However, these second-order controls explain much of the functional patterns that determine flow regimes. Mushiake et al. (1981) observed that bedrock lithology plays a key role in the natural variations of functional patterns in Japanese catchments. They showed that catchments on older sedimentary rocks formed in the Paleozoic and Mesozoic exhibit flashy hydrographs and a low baseflow component compared to younger catchments on volcanic rock. They also found that catchments underlain by younger volcanic rock formed in the Quaternary have a greater groundwater component compared to those on older volcanic rock formed in the Tertiary. Lohse and Dietrich (2005) found that vertical saturated hydraulic conductivity declines as volcanic soils in Hawaii develop with time, due to the formation of secondary clay mineral layers in those soil profiles. This evolution impedes rates of vertical percolation of water and increases shallow lateral subsurface flow. Jefferson et al. (2010) studied landscape and hydrological signatures in basalt catchments in the Oregon Cascades and reported that springs were absent in older catchments, which exhibit well-developed drainage networks and flashy hydrographs. They hypothesized that the differences in hydrological response between young and old catchments was the result of changes in flow pathways, from vertical recharge and groundwater-dominated flow to shallow subsurface lateral flow with sufficient erosive power to create channel networks through landscape incision.

Prior work has suggested that climate controls variations in hydrologic regimes in two ways. One is direct control on the water balance by function of available water and energy (e.g., Budyko, 1974; Milly, 1994; Troch et al., 2009; Sivapalan et al., 2011). The other is indirect control on the hydrologic regime as one of four independent factors (climate, bedrock weatherability, tectonic uplift and time) that control the evolution of landscape properties (landforms, soils, and vegetation; Troch et al., 2015). For example, Gentine et al. (2012) found that catchment-scale rooting depth was strongly correlated with the seasonal distribution of water and energy availability in 431 catchments in the United States. Wang and Wu (2013) found a strong inverse correlation between perennial stream density and climate aridity index in 185 catchments across the United States. Both studies suggested that these catchments evolved by adapting to the long-term climate conditions.

From this overview it is clear that both climate and time influence spatial variations of hydrological response. In this study we investigated the degree of control of climate and time on the differences in hydrological response between catchments that have different internal organizations. More specifically, we asked the following questions: (1) how do landscape features and hydrological response differ with catchment age and climate? (2) What is the dominant control of catchment coevolution in basaltic landscapes? (3) What are the possible mechanisms related to the historical development of volcanic catchments? To address these questions we selected 14 catchments dominated by basaltic rock with different ages and from a range of climatic regions of Japan. The age of the catchments ranges from 0.225 to 82.2 Ma (million years). The aridity index (ratio of annual potential evaporation to annual precipitation) for these catchments ranges from 0.28 to 0.82. For each catchment we derived indices of landscape features (drainage density and slope–area relationship) and hydrological response (annual water balance, baseflow index, and flow–duration curves). We then used regression analysis to explore possible relationships between these features and catchment age and climate. Based on the observed relationships, we proposed a hypothesis that explains as many observations in volcanic catchments, both these and others, as possible.

2 Study area and materials

2.1 Study catchments

The study area is located within the Japanese archipelago, where 28 % of the land is underlain by volcanic rock. The makeup of the volcanic landscape in Japan with respect to time since formation, or geological age, is as follows: 11 % was formed in the Quaternary, 12 % in the Neogene, 1 % in the Paleogene, and 4 % in the Cretaceous (Murata and Kano, 1995). The volcanic landscapes formed in the Quaternary and Neogene are mostly in northeastern and southwestern Japan, and those formed in the Paleogene and Cretaceous are predominantly in central and western Japan. The majority of the basement rock of Japan was formed in the Jurassic to Paleogene at the margin of the Asian continent and moved toward the current position during the spreading of the Japan Sea basin in the Neogene (Taira, 2001). The volcanic rocks formed in the Cretaceous, which are prevalent in western Japan, were formed at the margin of the Asian continent and have been exposed to the atmosphere since their formation (Taira, 2001). The volcanic rocks of the Quaternary and Neogene age, on the other hand, were formed after the spreading of the Japan Sea basin and thus have lain in their current position since their formation.

The catchments selected for the study are among the catchments listed in the Japanese Dam Database (National Institute for Land and Infrastructure Management, Japan, 2015). Out of 130 catchments listed in the database, we selected 14 that contain more than 50 % volcanic rock coverage (Table 1). Basalt is the dominant rock type in all catchments, while rhyolite is found in five of the catchments (GOS, YUD, AIM, IKR, and KWM), but at less than 15 % of the volcanic rock area. The catchment areas range from 30.9 to 635.0 km^2. Daily streamflow observed at each dam was obtained from the Japanese Dam Database. The streamflow observations are inflow to the dams and no major dams are located upstream; therefore, the streamflows examined are not affected by human regulations.

The catchment ages were assigned from the average of the ages of different volcanic rocks, weighted by area of coverage, which was obtained from the Seamless Geological Map of Japan (Geological Survey of Japan, 2012). The map is the product of the merged 1 : 200 000 geologic map sheets covering the entire country of Japan. The age of the catchments ranged from 0.225 to 82.2 Ma. The youngest catchment (SNK) is located in central Japan (Fig. 1). The oldest catchment (HAZ) is covered with basalt formed in the Late Cretaceous (Fig. 2). The other 12 catchments were formed in the Neogene period (2.7–18.5 Ma) and are clustered in northeastern Japan, with the exception of the Shimouke catchment in southern Japan.

To delineate catchments, digital elevation models (DEMs) of 10 m × 10 m were downloaded from the website of the Geographic Survey Institute, Japan (2015). The stream networks were derived from digitized geographical information system (GIS) maps from the National Land Numerical Information website (Ministry of Land, Infrastructure and Transportation, Japan, 2015). Tectonic uplift for each catchment was estimated from the map of uplift in the Quaternary with 250 m intervals (National Research Center for Disaster Prevention, Japan, 1973). We imposed the delineated catchments on the tectonic uplift map and determined the uplift from the dominant area of the catchment coverage on the map.

Table 1. List of study catchments

Catchment name	ID	Age (Ma)	Volcanic coverage (–)	Area (km^2)
Shinaki	SNK	0.225	0.994	30.9
Tamagawa	TMG	2.78	0.852	287.0
Aseishigawa	ASE	3.52	0.940	225.5
Kawamata	KWM	3.69	0.507	179.4
Shimouke	SIM	4.00	0.982	185.0
Gosho	GOS	6.39	0.750	635.0
Naruko	NAR	8.43	0.598	210.0
Ikari	IKR	10.7	0.653	271.2
Aimata	AIM	11.7	0.603	110.8
Ishibuchi	ISB	11.8	0.732	154.0
Kamafusa	KAM	12.8	0.621	195.3
Shichikashuku	SCK	13.9	0.619	230.9
Yuda	YUD	18.5	0.633	583.0
Hazi	HAZ	82.2	0.623	217.0

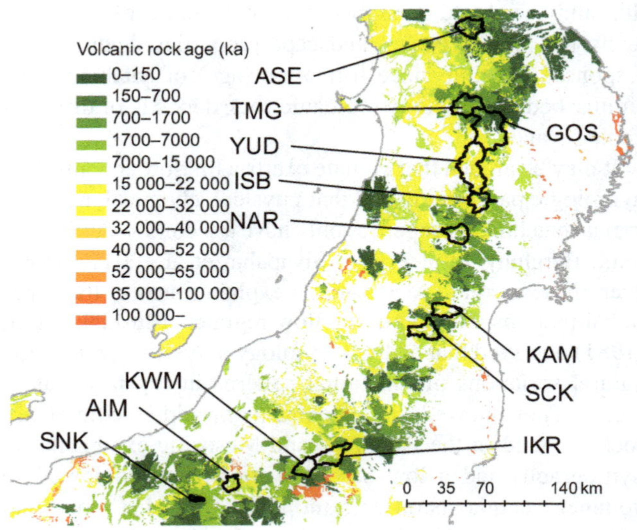

Figure 1. Study catchments in northeastern Japan.

2.2 Meteorological data

A 20-year (from 1993 through 2012) meteorological data set (including precipitation, temperature, humidity, and wind speed) was distilled from a gridded data set (hereafter referred to as the raw data set). The gridded data set covers all of Japan with grid cells of approximately 1 km × 1 km in size. The data interpolation was performed using observed data from the Automated Meteorological Data Acquisition System (AMeDAS) of Japan and the Mesh Climate Data 2000 (Japan Meteorological Agency, 2002). The Mesh Climate Data 2000 provides estimated of monthly climatic means from 1971 through 2000 for these grid cells by empirical relationships to account for orographic effects. The details of the raw data set are described in the Appendix.

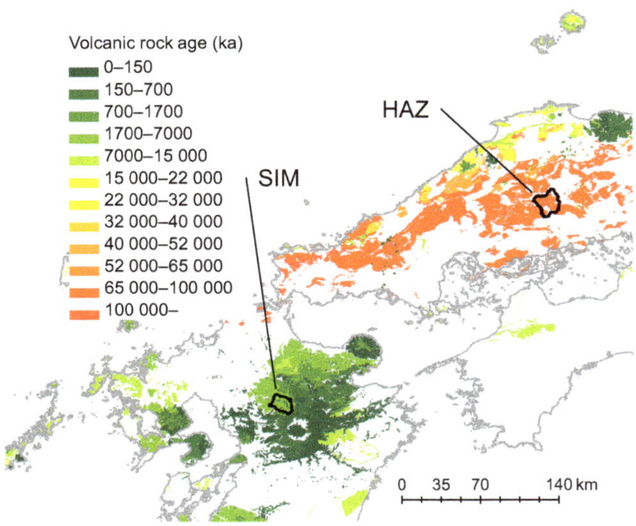

Figure 2. Study catchments in western Japan.

3 Methods

3.1 Correction of meteorological data

The precipitation in the raw data set accounts for orographic effects based on empirical relationships between precipitation and geographic features, such as elevation and distance from shoreline. It might not represent local orographic effects (e.g., higher rate of increase in precipitation on windward side of mountains), and does not correct for snow undercatch. Thus, we corrected the precipitation data as described in the Appendix. We assessed the validity of the raw and corrected data sets by means of the long-term water balance of the catchments. Given the mean of catchment-normalized annual precipitation for the raw data set P_{raw} (mm yr^{-1}) and precipitation-corrected data set P_{cor} (mm yr^{-1}), and streamflow Q (mm yr^{-1}), we estimated the mean annual evapotranspiration for each data set, E_{raw} and E_{cor} (mm yr^{-1}), as follows:

$$E_{raw} = P_{raw} - Q, \quad (1)$$
$$E_{cor} = P_{cor} - Q. \quad (2)$$

To cross-check the acceptability of E_{raw} and E_{cor}, catchment-scale actual evapotranspiration, E_a (mm yr^{-1}), was directly estimated with the complementary relation method (e.g., Bouchet, 1963; Morton, 1978; Brutsaert and Stricker, 1979). In this study, the complementary relationship method proposed by Otsuki et al. (1984a) was used to estimate E_a for each catchment. The method was used because it was rigorously validated with the long-term evapotranspiration in several experimental watersheds across Japan and does not require additional parameters (Otsuki et al., 1984b; Oue et al., 1992; Yoshida et al., 1994).

Thus, we have used three types of meteorological data sets to estimate the long-term water balance:

- Raw data set: P_{raw}, E_{raw};
- Corrected P data set: P_{cor}, E_{cor};
- Estimated E data set: P_{cor}, E_a.

The estimated data were validated against the Budyko hypothesis (Budyko, 1974). Budyko (1974) postulated an empirical relationship between the long-term evaporation index, E/P (E: evapotranspiration, P: precipitation), and the aridity index, PE/P (PE: potential evaporation) and argued that this relationship can be satisfactorily applied to most catchments. The large bias or discrepancy of estimated evaporation from the prediction of the Budyko curve was interpreted to derive from errors in the data set (Fig. 3).

3.2 Geomorphologic signatures

Two measures of landscape characteristics were used: drainage density and slope–area relationship. Drainage density is defined as the total length of streams divided by catchment area (km km^{-2}). To obtain the stream length, we used the length of the vectors defining blue-colored streamlines in the digitized version of topographic maps (Geographic Survey Institute, Japan, 2015). The scale of the map is 1 : 25 000. Because our study catchments are all located in humid regions where precipitation is evenly distributed, we assumed that the blue lines of our digital data set represents perennial streams.

To understand the timescale of landscape evolution with climate and geological age, we compared the relationship between local slope and contributing area for all the catchments. The slope–area relationship is the log–log relationship between the local slope and the upslope contributing area for all grid cells within a catchment. The local slope was estimated for each grid cell of a 10 m DEM with the D-infinity algorithm (Tarboton, 1997). The slope–area relationship can be used to discriminate between different geomorphologic regimes (Tarboton et al., 1992; Tucker and Bras, 1998; Bogaart and Troch, 2006). The slope–area relationship typically shows one or more inflection points, each of which can be interpreted as the threshold points where the dominant geomorphological process switches to other processes (Tucker and Bras, 1998; Bogaart and Troch, 2006). Hillslopes dominated by diffusional transport yield a positive slope–area relationship, i.e., slope increases with area, while hillslopes dominated by landslides or debris flows theoretically shows uniform slope gradient at a value equal to the threshold slope that induces the landslide (Tucker and Bras, 1998). Inflection points are found for a larger contributing area than those at the transition from hillslopes dominated by diffusional transport to the ones dominated by landslide/debris flow. The larger inflection points can be interpreted as the initiation of the hillslopes dominated by fluvial transport, which typically shows a negative slope–area relationship.

3.3 Hydrological signatures

Two signatures were used to represent the hydrological response of the study catchments. The first one is the baseflow index, the ratio of long-term baseflow component to total streamflow (e.g., Vogel and Kroll, 1992; Kroll et al., 2004). The higher the baseflow index the higher the groundwater contribution to catchment outflow. We separated the baseflow component from total streamflow as follows:

$$Q_b(t) = \epsilon Q_b(t-1) + \frac{1-\epsilon}{2}(Q(t) - Q(t-1)), \qquad (3)$$

where $Q_b(t) = Q(t)$ when $Q_b(t) > Q(t)$, $Q(t)$ is total streamflow at time t, $Q_b(t)$ is the estimated baseflow, and ϵ is a low-pass filter parameter (Arnold and Allen, 1999; Eckhardt, 2005). To compare hydrological responses among catchments consistently, the filter parameter ϵ was set at 0.925 for all catchments (Sawicz et al., 2011).

To quantify an index of flow variability, the slope of the flow–duration curve was used as the second signature. The flow–duration curve is the cumulative probability distribution of daily streamflow. The slope of the flow–duration curve is calculated between the 33rd and 66th streamflow percentiles, because this range represents a relatively linear part of the flow–duration curve on semi-log scale (Yadav et al., 2007; Zhang et al., 2008). This is calculated from

$$S_{\mathrm{fdc}} = \frac{Q_{33} - Q_{66}}{0.66 - 0.33}, \qquad (4)$$

where Q_{66} and Q_{33} are the streamflows that correspond to 66 % and 33 % exceedance probability, respectively. A high slope value indicates a variable flow regime, while a low slope value means a more damped response. A damped response can arise as a result of a combination of persistent (widespread and year-round) precipitation, dominance of groundwater contribution to streamflow, or a combination of the two.

4 Results

4.1 Initial data evaluation

To evaluate data consistency with long-term water partitioning, we plotted evaporation index versus aridity index for all catchments and compared the plotted values against the Budyko curve (Fig. 3). The plotted data points for the raw data set are all well below the Budyko curve. Catchment-scale actual evapotranspiration, E_a, estimated from the complementary relation method, on the other hand, showed good agreement with the Budyko curve, suggesting realistic representation of long-term evapotranspiration in each of the catchment.

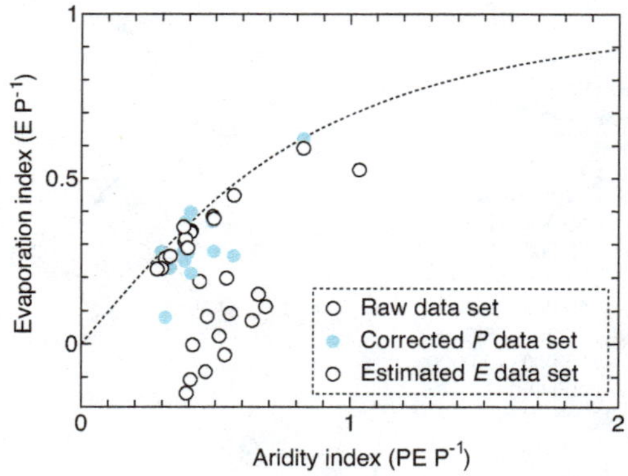

Figure 3. Raw, corrected P, and estimated E data sets plotted with Budyko curve.

Errors in the estimates of the raw data set could be caused by errors in discharge observations, trans-boundary groundwater flux, and catchment-averaged precipitation. The discharge, or dam inflow, was validated with the dam outflow, and the bias in the total inflow to total outflow was less than 2 % over 20 years. We thus assume that the dam inflow data are reliable. Another potential error is trans-boundary groundwater flux. Quantifying this flux is very difficult; however, by focusing on relatively large catchments ($> 30\,\mathrm{km}^2$), we argue that the influence of such a flux is negligible. It is more likely that the errors in annual evaporation estimated from the water balance are caused by underestimation of areal precipitation from the gridded data. Indeed, the corrected precipitation data set showed a substantial improvement from the raw data set compared to the Budyko hypothesis although some data still plot below the Budyko curve, likely caused by additional precipitation underestimation due to snow undercatch.

The aridity indices calculated with P_{cor} were used for subsequent analysis (Table 2). All the catchments lie in a energy-limited region (i.e., $\mathrm{PE}/P < 1$). Although no correlation between climate aridity indices and catchment ages was found ($R = 0.501$, $p = 0.068$), the oldest catchment lies in the driest region, while the youngest catchment lies in a humid region (Fig. 4). Thus, it is not possible to thoroughly investigate the controls of climate and time on catchment coevolution, and the regression of any signatures is substantially affected by the behavior of the youngest and oldest catchments.

4.2 Analysis by geological age: controls of time on catchment coevolution

All the landscape and hydrological indices obtained from the analysis are presented in Table 3. Baseflow index, drainage density, and slope of the flow–duration curve were all strongly correlated with catchment age (Figs. 5 and 6). Figure 5 shows a significant correlation between the baseflow index and catchment age ($R = -0.756$, $p = 0.0017$), and be-

Table 2. Estimated climate variables and indices.

ID	P_{cor} (mm yr^{-1})	PE (mm yr^{-1})	Q (mm yr^{-1})	E_{cor} (mm yr^{-1})	E_a (mm yr^{-1})	PE/P	E_{cor}/P
SNK	2454	932	1770	684	864	0.38	0.279
TMG	2937	916	2700	237	754	0.312	0.081
ASE	2215	903	1743	471	753	0.408	0.213
KWM	2333	947	1404	928	783	0.406	0.398
SIM	3112	1229	2267	844	899	0.395	0.271
GOS	2544	976	1908	636	780	0.384	0.25
NAR	2852	941	2202	649	752	0.33	0.228
IKR	1770	1006	1301	469	792	0.568	0.265
AIM	2652	1026	1679	973	839	0.387	0.367
ISB	3305	928	2541	763	743	0.281	0.231
KAM	2074	1024	1498	576	782	0.494	0.278
SCK	2100	1028	1326	774	807	0.49	0.369
YUD	3206	955	2316	890	727	0.298	0.278
HAZ	1513	1249	573	939	895	0.826	0.621

Table 3. Results of hydrometric and landscape analysis of study catchments

ID	Age	PE/P	E/P	Drainage density (km km^{-2})	Baseflow index (Q_b/Q)	Baseflow coefficient (Q_b/P)	Slope of FDC
SNK	0.225	0.380	0.352	1.080	0.835	0.602	0.322
TMG	2.78	0.312	0.257	0.882	0.713	0.656	0.850
ASE	3.52	0.408	0.340	0.887	0.775	0.610	0.585
KWM	3.69	0.406	0.336	0.898	0.737	0.444	0.827
SIM	4.00	0.395	0.289	0.859	0.585	0.426	0.880
GOS	6.39	0.384	0.307	1.000	0.684	0.513	0.870
NAR	8.43	0.330	0.264	0.736	0.707	0.546	0.668
IKR	10.7	0.568	0.448	0.867	0.708	0.520	0.776
AIM	11.7	0.387	0.316	0.771	0.712	0.451	1.071
ISB	11.8	0.281	0.225	0.969	0.666	0.512	1.034
KAM	12.8	0.494	0.377	0.707	0.680	0.491	0.735
SCK	13.9	0.490	0.384	0.793	0.710	0.448	0.754
YUD	18.5	0.298	0.227	0.948	0.663	0.479	1.011
HAZ	82.2	0.826	0.592	0.777	0.533	0.202	0.895

tween the drainage density and catchment age ($R = -0.603$, $p = 0.0222$), respectively. Because the catchment age spans 3 orders of magnitude, log (catchment age) was used in the analysis. No significant correlation was found between catchment area and the baseflow index ($R = -0.24$, $p = 0.42$) or drainage density ($R = 0.23$, $p = 0.43$).

The oldest catchment (HAZ) had a baseflow index of 0.533, while in the youngest catchment (SNK), the baseflow index of 0.835 was the highest of the study catchments, suggesting groundwater-dominated flow regime. The snow-dominated catchments tended to have a larger baseflow index, because we used a low-pass filter to exclude high-frequency signals from daily hydrographs to separate baseflow from total runoff. Two of our catchments (SIM and HAZ) are not snow affected, whereas the others are snow dominated as shown in Figs. A3 and A4. Plot for the catchments that receive little snow are negatively deviated from the regression line (Fig. 5).

The decrease in baseflow index with age was supported by the slope of the flow–duration curves (Fig. 6). The slopes of the flow–duration curve were positively correlated with catchment age ($R = 0.74$, $p = 0.002$), i.e., as catchments age, the slope of the flow–duration curve increases, indicating flashier runoff.

The flow–duration curves for all the study catchments are presented in the inset of Fig. 7, in which each line represents the data from a certain range in aridity index. It is clear that the aridity index controls the total amount of generated streamflow, i.e., the higher the aridity index is, the less streamflow generated by the catchment. Figure 7 depicts the flow–duration curves for the catchments, whose aridity index are in the range between 0.35 and 0.45. We used this range because it comprises the diversified catchments in terms of

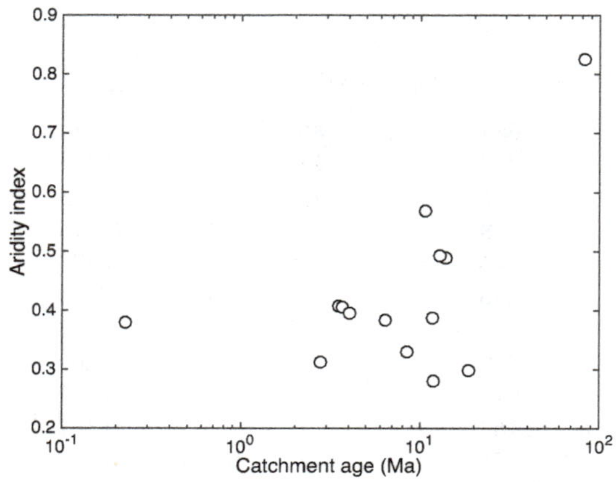

Figure 4. Relation between aridity index and age for the study catchments.

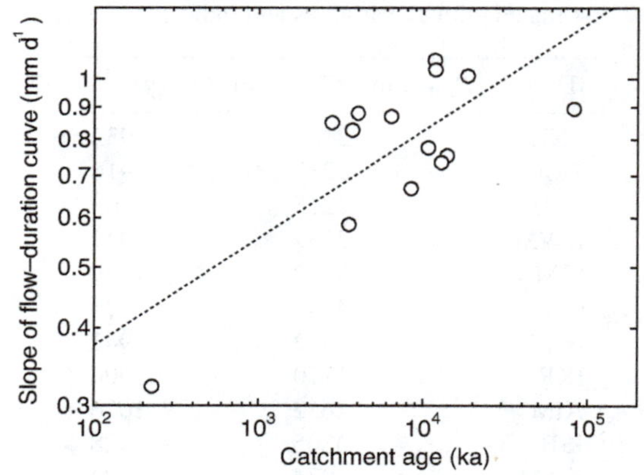

Figure 6. Slope of flow–duration curves with catchment age ($R = 0.74$, $p = 0.002$).

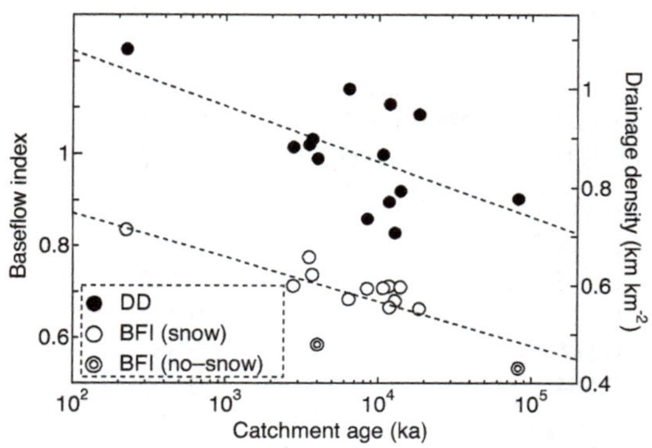

Figure 5. Influence of catchment age on baseflow index ($R = -0.756$, $p = 0.0017$) and drainage density ($R = -0.603$, $p = 0.0222$).

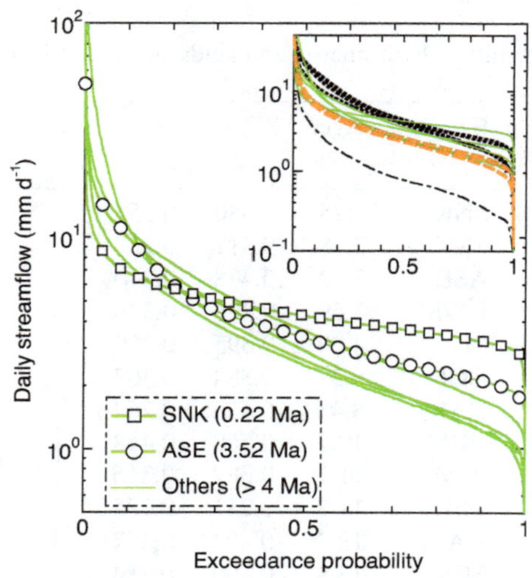

Figure 7. Flow–duration curves for the catchments, whose aridity index is in the range between 0.35 and 0.45. Inset of this figure illustrates flow–duration curves for all study catchments. Each colored line represents data from a certain range in aridity index (key to line colors is as follows: blue: $PE/P < 0.35$; green: $0.35 < PE/P < 0.45$; orange: $0.45 < PE/P < 0.65$; and black: $PE/P > 0.65$).

age (from 0.225 to 11.7 Ma). The slopes of the flow–duration curves of relatively young catchments, less than 4 Ma, were noticeably less than those of the older catchments. This implies that flow paths in older catchments with a similar aridity index have changed from deep groundwater to surface or near-surface flow paths.

It should be noted that there is no statistically significant log-linear fit between age and drainage density ($p = 0.28$) or age and slope of the flow–duration curve ($p = 0.30$) without the youngest catchment. We will later discuss the uniqueness of the youngest catchments and how it influences the interpretation of our results.

4.3 Analysis by climate: climate controls on catchment coevolution

The drainage density and baseflow index shows no significant correlation with aridity index (Fig. 8), although there are decreasing trends in both.

To highlight the seasonality of the intra-annual distribution of precipitation in each catchment, we calculated the seasonal index (SI) (Walsh and Lawler, 1981).

$$\mathrm{SI} = \frac{1}{P}\sum_{m=1}^{12}\left|X_m - \frac{P}{12}\right|, \tag{5}$$

where P is the mean annual precipitation and X_m is the mean precipitation of month m. The maximum value of SI is 1.83 if all the precipitation fell in a single month, while the minimum is 0 where the precipitation is perfectly evenly distributed throughout the year. The calculated SI of our catchments ranged from 0.182 to 0.506, which means they are classified as "very equable", "equable but with a definite wetter season" or "rather seasonal with a short drier season" (Walsh and Lawler, 1981). We therefore assumed the season-

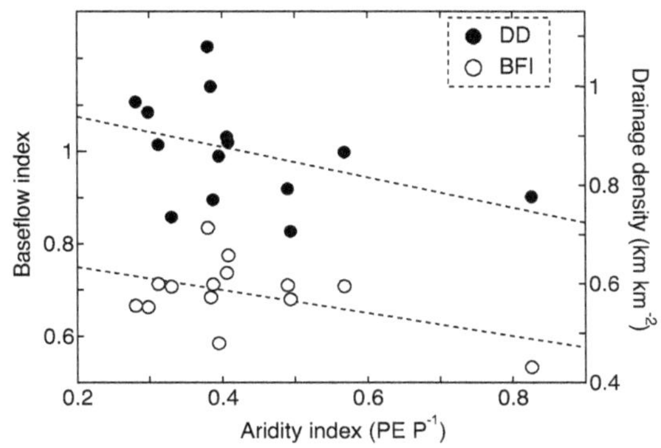

Figure 8. Climate controls on baseflow index and drainage density. The relationships with aridity index were not significant for both drainage density ($R = -0.451$, $p = 0.105$) and baseflow index ($R = -0.372$, $p = 0.190$).

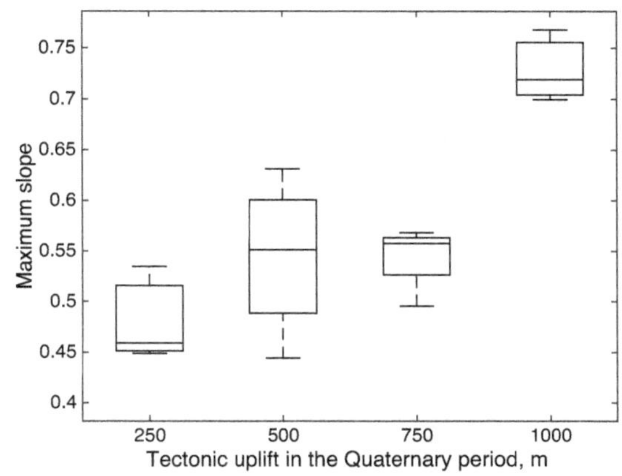

Figure 9. Tectonic uplift in the Quaternary period and maximum slope of mean slope–area relationship.

ality of precipitation of our catchments does not influence the results.

4.4 Geomorphological signatures of landscape evolution

The rate of landscape evolution is a function not only of time but also of climate, geology, and tectonics that drive the evolution process (Troch et al., 2015). To understand the timescale and governing processes of landscape evolution, an estimation of tectonic uplift is essential, especially in dynamic landscapes such as those of Japan. Figure 9 shows a strong correlation between the maximum slope captured in the mean slope–area relationship for the study catchments and the rate of tectonic uplift in the Quaternary period in Japan (National Research Center for Disaster Prevention, Japan, 1973). Numerical experiments with landscape evolution models also suggested that higher tectonic uplift rate yields steeper slopes (e.g., Yetemen, 2015). We thus assumed that the effect of the difference in tectonic uplift rate can be eliminated by comparing the catchments with similar maximum slope.

We then performed slope–area analysis (Tarboton, 1997; Jefferson et al., 2010) to infer the evolution pathways based on the current evidence. To quantify the activity level of climate on the landscape evolution, we used effective precipitation $P_{\mathrm{eff}} = P_{\mathrm{cor}} - E_{\mathrm{a}}$, because Jefferson et al. (2014) showed that mean annual precipitation has a strong link with erosion rate in basaltic islands and is the major driving factor for landscape evolution.

The mean slope–area relationship for three catchments with the smallest range of uplift rates, less than 250 m in the Quaternary, are presented in Fig. 10: SNK (0.225 Ma), ASE (3.52 Ma), and GOS (6.39 Ma). The mean slope–area relationship for the oldest catchment HAZ (82.2 Ma) is also shown for comparison. Note that the maximum slopes of the mean slope–area relationship were similar in SNK, ASE, and GOS as were the effective precipitation values; the mean slope–area relationship for HAZ differed in both respects. The smaller inflection points occurred at the contributing area value of approximately 300 m^2 in all four catchments in Fig. 10. The mean slope–area relationship in the younger two catchments (SNK and ASE) was relatively straight from the inflection points down to the outlet of the catchments, whereas a larger inflection point, which can be interpreted as the point of initiation of a fluvial channel, was found in the GOS catchment at a contributing area value of 4.4×10^4 m^2 (green dashed line in Fig. 10), suggesting the development of fluvial channels had occurred between 3 and 6 Ma since formation. Although the activity level of climate on the landscape evolution, i.e., effective precipitation, of HAZ is the lowest among the catchments in Fig. 10, the mean slope–area relationship for HAZ exhibited the gentlest slope among them where the contributing area is larger than 10^4 m^2.

Figure 11 depicts the mean slope–area relationship of catchments with uplift rate of 750 m in the Quaternary: TMG (2.78 Ma), NAR (8.43 Ma), ISB (11.8 Ma), YUD (18.5 Ma) and HAZ (82.2 Ma). The maximum slopes were similar in all of these catchments, while the effective precipitation value of HAZ was substantially less than those of the other catchments. The smaller inflection points were found at a contributing area value of approximately 300 m^2, which is similar to the catchments with the lowest uplift rate (Fig. 10), whereas clear larger inflection points were found in catchments except HAZ (Fig. 11). The larger inflection point in the youngest catchment in Fig. 11, TMG, at a contributing area of 2.1×10^6 m^2 (red dashed line) suggested that fluvial erosion had emerged earlier (less than 3 Ma of age) than in the catchments with the lowest uplift rate. With the only exception of ISB, the inflection points chronologically shifted from a lower position of the catchments (larger contributing areas) towards upper areas (smaller contributing areas); the inflection points emerged at 4.5×10^5 m^2 for NAR (dashed light blue line) and at 4.2×10^4 m^2 for YUD (dashed deep

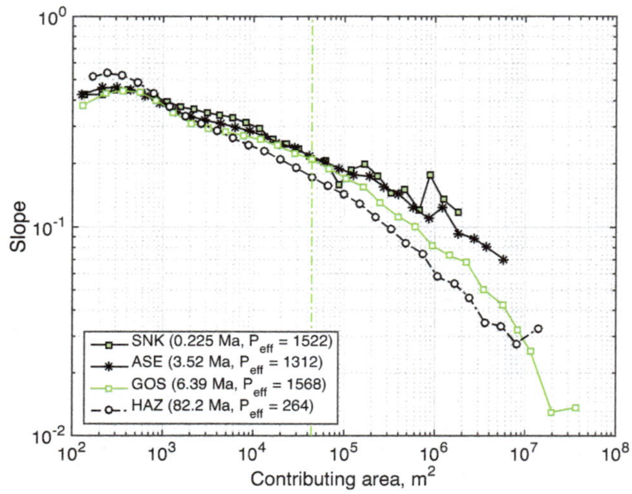

Figure 10. Mean slope–area relationship for the catchments with tectonic uplift of 250 m in the Quaternary. Vertical dashed lines represent visually determined larger inflection point for GOS (4.4×10^4 m^2).

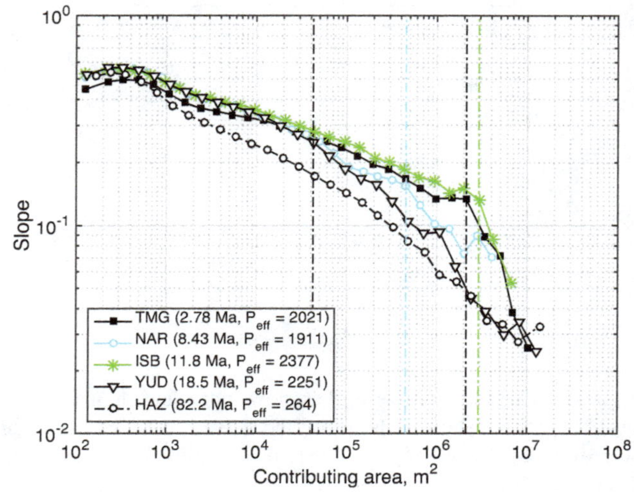

Figure 11. Mean slope–area relationship for the catchments with tectonic uplift of 750 m in the Quaternary. Vertical dashed lines represent visually determined larger inflection points for TMG (2.1×10^6 m^2, red), NAR (4.5×10^5 m^2, light blue), YUD (4.2×10^4 m^2, deep blue), and ISB (2.9×10^6 m^2, green), respectively.

blue line), respectively. Note that the contributing areas of the inflection points do not necessarily coincide with those of the averaged contributing areas for the channel heads of digitized stream networks: these were 5.2×10^6 m^2 for TMG, 9.2×10^5 m^2 for NAR, 5.3×10^5 m^2 for ISB, 7.7×10^5 m^2 for YUD, and 7.0×10^5 m^2 for HAZ.

The shift of the inflection points towards an upper position of the catchments implies the extension of channels dominated by fluvial erosion. Although there was no noticeable inflection point in the relationship for HAZ, the lowest slope in HAZ suggested that the catchment was substantially eroded. The inflection point for ISB was observed at a contributing area of 2.9×10^6 m^2, which was the largest value for the inflection point in all study catchments. The exceptional catchment, ISB, may invalidate our interpretation that the channels dominated by fluvial erosion extended chronologically.

Last, we compared the slope–area relationships for the catchments with tectonic uplift of 1000 m in the Quaternary: KWM (3.69 Ma), IKR (10.7 Ma), and AIM (11.7 Ma) (Fig. 12). Again, the slope–area relationship for the oldest catchment, HAZ, was added to the figure for reference. Clear inflection points were observed at contributing area values of 2.8×10^5 m^2 for KWM, 1.5×10^5 m^2 for IKR, and 3.7×10^5 m^2 for AIM.

Interpretation of slope–area relationship is not straightforward when tectonic uplift rates are incorporated. The inflection points for the catchments with the highest uplift rate do not corroborate the findings in the catchments with modest

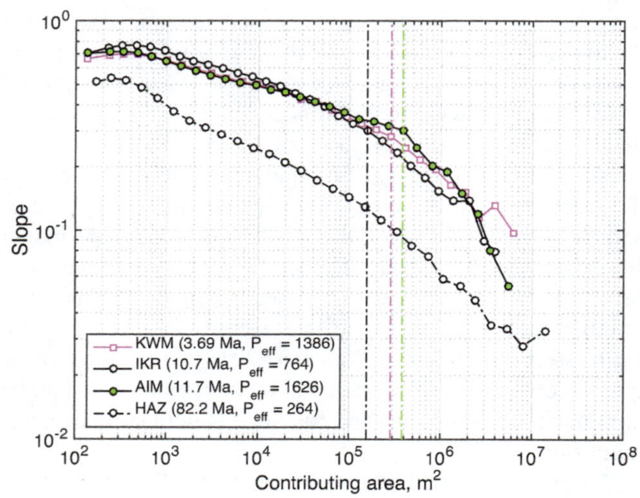

Figure 12. Mean slope–area relationship for the catchments with tectonic uplift of 1000 m in the Quaternary. Vertical dashed lines represent visually determined larger inflection points for KWM (2.8×10^5 m^2, pink), IKR (1.5×10^5 m^2, deep blue), and AIM (3.7×10^4 m^2, green), respectively.

uplift rates. One possible explanation for this is that the steep slopes in the high uplift catchments caused frequent landslides or debris flows, and subsequent rate of landscape evolution outpaced that based on chronological evolution at low to moderate rates of uplift.

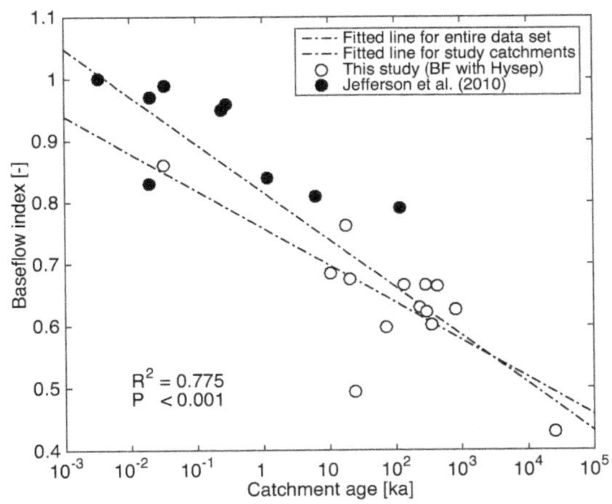

Figure 13. Baseflow index with catchment age. Dashed line is regressed line for the study catchments ($R = -0.756$, $p = 0.0017$). Dotted line is regressed line for the integrated data set of the catchments in Japan and the Oregon Cascades from Jefferson et al. (2010) ($R = -0.879$, $p < 0.001$).

5 Discussion

5.1 Relation between catchment age and hydrologic response

This study was predicated on the assumption that there is a connection between geological age, climatic conditions, and hydrological response.

Our results show that there are significant negative correlations between geologic age, baseflow index, and drainage density. Variability in the baseflow index and flow–duration curves suggest that younger catchments tend to have a larger groundwater flow component, while older catchments exhibit more flashy runoff. The decrease in baseflow in older catchments suggests that the major flow pathways have changed over time from deep groundwater aquifers to shallow subsurface flows.

The observed decrease of baseflow index with catchment age is consistent with the relationship observed in the Oregon Cascades in the western United States (Fig. 13, Jefferson et al., 2010). To compare the baseflow index values of our study with those presented by Jefferson et al. (2010), we separated the baseflow from total streamflow by using the Hysep procedure (Sloto and Crouse, 1996) presented in Jefferson et al. (2010). The combined data set of our catchments and with those of Jefferson et al. (2010) showed a steeper decline of the baseflow index with catchment age and more significant correlation between them ($R = -0.879$, $p < 0.001$) than those for the catchments in Japan alone. Jefferson et al. (2010) concluded that this decrease is the result of decreasing groundwater recharge due to a development of clay layers in the near subsurface. The emergent negative trend of baseflow index for the integrated data sets suggests that the clay layers have continued to develop over longer periods than studied by Jefferson et al. (2010).

The consistency in the rate of decline of baseflow index with time since formation suggests the similarity in catchment evolution rate in both studies. Troch et al. (2015) proposed the concept of hydrological age in catchment coevolution; i.e., "the age of a catchment is not only a function of time since formation, but also a function of how 'active' the climate, geology, and tectonics are to drive coevolution". The aridity index for the catchments in the Oregon Cascades ranged from 0.5 to 1.0 (Zomer et al., 2008), and annual precipitation from 1800 to 2700 mm (Jefferson et al., 2010). These conditions are in the same range of our study (aridity index: 0.281–0.826; annual precipitation: 1513–3305 mm), which allows the two studies to be compared. Studies addressing the tectonic uplift in the western Cascades suggests that the portions of the landscape are uplifted during the early Pliocene (5.3–3.6 Ma; Priest, 1990); and the cross section of the western Cascades suggested that the rate of tectonic uplift is on the order of 700 m in the period (Conrey et al., 2002), which corresponds to 350–500 m in the Quaternary. The rates of tectonic uplift of our catchments, ranging from 250 to 1000 m, suggests some of our study catchments are more dynamic than the catchments in the Oregon Cascades. However, the significant correlation between the catchment age and baseflow index for the combined data set implies that the differences in tectonic uplift rate do not exert significant influence on the rate of catchment coevolution.

No correlation was found between baseflow index and aridity index, not even in the United States (Fig. 14). These results indicate that the partitioning of total runoff into baseflow and quick flow was not dominantly controlled by climate, at least in the narrow range of aridity index of our catchments, suggesting that the catchment age is a strong control on hydrologic response, which is in agreement with the findings of Jefferson et al. (2010). Significant correlations between catchment age and hydrological responses allow us to enhance the predictive capabilities in ungaged environments (e.g., Sivapalan et al., 2011; Troch et al., 2013; Harman and Troch, 2014).

5.2 Reading the landscape

In addition to studying the pathways in terms of individual catchment forming factors independently, Troch et al. (2015) proposed that the landscapes could be read to infer their evolutional pathways from the current geomorphological evidence.

In the current set of catchments, we found a strong negative correlation between the drainage density and catchment age. This finding contradicts that of Jefferson et al. (2010), who found a positive correlation between drainage density and catchment age. The disparity may arise from the differences in the range of catchment ages: 0.225–82.2 Ma for our catchments, versus 0.017–15.5 Ma for the Oregon Cas-

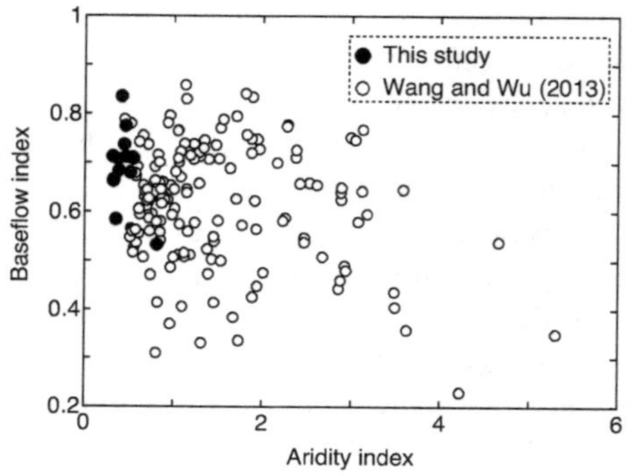

Figure 14. Relation between baseflow index and aridity index in Japan and MOPEX (Model Parameter Estimation Experiment) catchments in the US aridity index does not explain variability in baseflow index.

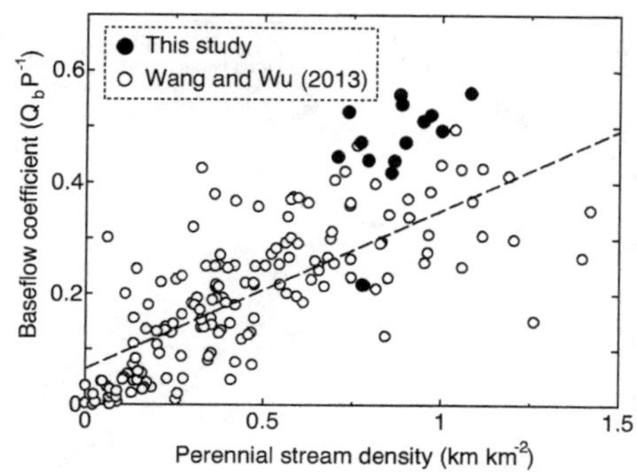

Figure 16. Perennial stream density and baseflow coefficient (Q_b/P). The baseflow, Q_b, in their study was estimated by using the same equation and parameter value as we used in Eq. (3). The correlation efficient between them was 0.74 without the study catchments (Wang and Wu, 2013) and 0.72 for the entire data set.

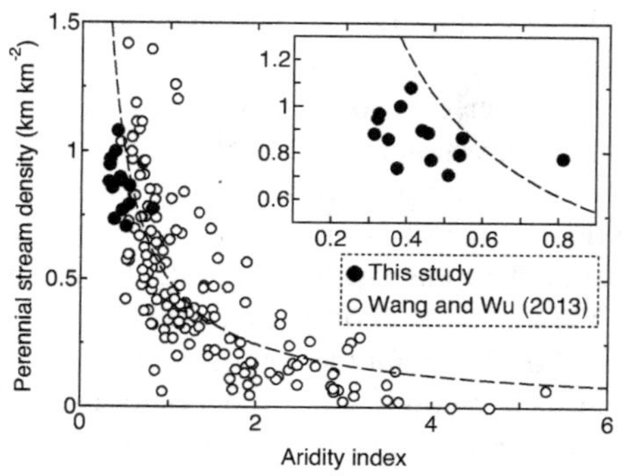

Figure 15. Perennial stream density (D_p) and aridity index (PE/P). The dashed line represents the fitted curve ($D_p = 0.444/(\text{PE}/P)$) from Wang and Wu (2013).

cades catchments of Jefferson et al. (2010). From the global data set of visible dissection on volcanic islands, Jefferson et al. (2014) observed that the visible dissection begins after about 0.5 Ma, that the dissection rate is much faster in humid landscapes, and that there is a transition from weak dissection to substantial dissection between 0.5 and 2 Ma, even in arid environments. Conversely, Suzuki (1969) observed that the dissection rate in volcanic landscapes in Japan declined markedly after 0.1 Ma after initiation. These data suggest that the increase in drainage density in the Oregon Cascades with age arose as a result of intense surface dissection, whereas the catchments of our study had possibly reached maturity in regard to dissection.

We thus hypothesize that the decline in drainage density with catchment age is caused by a mechanism other than the one described in Jefferson et al. (2010). Wang and Wu (2013) examined 185 catchments in the United States with aridity index ranging from 0.26 to 5.50. They found a systematic decrease in perennial stream density with increasing aridity index (Fig. 15). They hypothesized that the aridity index is the first-order indicator of the partitioning of precipitation into baseflow, and found a strong correlation between perennial stream density and the baseflow coefficient, Q_b/P, which follows the complementary Budyko curve (Fig. 16). We assume that our digital stream data set represents perennial streams because all the study catchments are located in a humid climate; hence, the drainage density can be compared with the perennial stream density of Wang and Wu (2013). However, the perennial stream density–aridity index curve derived by Wang and Wu (2013) does not explain the decreasing trend in drainage density of our catchments (Fig. 15, inset) even though the plotted data of our catchments cluster around the fitted curve (Fig. 15). Based on the general trend that catchments with lower baseflow coefficients tend to have lower drainage densities, we hypothesize the decline in drainage density with age therefore suggests that the proportion of precipitation that recharges groundwater aquifers declines with age.

However, plotting the drainage density of our study together with those observed in the Oregon Cascades leads to a different interpretation of the data (Fig. 17). First, the youngest catchment in our data set, SNK, has a much higher drainage density (1.08 km km^{-2}) than those of similar age in the Oregon Cascades. Second, if we remove SNK, no significant correlation between drainage density and catchment age can be found for the study catchments, although they still exhibit a negative trend ($R = -0.32$, $p = 0.28$). From lines (b) and (c) in Fig. 17, we can formulate an alternative hypothesis: the volcanic catchments reach maturity with respect to surface dissection at approximately 2 Ma of development, and the drainage density does not undergo significant change after that.

The mapping standards of the United States and Japan are both not available; hence, the difference in the mapping standards might be a reason for the anomalous drainage density of SNK. However, it is not convincing given the drainage densities of the catchments older than 2 Ma are similar. We

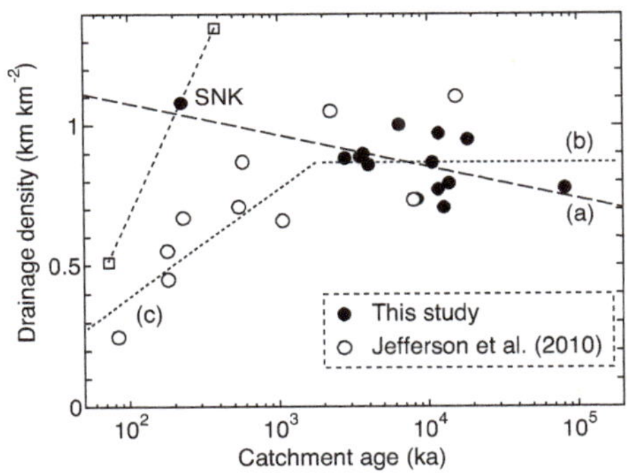

Figure 17. Drainage density with catchment age. Line **(a)** is the regression line for the study catchments ($R = -0.603$, $p = 0.0222$). Line **(b)** is the mean drainage density for the catchments older than 2 Ma in Japan and the Oregon Cascades (0.873 km km^{-2}). Line **(c)** is the regressed line for younger than 2 Ma in the Oregon Cascades. The drainage densities and basalt ages of the upper and lower part of SNK are plotted as squares (the upper part: drainage density is 0.51 km km^{-2} and basalt age is 72 ka; the lower part: 1.35 km km^{-2} and 380 ka).

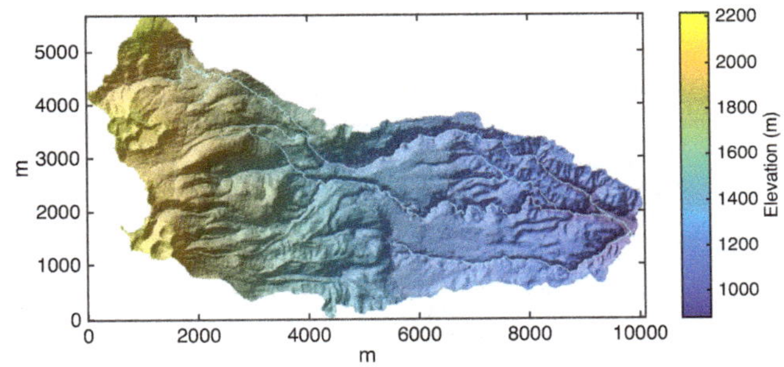

Figure 18. Topography and stream networks in SNK catchment.

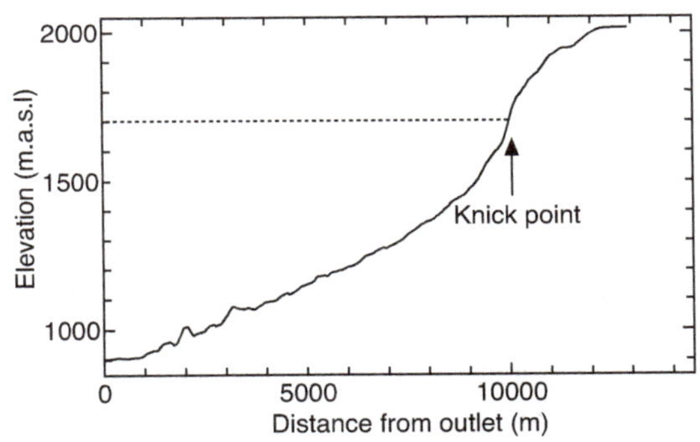

Figure 19. Longitudinal profile of the longest stream in SNK. Knick point is identified as the point at which the stream slope is the steepest.

thus assume that the difference in the mapping standards are negligible for comparison although there still remains uncertainty.

To test the alternative hypotheses, we investigated why the SNK catchment has such a high drainage density. The digital elevation model of SNK reveals distinct landscape differences between the upper and lower part of the catchment; i.e., less dissected plateau in the upper and highly dissected valleys in the lower catchments (Fig. 18). The longitudinal profile of the longest stream in SNK shows a point of convexity, or a knick point, defined as a steep reach between two low-gradient reaches (Fig. 19). The knick point is interpreted as a hallmark of the transient state of a river system caused by changes in the lithology across which the river flows, changes in the rate of tectonic uplift or spatial variability in climate forces (Anderson and Anderson, 2010). A detailed investigation of the cause of the knick point is beyond the scope of this paper; however, the basalt in the upper catchment is younger than that in the lower catchment. By dividing the upper and lower catchment along the 1700 m contour line, which is where the stream gradient is steepest (Fig. 19), we obtain the weighted average of the basalt age as 72 ka for the upper and 380 ka for the lower catchment areas and the drainage densities for those areas as 0.51 and 1.35 km km^{-2}, respectively. The plotted drainage density for the upper part of SNK is comparable with those observed in the Oregon Cascades; however, the high drainage density for the lower part of the catchment does not fit with either relationship (squares associated with SNK in Fig. 17). One possible explanation for the high drainage density in the lower part of the catchment is the acidity of stream water. The pH of the stream water in this catchment is as low as 2–3 due to its high sulfuric acid content (Ossaka et al., 1980). Acid precipitation significantly promotes weathering and soil formation by the input of rainfall and H^+ (Huang et al., 2013). Acidic stream water is neutralized by the chemical weathering of primary silicate minerals in humid forested watersheds (Johnson et al., 1981). The lower part of the SNK catchment is known for numerous hot springs of high acidity, which may have dissected at a much faster rate than neutral water. If the extremely high drainage density in the lower part of SNK has been caused by weathering due to acidic streamflow, it supports the latter hypothesis that the drainage density reaches maturity and does not change over time, but it does not reject the former on that the drainage density declines with age.

From the above analysis, the drainage densities reached a steady state or slightly decreasing trend 2 Ma after catchment formation. However, this contradicts the slope–area relationship, which indicates that the channels dominated by fluvial erosion emerged between 2 and 6 Ma since their formation, depending on the tectonic uplift rate, and extended towards the upper portion of the catchments until they become substantially eroded gentle hillslopes. It should also be noted that the contributing areas of the inflection points in mean slope–area relationship are generally smaller than those of the channel heads defined by the digitized stream lines. In other words, the signatures of fluvial erosion emerge in the upper portion of the catchments. The discrepancy be-

Table 4. Summary of hydrological and geomorphological analysis.

Catchment age	Slope–area relationship	Drainage densities	Hydrological responses
< 2 Ma		Increase of drainage density with surface dissection (Jefferson et al., 2010)	Younger catchments tended to have a larger groundwater flow component, while older catchments exhibit more flashy runoff
2–6 Ma	Dominant sediment transport shifted from landslides/debris flows to fluvial erosion, depending on the rate of tectonic uplift	Drainage densities reached an equilibrium, or slightly decreased with age	Continuous decrease of the baseflow index and increase of slope of the flow duration curve were observed
6–20 Ma	Extension of fluvial channels		
–80 Ma	Significantly eroded and gentle slope for entire catchment		

tween the inflection points and the channel heads implies that some portion of the fluvial channels are ephemeral rather than perennial. The extension of ephemeral streams in mature catchments supports the argument that the major flow pathways have changed over time from deep groundwater to shallow subsurface flow.

5.3 Hypothesis on mechanisms of coevolution in volcanic catchments

Table 4 summarizes all the hydrological and geomorphological signatures of the study catchments. From this summary, we propose a hypothesis that explains as many observations of catchment evolution in volcanic landscapes as possible.

In the early stages of development in volcanic landscapes, most of the infiltrated water flows vertically due to the high permeability of the young basalt, which typically contains many cracks and fissures (Lohse and Dietrich, 2005). In catchments younger than 2 Ma, subsurface flow is dominant to the extent that signatures of fluvial erosion in the slope–area relationship are weak (Fig. 20a). As time progresses, subsurface impermeable layers develop due to chemical weathering and mineral precipitation, and the major flow pathways change from vertical flow to shallow subsurface flow The slope–area relationship suggests that from 2 to 6 Ma of age, fluvial channels emerge in the lower part of catchments (Fig. 20b). As chemical weathering further continues (6–20 Ma), recharge to deep aquifers decreases, which would result in a shorter transit time of water and flashier hydrological responses (Fig. 20c). Increased pore pressure in steep slopes causes landslides. Mass wasting from landslides gradually decreases the local elevation until a threshold slope is reached below which landslides cannot be triggered. The lower gradient of the channel probably cannot support a high perennial stream density, leaving most of the stream network as ephemeral. The channels dominated by fluvial erosion extended towards the upper portion of the catchments until they become substantially eroded gentle

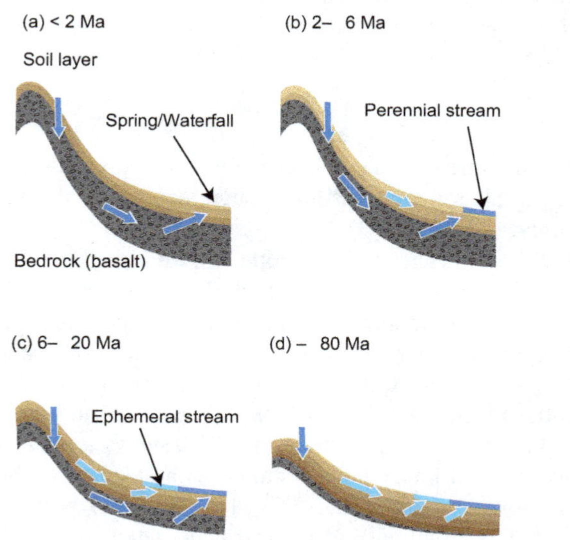

Figure 20. Conceptual model describing coevolution in volcanic catchments. In the early stages of evolution, deep groundwater flow in the bedrock (deep blue arrows) is the dominant flow pathways. As time progresses, subsurface impermeable layers develop (deepening of color in soil layer), and the major flow pathways change from vertical flow to shallow subsurface flow (light blue arrows).

hillslopes (Fig. 20d). As chemical weathering further continues, recharge to deep aquifers decreases, and further disconnection of the channel network from aquifers might result in a decrease in the extent of the perennial stream network.

It should be noted that the interconnected nature of landscape evolution and hydrologic coevolution make it difficult to determine causality. The limitation of our data set is also a confounding influence, as the oldest catchment lies in the driest region and the youngest catchment lies in a humid region. Separating the effects of climate and base rock geology is therefore difficult if we are restricted to empirical analysis. One way to test the hypothesis in such an environment is to use process-based hydrological models in an attempt to derive responses that are not observable with direct measurement. For instance, Troch et al. (2013) decoupled the landscape and climate properties using a hydrological model developed by Carrillo et al. (2011) and illustrated how catchment characteristics varied with climate gradients.

6 Conclusions

We examined 14 volcanic catchments of different ages, ranging from 0.225 to 82.2 Ma. We derived indices describing landscape features (drainage density and slope–area relationship) and hydrological responses (baseflow index, annual water balance, and slope of the flow–duration curve). The age of the volcanic rock was significantly related to the partitioning of deep groundwater recharge and shallow subsurface flow. Younger catchments tend to be associated with a larger component of groundwater flow, older ones with more flashy hydrographs. The decrease in baseflow with age suggests that the major flow pathways have changed from deep groundwater aquifers to shallow subsurface flows. The drainage densities of the catchments also decreased with catchment age. While aridity index does not explain the trend in drainage density given a quite narrow range of aridity index of our data set, catchment age has a more significant impact on the changes in drainage density; alternatively, drainage density has stayed constant in mature catchments. Our results suggests two hypotheses to be tested in future studies on the evolution of drainage density in matured catchments. One is that as catchments further evolve, hydrologically active channels retreat as less recharge leads to lower average aquifer levels and less baseflow; the other is that it does not significantly change after catchments reached maturity in terms of surface dissection.

Appendix A

Raw meteorological data

Required meteorological data (precipitation, temperature, relative humidity, and wind speed) were collected from 1976 through 2008 at existing observation stations. The meteorological data were obtained from the database of the Automated Meteorological Data Acquisition System (Japan Meteorological Agency). To estimate the spatial variation in the observed meteorological variables, the inverse distance weighting method was employed. In this method, the grid cell values are determined by calculating the weighted average of values observed at observation stations in the neighborhood of each grid cell. The closer a station is to the center of the cell being estimated, the greater its weight in the averaging.

Precipitation $p(x)$ (mm day^{-1}) in grid cell x was estimated by calculating the average of data from three observation stations, weighted according to the distance from x to the observation station. Then, the ratio of the observed precipitation to the estimated climatic value $r(i)$ (where $i = 1, 2, 3$) in the grid cell of station i was calculated as follows:

$$r(i) = p_{\mathrm{o}}(i)/p_{\mathrm{m}}(i), \tag{A1}$$

where $p_{\mathrm{o}}(i)$ (mm day^{-1}) is the observed precipitation at station i and $p_{\mathrm{m}}(i)$ (mm day^{-1}) is the climatic value derived from Mesh Climatic Data 2000, which is the monthly climatic precipitation for the corresponding basic grid square (third area partition) estimated from the observed spatial distribution of rainfall from 1971 through 2000 (Japan Meteorological Agency, 2002).

Then, ratio $r(i)$ is interpolated to each grid square of the model by using the inverse distance weighting method as follows:

$$r(x) = \sum_{i=1}^{3} \frac{w(i)r(i)}{\sum_{j=1}^{3} w(j)}, \tag{A2}$$

$$w(i) = \frac{1}{d(x,i)^2}, \tag{A3}$$

where $d(i)$ is the distance from grid cell x to observation station i.

Finally, $p(x)$ is estimated by multiplying $r(i)$ by $p_{\mathrm{m}}(i)$. Ideally, the estimated grid cell x should be located within a triangle formed by the three observation points; however, the same procedure was applied for grid cells outside of such a triangle. The values of other meteorological variables used for ET estimation were similarly interpolated by using the inverse distance method.

Correction of meteorological data

Precipitation corrected data set

Precipitation was corrected by accounting for local orographic effects and undercatch of snow due to wind. To account for the orographic effect in complex terrain, we collected the observed precipitation in mountainous areas from the database of Water Information System in Japan (available at http://www1.river.go.jp/). These observation data provide more reliable local orographic effects and potentially improve the estimated precipitation in the raw data set, P_{raw}. However, they are not consistent in the record length and often lacks data in winter due to heavy snow. Thus, we first

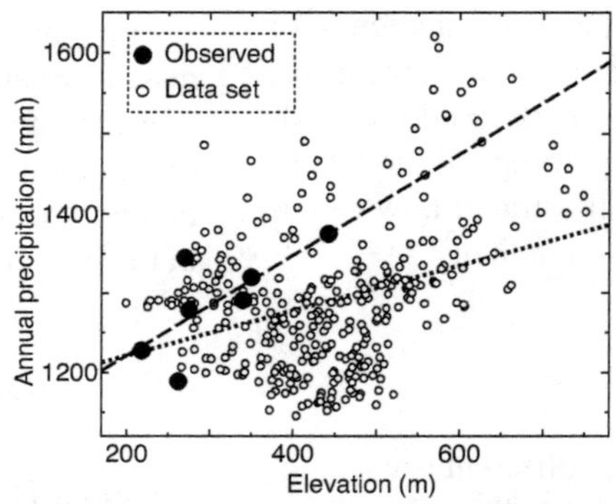

Figure A1. Altitudinal distribution of precipitation (HAZ catch-ment, annual).

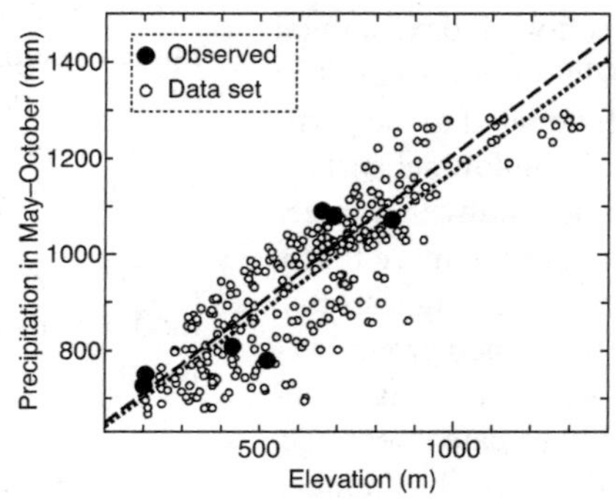

Figure A2. Altitudinal distribution of precipitation with elevation (ASE catchment, May–October).

sampled the precipitation data, P_{obs}, observed in each study catchment, whose record length is at least 5 years. We then aggregated P_{raw} and P_{obs} into annual-averaged values and plotted them with the elevation of the grid cells and observed stations, respectively. In snow dominant areas where precipitation data are not available in winter, we aggregated the data only in summer during which the data are available. In those cases, we define P_{obs} as the mean precipitation for the sampling periods.

The comparison of the regressed lines showed the degree of underestimation for precipitation in each catchment. Figure A1 illustrates P_{raw} and P_{obs} in the HAZ catchment, where no data were missing in winter. We corrected P_{raw} so that the regressed line of P_{raw} agrees to that of P_{obs}. The results of this correction are defined as a orographic-corrected data set, P_{orog}.

Figure A2 compares P_{raw} and P_{obs} that were calculated from May through October, when the observation data are available in the ASE catchment, a typical snow dominant catchment. We corrected P_{raw} for the entire period by using the relationship obtained in Fig. A2. This might be a reason for the underestimation of P_{cor} shown in Table 2.

Next, the snow undercatch is corrected as follows. The daily precipitation P_{orog} was classified into snowfall P_{snow} and rainfall P_{rain} depending on the daily average temperature T_{a}. The value of $T_{\text{a}} = 2\ 2\,^{\circ}\text{C}$ was used for the classification. The correction of snow undercatch was only applied with P_{snow} with undercatch ratio C_{r}:

$$C_{\text{r}} = \frac{1}{1+mU}, \tag{A4}$$

where m is a specific parameter that depends on the type of rain gauge (Yokoyama et al., 2003) and U is daily averaged wind speed ($\text{m}\,\text{s}^{-1}$). The typical value of m was set to 0.346 for for the AMeDAS rain gauges.

The precipitation data set corrected with the abovementioned procedures was defined as precipitation-corrected data set P_{cor}, calculated as follows:

$$P_{\text{cor}} = P_{\text{snow}}/C_{\text{r}} + P_{\text{rain}}. \tag{A5}$$

The mean annual evaporation calculated from the water balance with corrected precipitation P_{cor} was estimated as E_{cor}.

$$E_{\text{cor}} = P_{\text{cor}} - Q \tag{A6}$$

Evaluation with catchment water balance

We evaluated the monthly water balance for each of data set with a simple water balance model.

$$S(m) = S(m-1) + P(m) - \text{AET}(m) - Q(m), \tag{A7}$$

where $S(m)$ is catchment storage in month m (mm), $P(m)$ is monthly precipitation (mm), $\text{AET}(m)$ is monthly actual

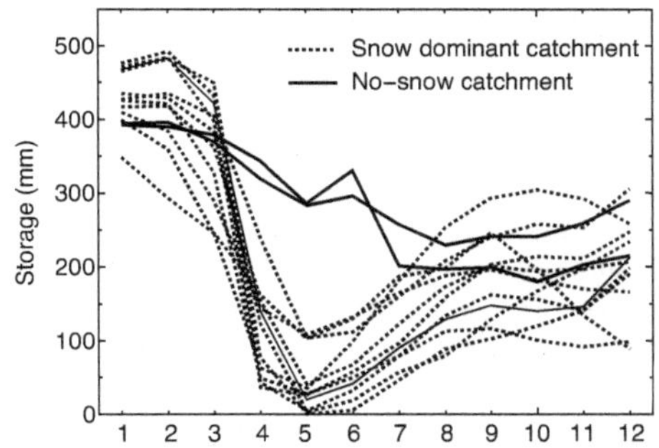

Figure A3. Seasonal water balance for raw data set.

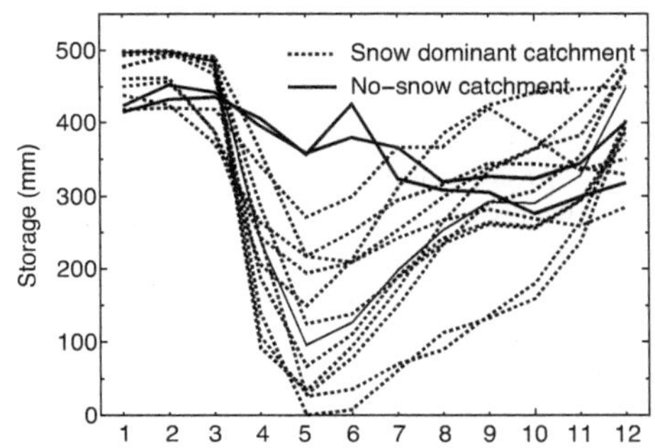

Figure A4. Seasonal water balance for precipitation corrected data set.

evapotranspiration (mm), and $Q(m)$ is monthly observed streamflow normalized by catchment area (mm). $\mathrm{AET}(m)$ was assumed to be $\mathrm{AET}(m) = \mathrm{PE}(m)S(m-1)/S_{\max}$, in which $\mathrm{PE}(m)$ is monthly potential evaporation estimated with the Penman–Monteith equation (mm) and $S_{\max}$ is parameter for the maximum storage (mm). Here, we assumed $S_{\max} = 500$ and set the initial value of S as 400 (mm) at the beginning of the year (i.e., we reset S on 1 January).

Figures A3 and A4 illustrate the monthly mean of calculated storages with raw data set P_{raw} and P_{cor}, respectively. The monthly storage calculated with P_{raw} declined at the end of each year, especially in snow dominant catchments (Fig. A3), while the monthly storage calculated with P_{cor} recovered to the initial state (i.e. 400 mm) at the end of year (Fig. A4), suggesting that the correction of precipitation was successful in terms of the annual water balance.

Acknowledgements. The first author acknowledges funding support by NARO Grant for Research Abroad to conduct this study. The authors acknowledge Anne Jefferson and Dingbao Wang for providing the data for this study. We are indebted to our reviewers, Anne Jefferson, Wouter Berghuijs, and an anonymous reviewer for their thoughtful reviews and comments. The authors would also like to thank Luke Pangle and Antônio Meira, whose feedback on a draft manuscript greatly improved the early version of this paper.

Edited by: D. Wang

References

Anderson, R. S. and Anderson, S. P.: Geomorphology: The Mechanics and Chemistry of Landscapes, Cambridge University Press, 2010.

Arnold, J. G. and Allen, P. M.: Automated methods for estimating baseflow and ground water recharge from streamflow records, J. Am. Water Resour. Assoc., 35, 411–424, 1999.

Berghuijs, W. R., Sivapalan, M., Woods, R. A., and Savenije, H. H. G.: Patterns of similarity of seasonal water balances: a window into streamflow variability over a range of time scales, Water Resour. Res., 50, 5638–5661, doi:10.1002/2014WR015692, 2014.

Blöschl, G., Sivapalan, M., and Wagener, T.: Runoff Prediction in Ungauged Basins: Synthesis Across Processes, Places and Scales, Cambridge University Press, 2013.

Bogaart, P. W. and Troch, P. A.: Curvature distribution within hillslopes and catchments and its effect on the hydrological response, Hydrol. Earth Syst. Sci., 10, 925–936, doi:10.5194/hess-10-925-2006, 2006.

Bouchet, R.: Evapotranspiration réelle et potentielle, signification climatique, IAHS-AISH P., 62, 134–142, 1963.

Brutsaert, W. and Stricker, H.: An advection-aridity approach to estimate actual regional evapotranspiration, Water Resour. Res., 15, 443–450, 1979.

Budyko, M.: Climate and life, Academic Press, New York, 508 pp., 1974.

Conrey, R. M., Taylor, E. M., Donnelly-Nolan, J. M., and Sherrod, D. R.: North-central Oregon Cascades: Exploring petrologic and tectonic intimacy in a propagating intra-arc rift, Field guide to geologic processes in Cascadia, Oregon Department of Geology and Mineral Industry: Salem, 36, 47–90, 2002.

Carrillo, G., Troch, P. A., Sivapalan, M., Wagener, T., Harman, C., and Sawicz, K.: Catchment classification: hydrological analysis of catchment behavior through process-based modeling along a climate gradient, Hydrol. Earth Syst. Sci., 15, 3411–3430, doi:10.5194/hess-15-3411-2011, 2011.

Eckhardt, K.: How to construct recursive digital filters for baseflow separation, Hydrol. Process., 19, 507–515, 2005.

Gentine, P., D'Odorico, P., Lintner, B. R., Sivandran, G., and Salvucci, G.: Interdependence of climate, soil, and vegetation as constrained by the Budyko curve, Geophys. Res. Lett., 39, L19404, doi:10.1029/2012GL053492, 2012.

Geographic Survey Institute, Japan: Geographic Data Download Service, available at: http://fgd.gsi.go.jp/download/, last access: 23 August 2015.

Geological Survey of Japan: Seamless Digital Geological Map of Japan 1 : 200 000, 2012 version, Research Information Database DB084, Geological Survey of Japan, Tsukuba, 2012.

Harman, C. and Troch, P. A.: What makes Darwinian hydrology "Darwinian"? Asking a different kind of question about landscapes, Hydrol. Earth Syst. Sci., 18, 417–433, doi:10.5194/hess-18-417-2014, 2014.

Huang, L.-M., Zhang, G.-L., and Yang, J.-L.: Weathering and soil formation rates based on geochemical mass balances in a small forested watershed under acid precipitation in subtropical China, Catena, 105, 11–20, 2013.

Japan Meteorological Agency: Mesh Climatic Data of Japan, Tokyo, 2002.

Jefferson, A. J., Grant, G., Lewis, S., and Lancaster, S.: Coevolution of hydrology and topography on a basalt landscape in the Oregon Cascade Range, USA, Earth Surf. Proc. Land., 35, 803–816, 2010.

Jefferson, A. J., Ferrier, K. L., Perron, J. T., and Ramalho, R.: Controls on the Hydrological and Topographic Evolution of Shield Volcanoes and Volcanic Ocean Islands, The Galapagos: A Natural Laboratory for the Earth Sciences, Geophys. Monogr. Ser., 204, 185–213, doi:10.1002/9781118852538, 2014.

Johnson, N. M., Driscoll, C. T., Eaton, J. S., Likens, G. E., and McDowell, W. H.: "Acid rain", dissolved aluminum and chemical weathering at the Hubbard Brook Experimental Forest, New Hampshire, Geochim. Cosmochim. Ac., 45, 1421–1437, 1981.

Kroll, C., Luz, J., Allen, B., and Vogel, R. M.: Developing a watershed characteristics database to improve low streamflow prediction, J. Hydrol. Eng., 9, 116–125, 2004.

Lohse, K. A. and Dietrich, W. E.: Contrasting effects of soil development on hydrological properties and flow paths, Water Resour. Res., 41, W12419, doi:10.1029/2004WR003403, 2005.

McDonnell, J., Sivapalan, M., Vaché, K., Dunn, S., Grant, G., Haggerty, R., Hinz, C., Hooper, R., Kirchner, J., Roderick, M. L., Selker, J., and Weiler, M.: Moving beyond heterogeneity and process complexity: a new vision for watershed hydrology, Water Resour. Res., 43, W07301, doi:10.1029/2006WR005467, 2007.

Milly, P.: Climate, soil water storage, and the average annual, Water Resour. Res., 30, 2143–2156, 1994.

Ministry of Land, Infrastructure and Transportation, Japan: National Land Numerical Information Download Service, available at: http://nlftp.mlit.go.jp/ksj-e/index.html, last access: 23 August 2015.

Morton, F. I.: Estimating evapotranspiration from potential evaporation: practicality of an iconoclastic approach, J. Hydrol., 38, 1–32, 1978.

Murata, Y. and Kano, K.: The areas of the geologic units comprising the Japanese islands, calculated by using the Geological Map of Japan 1 : 1,000,000, 3rd Edn., CD-Rom version, Chishitsu News, 493, 26–29, 1995.

Mushiake, K., Takahashi, Y., and Ando, Y.: Effects of basin geology on river flow regime in mountainous areas of Japan, Proc. Jpn. Soc. Civ. Eng., 1981, 51–62, 1981.

National Institute for Land and Infrastructure Management, Japan: Japanese Dam Database, available at: http://dam5.nilim.go.jp/dam/, last access: 23 February 2015.

National Research Center for Disaster Prevention, Japan: Explanatory Text of The Quaternary Tectonic Map of Japan, 1973.

Ossaka, J., Ozawa, T., Nomara, T., Ossaka, T., Hirabayashi, J., Takaesu, A., and Hayashi, T.: Variation of chemical compositions in volcanic gases and water at Kusatsu-Shirane Volcano and its activity in 1976, B. Volcanol., 43, 207–216, 1980.

Otsuki, K., Mitsuno, T., and Maruyama, T.: Comparison between water budget and complementary relationship estimates of catchment evapotranspiration – studies on the estimation of actual evapotranspiration (2), Transact. Japanese Soc. Irrig. Drai. Rural Eng., 112, 17–23, 1984a.

Otsuki, K., Mitsuno, T., and Maruyama, T.: Evaporation in Japan estimated from meteorological data – Studies on the estimation of actual evapotranspiration (3)-, Transact. Japanese Soc. Irrig. Drai. Rural Eng., 112, 25–32, 1984b.

Oue, H., Otsuki, K., Ohara, Y., and Maruyama, T.: Application on a complementary relationship for estimating evapotranspiration in a small area, Transact. Japanese Soc. Irrig. Drai. Rural Eng., 161, 45–50, 1992.

Priest, G. R.: Volcanic and tectonic evolution of the Cascade volcanic arc, central Oregon, J. Geophys. Res.-Sol. Ea., 95, 19583–19599, 1990.

Sawicz, K., Wagener, T., Sivapalan, M., Troch, P. A., and Carrillo, G.: Catchment classification: empirical analysis of hydrologic similarity based on catchment function in the eastern USA, Hydrol. Earth Syst. Sci., 15, 2895–2911, doi:10.5194/hess-15-2895-2011, 2011.

Sivapalan, M.: Process complexity at hillslope scale, process simplicity at the watershed scale: is there a connection?, Hydrol. Process., 17, 1037–1041, 2003.

Sivapalan, M., Yaeger, M. A., Harman, C. J., Xu, X., and Troch, P. A.: Functional model of water balance variability at the catchment scale: 1. Evidence of hydrologic similarity and space-time symmetry, Water Resour. Res., 47, W02522, doi:10.1029/2010WR009568, 2011.

Sivapalan, M., Savenije, H. H., and Blöschl, G.: Socio-hydrology: a new science of people and water, Hydrol. Process., 26, 1270–1276, 2012.

Sloto, R. A. and Crouse, M. Y.: HYSEP, a computer program for streamflow hydrograph separation and analysis, US Department of the Interior, US Geological Survey, 1996.

Suzuki, T.: Rate of erosion in strato-volcanoes in Japan, Bull. Volcanol. Soc. Jpn., 14, 133–147, 1969.

Taira, A.: Tectonic evolution of the Japanese island arc system, Annu. Rev. Earth Pl. Sc., 29, 109–134, 2001.

Tarboton, D. G.: A new method for the determination of flow directions and upslope areas in grid digital elevation models, Water Resour. Res., 33, 309–319, 1997.

Tarboton, D. G., Bras, R. L., and Rodriguez-Iturbe, I.: A physical basis for drainage density, Geomorphology, 5, 59–76, 1992.

Troch, P. A., Martinez, G. F., Pauwels, V., Durcik, M., Sivapalan, M., Harman, C., Brooks, P. D., Gupta, H., and Huxman, T.: Climate and vegetation water use efficiency at catchment scales, Hydrol. Process., 23, 2409–2414, 2009.

Troch, P. A., Carrillo, G., Sivapalan, M., Wagener, T., and Sawicz, K.: Climate-vegetation-soil interactions and long-term hydrologic partitioning: signatures of catchment co-evolution, Hydrol. Earth Syst. Sci., 17, 2209–2217, doi:10.5194/hess-17-2209-2013, 2013.

Troch, P. A., Lahmers, T., Meira, A., Mukherjee, R., Pedersen, J. W., Roy, T., and Valdés-Pineda, R.: Catchment coevolution: a useful framework for improving predictions of hydrological change?, Water Resour. Res., 51, 4903–4922, doi:10.1002/2015WR017032, 2015.

Tucker, G. E. and Bras, R. L.: Hillslope processes, drainage density, and landscape morphology, Water Resour. Res., 34, 2751–2764, 1998.

Vogel, R. M. and Kroll, C. N.: Regional geohydrologic–geomorphic relationships for the estimation of low-flow statistics, Water Resour. Res., 28, 2451–2458, 1992.

Wagener, T., Sivapalan, M., Troch, P., and Woods, R.: Catchment classification and hydrologic similarity, Geogr. Compass, 1, 901–931, doi:10.1111/j.1749-8198.2007.00039.x, 2007.

Wagener, T., Blöschl, G., Goodrich, D., Gupta, H., Sivapalan, M., Tachikawa, Y., Troch, P., and Weiler, M.: A synthesis framework for runoff predictions in ungauged basins, in: chapt. 2, Runoff Predictions in Ungauged Basins, edited by: Blöschl, G., Sivapalan, M., Wagener, T., Viglione, A., and Savenije, H., Cambridge University Press, Cambridge, UK, 11–28, 2013.

Walsh, P. and Lawler, D.: Rainfall seasonality: description, spatial patterns and change through time, Weather, 36, 201–208, 1981.

Wang, D. and Wu, L.: Similarity of climate control on base flow and perennial stream density in the Budyko framework, Hydrol. Earth Syst. Sci., 17, 315–324, doi:10.5194/hess-17-315-2013, 2013.

Yadav, M., Wagener, T., and Gupta, H.: Regionalization of constraints on expected watershed response behavior for improved predictions in ungauged basins, Adv. Water Resour., 30, 1756–1774, 2007.

Yetemen, O.: Modeling the ecohydrologic role of solar radiation on catchment development in semi-arid ecosystems, PhD thesis, University of Washington, Seattle, WA, USA, 2015.

Yokoyama, K., Ohno, H., Kominami, Y., Inoue, S., and Kawakata, T.: Performance of Japanese precipitation gauges in winter, J. Jpn. Soc. Snow Ice, 65, 303–316, 2003.

Yoshida, H., Hashino, M., Saka, K., and Muraoka, K.: Estimation of basin transpiration based on heat pulse velocity and micrometeorological data and comparison with one estimated by water budget method, P. Hydraul. Eng., 38, 131–136, 1994.

Zhang, X., Zhang, L., Zhao, J., Rustomji, P., and Hairsine, P.: Responses of streamflow to changes in climate and land use/cover in the Loess Plateau, China, Water Resour. Res., 44, W00A07, doi:10.1029/2007WR006711, 2008.

Zomer, R. J., Trabucco, A., Bossio, D. A., and Verchot, L. V.: Climate change mitigation: a spatial analysis of global land suitability for clean development mechanism afforestation and reforestation, Agr. Ecosyst. Environ., 126, 67–80, 2008.

5

Hillslope-scale experiment demonstrates the role of convergence during two-step saturation

A. I. Gevaert[1,2,*], A. J. Teuling[1], R. Uijlenhoet[1], S. B. DeLong[2,**], T. E. Huxman[2,3], L. A. Pangle[2], D. D. Breshears[2,4], J. Chorover[2,5], J. D. Pelletier[2,6], S. R. Saleska[2,7], X. Zeng[2,8], and P. A. Troch[2,9]

[1]Hydrology and Quantitative Water Management Group, Wageningen University, Wageningen, the Netherlands
[2]Biosphere 2 – College of Science, University of Arizona, Tucson, Arizona, USA
[3]Department of Ecology and Evolutionary Biology, University of California, Irvine, California, USA
[4]School of Natural Resources and the Environment, University of Arizona, Tucson, Arizona, USA
[5]Department of Soil, Water and Environmental Science, University of Arizona, Tucson, Arizona, USA
[6]Department of Geosciences, University of Arizona, Tucson, Arizona, USA
[7]Department of Ecology and Evolutionary Biology, University of Arizona, Tucson, Arizona, USA
[8]Department of Atmospheric Sciences, University of Arizona, Tucson, Arizona, USA
[9]Department of Hydrology and Water Resources, University of Arizona, Tucson, Arizona, USA
[*]now at: Department of Earth Sciences, VU University, Amsterdam, the Netherlands
[**]now at: US Geological Survey, Menlo Park, California, USA

Correspondence to: A. I. Gevaert (a.i.gevaert@vu.nl)

Abstract. Subsurface flow and storage dynamics at hillslope scale are difficult to ascertain, often in part due to a lack of sufficient high-resolution measurements and an incomplete understanding of boundary conditions, soil properties, and other environmental aspects. A continuous and extreme rainfall experiment on an artificial hillslope at Biosphere 2's Landscape Evolution Observatory (LEO) resulted in saturation excess overland flow and gully erosion in the convergent hillslope area. An array of 496 soil moisture sensors revealed a two-step saturation process. First, the downward movement of the wetting front brought soils to a relatively constant but still unsaturated moisture content. Second, soils were brought to saturated conditions from below in response to rising water tables. Convergent areas responded faster than upslope areas, due to contributions from lateral subsurface flow driven by the topography of the bottom boundary, which is comparable to impermeable bedrock in natural environments. This led to the formation of a groundwater ridge in the convergent area, triggering saturation excess runoff generation. This unique experiment demonstrates, at very high spatial and temporal resolution, the role of convergence on subsurface storage and flow dynamics. The results bring into question the representation of saturation excess overland flow in conceptual rainfall-runoff models and land-surface models, since flow is gravity-driven in many of these models and upper layers cannot become saturated from below. The results also provide a baseline to study the role of the co-evolution of ecological and hydrological processes in determining landscape water dynamics during future experiments in LEO.

1 Introduction

Understanding hillslope runoff response to extreme rainfall events is an important topic in hydrology, key to correct prediction of extreme streamflow, erosion and/or landslides, and important to integrated studies of landscapes where such processes affect vegetation dynamics, biogeochemical cycling and biosphere–atmosphere exchanges. In humid regions, saturation excess is one of the dominant mechanisms of overland flow generation (Dunne, 1978; Ward, 1984). Saturation excess occurs when the amount of incoming water exceeds the soil storage capacity at a certain location. Water can enter the soil reservoir through vertical infiltration or lateral subsurface flow (Freeze, 1972; Fiori et al., 2007). The development of saturated areas in catchments is central to the variable source area concept, which states that the bulk of catchment runoff is generated from a relatively small fraction of the total surface area of the system (Dunne and Black, 1970; Freeze, 1974). This source area is generally concentrated around a stream bed and can expand upslope into dry channels and laterally up hillslopes. The source areas expand and contract with the seasons (Dunne and Black, 1970) as well as during and after an intense rainfall event (Dunne et al., 1975; Bernier, 1985).

Many factors can influence the development of variable source areas. Firstly, soil hydraulic characteristics play an important role (Dunne and Black, 1970). For example, studies have shown that the presence of a capillary fringe, the zone of the soil profile above the groundwater table that is saturated at negative pore water pressures (Abdul and Gillham, 1984), is critical in the formation of variable source areas (Abdul and Gillham, 1984; Abdul and Gillham, 1989). Other important factors include antecedent moisture conditions (Beven, 1977; Phi et al., 2013), rainfall characteristics (Dunne and Black, 1970) and catchment geomorphology (Beven and Kirkby, 1979). Analytical studies of the effect of slope shape on saturated areas showed that the convergence of subsurface flow generates more saturated areas than planar or divergent alternatives (O'Loughlin, 1981; Troch et al., 2003). In field studies, however, the lack of a sufficiently dense array of subsurface sensors, unknown variability of soil properties and initial/boundary conditions, and the influence of multiple, competing water loss processes and rainfall input dynamics complicate the study of the role of convergence during saturation excess runoff generation.

A new experimental facility at the University of Arizona provides an opportunity to address some of these limitations, at least in part. Data from the hillslopes at the Landscape Evolution Observatory (LEO) at Biosphere 2 in Arizona provide an opportunity to study hillslope hydrological processes under highly controlled conditions. LEO was designed to improve understanding of the evolution of landscapes by studying the interactions between hydrology, ecology and soil science through years of experimentation (Hopp et al., 2009). The main focus of the research on the artificial hillslopes is not to mimic natural conditions, but to study underlying hydrological processes in great detail and under simplified and/or controlled conditions – features that constrain our ability to generate knowledge in natural hillslope settings (Huxman et al., 2009). In the first years of the project, the hillslope will remain devoid of vegetation to limit the relevant processes, but vegetation will be added once the initial set of experiments is completed. The unique experimental setup permits constant rainfall rates and known initial and boundary conditions at hillslope scale and the dense sensor network in the hillslope offers the potential to observe the hydrological response at a high spatial and temporal resolution.

The first experiment in LEO consisted of a single extreme rainfall event that saturated part of the hillslope and led to unintended gully erosion. Previously, Niu et al. (2014) analyzed the hydrological response of this experiment using a physically based model and found that incipient subsurface heterogeneity at the lower end of the hillslope was important to reconcile model based dynamics with measurements. Here, we analyze sensor data collected within the hillslope, focusing on soil moisture data. In Sect. 2, we will describe the instrumentation of the hillslope and the characteristics of the rainfall event. The results of the data analysis with special attention for the effect of convergence during the formation of saturation excess runoff are presented in Sect. 3. The saturation process, groundwater response, and overland flow generation mechanisms are discussed in Sect. 4. In Sect. 5, we will conclude with a short summary that speaks to specific hydrological phenomena, along with a description of hypotheses to guide future experimentation.

2 Experimental setup

Our study focused on the first of three LEO hillslopes at Biosphere 2. The hillslope is 11 × 30 m, and has an overall slope of 10°. The slope has a convergent shape with a central trough running from the toe of the slope to 18 m upslope (Fig. 1). The overall slope between the central trough and the far sides of the hillslope is 7°. The upslope, bottom and side boundaries are impermeable. At the toe of the slope a 0.5 m wide gravel section and a perforated plate followed by an open trough provide seepage face boundary flow conditions. The hillslope has been filled to a constant depth of 1 m using granular basaltic tephra ground to the texture of loamy sand. During construction of the hillslope, loose material was spread over a cross-slope strip of the hillslope and subsequently compacted to a specified depth. This process was repeated for several vertical layers and horizontal strips of the hillslope moving from the toe of the slope to the upper end. Soil cores from a barrel that was filled and compacted by the same method were used to measure relevant soil characteristics. The cores were taken from the barrel rather than the hillslope itself to limit disturbance of the hillslope. The material has a porosity of 0.39, a bulk density of 1.59 g cm^{-3} and a capillary fringe of approximately 30 cm. The van Genuchten curve was determined using Tempe cells and a WP4-T Dewpoint Potentiometer for the wet and dry ends, respectively (curve and relevant parameters provided in Fig. 2). The saturated hydraulic conductivity of the material based on column experiments is 1.92 m d^{-1} (2.2×10^{-5} m s^{-1}). However, based on the particle size distribution, the saturated hydraulic conductivity is 0.67 m d^{-1} (7.8×10^{-6} m s^{-1}), and later model calibration suggested that the effective value at hillslope scale is closer to 12.10 m d^{-1} (1.4×10^{-4} m s^{-1}) (Niu et al., 2014). We do not expect that the saturated hydraulic conductivity was significantly anisotropic during this

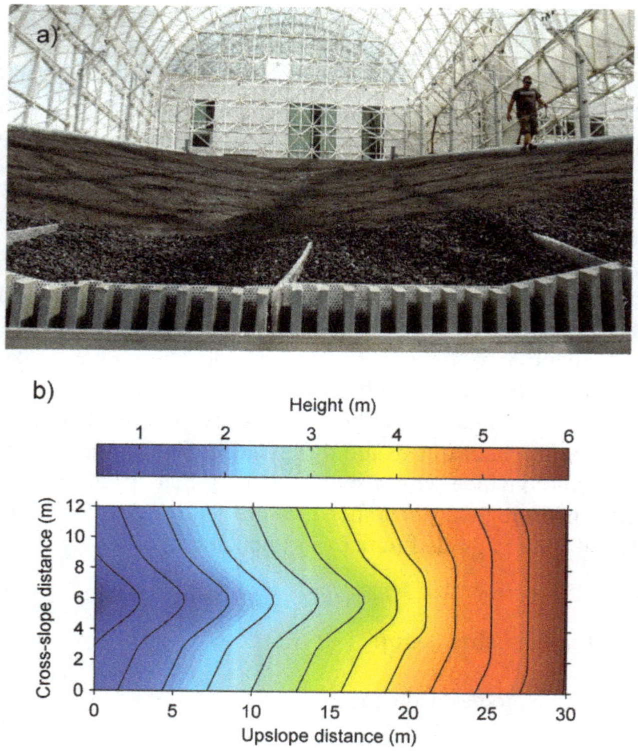

Figure 1. Hillslope design. **(a)** Photo of one of the Biosphere 2 hillslopes. **(b)** Topographic map of the hillslope with contours (solid lines) drawn every 0.5 m. Note the convergent trough in the center. The soil has a constant depth of 1 m.

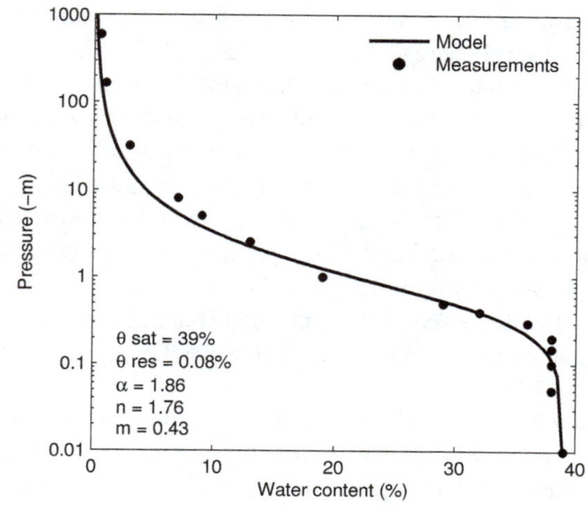

Figure 2. Laboratory measurements of the water retention curve and the van Genuchten model that best fits the measurements. The parameters describing the curve are shown in the plot.

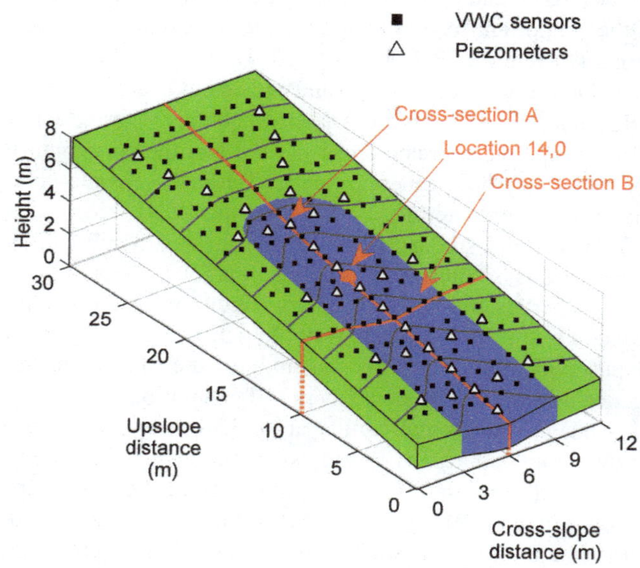

Figure 3. Schematization of the instrumentation of the hillslope with a vertical exaggeration factor of 1.5. The figure shows contour lines (grey), locations of soil moisture sensors, piezometers, and cross-sections used for analysis. The convergent zone of the slope, defined as all locations within 3 m of the central trough, is shown in blue, and the upslope area in green.

experiment because the material and compaction are homogeneous and this was the first experiment to be performed on the hillslope. In this paper, the hillslope is divided into a convergent and upslope area to study the effect of convergence (Fig. 3).

Subsurface flow is collected along the lower end of the hillslope, which is divided into six sections. From each section, the flow is routed to an electromagnetic flow meter and a tipping bucket, installed in series. A composite of measurements from these instruments gives accurate estimates of flow ($R^2 = 0.99$) when compared to known flow rates. Though instrumentation to measure overland flow is absent, estimates were made by two different methods. Before the rainfall was turned off, estimates were based on water balance analysis using measured precipitation, storage and subsurface flow data. The evaporation term was neglected because this period was during the night. After the rainfall was turned off, overland flow was collected every half hour and the flow rate was calculated based on the time that was needed to fill a fixed volume.

Within the hillslope area, an array of 496 5TM Decagon (Pullman, WA, USA) soil moisture sensors recorded volumetric water content (VWC) with an error margin of $\pm 2\,\%$. These are located on a regular grid of 154 vertical transects (Fig. 3). At each transect, sensors are located at 3 or 4 depths between 5 and 85 cm below the soil surface. In addition, the groundwater table is measured by 34 vibrating wire piezometers (GeoInstruments, San Francisco, CA, USA) installed at the bottom of the soil profile. These are placed along the central trough and along several cross-sections of the hillslope (Fig. 3). Also, ten load cells measure the total system mass, which can detect mass change equivalent to less than 1 cm of water ($\pm 0.05\,\%$).

Rainfall was applied to the hillslope by means of a sprinkler network, and was measured by an electromagnetic flow meter in the irrigation line. The experiment consisted of a single continuous rainfall event with a constant intensity of $12\,\text{mm}\,\text{h}^{-1}$. This experiment was intended to bring the hillslope to hydrologic steady state rather than to be in line with a specific climate, and the rainfall rate was chosen due to its relatively even spatial distribution. However, after 22 h, overland flow and erosion were observed and rainfall was stopped to prevent further unplanned changes to the topography of the hillslope. The total rainfall depth of the resulting event is similar to events that can trigger discharge extremes and/or landslides in natural environments (Turner et al., 2010; Brauer et al., 2011; Nguyen et al., 2013). Initial conditions were relatively dry, with volumetric water content of 8–11 % in most of the hillslope, except for the bottom of soil near the central trough where conditions were wetter due to rainfall system delivery testing several weeks prior to the current experiment (Fig. 4a). Data were collected from all instruments every 15 min.

3 Results

The rainfall event unexpectedly saturated part of the hillslope to the surface in and close to the central trough. Soil moisture time series show that the saturation process can be described by a sequential step-wise process rather than by a gradual process at all depths: three relatively stable phases (1–3) were separated by two rather abrupt steps. These steps are visible at individual locations (see examples in Fig. 5a) as well as in horizontally averaged data over the whole hillslope, over the convergent area only, and over upslope area only (example for 35 cm depth in Fig. S1 in the Supplement).

The time series in Fig. 5a as well as the time series of all other sensors that reached saturated conditions during the experiment show volumetric water contents exceeding the maximum porosity determined in the laboratory. The sensors were calibrated before they were installed in the hillslope by exposing the sensors to typical soil moisture values in the same soil material. The overshoot was not observed during this process. Soil moisture values closer to saturation were included in a second calibration after the experiment, but again no overshoot was observed and there were no significant changes in the calibration curves. Further testing of soil moisture sensors in large barrels under control conditions showed that the sensors read values exceeding the porosity of the soil when influenced by a capillary fringe or groundwater table. However, the relation between the overshoot and the depth under the groundwater table was not clear. Therefore, saturation is assumed when measured volumetric water contents exceed the porosity determined in the laboratory. This assumption is further justified by comparing storage estimates based on spatial averaging of soil moisture data and load cell measurements. Spatial averaging of raw soil

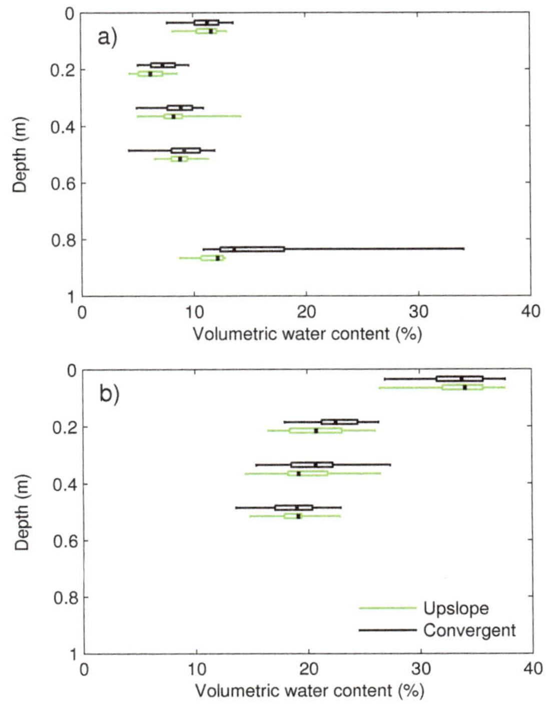

Figure 4. Box-and-whisker plots show the volumetric water content in phase 1 **(a)** and phase 2 **(b)** for the convergent and upslope areas and for each sensor depth. The whiskers show the 5th and 95th percentiles. Sensor depths are slightly offset to improve the visibility of the data.

moisture data resulted in significant overestimation of storage compared to the change in system mass measured by the load cells. This estimate was also significantly larger than the cumulative rainfall. When a maximum moisture content of 39 % was used, however, estimates compared well to load cell data (Fig. 5b).

Unfortunately, the piezometer data showed sensitivity to ambient temperature fluctuations. Due to this, piezometers read both negative values as well as values exceeding the soil depth of the hillslope (Fig. 6). Under normal applications these piezometers hang in deep wells where temperatures are more or less constant. In our application, the piezometers are mounted on the bottom of the hillslope and are subject to diurnal temperature fluctuations. It was impossible to find a reliable correction method, so values are reported relative to the initial conditions, when there was no groundwater table. While the values should not be interpreted as accurate representations of actual groundwater levels, the data nonetheless do provide a qualitative assessment of groundwater table dynamics.

The first step in the saturation process marked a sudden increase from initially dry conditions in the first phase to wetter, but still unsaturated conditions in the second phase (Fig. 5a) corresponding to the arrival of the infiltration front. After the passage of the infiltration front, the soil moisture

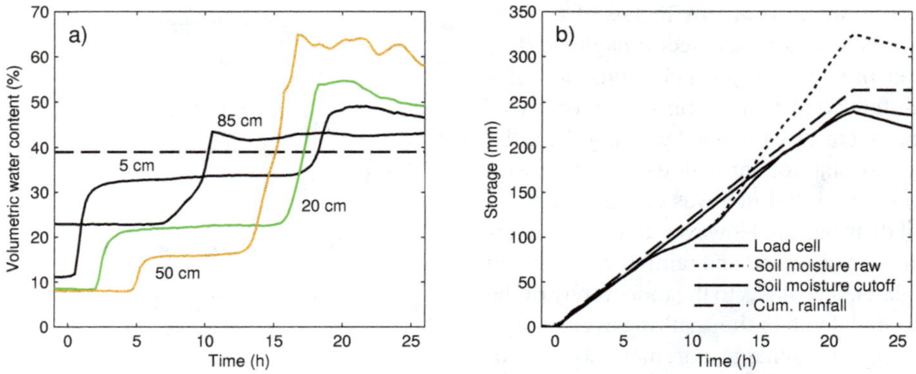

Figure 5. Volumetric water content **(a)** and storage estimates **(b)** in time relative to the start of the rainfall event. The volumetric water content is shown for four sensors along a vertical transect at a location (14, 0), where depths are relative to the surface. A horizontal line indicates the maximum porosity, above which locations are considered to be saturated. Storage estimates **(b)** relative to initial storage are derived from (i) load cell data, (ii) spatial averaging of raw soil moisture data and (iii) spatial averaging of soil moisture data with a cutoff of 39 %. Cumulative rainfall is added to the plot as a reference.

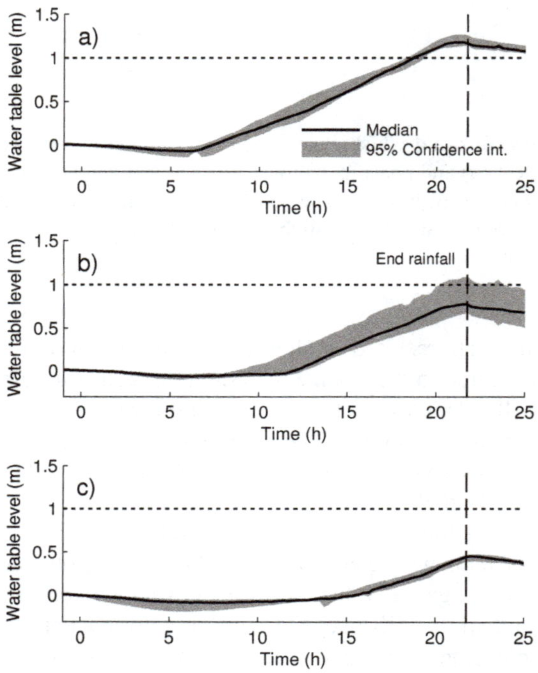

Figure 6. Median groundwater levels and the 95 % bootstrap confidence intervals are shown for piezometers in the central trough **(a)**, convergent area **(b)** and upslope area **(c)**. Plotted water table levels are relative to initial conditions, when no groundwater table was present. The dashed line marks the end of the rainfall event and the dotted line represents the soil surface.

content stayed steady in time but decreased with depth (Figs. 5a and 4b). The mean and median moisture contents at each depth in phase 2 were significantly different ($p < 0.01$). The second step marked a transition from unsaturated but wet conditions in the second phase to saturated conditions in the third phase, marking the arrival of the saturation front. Sensors at 85 cm depth behave differently, and show a single step from phase 1 (initial conditions) directly to phase 3 (saturated conditions).

Phase 2 was observed first at the shallower sensors hours after the start of the rainfall event ($T = 5.5$ h), and propagated downwards to the bottom of the soil profile (cross-sections A and B in Fig. 7). The third phase was observed first at the lower sensor depths and moved to the surface. Although the propagation of the second phase was relatively even across the hillslope, that of the third phase was not. The third phase reached the soil surface in and near the central trough ($T = 16.5$–22 h), while at the far sides and top of the hillslope it did not reach above 50 cm depth. The difference in expansion in the convergent and upslope areas formed a groundwater ridge in the cross-slope direction. The groundwater ridge was observed in both soil moisture data (Fig. 7b) and piezometer data (Fig. 8). The development of the three saturation phases at hillslope scale is shown in Animation S1.

The difference in the level of saturation in the convergent and upslope areas was further analyzed by means of simple column storage calculations. A maximum value equal to the porosity was applied to the data to prevent the observed overshoot in soil moisture values (Fig. 5a) from influencing the calculations. The results of the calculations show that soil columns in the convergent area saturated sooner than would be expected based on the rainfall rate and initial soil moisture conditions alone. In the central trough, the difference was equivalent to 3–4 h of rainfall. On the other hand, storage in upslope areas was less than would be expected. This indicates that there was an additional net incoming flux in the convergent area and a net outgoing flux in the upslope area.

Subsurface flow at the bottom of the hillslope started after 13 h of rainfall. Overland flow measurements are uncertain, but started after 14–20 h and continued until approximately

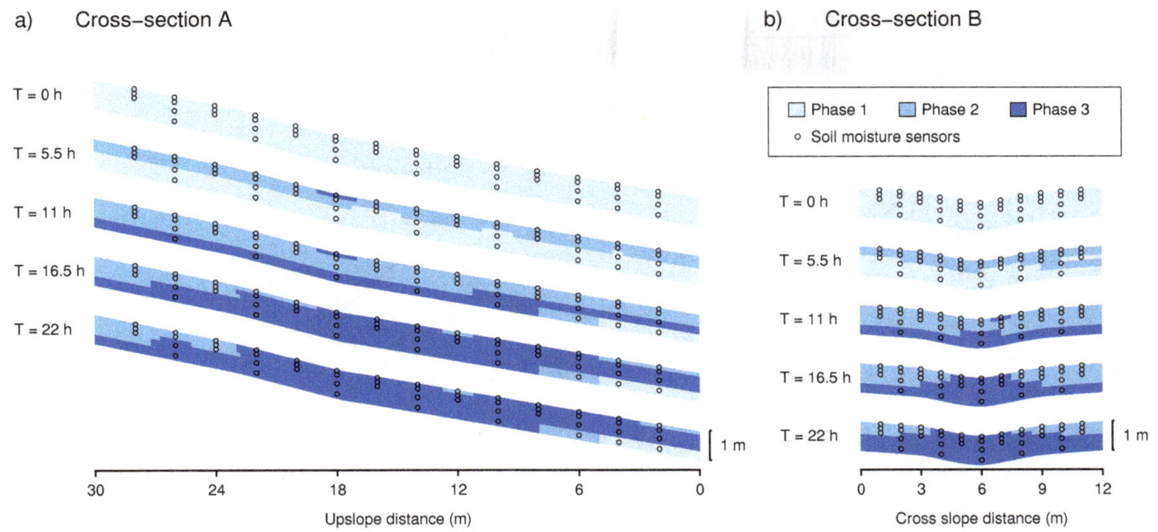

Figure 7. The development of the two-step saturation between the start ($T = 0$ h) and end ($T = 22$ h) of the rainfall event for cross-sections A **(a)** and B **(b)**. Open circles show soil moisture sensor locations.

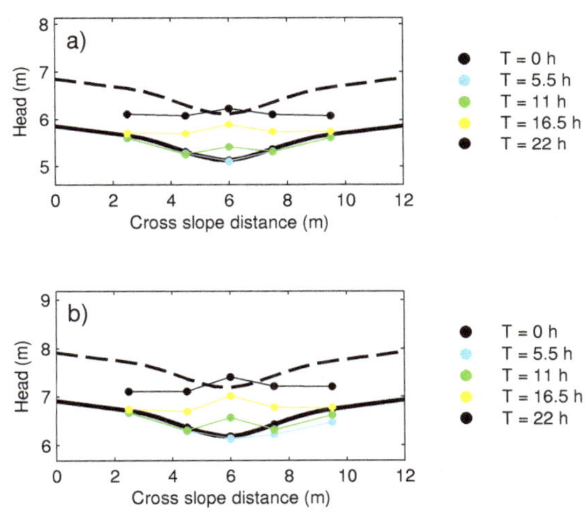

Figure 8. Piezometric data for two cross-slope transects at 7 **(a)** and 13 m **(b)** from the toe of the slope are shown at several time steps between the start and end of the rainfall event. The solid line represents the impermeable boundary at the bottom of the hillslope and the dashed line represents the soil surface.

24 h after the rainfall was turned off (Fig. 9a). According to soil moisture data, saturation first reached the surface after 18–19 h (Fig. 9b). Once the groundwater ridge reached the surface, it expanded along the side slopes of the central trough. The ridge created a slightly reversed hydraulic gradient from the ridge in the direction of the side slopes. This gradient was strongest when the ridge had just reached the surface and then slightly decreased in time (Fig. 7). Longitudinal cross-sections shown in Fig. 7a reveal increasing saturation in the downslope direction along the upper part of the hillslope. Saturation at the surface occurred along the lowest 20 m.

The hillslope is divided into a convergent and upslope part to compare the timing and speed of the saturation and subsequent drying processes. The timing of step 1 is the time at which the soil moisture starts to increase and marks the transition from phase 1 into phase 2. In the same way, the timing of step 2 marks the transition from phase 2 to phase 3. The relationship between sensor depth and the timing of step 1 was approximately linear for both parts (Fig. 9b). The speed of the infiltration front based on this relationship was 7.8 cm h^{-1} in the convergent area and 6.7 cm h^{-1} in the upslope area. However, the difference in timing was not significant based on the bootstrap confidence interval ($p = 0.15$).

Once the infiltration front reached the bottom of the soil profile, a water column quickly formed and rose upwards (Fig. 9b). In contrast to the infiltration front, the development of saturation in the soil profile was significantly different between the two areas. First, at each soil depth step 2 was observed significantly sooner in the convergent area than in the upslope area. In addition, step 2 was not observed near the surface in the upslope area, meaning that these areas did not reach phase 3. This is supported by the piezometer data, where groundwater table levels reached the surface in the central trough but stayed lower in the upslope area (Fig. 6). However, the speed of the propagation of step 2 was similar in the convergent and upslope areas, at 15.5 and 15.0 cm h^{-1}, respectively, or about two times faster than the infiltration front. The uncertainty in groundwater level values prevents us from accurately estimating the speed of groundwater table rise.

The difference between the convergent and upslope areas was also evident during the subsequent drying phase. The

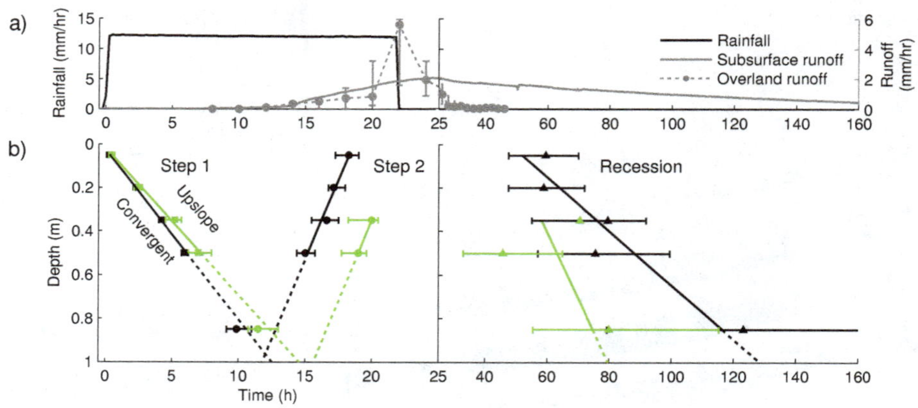

Figure 9. The rainfall-runoff characteristic **(a)** and the timing of different phases of saturation and recession for the convergent and upslope areas **(b)**. Time is relative to the start of the rainfall event and depth is relative to the soil surface. Overland flow estimates **(a)** are based on water balance analysis and volumetric measurements, and error bars express operator's uncertainty. In **(b)**, the median timing of each phase is shown with error bars representing the 95 % bootstrapping confidence interval of the median. Lines show best fits through the medians and are solid within the range of fitting and dotted when extrapolated.

recession start time is defined as the time at which the soil moisture in a saturated location first drops below the maximum porosity. This phase was slower than the wetting phase and started in relatively shallow or upslope locations before moving on to deeper locations and the convergent area (Fig. 9b). Subsurface flow at the toe of the slope peaked just after the end of the rainfall event and had a long recession tail (Fig. 9a). Overland flow peaked at the time rainfall was turned off, though the size of this peak is very uncertain. Runoff over the surface then continued for more than 24 h after the sprinklers were turned off, causing erosion in the central trough. This erosion formed a gully extending 18 m upslope, and changed the topography of the hillslope.

4 Discussion

Our results increase understanding of the behavior of saturation excess overland flow, and thereby can contribute to the development of how robust hydrological models and Earth system processes are impacted by surface and subsurface flow behavior. LEO represents an opportunity to experimentally approach these issues to refine models and then connect the processes to real settings. It is important to note that LEO's experimental setup is more typical of a zero-order basin than a hillslope due to the impermeable lateral boundaries of the artificial hillslope. Nonetheless, we expect that this system would behave similarly without these boundaries, and results of this experiment should be relevant for both. Rainfall falling on any part of the hillslope is eventually routed towards the central trough due to the convergent topography of the hillslope (Fig. 3). The topography of the bottom impermeable boundary or bedrock rather than the surface is the main driver of this convergent flow. Due to the convergence, flow over the lateral boundaries would only be relevant once the groundwater table extends to the sides of the hillslope. In natural hillslopes, water would then move across those boundaries to adjacent hillslopes. However, in natural rain events, adjacent hillslopes will receive similar amounts of rainfall and thus the groundwater table may rise in a similar fashion, sustaining the no-flow boundary at the topographic divide. Our results provide insights related to a two-step saturation, groundwater ridging, and overland flow.

4.1 Two-step saturation

The results show that the hillslope is saturated by a stepwise process, as has been previously observed in small-scale experiments on slopes with planar geometry (Phi et al., 2013). Identification of the steps and phases is aided by the experimental setup with constant rainfall rates, but is likely more challenging in natural catchments due to the rare nature of intense storms and the fact that soil moisture signals will also reflect the effect of varying rain rates. The steps were not only visible at individual locations (example in Fig. 5a), but also in aggregated data to the scale of the convergent area, upslope area and the entire hillslope (Fig. S1 in the Supplement). This demonstrates that the two-step process is relevant to both fundamental hydrological understanding and land surface modeling in weather, climate, and hydrological simulations that treat heterogeneous surfaces (such as the whole area in LEO) as a single grid box.

In our hillslope-scale experiment, the steps are very consistent throughout the hillslope, apart from a few individual locations that seem to show different behavior. One of these is at the bottom of the soil profile at the toe of the slope, where sensors appear to stay in phases 1 and 2 and therefore do not reach saturation (Fig. 7). However, the soil moisture content in these locations reaches over 36–37 %. Taking the ±2 % error margin of the sensors into account, the moisture content comes quite close to the maximum porosity, and we expect that these locations were likely saturated.

Another local deviation is observed in the slightly asymmetrical infiltration front (Fig. 7), which would not be expected in a homogeneous soil. However, at hillslope scale, there is very little scatter around the linear relationship between timing of the first step and sensor depth (Fig. 9). This suggests that the first observation is due to small-scale variations resulting from the indicated time and location of the cross-section, while at hillslope scale the soil is relatively homogeneous compared to natural hillslopes. Although the scale of the hillslope prevents truly homogeneous conditions, the absence of vegetation and other disturbances excludes the formation of instabilities, which can have a large influence on flow paths, such as finger flow or macropore flow (e.g., Beven and Germann, 2013). Soil cracks can form due to swelling and shrinking as minerals absorb and release water, but this is unlikely in LEO because the soil does not contain the clay minerals required for this process. However, in the future, we expect that the subsurface structure of the hillslope will constantly change as more experiments are executed and hydrological pathways develop, especially after vegetation is introduced.

The two-step saturation process supports existing theories of water movement in hillslopes in which water first moves vertically through the unsaturated zone, contributes to the formation of a groundwater table, and subsequently moves downslope. This concept is an important assumption in previous physically based modeling studies (e.g., Robinson and Sivapalan, 1996). In fact, due to the dense sensor network and highly controlled conditions, LEO is well suited to testing assumptions and concepts in hydrological models.

Conceptual models can simulate two-step saturation by including a tension water reservoir and a free water reservoir, as in the Sacramento model (Burnash et al., 1973; Duan et al., 2001) and the Xinanjiang model (Zhao, 1992). In these models, water infiltrates into the tension reservoir, which represents available soil water. Once the tension reservoir has been filled, water is routed to the free water reservoir. The free water reservoir represents gravitational soil water and is therefore likely to fill from the bottom upwards as was observed in this experiment. Another important characteristic of the saturation process is the role of convergence. In this experiment, the difference in timing and degree of saturation as well as the timing of the recession period indicate there was flow from the upslope area to the convergent area forced by the topography of the bedrock. The results of column storage calculations also support the importance of lateral subsurface flow. However, these and other conceptual models only allow for one-way interaction between reservoirs, and do not allow for lateral redistribution of water (e.g., Duan et al., 2001; van Esse et al., 2013; Zhao, 1992), which was a crucial mechanism of overland flow generation in this experiment, due to the relatively shallow soil.

Field studies on saturation and overland flow generation have shown similar development of saturated areas (Dunne and Black, 1970; Wilson and Dietrich, 1987) as observed in phase 3 of the saturation process. One study focusing on the role of topography in throughflow generation on a hillslope with similar characteristics to the artificial hillslope in this study also suggested the importance of convergent soil water flow in the formation of the saturated wedge (Anderson and Burt, 1978). However, the resolution of data collected during field studies is often limited (Anderson and Burt, 1978; Dunne and Black, 1970), and results are affected by environmental factors such as varying rainfall rates (Anderson and Burt, 1978) or bedrock permeability (Wilson and Dietrich, 1987). The present study of the saturation process uniquely combines high-resolution data and controlled conditions with near-field scales.

Although the main concept of the two-step saturation process is in line with existing theories in hillslope hydrology, some observations are not easily explained. In the case of a constant infiltration rate and a homogeneous soil, one would expect the equilibrium moisture content of phase 2 to be similar at all depths based on Richards' equation. Instead, our data show a significant decrease in soil moisture content with depth in the second phase in both convergent and upslope areas at hillslope scale (Fig. 4b). Air entrapment is a possible explanation for this observation, but is not deemed likely for a few reasons. First, air entrapment is observed when there is ponding at the surface (Culligan et al., 2000; Starr et al., 1978), for example during surface irrigation (Dixon and Linden, 1972). During this experiment, we did not observe ponding on the surface during the second phase. Furthermore, we observe the decrease in moisture content at times when none of the overlying soil is saturated, leaving enough empty pores for pressurized air to escape. Finally, the rainfall intensity distribution was not completely uniform over the hillslope and the scale of the hillslope makes air entrapment unlikely. A second possible explanation for the decreasing water content, yet constant specific infiltration rate, is decreasing porosity with depth. However, if we assume a significant decrease in maximum porosity, storage estimates by spatial averaging of soil moisture data significantly underestimate storage estimates based on load cell data shown in Fig. 5b. Another possibility is that there are changes in water retention characteristics with depth. In the relevant part of the retention curve, small changes in soil water pressure can lead to relatively large differences in soil moisture content. Finally, an increasing hydraulic conductivity with depth could also explain the observations, though this is deemed unlikely to develop from an initially homogeneous soil profile. The true explanation for these observations cannot be determined based on the collected data, and requires further research into the role of vertical heterogeneity in soil hydraulic properties. A hypothesis that can be tested after the introduction of vegetation in LEO to learn more about this phenomenon is that the presence of roots will reduce soil moisture values at shallower soil depths due to increased potential for preferential flow.

Other interesting observations concern the speeds of the downward propagating infiltration front and the upward propagating saturation front. The infiltration front speed as reported in the Results section (6–7 $\mathrm{cm\,h^{-1}}$) is more than five times faster than the applied rainfall rate (1.2 $\mathrm{cm\,h^{-1}}$). The fact that the moisture content is relatively stable in time after the passage of the wetting front suggests that the specific flux is constant with depth. At the same time, this flux must be equal to the rainfall rate at the surface. This suggests that a limited portion of the available pore space in the soil is used in water transport. The saturation front speed is higher than the infiltration front speed, which can be linked to the smaller specific yield between the second and third phases than between the first and second phases. We would expect the rise of the saturation front to accelerate due to decreasing available pore space (Fig. 4b), but we do not observe this at hillslope scale. Decreasing porosity with depth could explain this observation. However, as mentioned previously, this would result in an underestimation of system storage, and is therefore unlikely. Finally, we would expect the speed of the saturation front to be higher in the convergent zone than in the upslope area in the presence of flow convergence. Instead, the speeds in both areas are very similar (Fig. 9b). On the other hand, the error bars do not exclude the possibility of accelerated groundwater table rise or faster groundwater table rise in the convergent area than the upslope area. Furthermore, the number of locations that reach phase 3 at 35 cm depth in the upslope area is limited, reducing the power of the observed rate of groundwater table rise in the upslope area.

4.2 Groundwater ridging

Piezometer data (Fig. 8) indicate that a groundwater ridge formed during the rainfall event. Though soil moisture data indicate saturation rather than the phreatic surface, the soil moisture data support the piezometer data due to the similarity in ridge development in both data sets (Fig. 7b). The piezometer and soil moisture data are not co-located, but the level of the ridge in the soil moisture data appears higher than that in piezometer data. At first glance this could be explained by the capillary fringe. However, it is important to remember that the groundwater table values are not accurate, as shown by the negative values and the values extending above the soil surface (Fig. 6). The reversed hydraulic gradient was highest when the ridge reached the surface and then decreased due to continued rainfall and subsurface flow along the gradient (Cloke et al., 2006).

Several studies have linked the formation of a groundwater ridge to the presence of a capillary fringe (Gillham, 1984; Abdul and Gillham, 1984; Abdul and Gillham, 1989). Theory predicts that when rainfall is added to a soil where the capillary fringe reaches the land surface, the groundwater table rises rapidly and a groundwater ridge is formed. However, studies have not clearly demonstrated this mechanism in practice. In studies by Gillham (1984) and Novakowski et al. (1988), groundwater tables did not rise to the surface in response to rainfall events even though the capillary fringe extended to the surface. Another study in Canada described the development of a groundwater mound in response to infiltrating snowmelt, but observed gradual rather than rapid groundwater table rise (Buttle and Sami, 1992). In addition to the height of the capillary fringe, studies have identified several other factors that influence groundwater ridging. Waswa et al. (2013) found that the magnitude of the groundwater ridging response was linearly related to the maximum rainfall intensity of an event. A numerical study by Cloke et al. (2006) also showed that rainfall intensity influenced the development of a groundwater ridge, but found other factors such as the slope, saturated hydraulic conductivity and initial water table height to be more important.

In many studies of groundwater ridging, initial groundwater tables were (nearly) horizontal (Abdul and Gillham, 1984; Novakowski et al., 1988; Abdul and Gillham, 1989; Buttle and Sami, 1992). As a consequence, the distance to the groundwater table was lower near the stream than farther upslope and the capillary fringe often reached the surface for only a limited difference from the stream. Linked to this, the specific yield was lower near the stream than upslope. Under these conditions, areas near the stream respond quickly to rainfall, while upslope areas respond more slowly, allowing a groundwater ridge to form. In the present study, there was no initial groundwater table and the soil depth is uniform, limiting the influence of the capillary fringe in groundwater ridge development. The specific yield was not completely uniform in the hillslope due to differences in initial conditions of the lowest soil layer (Fig. 4). However, the column storage calculations indicate that there was net flow from the upslope area to the convergent area despite the reversed hydraulic gradient. This suggests that the formation of the groundwater ridge in this study is affected by subsurface flow driven by the convergent topography of the bedrock.

4.3 Overland flow

The estimation of overland flow based on the water balance indicates that overland flow may have started as early as 14 h after the start of the rainfall event (Fig. 9). Overland flow starting before saturation reached the surface would suggest Hortonian overland flow rather than Dunnian overland flow. However, the error bars extend to zero until 20 h into the event due to uncertainties in the water balance analysis, and we do not expect Hortonian flow to have occurred for several reasons. First, the constant rain rate is lower than the saturated hydraulic conductivity, both as determined in the laboratory as determined by model calibration. Second, overland flow was not observed for the first eight hours of the experiment and once overland flow started it was limited to the central trough. Third, in subsequent experiments with the same rainfall rate, but a shorter duration, no ponding or overland flow was observed. Finally, model simulations based on the 3-D Richards equation do not show overland flow due to infiltration excess, but confirm overland flow due to saturation excess in the central trough (Niu et al., 2014).

A possible explanation for the overland flow is that tension saturation at the surface caused saturation excess runoff to occur. However, the continuation of overland flow after rainfall had stopped indicates that lateral subsurface flow was a major contributor to overland flow generation in this experiment. Piezometric data can help justify this interpretation. The data should not be used as accurate portrayals of groundwater levels; however, the data (Fig. 6) support the timing of water table rise (Fig. 9) and suggest that the water table reached the surface in the central trough. The continuation of overland flow for a long period after rainfall had stopped also signals a persistent hydrologic connectivity between upslope and convergent areas. Previously, Sklash and Farvolden (1979) demonstrated that runoff in their study area could be dominated by either event or pre-event water, depending on initial conditions. Under wetter conditions, such as was the case at the onset of runoff in the present study, overland flow and streamflow hydrographs were dominated by groundwater (Sklash and Farvolden, 1979).

Initially, the subsurface component of overland flow in the present experiment was caused by the groundwater ridge alone. However, gully formation and expansion influenced flow paths in the hillslope. The erosion gully formed by the overland flow was limited to the central trough in the hillslope. A study of rill formation in the artificial Chicken Creek catchment showed a larger network of rills, but the longer and deeper main rill was similarly located along the convergent axis of the catchment (Hofer et al., 2012). It is expected that the steep sides of the gully in the present study increased local hydraulic gradients, increasing runoff generation and the related erosion while groundwater levels remained high on either side of the gully. In this way, the subsurface component of overland flow was enhanced by morphologic changes caused by erosion.

5 Conclusions and outlook

The Landscape Evolution Observatory at Biosphere 2 consists of a hillslope-scale experiment in a highly controlled setting, where the first rainfall experiment resulted in overland flow and soil erosion. This experiment provided a unique opportunity to study the importance of convergence on the development of saturation through the use of the high-density sensor array in the system. Data collected from 496 soil moisture sensors at a high temporal resolution show a two-step saturation process: a first step related to the downward propagation of the infiltration front and a second step characterized by saturation of the soil from below in response to rising water tables. Soil profiles in convergent areas responded sooner than soil profiles in upslope areas. In addition, soil profiles in the convergent area saturated completely, while the soil surface in the upslope area remained unsaturated. This difference created a groundwater ridge. Due to the uniform soil depth and lack of soil heterogeneity, the difference between the two areas can be attributed solely to lateral subsurface flow in the saturated zone of the soil profiles driven by the convergent topography of the bedrock.

Our experimental data demonstrate the importance of convergence in the context of existing theories describing hillslope hydrology. Convergence influences subsurface flow and storage dynamics at hillslope scale, which alters the timing of the two-step process by location, and the degree of saturation in different topographic positions. Several observations at hillslope scale are not easily explained by existing theory and likely emerge from the large-scale nature of the experiment or the important differences in system complexity associated with the lack of vegetation and spatial heterogeneity. One important question is how to explain the observed decreasing moisture content with depth despite constant infiltration rates and a homogeneous soil. Further research into vertical heterogeneity and small-scale processes is needed to explain the mechanisms behind these observations. Insight in these mechanisms is an important step to improve understanding of saturation excess overland flow generation and related natural hazards such as flash floods and landslides, and their representation in land surface models, many of which currently do not allow for lateral and/or upward movement of water. The results of this experiment also set the stage for hypotheses concerning the role of vegetation in altering wetting dynamics that can be tested in LEO. One such hypothesis is that the resulting macro-pore structure should create more homogeneous changes in water content by facilitating water flow to soil depths. Comparison of the results of this experiment and such future experiments can highlight the role of the co-evolution of ecological and hydrological processes in determining landscape water dynamics.

Acknowledgements. We thank all of the technicians and graduate students involved in preparing and performing this experiment. Of these, we are especially grateful to Michael Sibayan, John Adams and Nate Abramson for technical support and Yvonne Smit for providing data on soil properties. We also thank George Waswa, reviewer Erwin Zehe and two anonymous reviewers for their constructive comments, which greatly helped improve the manuscript. In addition, we thank Murugesu Sivapalan, Ciaran Harman and William Dietrich for their thoughts about an earlier version. Finally, we are grateful to Paul Ingram for providing the hillslope photograph. Primary support for LEO is from the Philecology Foundation (Forth Worth, TX, USA).

Edited by: F. Tian

References

Abdul, A. S. and Gillham, R. W.: Laboratory studies of the effects of the capillary fringe on streamflow generation, Water Resour. Res., 20, 691–698, 1984.

Abdul, A. S. and Gillham, R. W.: Field studies of the effects of the capillary fringe on streamflow generation, J. Hydrol., 112, 1–18, 1989.

Anderson, M. G. and Burt, T. P.: The role of topography in controlling throughflow generation, Earth Surf. Process., 3, 331–344, 1978.

Bernier, P. Y.: Variable source areas and storm-flow generation: An update of the concept and a simulation effort, J. Hydrol., 79, 195–213, 1985.

Beven, K.: Hillslope hydrographs by the finite element method, Earth Surf. Process., 2, 13–28, 1977.

Beven, K. and Germann, P.: Macropores and water flow in soils revisited, Water Resour. Res., 49, 3071–3092, 2013.

Beven, K. and Kirkby, M. J.: A physically based, variable contributing area model of basin hydrology, Hydrolog. Sci. Bull., 24, 43–69, 1979.

Brauer, C. C., Teuling, A. J., Overeem, A., van der Velde, Y., Hazenberg, P., Warmerdam, P. M. M., and Uijlenhoet, R.: Anatomy of extraordinary rainfall and flash flood in a Dutch lowland catchment, Hydrol. Earth Syst. Sci., 15, 1991–2005, doi:10.5194/hess-15-1991-2011, 2011.

Burnash, R. J., Ferral, R. L., and McGuire, R. A.: A generalized streamflow simulation system: conceptual modeling for digital computers, US Department of Commerce National Weather Service and State of California Department of Water Resources, Sacramento, California, USA, 1973.

Buttle, J. M. and Sami, K.: Testing the groundwater ridging hypothesis of streamflow generation during snowmelt in a forested catchment, J. Hydrol., 135, 53–72, 1992.

Cloke, H. L., Anderson, M. G., McDonnell, J. J., and Renaud, J.-P.: Using numerical modeling to evaluate the capillary fringe groundwater ridging hypothesis of streamflow generation, J. Hydrol., 316, 141–162, 2006.

Culligan, P. J., Barry, D. A., Parlange, J. Y., Steenhuis, T. S., and Haverkamp, R.: Infiltration with controlled air escape, Water Resour. Res., 36, 781–785, 2000.

Dixon, R. M. and Linden, D. R.: Soil air pressure and water infiltration under border irrigation, Soil Sci. Soc. Am. J., 36, 948–953, 1972.

Duan, Q., Schaake, J., and Koren, V.: A priori estimation of land surface model parameters, in: Land Surface Hydrology, Meteorology, and Climate: Observation and Modeling, Water Sci. and Appl. 3, edited by: Lakshmi, V., Albertson, J., and Schaake, J., AGU, Washington, D.C., 77–94, 2001.

Dunne, T.: Field studies of hillslope flow processes, in: Hillslope Hydrology, edited by: Kirkby, M. J. and Chorley, R. J., Wiley, Chichester, 227–293, 1978.

Dunne, T. and Black, R. D.: Partial area contributions to storm runoff in a small New England watershed, Water Resour. Res., 6, 1296–1311, 1970.

Dunne, T., Moore, T. R., and Taylor, C. H.: Recognition and prediction of runoff-producing zones in humid regions, Hydrolog. Sci. Bull., 20, 305–327, 1975.

Fiori, A., Romanelli, M., Cavalli, D. J., and Russo, D.: Numerical experiments of streamflow generation in steep catchments, J. Hydrol., 339, 183–192, 2007.

Freeze, R. A.: Role of subsurface flow in generating surface runoff: 2. Upstream source areas, Water Resour. Res., 8, 1272–1283, 1972.

Freeze, R. A.: Streamflow generation, Rev. Geophys., 12, 627–647, 1974.

Gillham, R. W.: The capillary fringe and its effect on water table response, J. Hydrol., 67, 307–324, 1984.

Hofer, M., Lehmann, P., Stähli, M., Seifert, S., and Krafczyk, M.: Two approaches to modeling the initiation and development of rills in a man-made catchment, Water Resour. Res., 48, W01531, doi:10.1029/2011WR010719, 2012.

Hopp, L., Harman, C., Desilets, S. L. E., Graham, C. B., McDonnell, J. J., and Troch, P. A.: Hillslope hydrology under glass: confronting fundamental questions of soil-water-biota coevolution at Biosphere 2, Hydrol. Earth Syst. Sci., 13, 2105–2118, doi:10.5194/hess-13-2105-2009, 2009.

Huxman, T., Troch, P. A., Chorover, J., Breshears, D. D., Saleska, S., Pelletier, J., and Zeng, X.: The Hills are Alive: Earth Science in a Controlled Environment, EOS Trans. AGU, 90, 120–120, doi:10.1029/2009EO140003, 2009.

Nguyen, H. T., Wiatr, T., Fernández-Steeger, T. M., Reicherter, K., Rodrigues, D. M. M., and Azzam, R.: Landslide hazard and cascading effects following the extreme rainfall event on Madeira Island (February 2010), Nat. Hazards, 65, 635–652, 2013.

Niu, G.-Y., Pasetto, D., Scudeler, C., Paniconi, C., Putti, M., Troch, P. A., DeLong, S. B., Dontsova, K., Pangle, L., Breshears, D. D., Chorover, J., Huxman, T. E., Pelletier, J., Saleska, S. R., and Zeng, X.: Incipient subsurface heterogeneity and its effect on overland flow generation – insight from a modeling study of the first experiment at the Biosphere 2 Landscape Evolution Observatory, Hydrol. Earth Syst. Sci., 18, 1873–1883, doi:10.5194/hess-18-1873-2014, 2014.

Novakowski, K.-S. and Gillham, R. W.: Field investigations of the nature of water-table response to precipitation in shallow water-table environments, J. Hydrol., 97, 23–32, 1988.

O'Loughlin, E. M.: Saturation regions in catchments and their relations to soil and topographic properties, J. Hydrol., 53, 229–246, 1981.

Phi, S., Clarke, W., and Li, L.: Laboratory and numerical investigations of hillslope soil saturation development and runoff generation over rainfall events, J. Hydrol., 493, 1–15, 2013.

Robinson, J. S. and Sivapalan, M.: Instantaneous response functions of overland flow and subsurface stormflow for catchment models, Hydrol. Process., 10, 845–862, 1996.

Sklash, M. G. and Farvolden, R. N.: The role of groundwater in storm runoff, J. Hydrol, 43, 45–65, 1979.

Starr, J. L., DeRoo, H. C., Frink, C. R., and Parlange, J.-Y.: Leaching characteristics of a layered field soil, Soil Sci. Soc. Am. J., 42, 386–391, 1978.

Troch, P. A., Paniconi, C., and van Loon, E. E.: The hillslope-storage Boussinesq model for subsurface flow and variable source areas along complex hillslopes: 1. Formulation and characteristic response, Water Resour. Res., 39, 1316, doi:10.1029/2002WR001728, 2003.

Turner, T. R., Duke, S. D., Fransen, B. R., Reiter, M. L., Kroll, A. J., Ward, J. W., Bach, J. L., Justice, T. E., and Bilby, R. E.: Landslide densities associated with rainfall, stand age, and topography on forested landscapes, southwestern Washington, USA, J. Hydrol., 259, 2233–2247, 2010.

van Esse, W. R., Perrin, C., Booij, M. J., Augustijn, D. C. M., Fenicia, F., Kavetski, D., and Lobligeois, F.: The influence of conceptual model structure on model performance: a comparative study for 237 French catchments, Hydrol. Earth Syst. Sci., 17, 4227–4239, doi:10.5194/hess-17-4227-2013, 2013.

Ward, R. C.: On the response to precipitation of headwater streams in humid areas, J. Hydrol., 74, 171–189, 1984.

Waswa, G. W., Clulow, A. D., Freese, C., Le Roux, P. A. L., and Lorentz, S. A.: Transient pressure waves in the vadose zone and the rapid water table response, Vadose Zone J., 12, doi:10.2136/vzj2012.0054, 2013.

Wilson, C. J., and Dietrich, W. E.: The contribution of bedrock groundwater flow to storm runoff and high pore pressure development in hollows, Erosion and Sedimentation in the Pacific Rim, Int. Assoc. Hydrol. Sciences Pub. No. 165, IAHS Press, Wallingford, UK, 49–59, 1987.

Zhao, R. J.: The Xinanjiang model applied to China, J. Hydrol., 135, 371–381, 1984.

6

Multiresponse modeling of variably saturated flow and isotope tracer transport for a hillslope experiment at the Landscape Evolution Observatory

Carlotta Scudeler[1,2], Luke Pangle[3], Damiano Pasetto[4], Guo-Yue Niu[5,6], Till Volkmann[6], Claudio Paniconi[1], Mario Putti[2], and Peter Troch[5,6]

[1]Institut National de la Recherche Scientifique, Centre Eau Terre Environnement (INRS-ETE), Université du Québec, Quebec City, Canada
[2]Department of Mathematics, University of Padua, Padua, Italy
[3]Department of Geosciences, Georgia State University, Atlanta, USA
[4]Laboratory of Ecohydrology, École Polytechnique Fédérale de Lausanne, Lausanne, Switzerland
[5]Biosphere 2 – Earth Science, University of Arizona, Tucson, USA
[6]Department of Hydrology and Water Resources, University of Arizona, Tucson, USA

Correspondence to: Carlotta Scudeler (carlotta.scudeler@ete.inrs.ca)

Abstract. This paper explores the challenges of model parameterization and process representation when simulating multiple hydrologic responses from a highly controlled unsaturated flow and transport experiment with a physically based model. The experiment, conducted at the Landscape Evolution Observatory (LEO), involved alternate injections of water and deuterium-enriched water into an initially very dry hillslope. The multivariate observations included point measures of water content and tracer concentration in the soil, total storage within the hillslope, and integrated fluxes of water and tracer through the seepage face. The simulations were performed with a three-dimensional finite element model that solves the Richards and advection–dispersion equations. Integrated flow, integrated transport, distributed flow, and distributed transport responses were successively analyzed, with parameterization choices at each step supported by standard model performance metrics. In the first steps of our analysis, where seepage face flow, water storage, and average concentration at the seepage face were the target responses, an adequate match between measured and simulated variables was obtained using a simple parameterization consistent with that from a prior flow-only experiment at LEO. When passing to the distributed responses, it was necessary to introduce complexity to additional soil hydraulic parameters to obtain an adequate match for the point-scale flow response. This also improved the match against point measures of tracer concentration, although model performance here was considerably poorer. This suggests that still greater complexity is needed in the model parameterization, or that there may be gaps in process representation for simulating solute transport phenomena in very dry soils.

1 Introduction

Simulation models of water and solute interaction and migration through complex geologic media are essential tools for addressing fundamental and practical problems, ranging from basic scientific understanding of critical zone processes (Brooks et al., 2015) to improving the management of our freshwater resources (Gorelick and Zheng, 2015). Physically based distributed numerical models require a careful definition of spatially variable parameters and time variable boundary conditions, and can produce information for numerous response variables at different levels of spatio-temporal aggregation. It is increasingly acknowledged that proper implementation and verification of these models, in terms of both process representation and parameter identification, require detailed, multiresponse field or laboratory data, in contrast to traditional model evaluation based on a single, integrated response variable such as total discharge (Paniconi and Putti, 2015). However, multiobjective parameter estimation for nonlinear or coupled models with a high number of degrees of freedom is very challenging (Anderman and Hill, 1999; Keating et al., 2010), since classical techniques developed for simpler hydrological models (e.g., Gupta et al., 1998; Fenicia et al., 2007) are not readily extendable, in terms of robustness and efficiency, to more complex models. Traditional challenges, on both the experimental and modeling sides, are associated with soil heterogeneity, variability in parameters, and variably saturated conditions (e.g., Binley et al., 1989; Woolhiser et al., 1996; Neuweiler and Cirpka, 2005). An added source of complexity arises when passing from flow modeling to flow and transport modeling (e.g., Ghanbarian-Alavijeh et al., 2012; Russo et al., 2014).

While many hydrologic model assessment studies have reported good agreement between simulated and observed data when performance is measured against a single response variable, there are comparatively few studies that have made use of observation data from multiple response variables. Brunner et al. (2012), for instance, examined the performance of a one-dimensional (1-D) unsaturated zone flow model when water table measurements were supplemented by evapotranspiration and soil moisture observations. Sprenger et al. (2015) assessed the performance of three inverse modeling strategies based on the use of soil moisture and porewater isotope concentration data for a 1-D unsaturated flow and transport model. Kampf and Burges (2007) obtained encouraging results for a 2-D Richards equation flow model using integrated (subsurface outflow) and internal (piezometric water level and volumetric water content) measurements from a hillslope-scale experiment. Kumar et al. (2013) used multiple discharge measurements to calibrate and apply a distributed hydrologic model to 45 subcatchments of a river basin in Germany. Investigations based on hypothetical experiments are more common. Mishra and Parker (1989), for example, obtained smaller errors for simultaneous estimation of flow and transport parameters than for sequential estimation based on synthetically generated observations of water content, pressure head, and concentration.

In this study we perform a modeling analysis of the experimental data collected from an intensively measured hillslope at the Landscape Evolution Observatory (LEO) of the Biosphere 2 facility (Hopp et al., 2009). The simulations were conducted with the CATHY (CATchment HYdrology) model (Camporese et al., 2010; Weill et al., 2011), a physics-based numerical code that solves the 3-D Richards and advection–dispersion equations and includes coupling with surface routing equations. The availability of extensive observational data sets from detailed multidisciplinary experiments (recent examples in addition to LEO include the TERENO network of experimental catchments (Zacharias et al., 2011) and the Chicken Creek artificial catchment (Hofer et al., 2012)) can contribute vitally to testing and improving the current generation of integrated (surface–subsurface) hydrological models (Sebben et al., 2013; Maxwell et al., 2014).

Two experiments have been conducted to date at LEO, a rainfall and drainage test in February 2013 (Gevaert et al., 2014; Niu et al., 2014), which featured both subsurface and overland flow, and an isotope tracer test in April 2013 (Pangle et al., 2015), run under drier soil conditions and with reduced rainfall rates to avoid occurrence of surface runoff. Both of these experiments were performed on the first of the three hillslopes at LEO to be commissioned, hereafter referred to as LEO-1.

Using both integrated (load cell and seepage face) and distributed (point-scale soil moisture and concentration) data collected during the tracer experiment, the objective of this study is to explore the challenges of multiresponse performance assessment for a 3-D variably saturated flow and solute transport model. In a first step we consider only integrated flow responses, and the CATHY model is initially parameterized according to analyses of the February 2013 experiment. As integrated transport and point-scale flow and transport observations are progressively introduced in the analysis, the impact of different configurations (spatially uniform vs. spatially variable parameters, treatment of initial and boundary conditions, etc.) on the model's ability to capture the expanding and increasingly detailed response variables is examined. The boundary condition configurations, for instance, include a sink-based treatment of isotope fractionation to allow only a portion of the tracer to evaporate with the water.

2 Study site: Biosphere 2 Landscape Evolution Observatory

LEO is a large-scale community-oriented research infrastructure managed by the University of Arizona at the Biosphere 2, Oracle, USA (Hopp et al., 2009; Huxman et al., 2009; Pangle et al., 2015). It consists of three identical convergent artificial landscapes (or hillslopes) constructed with the aim of advancing our predictive understanding of the coupled physical, chemical, biological, and geological processes at Earth's surface in changing climates. For the first years of LEO operation, vegetation is not present and the research is focused on the characterization of the hydrological response of the hillslopes in terms of water transit times, generation of seepage and overland flow, internal dynamics of soil moisture, and evaporation. The three hillslopes are 30 m long and 11 m wide and of 10° average slope. The local slope varies from upslope positions to the convergence zone, with a maximum slope of 17° near the convergence zone. The landscapes are filled with 1 m of basaltic tephra ground to homogeneous loamy sand, chosen mainly for its primary elemental composition that includes critical nutrients for plant growth. The three landscapes are housed in a 2000 m^2 environmentally controlled facility. Each landscape contains a sensor and sampler network capable of resolving meter-scale lateral heterogeneity and submeter-scale vertical heterogeneity in water, energy, and carbon states and fluxes. The density of sensors and the frequency at which they can be polled allows for a monitoring intensity that is impossible to achieve in natural field settings. Additionally, each landscape has 10 load cells embedded into the structure that allow measurement of changes in total system mass and an engineered rain system that allows application of precipitation at rates between 2 and 40 $mm\,h^{-1}$. Each landscape at LEO has five independent plumbing circuits, each including a different array of sprinkler heads, and therefore generating a different rain flux. Tracers can be introduced into the system via the rainfall simulator at a constant or time-varying rate. The embedded soil water solution and soil gas samplers facilitate the use of these tracers to study water and solute movement within the hillslopes at a very dense spatial scale.

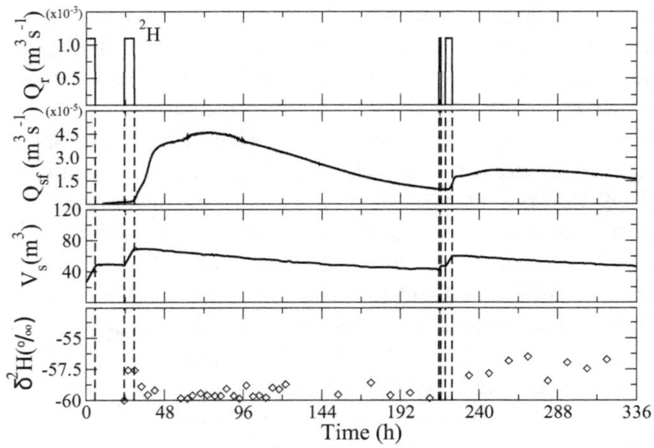

Figure 1. Hydrological response to the tracer experiment at the LEO-1 hillslope. From top: measured rain input pulses Q_r (the red pulse is deuterium-enriched); seepage face flow Q_{sf}; total water storage V_S; and mean δ^2H values at the seepage face. Time 0 corresponds to 09:30, 13 April 2013. The vertical dashed lines indicate the timing of the three pulses of rain (red when the water is deuterium-enriched and blue when it is not).

3 Methodology

3.1 Isotope tracer experiment

The first tracer experiment performed at the LEO-1 hillslope began at 09:30 on 13 April 2013. The experiment consisted of three rainfall events that were applied over 10 days (Fig. 1). During each event the rainfall was applied at a rate of 12 $mm\,h^{-1}$ for durations, respectively, of 5.5, 6, and 5.25 h. Rainfall was interrupted for 2.75 h during the third event (1.25 h from the start) due to necessary equipment maintenance, then restarted. During the second event deuterium-enriched water was introduced into the rain system. The enriched water had a hydrogen isotopic composition (expressed using the delta notation as δ^2H) of approximately 0 ‰, which corresponds to an enrichment of approximately 60 ‰ compared to typical (non-enriched) rainfall source water.

At the time of this experiment we consistently used one plumbing circuit because the spatial distribution of rainfall produced by this circuit had been well characterized by in situ testing. This allowed us to examine the possible influence of spatially heterogeneous rain patterns on flow and transport. The purpose of the first rain application was to increase the average moisture content of the landscape, which had received no rain for more than 40 days prior. The second rainfall application was used to introduce the deuterium tracer. No additional rain was applied for multiple days so that the tracer transport within, and out of the landscape, would be affected by soil moisture redistribution and evaporation. The third and final rainfall application was applied with the intention of forcing additional tracer mass beyond the seepage face boundary, to reveal additional detail in the measured breakthrough curve. In retrospect, and following laboratory analysis that spanned several weeks, we only observed the initiation of the tracer breakthrough curve at the seepage face.

The initial conditions of the system were very dry. The estimated total initial volume of water was about 26 m^3 (the total water storage capacity of the hillslope is approximately 130 m^3). All the rain water applied infiltrated into the soil. Seepage face outflow at the downslope vertical plane started 5 h after the beginning of the experiment. Two outflow peaks were observed: the first one after the second pulse of rain, with a peak of $4.5 \times 10^{-5}\,m^3\,s^{-1}$, and the second one after the final pulse, with a peak of $2.1 \times 10^{-5}\,m^3\,s^{-1}$. Temporal changes in total soil water storage were monitored via the load cell measurements, flow from the seepage face boundary was measured with electronic flow meters and tipping bucket gauges, and matric potential and water content were measured at 496 locations with, respectively, MPS-2 and 5TM Decagon sensors installed at depths of 5, 20, 50, and 85 cm from the landscape surface. Cumulative fluxes and instantaneous state variables were recorded at 15 min intervals. The estimated evaporation rate, derived from the seepage face measurements and load cell data and calculated as the difference between the change in water volume and the cumulative volume flowing out from the seepage face over the selected time interval, was, on average, $1.9 \times 10^{-5}\,m^3\,s^{-1}$ (5.0 $mm\,d^{-1}$) between rain pulses and $1.5 \times 10^{-5}\,m^3\,s^{-1}$ (3.9 $mm\,d^{-1}$) after the third rain pulse.

The movement of the deuterium-enriched water within and out of the landscape was monitored through manual sampling and subsequent laboratory analysis. Prenart quartz water sampling devices were used to extract soil water samples

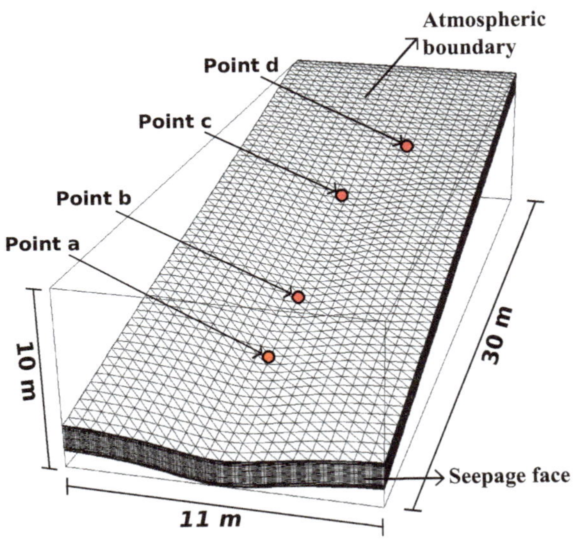

Figure 2. The 3-D numerical grid for the LEO landscape. Points a, b, c, and d are the locations where samples were extracted during the experiment for subsequent laboratory analysis.

periodically throughout the experiment. Data reported in this manuscript include samples collected at 5, 20, 50, and 85 cm depths from the surface at the four locations shown in Fig. 2. Flow from the seepage face boundary was collected with a custom autosampler (sampling cylinders of 5 cm length and 3 cm circumference). The deuterium concentration within all water samples was measured via laser spectroscopy (LGR LWIA Model DLT-100) at the University of Arizona. Analytical precision was better than 0.5‰ for $\delta^2 H$. All isotopic data are expressed relative to the VSMOW international reference or the VSMOW-SLAP scale. The seepage face isotopic data indicate that the residual soil water in the landscape prior to the experiment had become enriched in deuterium (compared to the rainfall water) during evaporation. In fact, during evaporation, hydrogen preferentially goes into the vapor phase compared to deuterium, so that the liquid phase remaining in the soil easily becomes deuterium-enriched. Thus, the $\delta^2 H$ values in the early seepage face flow may reflect some mixing of the new rain water with the evaporatively enriched water. This slight enrichment disappears in the seepage flow at later times because of the dilution by the newly infiltrating water.

3.2 Hydrological model

The CATHY (CATchment HYdrology) model (Camporese et al., 2010) used to simulate the isotope tracer experiment has been previously implemented for LEO to study coupled surface and subsurface flow (Niu et al., 2014) and sensor performance (Pasetto et al., 2015). The description here will thus be limited to aspects pertaining particularly to the implementation for LEO of the solute transport component of the model. The numerical solver for the advection–dispersion transport equation is described in detail in Putti and Paniconi (1995), and, like the flow solver, is based on a three-dimensional finite element discretization in space and a weighted finite difference discretization in time. The velocity field and nodal saturation values computed by the flow solver are passed as input at given time steps to the transport solver. The governing equations for the flow and transport solvers are

$$S_\mathrm{w} S_\mathrm{s} \frac{\partial \psi}{\partial t} + n \frac{\partial S_\mathrm{w}}{\partial t} = \nabla \cdot \left[K_\mathrm{r}(\psi) \mathbf{K}_\mathrm{s} \left(\nabla \psi + \boldsymbol{\eta}_z \right) \right] + q, \tag{1}$$

$$\frac{\partial \left(n S_\mathrm{w} c \right)}{\partial t} = \nabla \cdot (\mathbf{D} \nabla c) - \nabla \cdot (\boldsymbol{v} c) + f_\mathrm{c}, \tag{2}$$

where $S_\mathrm{w} = \theta / \theta_\mathrm{s}$ is the water saturation (–), θ is the volumetric moisture content (–), θ_s is the saturated moisture content (–) (generally equal to the porosity n (–)), S_s is the aquifer-specific storage coefficient (L^{-1}), ψ is the pressure head (L), t is the time (T), ∇ is the gradient operator (L^{-1}), $K_\mathrm{r}(\psi)$ is the relative hydraulic conductivity function (–), $\mathbf{K}_\mathrm{s}$ is the hydraulic conductivity tensor ($\mathrm{L T}^{-1}$) (considered to be diagonal, with k_s the saturated hydraulic conductivity parameter for the isotropic case and k_v and k_h, respectively, the vertical and horizontal hydraulic conductivity parameters for the anisotropic case), $\boldsymbol{\eta}_z = (0, 0, 1)^\mathrm{T}$, z is the vertical coordinate directed upward (L), q is a source (when positive) or sink (when negative) term (T^{-1}), c is the solute concentration ($\mathrm{M L}^{-3}$), $\mathbf{D}$ is the dispersion tensor ($\mathrm{L}^2 \mathrm{T}^{-1}$), $\boldsymbol{v} = (v_1, v_2, v_3)^\mathrm{T}$ is the Darcy velocity vector ($\mathrm{L T}^{-1}$), and f_c is a correction term ($\mathrm{M T}^{-1} \mathrm{L}^{-3}$) used in the treatment of the surface boundary condition for the transport equation during evaporation. The velocity vector is obtained from the flow equation as $\boldsymbol{v} = -K_\mathrm{r} \mathbf{K}_\mathrm{s} (\nabla \psi + \boldsymbol{\eta}_z)$, while the dispersion tensor can be expressed as

$$D_{ij} = n S_\mathrm{w} \widetilde{D}_{ij} = \alpha_\mathrm{t} |v| \boldsymbol{\delta}_{ij} + (\alpha_\mathrm{l} - \alpha_\mathrm{t}) \frac{v_i v_j}{|v|} + n S_\mathrm{w} D_\mathrm{o} \tau \boldsymbol{\delta}_{ij} \quad i, j = 1, 2, 3, \tag{3}$$

where $|v| = \sqrt{v_1^2 + v_2^2 + v_3^2}$, α_l is the longitudinal dispersivity (L), α_t is the transverse dispersivity (L), $\boldsymbol{\delta}_{ij}$ is the Kronecker delta (–), D_o is the molecular diffusion coefficient ($\mathrm{L}^2 \mathrm{T}^{-1}$), and τ is the tortuosity (we assume $\tau = 1$) (–). The evaluation of integrals arising in finite element discretization of the dispersion fluxes is performed using a rotated reference system spanned by the unit vectors (x_1, x_2, x_3) that are aligned with the principal directions of anisotropy of $\mathbf{D}$, whereby $x_1 = v/|v|$. Within this reference system, $\mathbf{D}$ becomes diagonal, with the three components defined as

$$D_{11} = \alpha_\mathrm{l} |v| + n S_\mathrm{w} D_\mathrm{o} \tau, \tag{4}$$

$$D_{22} = D_{33} = \alpha_\mathrm{t} |v| + n S_\mathrm{w} D_\mathrm{o} \tau. \tag{5}$$

The soil moisture–pressure head and relative conductivity–pressure head dependencies are described by the van

Genuchten (1980) relationship:

$$S_e = \left[1 + \left(\frac{|\psi|}{|\psi_{sat}|}\right)^{n_{VG}}\right]^{-m}, \tag{6}$$

$$K_r(\psi) = S_e^{0.5}\left[1 - \left(1 - S_e^{\frac{1}{m}}\right)^m\right]^2, \tag{7}$$

where $S_e = (S_w - S_{wr})/(1 - S_{wr})$ is the effective saturation (–), S_{wr} is the residual water saturation (–), $m = (1 - 1/n_{VG})$, n_{VG} is a fitting parameter ranging between 1.25 and 6 (–), and ψ_{sat} is related to the air entry suction (L).

The transport equation (2) is solved in its conservative form, i.e., without applying the chain rule to the advective and storage terms. Using Euler time stepping, the resulting discretized system is

$$\left([\mathbf{A}+\mathbf{B}]^{k+1} + \frac{1}{\Delta t_k}\mathbf{M}^{k+1}\right)\hat{\boldsymbol{c}}^{k+1} = \frac{1}{\Delta t_k}\mathbf{M}^k\hat{\boldsymbol{c}}^k - \boldsymbol{b}^{t,k+1}, \tag{8}$$

where k is the time counter, $\hat{\boldsymbol{c}}$ is the vector of the numerical approximation of c at each node of the grid, and the coefficients of the, respectively, dispersion, advection, and mass matrices are

$$a_{ij} = \int_\Omega \mathbf{D}\nabla\phi_i \nabla\phi_j d\Omega, \tag{9}$$

$$b_{ij} = \int_\Omega \nabla(\boldsymbol{v}\phi_j)\phi_i d\Omega, \tag{10}$$

$$m_{ij} = \int_\Omega nS_w\phi_i\phi_j d\Omega, \tag{11}$$

where $i, j = 1, \ldots, N$, with N the number of nodes, Ω is the discretized domain, and ϕ are the basis functions of the Galerkin finite element scheme. The boundary condition vector for the discretized transport equation is

$$b_i^t = \int_{\Gamma^t} (-\mathbf{D}\nabla c)\cdot \boldsymbol{\nu}\phi_i d\Gamma^t = \int_{\Gamma^t} q_n^t \phi_i d\Gamma^t, \tag{12}$$

where Γ^t is the boundary of the domain Ω, q_n^t ($\mathrm{ML^{-2}T^{-1}}$) is the Neumann (dispersive) flux, and $\boldsymbol{\nu}$ is the outward normal vector to the boundary. Cauchy, or mixed, boundary conditions can be easily implemented as variations of Eq. (12), involving an additional term in the system matrix implementing the advective component of the Cauchy condition.

3.3 Model setup for the LEO tracer experiment

We discretized the 30 m × 11 m × 1 m LEO hillslope into 60 × 22 grid cells in the lateral direction and 30 layers in the vertical direction (Fig. 2). The resulting surface mesh consists of 1403 nodes and 2640 triangular elements. This horizontal discretization was chosen in order to have the nodes of the computational mesh aligned with the sensor and sampler locations, thereby allowing us to directly compare simulated and measured distributed responses. This same principle was used to guide the vertical discretization (the interface between two layers is set at the sensor and sampler heights). The surface mesh was projected vertically to form a 3-D tetrahedral mesh with parallel layers of varying thicknesses, with the thinnest layers assigned to the surface and bottom layers. This allows the numerical model to accurately capture infiltration/evaporation processes at the surface and the formation of base flow at the bottom of the domain. From top to bottom the thicknesses of the 30 layers are 0.01 m for the first five layers, 0.025 m from layer 6 to layer 9, 0.05 m for layer 10, 0.06 m from layer 11 to layer 20, 0.05 m for layer 21, 0.025 m from layer 22 to layer 25, and 0.01 m from layer 26 to layer 30.

Measurements showed that the average δ^2H of the rain source water at LEO was −60 ‰. For the transport model, we used a normalized concentration defined as

$$c = \frac{\delta^2 H_{ref} - \delta^2 H}{\delta^2 H_{ref}}, \tag{13}$$

where $\delta^2 H_{ref} = -60$ ‰ and $\delta^2 H$ is the actual value. Thus the initial conditions, as well as the concentrations of the first and third pulses, were $c = 0$, while the second pulse had an imposed concentration of $c = 1$. Note that, with this transformation, the dimension of the term f_c of Eq. (2) becomes $\mathrm{T^{-1}}$.

A careful treatment of boundary conditions was essential to modeling the isotope tracer experiment, in particular at the land surface where three different cases needed to be considered. These cases are schematically summarized in Table 1, in relation to the simulations performed, and further noted here: (1) rain with ^2H-enriched water, handled as a Neumann prescribed flux condition for flow ($q_n^f = -\mathbf{K}_s K_r(\psi)(\nabla\psi + \boldsymbol{\eta}_z)\cdot\boldsymbol{\nu} = \boldsymbol{v}\cdot\boldsymbol{\nu}$) and a Cauchy prescribed advective flux condition for transport ($q_c^t = (\boldsymbol{v}c - \mathbf{D}\nabla c)\cdot\boldsymbol{\nu} = \boldsymbol{v}\cdot\boldsymbol{\nu}c^*$); (2) rain with no ^2H-enriched water, handled with the same Neumann condition as case (1) for flow and a zero Cauchy prescribed total flux condition for transport ($q_c^t = (\boldsymbol{v}c - \mathbf{D}\nabla c)\cdot\boldsymbol{\nu} = 0$ with the concentration values at the surface nodes computed by the model); (3) evaporation, handled with the same Neumann condition as case (1) for flow and a zero Neumann prescribed dispersive flux condition for transport ($q_n^t = -\mathbf{D}\nabla c\cdot\boldsymbol{\nu} = 0$ with the concentration values at the surface nodes computed by the model). With the zero dispersive flux condition of case (3), all the isotopic mass in solution with the evaporating water leaves the domain by advection.

In addition to this "base case" treatment of rainfall and evaporation, we also introduced some variations in the surface boundary conditions. For rainfall (cases 1 and 2 above), we tested both uniform and variable spatial distributions. For the latter, a rainfall pattern with slightly higher rates towards the center of the landscape was used, as indicated by measurements taken during testing of the engineered rain sys-

Table 1. Treatment of boundary conditions at the land surface during the rainfall and evaporation periods for the flow and transport models.

Simulation (see Tables 2 and 3)	Rain with ^2H-enriched water (second pulse)		Rain with no ^2H-enriched water (first and third pulses)		Evaporation (between rain pulses and after the third pulse)	
	Flow	Transport	Flow	Transport	Flow	Transport
a–f, g–i	$q_n^{\mathrm{f}} = -12\,\mathrm{mm\,h^{-1}}$	$q_c{}^{\mathrm{t}} = \boldsymbol{v} \cdot \boldsymbol{\nu} c^*$, $c^* = 1$	$q_n^{\mathrm{f}} = -12\,\mathrm{mm\,h^{-1}}$	$q_c^{\mathrm{t}} = 0$	$q_n^{\mathrm{f}} = 5$ or $3.9\,\mathrm{mm\,h^{-1}}$	$q_c^{\mathrm{t}} = 0$ $q_c^{\mathrm{t}} = 0$
j	$q_n^{\mathrm{f}} = -12\,\mathrm{mm\,h^{-1}}$	$q_c^{\mathrm{t}} = \boldsymbol{v} \cdot \boldsymbol{\nu} c^*$, $c^* = 1$	$q_n^{\mathrm{f}} = -12\,\mathrm{mm\,h^{-1}}$	$q_c^{\mathrm{t}} = 0$	Sink q (Fig. 3)	Source f_c (Fig. 3)
k	$q_n^{\mathrm{f}} = -12\,\mathrm{mm\,h^{-1}}$	$q_c^{\mathrm{t}} = \boldsymbol{v} \cdot \boldsymbol{\nu} c^*$, $c^* = 1$	$q_n^{\mathrm{f}} = -12\,\mathrm{mm\,h^{-1}}$	$q_c^{\mathrm{t}} = 0$	Sink q (Fig. 3)	Source f_c, $f_c = -qc$

tem. This pattern was generated in such a way that the mean rainfall rate and the total volume of water injected were preserved. For evaporation, since there were no measurements of soil evaporation isotopic composition at the LEO landscape, we tested two other hypotheses — that none or only a portion (fractionation) of the isotope tracer evaporated — in addition to the zero dispersive flux condition of case (3).

To prevent an isotope tracer from leaving the system through the landscape surface, we treated the evaporation as a sink term in the flow model, distributed exponentially from the surface to a depth of 38 cm, rather than as a Neumann boundary condition. In generating the sink term, we ensured that the total volume of water evaporated was the same as in the Neumann boundary condition treatment. The sink term function q in Eq. (1) applied to each layer i ($i = 1, \ldots, 13$ for a total depth of 38 cm) is

$$q_i = \frac{F_{\mathrm{ev}}}{\sum_{i=1}^{13}\left(e^{-\lambda z_i}\Delta z_i\right)} e^{-\lambda z_i}, \tag{14}$$

where q_i is applied to each tetrahedron of layer i, λ ($\mathrm{L^{-1}}$) is a parameter set to $1\,\mathrm{m^{-1}}$ in this case, z_i is the depth from the surface to the center of layer i, Δz_i is the thickness of layer i, and F_{ev} ($\mathrm{LT^{-1}}$) is the homogeneous evaporative flux used in the Neumann boundary condition case (with rates $-5.8 \times 10^{-8}\,\mathrm{m\,s^{-1}}$ between rain pulses and $-4.5 \times 10^{-8}\,\mathrm{m\,s^{-1}}$ after the third pulse). The applied sink fluxes are shown schematically in Fig. 3. Note that if the element reaches the residual water saturation, parameterized by its corresponding pressure head level, the evaporation process becomes soil-limited. When this occurs, the actual sink term function is automatically smaller than the imposed value. To ensure that all the tracer mass stays in the system, for the transport model we set the correction term f_c in Eq. (2) equal to $-qc$. In this way we inject back into the system the same amount of tracer mass that has exited with the sink term q.

Most land surface hydrological models still neglect fractionation, even though it can significantly influence the mass exchange at the land surface and the concentration profiles in the soil. Barnes and Allison (1988) examined isotope transport phenomena under both saturated and unsaturated conditions. In the latter case they experimentally observed that

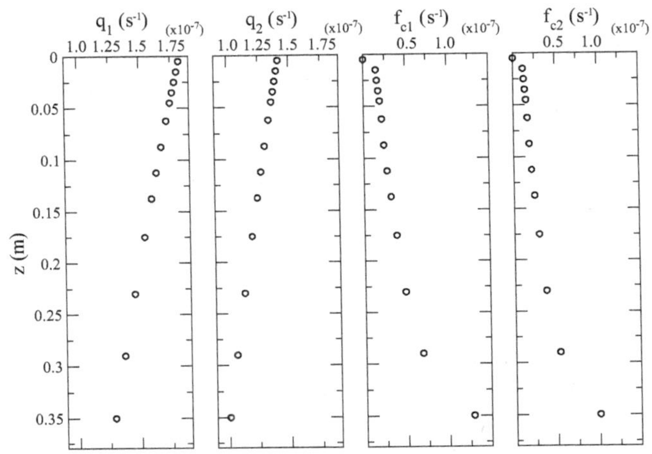

Figure 3. Sink term q and correction source term f_c over depth z added to the flow and transport equations, respectively. q_1 and f_{c1} are applied between rain pulses 1, 2, and 3, while q_2 and f_{c2} are applied after rain pulse 3.

at steady state the maximum concentration of the heavier isotope species (e.g., ^2H) occurs a short distance below the surface and decreases rapidly beyond that depth. The resulting profile can be explained as the result of vapor diffusion and isotopic exchange dominating the zone above the drying front and the balance between capillary and diffusive liquid water transport below the drying front (Craig and Gordon, 1965; Clark and Fritz, 1997; Horita et al., 2008). Alternative conceptualizations of the fractionation process have also been recently developed (e.g., Braud et al., 2009; Haverd and Cuntz, 2010). In this work the fractionation process was incorporated into the CATHY model using the sink term approach described above, setting 38 cm as the soil depth at which the maximum ^2H tracer concentration occurs (thereby assuming that the soil above is dominated by water vapor diffusion due to evaporation). The correction source term f_c introduced into the transport equation is now modified such that there is no tracer mass re-injection in the first layer, and the amount re-injected progressively increases from $qc/12$ to qc between layers 2 and 13 (Fig. 3). The reasoning here is that the rate at which a tracer evaporates increases with evaporation and water vapor diffusion close to the surface.

Besides the surface boundary, we set up a seepage face condition at the 23×30 nodes that constitute the downslope lateral boundary. For the transport equation the seepage face nodes have a zero Neumann (dispersive) assigned flux so that ^2H is allowed to exit the domain through advection with the outflowing water. All other LEO boundaries (the three other lateral boundaries and the base of the hillslope) were set to a zero Neumann condition for both the flow and transport equations (with a zero water flux this implies that the advective flux for the transport equation is also zero).

The time stepping for the flow model is adaptive (based on convergence of the iterative scheme used to linearize Richards' equation (1)) and we set the time step range between 10^{-4} and 90 s. The results in terms of velocity and saturation values were saved every 90 or 900 s, respectively, during and between the rain events. These were linearly interpolated in time and read as input by the transport model, which was run with a fixed time step of 90 s for the entire simulation.

3.4 Simulations performed

The model simulations were used to interpret the integrated and point-scale flow and transport responses of the LEO hillslope. The guiding idea was to assess the need to increase the complexity of the model in progressing from first trying to reproduce the integrated flow response, then the integrated transport response, and finally the point-scale flow and transport responses. With the requirement that each new parameterization still had to satisfy the observation data set from the previous level, the space of admissible solutions was progressively reduced. Initially the soil was assumed to be homogeneous and isotropic. The values of the van Genuchten parameters ($n_{VG} = 2.26$, $\psi_{sat} = -0.6$ m, and $\theta_r = 0.002$), the porosity ($n = 0.39$), the saturated hydraulic conductivity ($k_s = 1.4 \times 10^{-4}$ m s^{-1}), and the specific storage ($S_s = 5 \times 10^{-4}$ m^{-1}) were obtained from laboratory analyses and simulations of prior LEO experiments (Niu et al., 2014; Pasetto et al., 2015). From this base set of parameter values for the first simulations, anisotropy and other variations were progressively introduced in the model.

In the first step of this procedure (integrated flow response), we examined the influence of spatial variability and anisotropy in saturated hydraulic conductivity (different k_s at the seepage face and over the rest of the hillslope, on the basis of a clogging hypothesis from accumulation of fine particles (Niu et al., 2014); higher k_h than k_v, on the basis of a hypothesis of slight vertical compaction leading to enhanced flow in the horizontal direction), rainfall (spatially uniform; spatially variable), and initial conditions (uniform; generated from a steady-state simulation under drainage and evaporation; matching the soil moisture distribution at each sensor location). Six simulations were run in the first step. The configurations for each run are summarized in Table 2. For the initial conditions, in all three configurations (uniform for runs a through c, steady state for run d, and matching sensors for runs e and f), the same total initial water storage (26 m^3 as reported earlier) was used. For the atmospheric forcing, the spatially uniform rainfall rate (runs a through e) was the mean measured rate reported earlier (12 mm h^{-1}), while the spatially variable case (run f) was handled as described earlier. The evaporation rate, on the other hand, was kept spatially uniform for all six simulations and equal to the mean rate of 5.0 mm d^{-1} between the three pulses and 3.9 mm d^{-1} after the third pulse.

In the second step (integrated transport response), the effects of the dispersivity coefficients α_l and α_t and of isotope evaporation mechanisms on the amount of tracer at the seepage face outlet were explored. In the third step (flow point-scale data), the analysis focused on the soil moisture profiles obtained by averaging the observations and model results at specific depths (5, 20, 50, and 85 cm), and spatially variable (by layer) soil hydraulic properties (n_{VG}) were introduced. Finally, for the point-scale transport we compared the results obtained from some of the different parameterizations used in the previous steps.

The simulations performed are summarized in Table 3. Model performance was assessed against available observations using the coefficient of efficiency (CE) on seepage face flow Q_{sf} for the integrated flow response and the root mean squared error (RMSE) on concentration c at the seepage face for the integrated transport response and on averaged θ profiles for the flow point-scale response. The CE and RMSE metrics, also reported in Table 3, are calculated as in Dawson et al. (2007):

$$\mathrm{CE} = 1 - \frac{\sum_{i=1}^{n_o}\left(Q_i - \hat{Q}_i\right)^2}{\sum_{i=1}^{n_o}\left(Q_i - \bar{Q}\right)^2}, \quad (15)$$

$$\mathrm{RMSE} = \sqrt{\frac{\sum_{i=1}^{n_o}\left(Q_i - \hat{Q}_i\right)^2}{n}}, \quad (16)$$

where n_o is the total number of observed data available at the different times, Q_i and $\hat{Q}_i$ are the observed and modeled values, respectively, and $\bar{Q}$ is the observed average value.

4 Results

4.1 Integrated flow response

In the first set of simulations we attempt to reproduce two integrated flow responses of the LEO hillslope, the measured seepage face flow and the measured total water storage. The results of the six simulations are presented in Fig. 4. The water balance partitioning between seepage face flow and internal storage was found to be strongly affected by the introduction of anisotropy and variability in the hydraulic conductivity. We also found that the distribution of initial conditions determines the timing of the first simulated seepage

Table 2. Configurations for the six simulations of the integrated flow analysis.

Simulation	Saturated hydraulic conductivity ($m s^{-1}$)			Initial conditions	Rainfall
	Horizontal, k_h	Vertical, k_v	Seepage face, k_{sf}		
a	1.4×10^{-4}	1.4×10^{-4}	1.4×10^{-4}	Uniform	Spatially uniform
b	6×10^{-4}	1.4×10^{-4}	1.4×10^{-4}	Uniform	Spatially uniform
c	6×10^{-4}	1.4×10^{-4}	2.2×10^{-5}	Uniform	Spatially uniform
d	6×10^{-4}	1.4×10^{-4}	2.2×10^{-5}	Steady state	Spatially uniform
e	6×10^{-4}	1.4×10^{-4}	2.2×10^{-5}	Interpolated soil moisture measurements	Spatially uniform
f	6×10^{-4}	1.4×10^{-4}	2.2×10^{-5}	Interpolated soil moisture measurements	Spatially variable

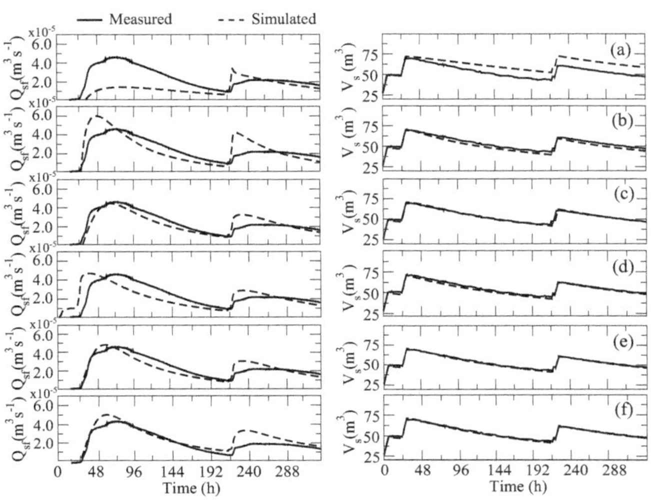

Figure 4. Results for the six simulations of the integrated flow response analysis (see Table 2). For each case the seepage face flow Q_{sf} (left) and total water storage V_s (right) are reported.

face peak and its shape. The spatial distribution of rain, on the other hand, was not found to have a significant impact on the model response. These general findings are described in more detail below.

In the first simulation (Fig. 4a), under the assumption of homogeneity, isotropy, uniform initial conditions, and spatially uniform rainfall and evaporation, the discrepancy between the simulated and observed response was large (a negative CE is reported in Table 3), with the first and second peaks of the discharge hydrograph, respectively, underestimated and overestimated by the model. In the second simulation, with the introduction of anisotropy (increasing the horizontal hydraulic conductivity k_h to $6 \times 10^{-4}\ m s^{-1}$), the overall model results for the seepage face flow improved notably compared to simulation a (CE passed from -0.62 to 0.64) and the match for the total water storage was improved significantly (Fig. 4b). Next, the introduction of low k_s at the seepage face lowered the hydrograph peaks and smoothed out its overall shape (Fig. 4c), moving the simulated hydrograph closer to the measured one (and increasing CE to 0.79). The effect of using distributed instead of uniform initial conditions is seen in comparing Figs. 4c, 4d, and 4e. Under uniform starting conditions the response was delayed in time, compared to the steady-state case (generated under a drainage and evaporation run from initially wet conditions), where the response to the first rain pulse was faster. This faster response resulted in increased drainage due to longer recession periods, adversely affecting the match for the second pulse but improving the result for the third pulse. The simulation for Fig. 4e, with initial conditions closest to the initial state of the hillslope, resulted in a further increase in CE to 0.82. For this run, the good match for the first hydrograph peak from simulation c of Table 2 was recovered, whilst retaining the good match for the second peak from simulation d. The simulated total water storage dynamics was already very well captured by simulation c and was not greatly affected by the initial conditions. The initial conditions from simulation e were used for all subsequent simulations discussed in this study. In the final simulation for the integrated flow response analysis, incorporating the spatial distribution of rainfall had a nominal impact on the results (Fig. 4f), with a slight increase in CE to 0.85. Thus the actual distribution of atmospheric forcing, so long as it is not highly variable (which was part of the experimental design for the LEO tracer experiment), is less important than capturing the correct mean rate and total volume of these hydrologic drivers.

4.2 Integrated transport response

The velocity field and saturation obtained from the sixth flow simulation (simulation f) of the preceding section were used as input to the transport model. Figure 5 and Table 3 show, respectively, the results for the average tracer concentration at the seepage face and the RMSE for different longitudinal dispersivity α_l values, namely 0.1, 0.01, and 0.001 m. The transverse dispersivity α_t was set 1 order of magnitude smaller than α_l. The three graphs and the RMSE values show that the discrepancy between the measured and simulated outflow concentrations decreases with α_l. The results show that

Table 3. Simulation descriptions, parameter configurations, and performance metrics (coefficient of efficiency CE and root mean squared error RMSE) for the integrated flow, integrated transport, and point-scale analysis steps.

Integrated flow analysis			Simulations	CE for Q_{sf}
Effect on seepage face flow Q_{sf}	Effect on total water storage V_s	a	Base case (Niu et al., 2014)	−0.62
		b	Anisotropy	0.64
		c	Low k_s at seepage face	0.79
		d	Initial conditions	0.28
		e	Initial conditions	0.82
		f	Rainfall distribution	0.85

Integrated transport analysis		α_l	Evaporation	RMSE
Effect on concentration c at the seepage face (flow configuration from simulation f)	g	0.1	All tracers	0.12
	h	0.01	All tracers	0.037
	i	0.001	All tracers	0.026
	j	0.001	Fractionation	0.03
	k	0.001	No tracer	0.045

Point-scale analysis									RMSE for averaged θ (at 5, 20, 50, and 85 cm depths)
Effect on averaged θ profiles	Effect on point-scale θ profiles	Effect on c point-scale profiles	f	Depth (cm)	5	20	50	85	
				n_{VG} (homogeneous)	2.26				10.36, 1.17, 1.73, 3.78
		(transport configuration from simulation i)	l	Depth (cm)	5	20	50	85	
				n_{VG} (layered)	1.8	2.26	2.0	1.9	5.61, 1.43, 0.95, 1.72

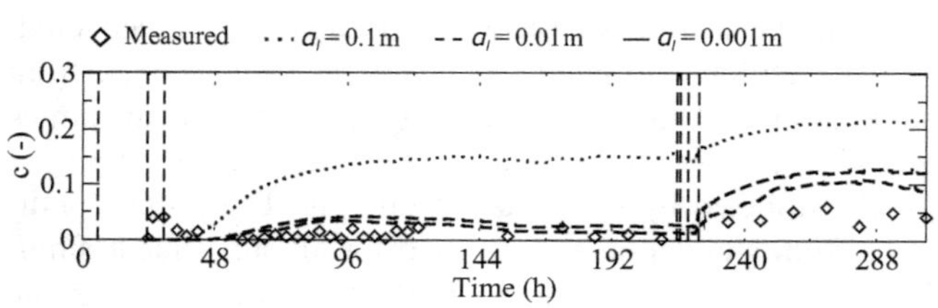

Figure 5. Results for the integrated transport response analysis for different values of dispersivity (simulations g, h, and i of Table 3). The vertical dashed lines indicate the timing of the three pulses of rain (red when the water is ^2H-enriched and blue when it is not).

the effect of the high dispersivity makes the tracer percolate down quickly to then flow out of the domain from the seepage face boundary. In fact, at the highest value, significant levels of ^2H-labeled water appeared in the outflow discharge after the second pulse, whereas in the measured data and in the model results for the smaller dispersivity values, the levels were much lower. In all three cases the model reproduced the increase in tracer concentration after the last pulse, but whereas for $\alpha_l = 0.1$ m the values were 4 times higher than the observed ones, for $\alpha_l = 0.01$ m and $\alpha_l = 0.001$ m they decreased significantly. The simulation using the lowest value of dispersivity was able to reproduce reasonably well the integrated measure of tracer response for the LEO hillslope.

To assess model accuracy, we report in Fig. 6 the mass balance results for the $\alpha_l = 0.001$ m case, in terms of a balance between the cumulative mass of deuterium that entered the hillslope (with the second rainfall pulse), that exited the system (through seepage face outflow and evaporation), and that remained in storage. The results show that for $\alpha_l = 0.001$ m and $\alpha_t = 0.0001$ m, at the end of the simulation (after 14 days), 52 % of the mass of ^2H injected has been lost through evaporation, about 4 % has seeped out, and the rest remained in storage, minus a cumulative mass balance error of about 2 % with respect to the total mass injected. The sudden mass balance error jump that occurs at the beginning of the third pulse of rain is most probably due to discontinuities in the time derivative of concentration and water saturation close to the surface (since the soil is very dry at this level and after the long evaporation period) as a consequence of the discontinuity in the atmospheric boundary condition. The high evaporative component computed by the model is a direct outcome of the zero dispersive flux surface boundary condition for the transport equation, through which any tracer in solution with evaporating water is advected away with the water. We examine next the impact of the sink term treatment of tracer exchange across the land surface boundary, preventing any isotope tracer from evaporating.

The results of the sink term simulation in terms of average seepage face tracer concentration and mass balance are reported, respectively, in Figs. 7 and 8. As expected, the seepage face concentration has now increased, but only slightly, compared to the previous simulation. In mass terms, the seep-

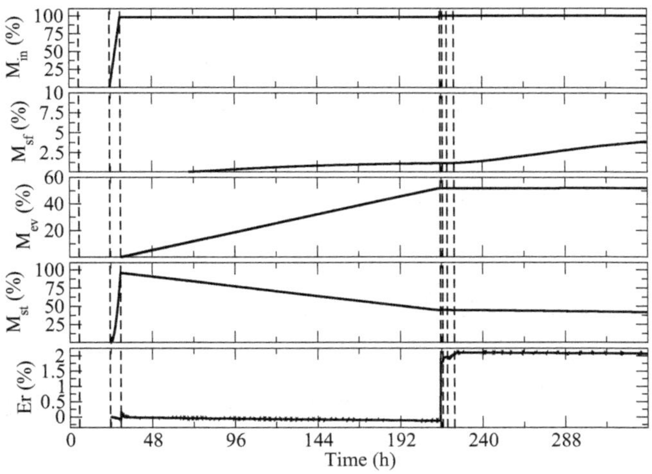

Figure 6. Simulated mass balance results for $\alpha_l = 0.001$ m (simulation i of Table 3). From top to bottom: ^2H mass that enters the system, M_{in} (normalized with respect to the total mass added to the system during the simulation); that exits through the seepage face, M_{sf}; that exits through evaporation, M_{ev}; and that remains in storage, M_{st}. The bottom graph shows the cumulative mass balance error $E_r = (M_{in} - M_{sf} - M_{ev} - M_{st})$. The vertical dashed lines indicate the timing of the three pulses of rain (red when the water is ^2H-enriched and blue when it is not).

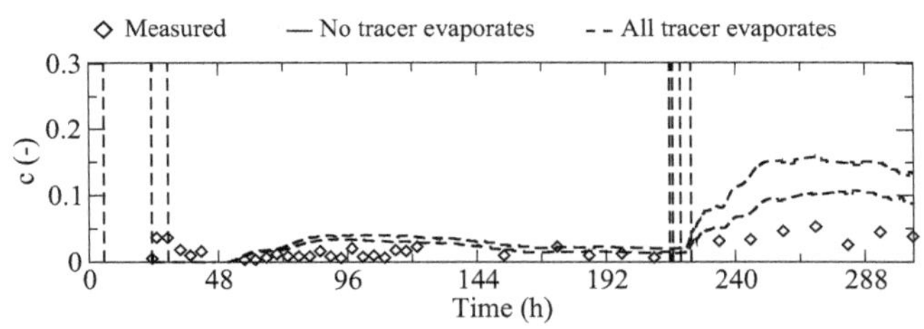

Figure 7. Measured and modeled average tracer concentration at the seepage face for the cases in which no tracer and all tracers leave the system with evaporation (simulations i and k of Table 3). Both simulations are run for $\alpha_l = 0.001$ m and $\alpha_t = 0.0001$ m. The vertical dashed lines indicate the timing of the three pulses of rain (red when the water is ^2H-enriched and blue when it is not).

age component has increased from 4 to 8 % by the end of the simulation. With no tracer mass now exiting via the landscape surface, it is found instead that much more of the mass has remained in storage (about 90 % compared to about 40 % when the tracer was allowed to evaporate with the water). This result strongly suggests that the tracer does not percolate far (deep) into the hillslope, perhaps as a result of the very dry conditions initially and during the whole experiment. A negative consequence of not allowing any tracer mass to evaporate, combined with low percolation, is an intense accumulation of the mass near the landscape surface, with tracer concentrations as high as 15. A compromise between allowing zero or all isotope tracers to leave the system via evaporation is to introduce isotopic fractionation processes into the model.

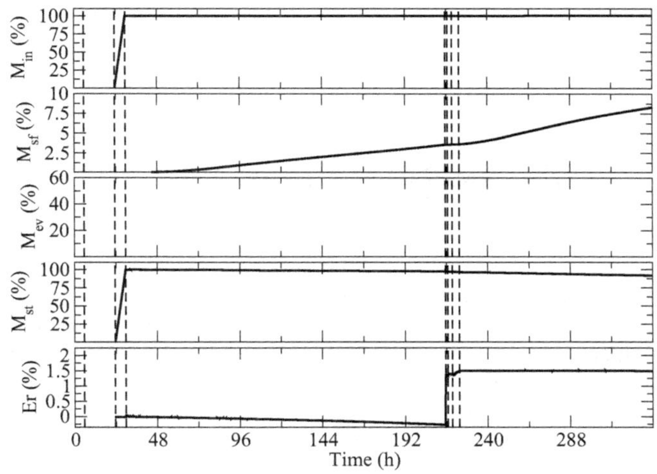

Figure 8. Simulated mass balance results for $\alpha_l = 0.001$ m when the sink term is used to perform evaporation and the correction term f_c added to the transport equation is used to force all the isotopic mass to stay in the system (simulation k of Table 3). From top to bottom: ^2H mass that enters the system, M_{in} (normalized with respect to the total mass added to the system during the simulation); that exits through the seepage face, M_{sf}; that exits through evaporation, M_{ev}; and that remains in storage, M_{st}. The bottom graph shows the cumulative mass balance error $E_r = (M_{in} - M_{sf} - M_{ev} - M_{st})$. The vertical dashed lines indicate the timing of the three pulses of rain (red when the water is ^2H-enriched and blue when it is not).

The results of the isotope fractionation simulation are reported in Figs. 9 and 10, respectively, for the average tracer concentration at the seepage face and the model mass balance results. The curve for the average concentration in Fig. 9 justly lies between the curves obtained by making all and no isotopes evaporate with water. The mass balance shows that at the end of the simulation, 6.5 % of the total mass injected has gone out through the seepage face, this result also falling between the previous simulations where zero or all isotope tracers in solution with the evaporating water was lost via evaporation. As expected, the evaporative mass loss is now significant (38 %) but not as high as obtained when evaporation was treated as a land surface Neumann boundary condition (52 %). The final mass balance error (0.75 %) is lower than for the two previous simulations, and the accumulation of isotope mass just below the land surface that occurred in the preceding case was not observed in this simulation.

4.3 Distributed flow response

For the distributed flow response analysis we first examined the behavior in time of the averaged soil water content value at the four depths of the sensor network (5, 20, 50, and 85 cm). That is, we compared the average of all sensor measurements at a given depth to the average of all simulated

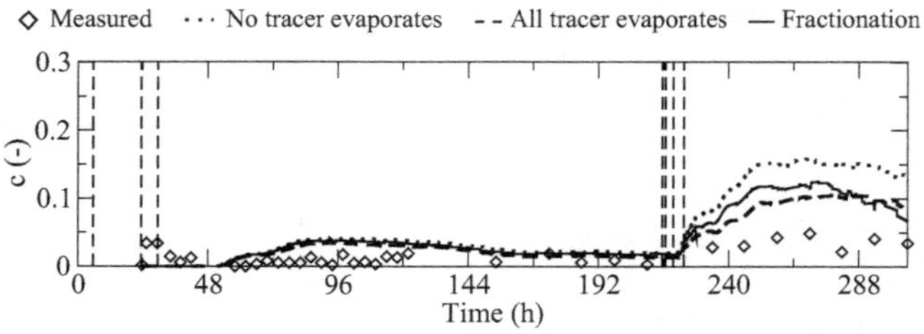

Figure 9. Measured and modeled average tracer concentration at the seepage face for the cases in which no tracer, all tracers, and some tracers (fractionation) leave the system with evaporation (simulations i, j, and k of Table 3). The three simulations are run for $\alpha_l = 0.001$ m and $\alpha_t = 0.0001$ m. The vertical dashed lines indicate the timing of the three pulses of rain (red when the water is ^2H-enriched and blue when it is not).

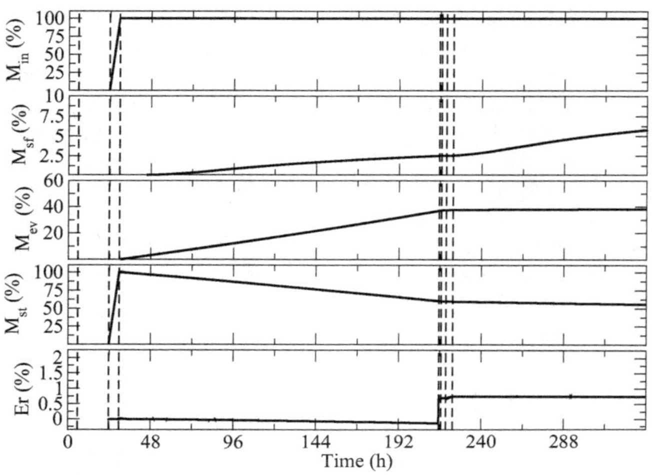

Figure 10. Simulated mass balance results for $\alpha_l = 0.001$ m when the correction source term f_c is added to the transport equation to perform isotopic fractionation (simulation k of Table 3). From top to bottom: ^2H mass that enters the system, M_{in} (normalized with respect to the total mass added to the system during the simulation); that exits through the seepage face, M_{sf}; that exits through evaporation, M_{ev}; and that remains in storage, M_{st}. The bottom graph shows the cumulative mass balance error $E_r = (M_{in} - M_{sf} - M_{ev} - M_{st})$. The vertical dashed lines indicate the timing of the three pulses of rain (red when the water is ^2H-enriched and blue when it is not).

nodal θ values at that depth. The graphs for the results of simulation f from Table 2 (the configuration that best retrieved the integrated flow response) are shown in Fig. 11, while the RMSE values are reported in Table 3. The results show that at 50 cm there is a small underestimation by the model and that the model does not perform well at 5 and 85 cm compared to the profile at 20 cm. At 85 cm depth the observed and calculated deviations from the mean are also completely different (for the model it is almost 0).

To address this problem we increased the retention capacity of the soil by reducing, selectively, the n_{VG} parameter of the van Genuchten hydraulic functions as reported in Ta-

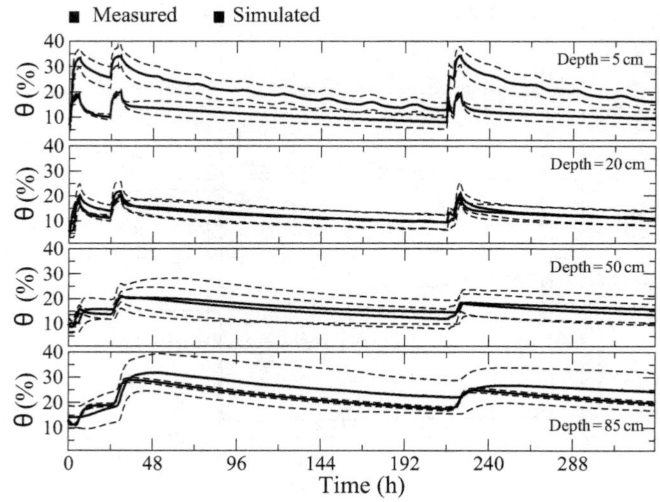

Figure 11. Averaged θ profiles at 5, 20, 50, and 85 cm depths from the surface. In each graph the deviation from the mean (1 standard deviation above and below) is shown as dashed lines. The results are obtained for simulation f reported in Table 3.

ble 3. We subdivided the soil profile into four strata encompassing the four sensor depths, and we decreased n_{VG} for the strata closest to the surface (from 0 to 10 cm, $n_{VG} = 1.8$), from 32 to 68 cm ($n_{VG} = 2.0$), and from 68 cm to the bottom ($n_{VG} = 1.9$). For the second stratum (from 10 to 32 cm) the retention parameter was left unaltered from all previous simulations ($n_{VG} = 2.26$) since the model already captured the observed response for the sensor at 20 cm depth quite well. The highest retention capacity (lowest n_{VG} value) was set in the first stratum since the observation data show that the water content close to the landscape surface remains quite high, both during infiltration and drainage. The n_{VG} values for the four strata reported here are the combination, from many trials, that best retrieved the observed averaged θ profiles. The results of this simulation are shown in Fig. 12 and reported in Table 3. Compared to the results of the homogeneous n_{VG} case, the model response improves significantly for the average profile at 5, 50, and 85 cm, even if the deviations at 85 cm are still very different.

To take the distributed flow response analysis further, in Fig. 13 we show the water content time series at the four specific points shown in Fig. 2, at 5, 20, 50, and 85 cm depths from the surface. Sensor data at each of the four points and for each of the four soil depths are compared against both the homogeneous n_{VG} case (simulation f from Table 2) and the layered n_{VG} case (different value for each of the four strata). Once again the more detailed parameterization (simulation l from Table 3, variable n_{VG}) gives better results, although for some of the soil depths (in particular at 50 and 85 cm) and for some of the points (in particular point c) the discrepancies between simulated and measured θ time series are quite marked. It should be remarked that we did not run, as we

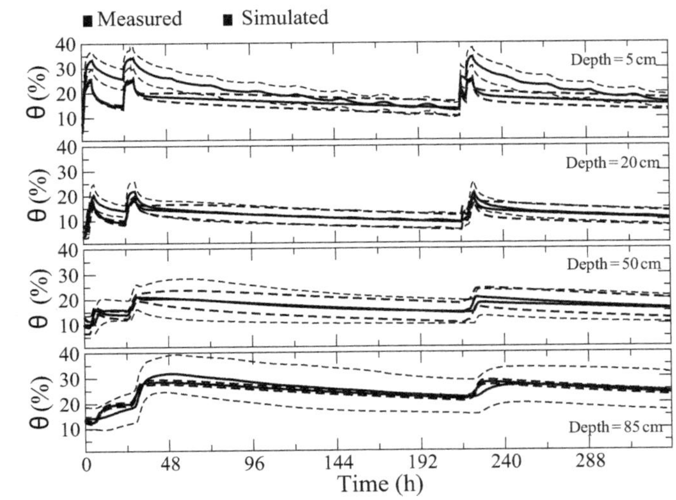

Figure 12. Averaged θ profiles at 5, 20, 50, and 85 cm depths from the surface. In each graph the deviation from the mean (1 standard deviation above and below) is shown as dashed lines. The results are obtained for simulation l reported in Table 3.

did for the simulation summarized in Fig. 12, repeated trials to find a best fit, so it may perhaps be possible to optimize the fits against both the averaged θ data and the point data (Fig. 13) by manipulating the soil retention capacity for the four strata. However, it seems more likely that in going from a distributed but nonetheless averaged response variable to a distributed, point-scale response variable, additional model parameter complexity is needed to obtain an adequate response for all individual response variables.

4.4 Distributed transport response

For the distributed transport response analysis we compared, as we did in Fig. 13 for the internal state flow response, the model results at individual points (a, b, c, d from Fig. 2) and individual soil depths (5, 20, 50, and 85 cm) for simulations using uniform (corresponding to configuration f from Table 2) and spatially variable (simulation l from Table 3) soil retention capacity. The results are shown in Fig. 14, and it can be seen that the model does not perform well at several locations within the hillslope (consistently at 20 cm depth, and at 5 cm depth for point b). Encouragingly, however, there is consistency with the previous distributed flow results, in that the variable n_{VG} run performs noticeably better than the spatially uniform case. For instance, with variable n_{VG} the results are improved at the bottom of the hillslope, at 50 cm (for points b and c the modeled response gets closer to the measurements, particularly after the third flush), and slightly at 5 cm (for point a).

For the distributed transport analysis we did not examine averaged concentration profiles at each of the four sensor depths (as we did for soil water content in Fig. 12) due to insufficient data. The sampling time and laboratory analysis costs for exhaustive measurement of isotopic compositions were prohibitive; thus, there are far fewer data available for the distributed transport analysis compared to the flow case. The data gaps are also evident in Fig. 14: there are no measurements for three of the graphs, and scarce data at 50 cm depth for points a and d. It is also important to note that no additional parameterization was attempted for the distributed transport analysis. The main explicit parameters in the transport equation are the dispersivity coefficients, and these were experimented with in the integrated transport analysis. The transport equation is of course strongly dependent on flow velocities, and thus implicitly on the conductivity and other soil hydraulic parameters that were assessed in the flow model analyses. These and other parameterization issues will be further discussed in the next section.

To complete the sequence of analyses from integrated flow and then transport to distributed flow and transport, we used the simulation results from the additional parameterization introduced for the distributed analyses (spatially variable soil retention capacity) to assess model performance against the integrated flow and transport responses. The results (Fig. 15) show that while the match against tracer concentration at the seepage face has somewhat improved (compare with Fig. 9), the match against both of the integrated flow responses (seepage outflow and total water storage) has significantly deteriorated (compare with simulation f of Fig. 4). This is not a surprising result, given that no attempt was made to parameterize the model in tandem against both integrated and point-scale observations (nor against joint flow and transport data); the implications will be discussed below.

5 Discussion

Mass transport in unsaturated soils is extremely important in the context of biosphere, critical zone, and Earth systems research because of exchanges of water and solutes that occur across the land surface interface. The study of hillslope transit time distributions (e.g., Fiori and Russo, 2008; Botter et al., 2010; Heidbüchel et al., 2013; Tetzlaff et al., 2014) is a good example of the need for a better understanding of such water and solute exchanges and the consequent subsurface flowpaths. The simulation of unsaturated zone mass transport phenomena is however known to be a particularly complex problem, compounded by any presence of heterogeneity. Wilson and Gelhar (1981), for instance, showed that spatial variations in moisture content affect solute plume spreading even without dispersive mixing, and that the rates of solute displacement are typically much smaller than the rates of moisture displacement. Birkholzer and Tsang (1997) demonstrated significant channeling effects (preferential solute pathways, with accompanying higher dispersion) at the extremes of very low saturation and full saturation. Studies that have combined comprehensive experimental observation with detailed subsurface simulation have also documented

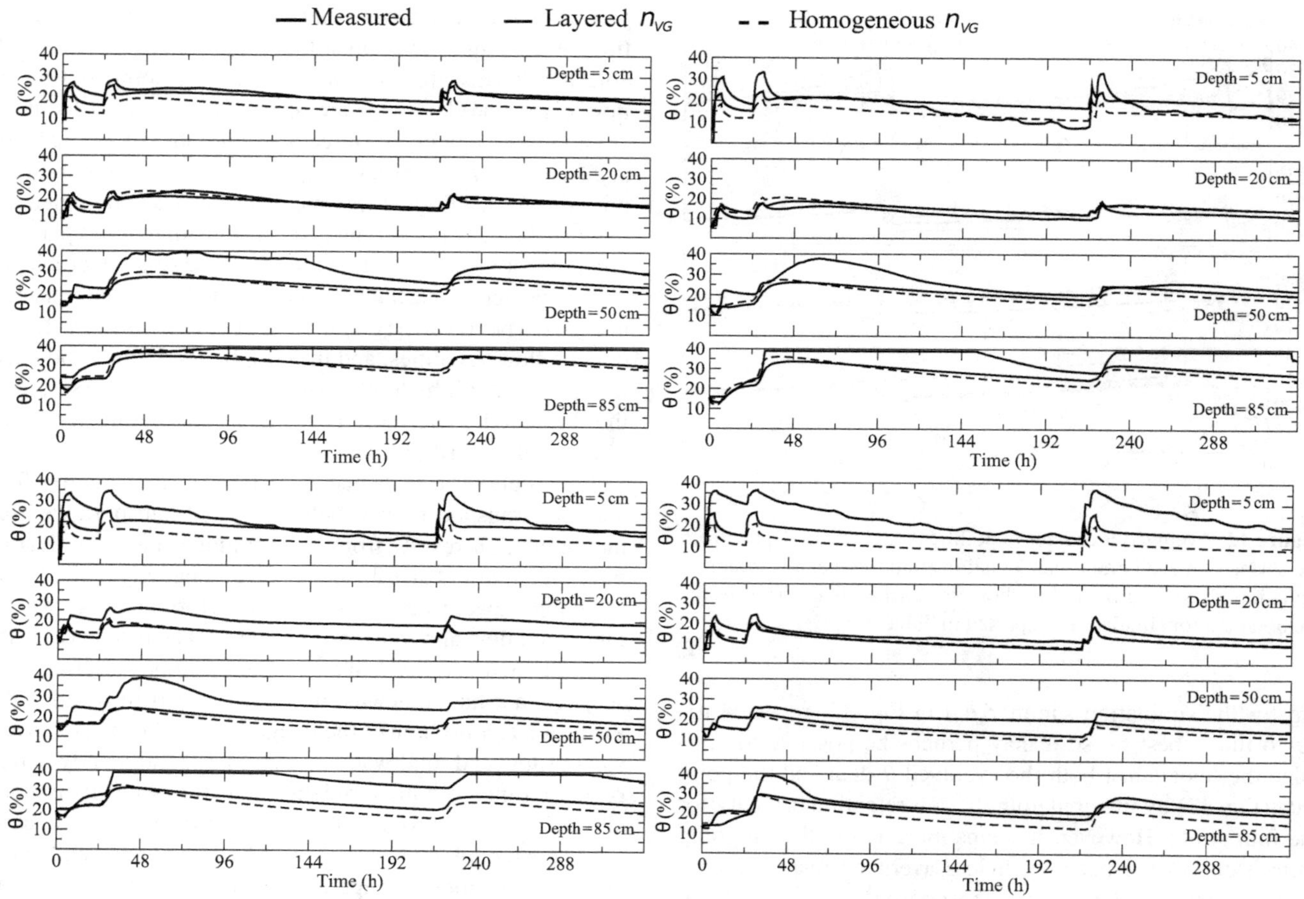

Figure 13. Distributed (internal state) hydrological response for the θ profiles at 5, 20, 50, and 85 cm depths from the surface for four locations on the LEO-1 hillslope: point a (top left), point b (top right), point c (bottom left), and point d (bottom right) of Fig. 2. The results are obtained for simulations f and l reported in Table 3.

some of the challenges faced in modeling solute transport under unsaturated and heterogeneous conditions (e.g., Haggerty et al., 2004; Zheng et al., 2011). In this context, for the tritium and bromide tracer experiments at the Las Cruces trench site, standard models gave good prediction of wetting front movement during infiltration but poor prediction of point soil water content and tracer transport (Hills et al., 1991; Wierenga et al., 1991). For the macrodispersion (MADE) experiment, Russo and Fiori (2009) found that heterogeneity further enhances solute spreading and breakthrough curve arrival times when the unsaturated zone is relatively dry or deep. In the present study, the additional complexity introduced for the point-scale responses (namely spatially variable soil retention capacity) did not match as favorably the integrated (flow) observation data set (Fig. 15). While this could perhaps be remedied using more rigorous or quantitative parameter estimation, the particular difficulties in capturing the point-scale concentration profiles, especially near the landscape surface, can be taken as further evidence of flaws or gaps in theoretical understanding and model formulation (process representation) for simulating solute transport phenomena in very dry, heterogeneous soils.

Various hypotheses have been invoked to explain possible factors that affect the migration and distribution of solutes under unsaturated, heterogeneous conditions, including turbulent mixing due to high rainfall (Havis et al., 1992); solute transfer between mobile and immobile water (De Smedt and Wierenga, 1984); mobile–immobile exchange and hysteresis (Butters et al., 1989; Russo et al., 1989a, b, 2014); lateral mixing due to velocity fluctuations (Russo et al., 1998); isotope effects (Barnes and Allison, 1988; LaBolle et al., 2008; Zhang et al., 2009); variable, state-dependent anisotropy (McCord et al., 1991); non-Gaussian early-time mean tracer plume behavior (Naff, 1990); non-Fickian solute migration at low water contents (Padilla et al., 1999) and for macroscopically homogeneous sand (Bromly and Hinz, 2004); and saturation-dependent dispersivity (Raoof and Hassanizadeh, 2013). In addition, Konikow et al. (1997) and Parker and van Genuchten (1984) discuss the importance of boundary condition treatment (e.g., water-solute injection,

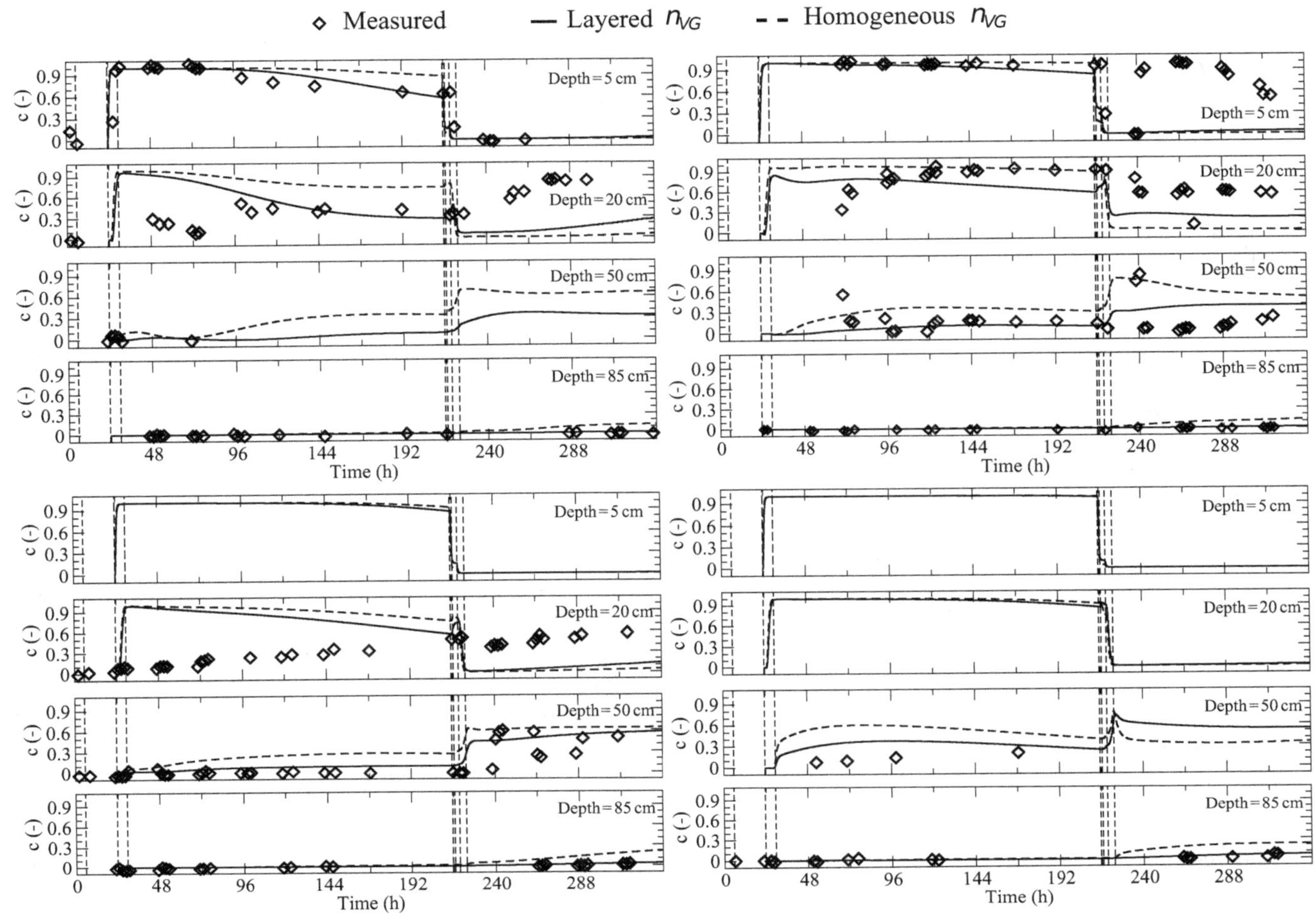

Figure 14. Distributed (internal state) hydrological response for the tracer concentration breakthrough curves at 5, 20, 50, and 85 cm depths from the surface for four locations on the LEO-1 hillslope: point a (top left), point b (top right), point c (bottom left), and point d (bottom right) of Fig. 2. There were no tracer concentration measurements at 5 cm depth for point c and at 5 and 20 cm depths for point d. The transport model is run for $\alpha_l = 0.001$ m and $\alpha_t = 0.0001$ m. The vertical dashed lines indicate the timing of the three pulses of rain (red when the water is ^2H-enriched and blue when it is not).

solute exchange between soil and atmosphere). Given the many open questions, for this first analysis of the LEO isotope tracer experiment the modeling was kept to the standard formulation of the Richards and advection–dispersion equations. Limitations encountered in the multiresponse performance assessment, from the standpoint of experimental procedure, model formulation, or numerical implementation, will inform follow-up studies at LEO. The simulation results from this tracer experiment, for instance, point to highly complex effects on plume migration of spatially variable water content in the dry soils that characterized the experiment, especially at early times.

The broad results of our study should be quite universal, particularly to deterministic numerical models based on the 3-D Richards and advection–dispersion equations. However, any model has its specific features and differs, for example, in the way equations are coded (e.g., choice of numerical solvers) or interface conditions are implemented (e.g., free-surface vs. boundary condition switching). For insights into the impact of specific model differences in the performance of CATHY-like models, see the intercomparison studies of Sulis et al. (2010) and Maxwell et al. (2014). These intercomparison studies have thus far focused only on flow processes, and there is an urgent need to extend the analyses to solute transport phenomena, in order to properly guide our assessment of the physical and numerical correctness of competing models as these models continue to increase in complexity. For instance, for this study there are aspects of the CATHY model related to how we implemented evaporation and fractionation that might be expected to negatively impact the generality of our findings, although in terms of isotope tracer mass exiting the seepage face the impact was quite small. But the implementation here was somewhat ad hoc, and more study is needed on the importance and proper

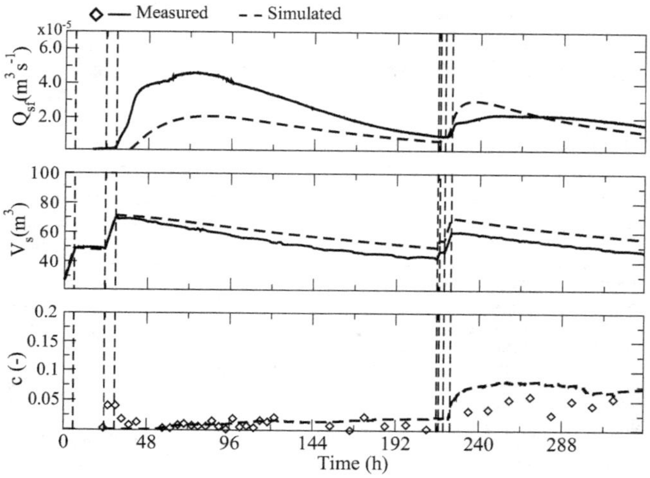

Figure 15. Performance of the model against integrated flow and transport responses (seepage face flow Q_{sf}, total water storage V_{s}, and average tracer concentration c at the seepage face) using the additional parameterization from the distributed analyses (spatially variable soil retention capacity, simulation 1 of Table 3). The vertical dashed lines indicate the timing of the three pulses of rain (red when the water is ^2H-enriched and blue when it is not).

representation of fractionation in solute transport models, especially under strongly unsaturated conditions.

6 Conclusions

In this study we have used multivariate observations (soil moisture, water and tracer outflow, breakthrough curves, and total water storage) culled from the first isotope tracer experiment at the LEO-1 hillslope of the Biosphere 2 facility to explore some of the challenges in modeling unsaturated flow and transport phenomena. Integrated (first flow and then transport) and distributed (again flow followed by transport) measurements were progressively introduced as response variables with which to assess the results from simulations with CATHY, a 3-D numerical model for variably saturated flow and advective–dispersive solute migration. Compared to the first flow experiment at LEO that was successfully modeled with CATHY (Niu et al., 2014), the modeling task for the tracer experiment was significantly more complicated due to joint simulation of both flow and transport processes; considerably drier initial conditions and reduced forcing; performance assessment against both system-wide and point-scale measurements; and multiple periods of water/tracer injection compared to a single rainfall episode. In some sense the previous flow study looked at the first-order response of the LEO hillslope, whereas the modeling exercise for the tracer experiment represents a first look at higher-order responses of the Biosphere 2 landscapes.

There are several findings from this first set of simulations of a LEO isotope tracer experiment. At the start of the exercise, where integrated flow measurements were used, we were able to obtain good matches for two response variables (total water storage and seepage face outflow) using parameter values and initial and boundary conditions that correspond quite closely to the actual experimental conditions and previous (flow experiment) model implementation (Niu et al., 2014). The same soil parameterization was successfully used to reproduce the integrated transport response. When passing to point-scale flow and finally point-scale transport, a refinement of the model setup (augmenting the degree of heterogeneity, mainly) was needed. Moreover, providing more information to the model (for example, the distribution of initial water storage rather than just the initial total volume) generally helped to improve the simulation results.

The effect of saturated hydraulic conductivity (heterogeneity and anisotropy) on the response of subsurface hydrologic models is well known, and was also borne out in this study. Also not surprisingly, the dispersivity parameter had a big impact on the transport simulations, with a clear trend to a better match against measured seepage face concentration as dispersivity was decreased. The spatial distribution of rainfall was not found to have a big impact on simulation results, and there was not much difference, in terms of isotope tracer mass exiting the seepage face, between the zero, partial, and no fractionation cases, suggesting that the injected tracer did not percolate very far into the hillslope, likely due to the very dry initial conditions.

We conclude with a few specific recommendations for alleviating some of the modeling and experimental limitations encountered during this study. On the modeling side, a more sophisticated treatment of solute transport phenomena beyond the standard advection–dispersion equation could start with incorporation of a mobile–immobile conceptualization and/or saturation-dependent dispersivity. Other upgrades to the CATHY model (e.g., Scudeler et al., 2016) will mitigate the grid Peclet constraint and provide more reliable flow velocity calculations, essential to maintaining low mass balance errors and high accuracy in solute transport. On the experimental side, higher tracer concentrations (including labeled tracers), wetter initial conditions, and more intensive direct or indirect measures of total tracer mass could help address the high sensitivity of solute transport to small-scale heterogeneity under dry soil conditions. Any experiments that provide spatially detailed observations of both flow and transport response variables that are then jointly used in estimating, for instance, conductivity and other soil hydraulic parameters (traditionally identified based solely on flow responses), would be critical to advancing the present state of hydrologic model verification, given the high impact that Darcy velocities, which are directly dependent on such parameters, have on solute mixing processes. Finally, future LEO isotope tracer experiments that also generate some surface runoff would offer valuable benchmark data for improving integrated surface–subsurface models.

7 Data availability

Access to the data for this study is available by contacting the corresponding author.

Acknowledgements. This work was supported in part by the National Science Foundation (NSF grants EAR-1344552 and EAR-1417097). The authors wish to thank Nate Abramson, Anouk Gevaert, Ed Hunt, Michael Pohlmann, Michael Sibayan, and Yvonne Smit for technical support. The Biosphere 2 facility and the capital required to conceive and construct LEO were provided through a charitable donation from the Philecology Foundation and its founder, Edward Bass. We gratefully acknowledge this charitable donation.

Edited by: P. Saco

References

Anderman, E. R. and Hill, M. C.: A new multistage groundwater transport inverse method: presentation, evaluation, and implications, Water Resour. Res., 35, 1053–1063, doi:10.1029/1998WR900114, 1999.

Barnes, C. J. and Allison, G. B.: Tracing of water movement in the unsaturated zone using stable isotopes of hydrogen and oxygen, J. Hydrol., 100, 143–176, doi:10.1016/0022-1694(88)90184-9, 1988.

Binley, A., Elgy, J., and Beven, K.: A physically based model of heterogeneous hillslopes: 1. Runoff production, Water Resour. Res., 25, 1219–1226, doi:10.1029/WR025i006p01219, 1989.

Birkholzer, J. and Tsang, C.-F.: Solute channeling in unsaturated heterogeneous porous media, Water Resour. Res., 33, 2221–2238, doi:10.1029/97WR01209, 1997.

Botter, G., Bertuzzo, E., and Rinaldo, A.: Transport in the hydrologic response: Travel time distributions, soil moisture dynamics, and the old water paradox, Water Resour. Res., 46, W03514, doi:10.1029/2009WR008371, 2010.

Braud, I., Biron, P., Bariac, T., Richard, P., Canale, L., Gaudet, J. P., and Vauclin, M.: Isotopic composition of bare soil evaporated water vapor. Part I: RUBIC IV experimental setup and results, J. Hydrol., 369, 1–16, doi:10.1016/j.jhydrol.2009.01.034, 2009.

Bromly, M. and Hinz, C.: Non-Fickian transport in homogeneous unsaturated repacked sand, Water Resour. Res., 40, W07402, doi:10.1029/2003WR002579, 2004.

Brooks, P. D., Chorover, J., Fan, Y., Godsey, S. E., Maxwell, R. M., McNamara, J. P., and Tague, C.: Hydrological partitioning in the critical zone: Recent advances and opportunities for developing transferable understanding of water cycle dynamics, Water Resour. Res., 51, 6973–6987, doi:10.1002/2015WR017039, 2015.

Brunner, P., Doherty, J., and Simmons, C. T.: Uncertainty assessment and implications for data acquisition in support of integrated hydrologic models, Water Resour. Res., 48, W07513, doi:10.1029/2011WR011342, 2012.

Butters, G. L., Jury, W. A., and Ernst, F. F.: Field scale transport of bromide in an unsaturated soil: 1. Experimental methodology and results, Water Resour. Res., 25, 1575–1581, doi:10.1029/WR025i007p01575, 1989.

Camporese, M., Paniconi, C., Putti, M., and Orlandini, S.: Surface-subsurface flow modeling with path-based runoff routing, boundary condition-based coupling, and assimilation of multisource observation data, Water Resour. Res., 46, W02512, doi:10.1029/2008WR007536, 2010.

Clark, I. D. and Fritz, P.: Environmental Isotopes in Hydrogeology, CRC Press, Boca Raton, Florida, 1997.

Craig, H. and Gordon, L. I.: Deuterium and oxygen 18 variations in the ocean and the marine atmosphere, in: Stable Isotopes in Oceanographic Studies and Paleotemperatures, edited by: Tongiorgi, E., Laboratorio di geologia nucleare, Pisa, 9–130, 1965.

Dawson, C. W., Abrahart, R. J., and See, L. M.: HydroTest: a web-based toolbox of evaluation metrics for the standardised assessment of hydrological forecasts, Environ. Model. Softw., 22, 1034–1052, doi:10.1016/j.envsoft.2006.06.008, 2007.

De Smedt, F. and Wierenga, P. J.: Solute Transfer through columns of glass beads, Water Resour. Res., 20, 225–232, doi:10.1029/WR020i002p00225, 1984.

Fenicia, F., Savenije, H. H. G., Matgen, P., and Pfister, L.: A comparison of alternative multiobjective calibration strategies for hydrological modeling, Water Resour. Res., 43, W03434, doi:10.1029/2006WR005098, 2007.

Fiori, A. and Russo, D.: Travel time distribution in a hillslope: Insight from numerical simulations, Water Resour. Res., 44, W12426, doi:10.1029/2008WR007135, 2008.

Gevaert, A. I., Teuling, A. J., Uijlenhoet, R., DeLong, S. B., Huxman, T. E., Pangle, L. A., Breshears, D. D., Chorover, J., Pelletier, J. D., Saleska, S. R., Zeng, X., and Troch, P. A.: Hillslope-scale experiment demonstrates the role of convergence during two-step saturation, Hydrol. Earth Syst. Sci., 18, 3681–3692, doi:10.5194/hess-18-3681-2014, 2014.

Ghanbarian-Alavijeh, B., Skinner, T. E., and Hunt, A. G.: Saturation dependence of dispersion in porous media, Phys. Rev. E, 86, doi:10.1103/PhysRevE.86.066316, 066316, 2012.

Gorelick, S. M. and Zheng, C.: Global change and the groundwater management challenge, Water Resour. Res., 51, 3031–3051, doi:10.1002/2014WR016825, 2015.

Gupta, H. V., Sorooshian, S., and Yapo, O. P.: Toward improved calibration of hydrologic models: Multiple and noncommensurable measures of information, Water Resour. Res., 34, 751–763, doi:10.1029/97WR03495, 1998.

Haggerty, R., Harvey, C. F., Freiherr von Schwerin, C., and Meigs, L. C.: What controls the apparent timescale of solute mass transfer in aquifers and soils? A comparison of experimental results, Water Resour. Res., 40, W01510, doi:10.1029/2002WR001716, 2004.

Haverd, V. and Cuntz, M.: Soil-Litter-Iso: A one-dimensional model for coupled transport of heat, water and stable isotopes in soil with a litter layer and root extraction, J. Hydrol., 388, 438–455, doi:10.1016/j.jhydrol.2010.05.029, 2010.

Havis, R. N., Smith, R. E., and Adrian, D. D.: Partitioning solute transport between infiltration and overland flow under rainfall, Water Resour. Res., 28, 2569–2580, doi:10.1029/92WR01366, 1992.

Heidbüchel, I., Troch, P. A., and Lyon, S. W.: Separating physical and meteorological controls of variable transit times in zero-order catchments, Water Resour. Res., 49, 7644–7657, doi:10.1002/2012WR013149, 2013.

Hills, R. G., Wierenga, P. J., Hudson, D. B., and Kirkland, M. R.: The second Las Cruces trench experiment: Experimental results and two-dimensional flow predictions, Water Resour. Res., 27, 2707–2718, doi:10.1029/91WR01538, 1991.

Hofer, M., Lehmann, P., Stähli, M., Seifert, S., and Krafczyk, M.: Two approaches to modeling the initiation and development of rills in a man-made catchment, Water Resour. Res., 48, W01531, doi:10.1029/2011WR010719, 2012.

Hopp, L., Harman, C., Desilets, S. L. E., Graham, C. B., McDonnell, J. J., and Troch, P. A.: Hillslope hydrology under glass: confronting fundamental questions of soil-water-biota coevolution at Biosphere 2, Hydrol. Earth Syst. Sci., 13, 2105–2118, doi:10.5194/hess-13-2105-2009, 2009.

Horita, J., Rozanski, K., and Cohen, S.: Isotope effects in the evaporation of water: a status report of the Craig-Gordon model, Isot. Environ. Healt. S., 44, 23–49, doi:10.1080/10256010801887174, 2008.

Huxman, T., Troch, P. A., Chorover, J., Breshears, D. D., Saleska, S., Pelletier, J., Zeng, X., and Espeleta, J.: The hills are alive: interdisciplinary Earth science at Biosphere 2, EOS T. Am. Geophys. Un., 90, 120 pp., 2009.

Kampf, S. K. and Burges, S. J.: Parameter estimation for a physics-based distributed hydrologic model using measured outflow fluxes and internal moisture states, Water Resour. Res., 43, W12414, doi:10.1029/2006WR005605, 2007.

Keating, E. H., Doherty, J., Vrugt, J. A., and Kang, Q.: Optimization and uncertainty assessment of strongly nonlinear groundwater models with high parameter dimensionality, Water Resour. Res., 46, W10517, doi:10.1029/2009WR008584, 2010.

Konikow, L. F., Sanford, W. E., and Campbell, P. J.: Constant-concentration boundary condition: lessons from the HYDROCOIN variable-density groundwater benchmark problem, Water Resour. Res., 33, 2253–2261, doi:10.1029/97WR01926, 1997.

Kumar, R., Samaniego, L., and Attinger, S.: Implications of distributed hydrologic model parameterization on water fluxes at multiple scales and locations, Water Resour. Res., 49, 360–379, doi:10.1029/2012WR012195, 2013.

LaBolle, E. M., Fogg, G. E., Eweis, J. B., Gravner, J., and Leaist, D. G.: Isotopic fractionation by diffusion in groundwater, Water Resour. Res., 44, W07405, doi:10.1029/2006WR005264, 2008.

Maxwell, R. M., Putti, M., Meyerhoff, S., Delfs, J.-O., Ferguson, I. M., Ivanov, V., Kim, J., Kolditz, O., Kollet, S. J., Kumar, M., Lopez, S., Niu, J., Paniconi, C., Park, Y.-J., Phanikumar, M. S., Shen, C., Sudicky, E. A., and Sulis, M.: Surface-subsurface model intercomparison: A first set of benchmark results to diagnose integrated hydrology and feedbacks, Water Resour. Res., 50, 1531–1549, doi:10.1002/2013WR013725, 2014.

McCord, J. T., Stephens, D. B., and Wilson, J. L.: Hysteresis and state-dependent anisotropy in modeling unsaturated hillslope hydrologic processes, Water Resour. Res., 27, 1501–1518, doi:10.1029/91WR00880, 1991.

Mishra, S. and Parker, J. C.: Parameter estimation for coupled unsaturated flow and transport, Water Resour. Res., 25, 385–396, doi:10.1029/WR025i003p00385, 1989.

Naff, R. L.: On the nature of the dispersive flux in saturated heterogeneous porous media, Water Resour. Res., 26, 1013–1026, doi:10.1029/WR026i005p01013, 1990.

Neuweiler, I. and Cirpka, O. A.: Homogenization of Richards equation in permeability fields with different connectivities, Water Resour. Res., 41, W02009, doi:10.1029/2004WR003329, 2005.

Niu, G.-Y., Pasetto, D., Scudeler, C., Paniconi, C., Putti, M., Troch, P. A., DeLong, S. B., Dontsova, K., Pangle, L., Breshears, D. D., Chorover, J., Huxman, T. E., Pelletier, J., Saleska, S. R., and Zeng, X.: Incipient subsurface heterogeneity and its effect on overland flow generation – insight from a modeling study of the first experiment at the Biosphere 2 Landscape Evolution Observatory, Hydrol. Earth Syst. Sci., 18, 1873–1883, doi:10.5194/hess-18-1873-2014, 2014.

Padilla, I. Y., Yeh, T.-C. J., and Conklin, M. H.: The effect of water content on solute transport in unsaturated porous media, Water Resour. Res., 35, 3303–3313, doi:10.1029/1999WR900171, 1999.

Pangle, L. A., DeLong, S. B., Abramson, N., Adams, J., Barron-Gafford, G., Breshears, D. D., Brooks, P., Chorover, J., Dietrich, W. E., Dontsova, K., Durcik, M., Espeleta, J., Ferre, T. P. A., Ferriere, R., Henderson, W., Hunt, E., Huxman, T. E., Millar, D., Murphy, B., Niu, G.-Y., Pavao-Zuckerman, M., Pelletier, J. D., Rasmussen, C., Ruiz, J., Saleska, S., Schaap, M., Sibayan, M., Troch, P., Tuller, M., van Haren, J., and Zeng, X.: Landscape Evolution Observatory: A large-scale controllable infrastructure to study coupled Earth-surface processes, Geomorphology, 244, 190–203, doi:10.1016/j.geomorph.2015.01.020, 2015.

Paniconi, C. and Putti, M.: Physically based modeling in catchment hydrology at 50: Survay and outlook, Water Resour. Res., 51, 7090–7129, doi:10.1002/2015WR017780, 2015.

Parker, J. C. and van Genuchten, M. T.: Flux-averaged and volume-averaged concentrations in continuum approaches to solute transport, Water Resour. Res., 20, 866–872, doi:10.1029/WR020i007p00866, 1984.

Pasetto, D., Niu, G.-Y., Pangle, L., Paniconi, C., Putti, M., and Troch, P. A.: Impact of sensor failure on the observability of flow dynamics at the Biosphere 2 LEO hillslopes, Adv. Water Resour., 86, 327–339, doi:10.1016/j.advwatres.2015.04.014, 2015.

Putti, M. and Paniconi, C.: Picard and Newton linearization for the coupled model of saltwater intrusion in aquifers, Adv. Water Resour., 18, 159–170, doi:10.1016/0309-1708(95)00006-5, 1995.

Raoof, A. and Hassanizadeh, S. M.: Saturation-dependent solute dispersivity in porous media: Pore-scale processes, Water Resour. Res., 49, 1943–1951, doi:10.1002/wrcr.20152, 2013.

Russo, D. and Fiori, A.: Stochastic analysis of transport in a combined heterogeneous vadose zone-groundwater flow system, Water Resour. Res., 45, W03426, doi:10.1029/2008WR007157, 2009.

Russo, D., Jury, W. A., and Butters, G. L.: Numerical analysis of solute transport during transient irrigation: 1. The effect of hysteresis and profile heterogeneity, Water Resour. Res., 25, 2109–2118, doi:10.1029/WR025i010p02109, 1989a.

Russo, D., Jury, W. A., and Butters, G. L.: Numerical analysis of solute transport during transient irrigation: 2. The effect of immobile water, Water Resour. Res., 25, 2119–2127, doi:10.1029/WR025i010p02119, 1989b.

Russo, D., Zaidel, J., and Laufer, A.: Numerical analysis of flow and transport in a three-dimensional partially saturated heterogeneous soil, Water Resour. Res., 34, 1451–1468, doi:10.1029/98WR00435, 1998.

Russo, D., Laufer, A., Gerstl, Z., Ronen, D., Weisbrod, N., and Zentner, E.: On the mechanism of field-scale solute transport: Insights from numerical simulations and field observations, Water Resour. Res., 50, 7484–7504, doi:10.1002/2014WR015514, 2014.

Scudeler, C., Putti, M., and Paniconi, C.: Mass-conservative reconstruction of Galerkin velocity fields for transport simulations, Adv. Water Resour., 94, 470–485, doi:10.1016/j.advwatres.2016.06.011, 2016.

Sebben, M. L., Werner, A. D., Liggett, J. E., Partington, D., and Simmons, C. T.: On the testing of fully integrated surface-subsurface hydrological models, Hydrol. Process., 27, 1276–1285, doi:10.1002/hyp.9630, 2013.

Sprenger, M., Volkmann, T. H. M., Blume, T., and Weiler, M.: Estimating flow and transport parameters in the unsaturated zone with pore water stable isotopes, Hydrol. Earth Syst. Sci., 19, 2617–2635, doi:10.5194/hess-19-2617-2015, 2015.

Sulis, M., Meyerhoff, S. B., Paniconi, C., Maxwell, R. M., Putti, M., and Kollet, S. J.: A comparison of two physics-based numerical models for simulating surface water–groundwater interactions, Adv. Water Resour., 33, 456–467, doi:10.1016/j.advwatres.2010.01.010, 2010.

Tetzlaff, D., Birkel, C., Dick, J., Geris, J., and Soulsby, C.: Storage dynamics in hydropedological units control hillslope connectivity, runoff generation, and the evolution of catchment transit time distributions, Water Resour. Res., 50, 969–985, doi:10.1002/2013WR014147, 2014.

van Genuchten, M. T.: A closed-form equation for predicting the hydraulic conductivity of unsaturated soils, Soil Sci. Soc. Am. J., 44, 892–898, doi:10.2136/sssaj1980.03615995004400050002x, 1980.

Weill, S., Mazzia, A., Putti, M., and Paniconi, C.: Coupling water flow and solute transport into a physically-based surface-subsurface hydrological model, Adv. Water Resour., 34, 128–136, doi:10.1016/j.advwatres.2010.10.001, 2011.

Wierenga, P. J., Hills, R. G., and Hudson, D. B.: The Las Cruces trench site: Characterization, experimental results, and one-dimensional flow predictions, Water Resour. Res., 27, 2695–2705, doi:10.1029/91WR01537, 1991.

Wilson, J. L. and Gelhar, L. W.: Analysis of longitudinal dispersion in unsaturated flow: 1. The analytical method, Water Resour. Res., 17, 122–130, doi:10.1029/WR017i001p00122, 1981.

Woolhiser, D. A., Smith, R. E., and Giraldez, J.-V.: Effects of spatial variability of saturated hydraulic conductivity on Hortonian overland flow, Water Resour. Res., 32, 671–678, doi:10.1029/95WR03108, 1996.

Zacharias, S., Bogena, H., Samaniego, L., Mauder, M., Fuß, R., Pütz, T., Frenzel, M., Schwank, M., Baessler, C., Butterbach-Bahl, K., Bens, O., Borg, E., Brauer, A., Dietrich, P., Hajnsek, I., Helle, G., Kiese, R., Kunstmann, H., Klotz, S., Munch, J. C., Papen, H., Priesack, E., Schmid, H. P., Steinbrecher, R., Rosenbaum, U., Teutsch, G., and Vereecken, H.: A network of terrestrial environmental observatories in Germany, Vadose Zone J., 10, 955–973, doi:10.2136/vzj2010.0139, 2011.

Zhang, X.-P., Yang, Z.-L., Niu, G.-Y., and Wang, X.-Y.: Stable water isotope simulation in different reservoirs of Manaus, Brazil, by Community Land Model incorporating stable isotopic effect, Int. J. Climatol., 29, 619–628, doi:10.1002/joc.1740, 2009.

Zheng, C., Bianchi, M., and Gorelick, S. M.: Lessons learned from 25 years of research at the MADE site, Ground Water, 49, 649–662, doi:10.1111/j.1745-6584.2010.00753.x, 2011.

7

Quantification of the influence of preferential flow on slope stability using a numerical modelling approach

W. Shao[1], T. A. Bogaard[1], M. Bakker[1], and R. Greco[2]

[1]Water Resources Section, Faculty of Civil Engineering and Geosciences, Delft University of Technology, Stevinweg 1, P.O. Box 5048, 2600 GA Delft, the Netherlands

[2]Dipartimento di Ingegneria Civile Design Edilizia e Ambiente, Seconda università di Napoli, via Roma 29, 81031 Aversa (CE), Italy

Correspondence to: W. Shao (w.shao@tudelft.nl)

Abstract. The effect of preferential flow on the stability of landslides is studied through numerical simulation of two types of rainfall events on a hypothetical hillslope. A model is developed that consists of two parts. The first part is a model for combined saturated/unsaturated subsurface flow and is used to compute the spatial and temporal water pressure response to rainfall. Preferential flow is simulated with a dual-permeability continuum model consisting of a matrix domain coupled to a preferential flow domain. The second part is a soil mechanics model and is used to compute the spatial and temporal distribution of the local factor of safety based on the water pressure distribution computed with the subsurface flow model. Two types of rainfall events were considered: long-duration, low-intensity rainfall, and short-duration, high-intensity rainfall. The effect of preferential flow on slope stability is assessed through comparison of the failure area when subsurface flow is simulated with the dual-permeability model as compared to a single-permeability model (no preferential flow). For the low-intensity rainfall case, preferential flow has a positive effect on drainage of the hillslope resulting in a smaller failure area. For the high-intensity rainfall case, preferential flow has a negative effect on the slope stability as the majority of rainfall infiltrates into the preferential flow domain when rainfall intensity exceeds the infiltration capacity of the matrix domain, resulting in larger water pressure and a larger failure area.

1 Introduction

Landslides are commonly triggered by rainfall events. Hydrological models can be integrated with slope stability analysis methods to calculate the factor of safety and predict the time and magnitude of landslides (Crosta and Frattini, 2008; Shuin et al., 2012; Aleotti and Chowdhury, 1999; Westen et al., 2006). Combined hydro-mechanical models can roughly be divided into two types: simplified conceptual models (Montrasio and Valentino, 2008; Dai et al., 2002) and numerical models (Stead et al., 2001; Jing, 2003; Brinkgreve et al., 2010; Pastor et al., 2008), and have different levels of complexity depending on the scale and the research purpose.

The limit equilibrium method or infinite slope stability approach is frequently integrated with Richards' equation (Lanni et al., 2013; Ng and Shi, 1998; Godt et al., 2008; Shuin et al., 2012; Wilkinson et al., 2002; Talebi et al., 2008; Greco et al., 2013) or conceptual models (Arnone et al., 2011; Simoni et al., 2008; Qiu et al., 2007) for landslide hazard evaluation. The limit equilibrium method and infinite slope approach assume or search for a potential failure surface. The factor of safety is defined as the ratio between the maximum retaining force and the driving force (e.g. Lanni et al., 2013; Lu et al., 2012). Although the underlying assumptions of the slope failure mechanism have limitations (Huang and Jia, 2009; Griffiths et al., 2011), the simplified slope stability analysis method has low computational demand and is widely used for geotechnical analyses at the slope scale (Talebi et al., 2008; Tsai and Yang, 2006; Abramson, 2002), watershed scale, and catchment scale (Borga et al., 2002a, b; Baum et al., 2010; Wilkinson et al., 2002).

The strength reduction method (Griffiths and Lu, 2005; Huang and Jia, 2009) or local factor of safety method (Lu et al., 2012) can result in similar factor of safety values and locations of the critical slip surface as the limit equilibrium method, while no assumption is needed for the critical failure surface (Griffiths and Lu, 2005; Hammouri et al., 2008; Kim et al., 1999). The location, shape, and magnitude of the plastic deformation area are used to quantify the slip surface and factor of safety (Griffiths and Lane, 1999). Geotechnical engineering software and numerical models – such as FLAC® (Itasca, 2002) or PLAXIS (Brinkgreve et al., 2010, based on the strength reduction method) – have been widely applied for slope stability analysis under the influence of transient hydrological conditions, such as rainstorms (Mukhlisin et al., 2008; Hamdhan and Schweiger, 2011) and reservoir water level variations (Huang and Jia, 2009; Zhou et al., 2014).

The Darcy–Richards equation is the most widely used approach in current software packages, but cannot effectively simulate preferential flow resulting in rapid infiltration (Nieber and Sidle, 2010; Beven and Germann, 2013). In highly heterogeneous slopes, preferential flow and transport can fundamentally influence subsurface flow (Jarvis, 2007; Hendrickx and Flury, 2001) and contaminant transport (Köhne et al., 2009; Allaire et al., 2009; Debieche et al., 2012; Zehe et al., 2001). A chain of connected macropores is commonly found in various types of soils, including forest soil and semiarid land (Uchida et al., 2001; Jarvis, 2007; Flury et al., 1994). For example, an earthworm burrow can extend from the surface deep into the soil, as can decayed plant roots or soil cracks (Jarvis, 2007; Beven and Germann, 1982; Hendrickx and Flury, 2001). The self-organising preferential flow network will become active and hydraulically connected with an increase in soil saturation (Nieber and Sidle, 2010). The saturated hydraulic conductivity of preferential flow paths is significantly larger than that of the soil matrix (Beven and Germann, 1982; Köhne et al., 2009). A significant portion of subsurface stormflow (Uchida et al., 2004; Zhang et al., 2006; Beven, 1981) is transmitted via preferential flow paths (Nieber and Sidle, 2010). Preferential flow through macropores, fractures, and other local high-permeability zones is extremely rapid, and contributes instantly to high porewater pressures in deep soils (Jarvis, 2007).

Quantification of landslide triggering mechanisms is an essential step in landslide forecasting. Some studies have shown that preferential flow is one of the major mechanisms affecting the timing and location of landslides (Sharma and Nakagawa, 2010; Uchida, 2004; Verachtert et al., 2013). In hillslopes, preferential flow paths, such as macropores, soil pipes, and fissures, have been associated with slope stability (Hencher, 2010; McDonnell, 1990; Uchida et al., 2001; Krzeminska et al., 2012; Debieche et al., 2012). Besides the fact that internal erosion in preferential flow paths deteriorates the slope mass and reduces the soil shear strength, the occurrence of preferential flow can give rapid access to the deeper soil and groundwater system, reduce soil shear strength (due to pore pressure changes), and influence the timing and frequency of landslides (Wienhöfer et al., 2011; Uchida, 2004; Verachtert et al., 2013).

Preferential flow and solute transport have been simulated at various scales including the scales of pores, soil columns, hillslopes, and catchments (Šimůnek et al., 2003; Gerke, 2006; Köhne et al., 2009) using increasingly sophisticated models such as the dual-porosity/dual-permeability model (Gerke and van Genuchten, 1993a; Jarvis et al., 1991; Larsbo and Jarvis, 2003), the multi-permeability model (Wu et al., 2004; Greco, 2002; Gwo et al., 1995), and conceptual models (Armstrong et al., 2000; Weiler, 2005; Mulungu et al., 2005). The dual-permeability model is widely used because of its clear physical concept and powerful simulating ability (Roulier and Jarvis, 2003; Kodešová et al., 2005; Gerke and Köhne, 2004; Köhne et al., 2006; Christiansen et al., 2004; Weiler, 2005; Therrien and Sudicky, 2005; Vogel et al., 2000). The dual-permeability model assumes that the soil consists of two interacting, overlapping pore domains. The matrix domain with relatively low permeability represents the soil micropores where flow is governed by Richards' equation. The preferential flow domain represents the highly permeable preferential flow paths, such as macropores, fractures, cracks, or large pores between soil aggregate. Preferential flow is described by Richards' equation (Šimůnek et al., 2008; Gerke and van Genuchten, 1993a) or the gravity-driven kinematic wave equation (Larsbo and Jarvis, 2003; Jarvis et al., 1991; Greco, 2002). The water exchange between the two domains is driven by the pressure head difference between the two domains (Pirastru and Niedda, 2010; Gerke and van Genuchten, 1993b). Dual-permeability models have proven to be effective for preferential flow simulation, but have not been incorporated into slope stability models.

The objective of this study is to quantify the temporal and the spatial effect of preferential flow on slope stability, and to analyse its underlying hydrological mechanisms using numerical experiments of rainfall-induced shallow landslides. This paper is organised as follows. The subsurface dual-permeability hydrological model is described. The subsurface hydrological model is sequentially coupled with a soil mechanics model and a stress-field-based local factor of safety slope stability method (Sect. 2.2). The numerical experiments and parameterisation are discussed in Sect. 3. The hydrological and geotechnical results are given in Sect. 4. The influence of preferential flow on subsurface hydrological processes and consequent slope stability is discussed in Sect. 5 by comparing the results of single- and dual-permeability models.

2 Methods

2.1 Subsurface flow model

The single-permeability model is described by one Richards' equation to represent flow in a homogenous soil. The dual-permeability model divides the flow domain into two overlapping and interacting continua, where two coupled Richards' equations are used to describe the matrix flow and preferential flow (Gerke and van Genuchten, 1993a):

$$[C_\mathrm{f}+\Theta_\mathrm{f}S_\mathrm{s}]\frac{\partial h_\mathrm{f}}{\partial t}=\nabla[K_\mathrm{f}(\nabla h_\mathrm{f}+\nabla z)]-\frac{\Gamma_\mathrm{w}}{w_\mathrm{f}},\tag{1}$$

$$[C_\mathrm{m}+\Theta_\mathrm{m}S_\mathrm{s}]\frac{\partial h_\mathrm{m}}{\partial t}=\nabla[K_\mathrm{m}(\nabla h_\mathrm{m}+\nabla z)]+\frac{\Gamma_\mathrm{w}}{w_\mathrm{m}},\tag{2}$$

where the subscript f indicates the preferential flow domain and the subscript "m" indicates the matrix domain. C is the differential water capacity ($\mathrm{d}\theta/\mathrm{d}h$) ($\mathrm{L}^{-1}$), Θ is the effective saturation (–), h is the pressure head (L), t is time (T), z is the vertical coordinate (positive upward), K is the isotropic hydraulic conductivity ($\mathrm{L\,T}^{-1}$), S_s is the specific storage (L^{-1}), w is the volumetric ratio of the preferential flow domain or the matrix domain over the total soil volume (–), and Γ_w is the water exchange term (T^{-1}) between the two domains.

The Brooks–Corey function is used to describe the hydraulic properties of both the matrix and preferential flow domains (Brooks and Corey, 1964):

$$\Theta=\frac{\theta-\theta_\mathrm{r}}{\theta_\mathrm{s}-\theta_\mathrm{r}}=\begin{cases}|\alpha_\mathrm{BC}h|^{n_\mathrm{BC}}, & \alpha_\mathrm{BC}h<-1\\ 1, & \alpha_\mathrm{BC}h\geq -1,\end{cases}\tag{3}$$

$$K=K_\mathrm{s}\Theta^{2/n_\mathrm{BC}+l_\mathrm{BC}+2}=K_\mathrm{s}|\alpha_\mathrm{BC}h|^{-2-n_\mathrm{BC}(l_\mathrm{BC}+2)},\tag{4}$$

$$C=-\frac{\mathrm{d}\theta}{\mathrm{d}|h|}=\begin{cases}\alpha_\mathrm{BC}n_\mathrm{BC}(\theta_\mathrm{s}-\theta_\mathrm{r})|\alpha_\mathrm{BC}h|^{-n_\mathrm{BC}-1}, & \alpha_\mathrm{BC}h<-1\\ 0, & \alpha_\mathrm{BC}h\geq -1,\end{cases}\tag{5}$$

where θ is the water content ($\mathrm{L}^3\,\mathrm{L}^{-3}$), subscripts s and r denote saturation and residual state, K_s is the saturated hydraulic conductivity ($\mathrm{L\,T}^{-1}$), and α_BC, l_BC, n_BC, are fitting parameters.

Γ_w is the water exchange rate between the two domains (Ray et al., 1997):

$$\Gamma_\mathrm{w}=\alpha_\mathrm{w}K_\mathrm{a}(h_\mathrm{f}-h_\mathrm{m}),\tag{6}$$

where α_w (L^{-2}) is the effective water transfer coefficient, and the relative hydraulic conductivity K_a ($\mathrm{L\,T}^{-1}$) is calculated by averaging the hydraulic conductivities of the two pore domains (Arora et al., 2011; Laine-Kaulio et al., 2014):

$$K_\mathrm{a}=\frac{K_\mathrm{f}+K_\mathrm{m}}{2}.\tag{7}$$

The volumetric ratio of the preferential flow domain and matrix domain sum up to 1:

$$w_\mathrm{f}+w_\mathrm{m}=1.\tag{8}$$

The total water content of the soil is the weighted average of the water contents of the two domains:

$$\theta=w_\mathrm{f}\theta_\mathrm{f}+w_\mathrm{m}\theta_\mathrm{m}.\tag{9}$$

The same holds for the total saturated hydraulic conductivity of the soil:

$$K_\mathrm{s}=w_\mathrm{f}K_\mathrm{sf}+w_\mathrm{m}K_\mathrm{sm}.\tag{10}$$

where K_sf and K_sm are saturated hydraulic conductivity of preferential flow domain and matrix domain, respectively. Boundary conditions may be specified for pressure head, specified flux, or mixed (Chui and Freyberg, 2009). In the case of a dual-permeability model, specified flux i (infiltration from rainfall) is divided between the matrix and preferential flow domains:

$$i=w_\mathrm{f}i_\mathrm{f}+w_\mathrm{m}i_\mathrm{m},\tag{11}$$

where i_m and i_f are boundary fluxes to the matrix and the preferential flow domains ($\mathrm{L\,T}^{-1}$), respectively. The two domains have an equal opportunity to receive rainfall and are initially equal to rainfall intensity R (Dusek et al., 2008):

$$R=i=i_\mathrm{f}=i_\mathrm{m}.\tag{12}$$

As the matrix domain has a larger volumetric ratio ($w_\mathrm{m}>w_\mathrm{f}$), the infiltration process is initially dominated by the matrix domain. The infiltration capacity of each domain is the product of saturated hydraulic conductivity and the pressure head gradient of each domain. Once the specified flux into the matrix is larger than its infiltration capacity, the boundary condition changes to specified pressure head and the specified flux for the preferential flow domain is increased to

$$i_\mathrm{f}=\frac{R-w_\mathrm{m}i_\mathrm{m}}{w_\mathrm{f}}.\tag{13}$$

Once the specified flux into the preferential flow domain is also larger than the infiltration capacity, the boundary conditions of both the matrix and the preferential flow domain are changed to a specified pressure head of zero and overland flow occurs.

2.2 Slope stability analysis method

The slope stability analysis is based on the local factor of safety approach (Lu et al., 2012). The plane-strain linear elasticity model is used to calculate the 2-D stress field (Abramson, 2002), which is governed by a momentum balance equation

$$\nabla(\boldsymbol{\sigma})+\gamma\boldsymbol{b}=0,\tag{14}$$

where $\boldsymbol{\sigma}$ is a stress tensor ($\mathrm{M\,L}^{-1}\,\mathrm{T}^{-2}$) with three independent stress variables in two-dimensional space, γ is the bulk unit weight of the slope material ($\mathrm{ML}^{-2}\,\mathrm{T}^{-2}$), and $\boldsymbol{b}$ is the

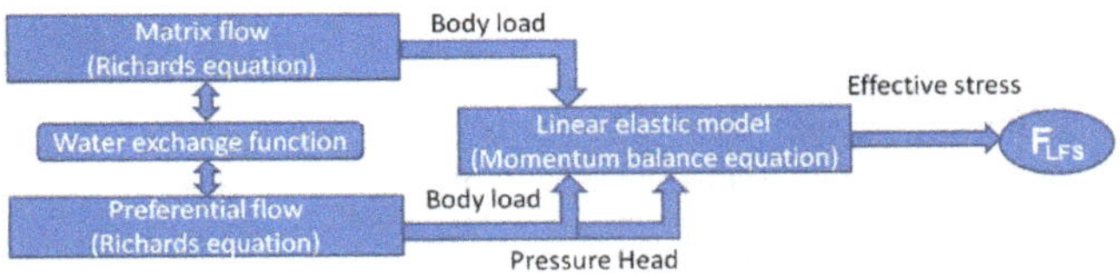

Figure 1. Structure of coupled dual-permeability model and soil mechanics model.

unit vector of body forces with two components. The effective stress equation can be formulated as

$$\boldsymbol{\sigma}' = \boldsymbol{\sigma} - \chi p_{\mathrm{w}}, \tag{15}$$

where $\boldsymbol{\sigma}'$ is the effective stress, p_{w} ($\mathrm{ML^{-1}T^{-2}}$) is the porewater pressure, and χ (–) is the matrix suction coefficient, which is usually approximated by the effective saturation (Lu et al., 2010).

The local factor of safety F_{LFS} is defined as the "ratio of the Coulomb stress at the current state of stress to the Coulomb stress of the potential failure state under the Mohr–Coulomb criterion" (Lu et al., 2012):

$$F_{\mathrm{LFS}} = \frac{\tau^*}{\tau}, \tag{16}$$

where τ^* is the limit Coulomb stress and τ is the actual shear stress ($\mathrm{ML^{-1}T^{-2}}$). Application of the Mohr–Coulomb failure criterion gives the local factor of safety at every point in the hillslope:

$$F_{\mathrm{LFS}} = \frac{2\cos\phi'}{\sigma_1' - \sigma_3'}\left[c' + \frac{\sigma_1' + \sigma_3'}{2}\tan\phi'\right], \tag{17}$$

where c' is the effective cohesion ($\mathrm{ML^{-1}T^{-2}}$), ϕ' is the friction angle, and $\boldsymbol{\sigma}_1'$ and $\boldsymbol{\sigma}_3'$ are the spatially varying first and third principal effective stress for the variably saturated soil ($\mathrm{ML^{-1}T^{-2}}$).

The influence of hydrology on slope stability is manifested in two ways. First, the unit weight function depends on the water content (Eq. 9). Second, the effective stress depends on the porewater pressure. In the dual-permeability model, the porewater pressure of the preferential flow domain is used in the computation of the effective stress.

Figure 1 summarises the structure of coupled dual-permeability and slope stability model. Two Richards' equations are coupled by the water exchange function. The hydrological results are sequentially coupled with a soil mechanics model without considering possible feedback of soil deformation on soil properties and the hydrological process.

In this study, COMSOL Multiphysics® is used to develop a coupled hydrological and slope stability model for both single-permeability and dual-permeability subsurface flow (Shao et al., 2014). For subsurface flow computations, COMSOL Multiphysics facilitates variable mesh resolution to optimise between coarse and fine mesh density to balance

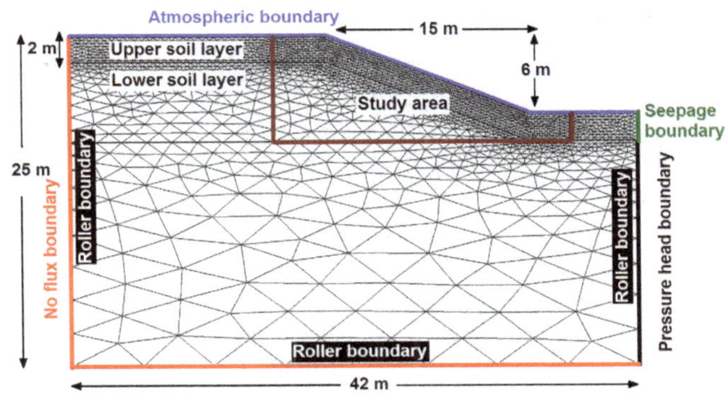

Figure 2. Computational mesh and boundary conditions.

computation time and improve the numerical solution where needed. The time discretization depends on the convergence of the numerical solution, which is automatically adjusted by the software. The technical details and model validation can be found in Shao et al. (2014).

3 Set-up of the numerical experiments

3.1 Slope geometry

Consider a slope of 23° consisting of fine-grained lithology such as clay shales with a more permeable weathered top soil layer (Bogaard, 2002; Berti and Simoni, 2012; Picarelli et al., 2006); this is a typical slope that is vulnerable to failure. The slope is 6 m high and 15 m long and consists of two layers with a 2 m thick homogeneous upper soil layer (see Fig. 2).

The model domain is 42 m by 25 m to reduce the effects of specified boundary conditions along the model boundary. The computational mesh and the boundary conditions are shown in Fig. 2. The boundary conditions of the subsurface flow model are atmospheric at the surface: the left-hand and bottom sides are no-flux boundaries; the right-hand side consists of a seepage boundary condition for the upper soil layer and a specified pressure head to mimic a constant groundwater table for the lower layer. For the soil mechanics model, the surface is a free boundary, the bottom boundary (only horizontal displacements) and the left- and right-hand sides (only vertical displacements) are all roller boundaries.

Since the pressure head in the surface area can change drastically during rainfall, a very dense mesh was used near the surface to accurately model the transient hydrological conditions. The mesh density of the upper layer is approximately 0.25 m (vertical) by 0.5 m (horizontal). A coarser mesh was defined in the lower part of the slope as a less dynamic condition will occur here.

3.2 Parameterisation

The volumetric ratio of the preferential flow domain w_{f} is 0.1; a typical range is $\sim$0.025–0.2 (Köhne et al., 2002). The

Table 1. Summary of parameters.

Symbol	Parameter name	Units	Upper layer (sandy loam)	Lower layer (clay)
θ_s	Saturated water content	(–)	0.412	0.385
θ_r	Residual water content	(–)	0.041	0.09
K_s	Saturated hydraulic conductivity	$(\mathrm{cm\,h^{-1}})$	2.59	0.06
K_{sf}	K_s of preferential flow domain	$(\mathrm{cm\,h^{-1}})$	23.76	0.214
K_{sm}	K_s of matrix domain	$(\mathrm{cm\,h^{-1}})$	0.2376	0.043
α_w	Water exchange coefficient	$(\mathrm{m^{-2}})$	0.2	0.2
α_{BC}	Brooks–Corey fitting parameter	$(\mathrm{cm^{-1}})$	0.068	0.027
n_{BC}	Brooks–Corey fitting parameter	(–)	0.322	0.131
l_{BC}	Brooks–Corey fitting parameter	(–)	1	1
γ_{dry}	Dry unit weight	$(\mathrm{kN\,m^{-3}})$	15.5	15.5
E	Young's modulus	(MPa)	10	10
ν	Poisson's ratio	(–)	0.35	0.35
ϕ'	Friction angle	(°)	35	35
c'	Effective cohesion (case 1)	(kPa)	5	5
c'	Effective cohesion (case 2)	(kPa)	3	6

pore-size distribution of the preferential flow domain allows unsaturated infiltration before the matrix domain is saturated (Dusek et al., 2008). A comparison is made between the hydrological results of the single-permeability and the dual-permeability models. The total weighted saturated hydraulic conductivity of the dual-permeability model is equal to the saturated hydraulic conductivity of the single-permeability model. The water exchange between the matrix and preferential flow domains depends on the hydraulic conductivity between the two domains K_a and the water exchange coefficient α_w (Eq. 6). Equilibrium between the preferential flow and matrix domains is reached quicker for smaller values (closer to 1) of K_{sf}/K_{sm} and larger values of α_w. Moderate values are used for K_{sf}/K_{sm} (100 in the upper layer and 5 in the lower layer) and for α_w ($0.2\,\mathrm{m^{-2}}$).

The soil hydraulic parameters are presented in Table 1. Preferential flow plays an important role in the upper soil layer where there is an abundance of macropores, but less so in the lower soil layer where macropores are almost non-existent (e.g. Bogner et al., 2013). In other words, the volumetric percentage of preferential flow domain is still the same, but in the lower layer the saturated hydraulic conductivity of macropores is more similar to the pores of the matrix. The more permeable top layer is sandy loam and the fine-grained lower layer is clay; the soil hydraulic parameters are taken from the unsaturated soil hydraulic database (UNSODA) (Nemes et al., 2001; Leij, 1996).

Parameterisation of soil hydraulic properties for dual-permeability models is difficult as the two conceptualised domains cannot be experimentally separated. In this paper, the pragmatic approach of Vogel et al. (2000) is adopted: the same hydraulic parameters are adopted for both domains, except for the saturated hydraulic conductivities.

The parameters of the soil mechanics model are also shown in Table 1. The slope stability is a function of two parameters: the friction angle and the effective cohesion. Decreasing the friction angle or increasing the effective cohesion leads to a higher factor of safety. The friction angle is physically constrained in a narrow range, and in this study is fixed. In numerical modelling, effective cohesion c' is scale dependent, and is usually defined as a linear function of the slope height to obtain identical values of the safety factor when applying it to different slope sizes (Griffiths and Lane, 1999; Lu et al., 2012). Very low values of effective cohesion result in initially unstable slopes, and very high values of effective cohesion results in slopes that are unconditionally stable. In this study, two sets of cohesion values were selected in the range that may result in rainfall-triggered landslides: a homogeneous case where the effective cohesion of both layers is 5 kPa and a case where the effective cohesion of the upper layer is smaller ($c'_1 = 3\,\mathrm{kPa}$) than the lower layer ($c'_2 = 6\,\mathrm{kPa}$).

Intermittent and variable rainfall may significantly influence the pressure response in the shallow layer soil, while the pressure response in the deeper layer soil is driven by percolation and less dynamic to intermittent rainfall. Two rainfall events are modelled: a low-intensity rainfall of $2\,\mathrm{mm\,h^{-1}}$ for 150 h and a high-intensity rainfall of $20\,\mathrm{mm\,h^{-1}}$ for 15 h. The two rainfall events are expected to result in distinctly different slope stabilities, even though the total rainfall of both events is equal. The initial condition is the steady pore-water pressure distribution obtained from running the model with a daily rainfall of $1.64\,\mathrm{mm\,day^{-1}}$ ($600\,\mathrm{mm\,year^{-1}}$) for 10 years.

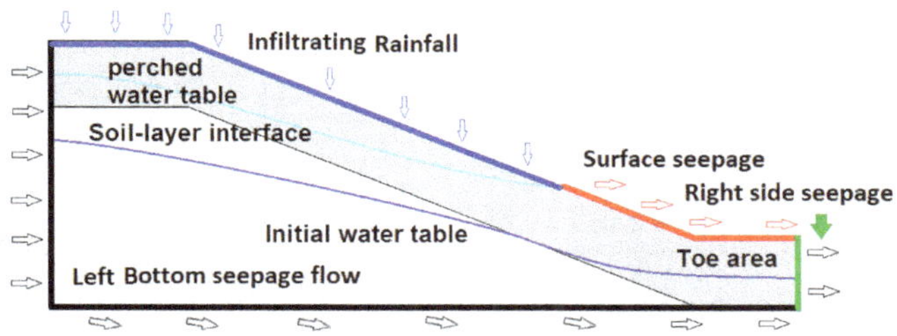

Figure 3. Flow component and water balance of study area.

4 Results

4.1 Subsurface flow

A schematic diagram of the subsurface flow components in the study area is shown in Fig. 3. Note that the study area is a small part of the model domain (Fig. 2). The main fluxes are the infiltration from rainfall (blue), the inflow/outflow along the left side and bottom (black), the seepage outflow along the surface (red), and the outflow along the right boundary (green).

Hydrological results for the single- and dual-permeability models are shown in Figs. 4 and 5, respectively. The graphs on the left are results for the long-duration, low-intensity rainfall case while the graphs on the right are results for the short-duration, high-intensity rainfall. Integrated fluxes, as shown in Fig. 3, are reported in $m^2 h^{-1}$.

For both models, all the rainfall infiltrates into the slope during the beginning of the rain event and when the soil becomes saturated infiltration decreases and saturation excess overland flow occurs. For the single-permeability model and low-intensity rainfall overland flow starts after 95 h (or 190 mm of rainfall) while for the high-intensity rainfall overland flow starts after 8.5 h (or 170 mm of rainfall) (Fig. 4a and b).

In the dual-permeability model (Fig. 5), the rainfall infiltration is divided over the two domains and additional rainfall infiltrates into the preferential flow domain when the matrix domain reaches infiltration capacity. Recall that the matrix domain is 90 % of the domain, and the preferential flow domain is 10 % of the domain. A smaller fraction of rainfall infiltrates into the preferential flow domain for the case of low-intensity rainfall (10–15 %) than for the case of high-intensity rainfall (50–85 %). Overland flow starts after 80 h (or 160 mm of rainfall) for the low-intensity case and after 60 h (or 120 mm of rainfall) for the high-intensity case.

The seepage outflow increases along the left, right, and bottom boundaries during the rainfall event (Figs. 4c, d, and 5c, d) and is smaller than the infiltration rate (storage is increasing). In the dual-permeability model and the low-intensity rain, outflow along the surface boundary starts after 115 h (or 230 mm of rainfall) while for the high-intensity rain outflow starts after 9 h (or 180 mm rainfall). The outflow rate along the surface boundary depends on the groundwater level in the upper layer. In the dual-permeability model, the outflow along the right boundary is approximately 10 times larger for the preferential flow domain than for the matrix domain, which is consistent with their volumetric ratio and their saturated hydraulic conductivity ratio. The water exchange between the two domains in the dual-permeability model is shown in Fig. 5e and f. For the low-intensity rainfall case, the water exchange from the preferential flow domain to the matrix domain increases during the first 100 h and then decreases, while the water exchange from the matrix domain to the preferential flow domain is almost always increasing (more negative). For the high-intensity rainfall case, the water exchange from the matrix to the preferential flow domain is negligible, while the water exchange from the preferential flow domain to the matrix domain reaches more than $0.3 m^2 h^{-1}$, which is similar to the infiltration into the preferential flow domain. After 5 h, approximately 75 % of infiltration into the matrix domain is water exchange from the preferential flow domain (Fig. 5f) and 25 % infiltration from the surface boundary (Fig. 5b).

4.2 Water balance

The integrated rainfall and water storage for the study area are shown for both models in Fig. 6. The water balance is obtained by integrating all flow components along the boundaries of the study area. The numerical water balance errors are between 2 and 3 %.

For all cases, the storage increase flattens out when the inflow decreases (Figs. 4 and 5). For the high-intensity rainfall, the dual-permeability model stores 8 % less water than the single-permeability model. The total storage after 150 h of low-intensity rainfall is less than after 15 h of high-intensity rainfall, probably caused by the longer time that water can drain from the study area under low-intensity rain.

For the dual-permeability model, the water exchange has a significant influence on the storage change in each domain. For the low-intensity rainfall, the storage in the preferential flow domain does not increase much after 6 h (Fig. 6). For the high-intensity rainfall, the storage in the preferential flow domain increases rapidly over the first 3 h as very little water infiltrates into the matrix domain due to the low infiltration capacity of the matrix. After 3 h, the preferential flow domain has almost reached full saturation and the large pressure difference between the preferential flow domain and matrix domain causes extensive water exchange (Fig. 5f).

4.3 Water content

The water content distribution in the study area is shown in Fig. 7 for both the single-permeability model (left-hand panels) and the dual-permeability model (centre and right-hand panels). The water exchange rate between the matrix and preferential flow domains of the dual-permeability

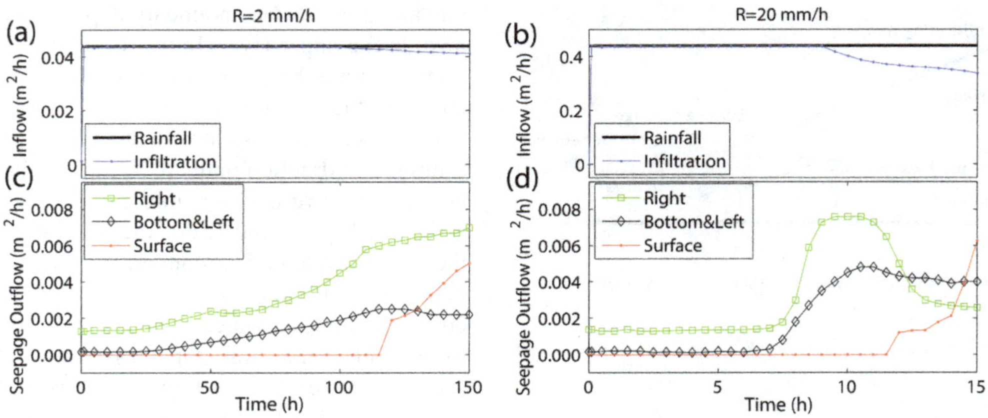

Figure 4. Integrated fluxes for single-permeability model and $2\,\mathrm{mm\,h^{-1}}$ (left panels) and $20\,\mathrm{mm\,h^{-1}}$ (right panels) rainfall. Rainfall and infiltration **(a, b)**, and outflow at the right, outflow at the left and bottom, and outflow at the surface **(c, d)**.

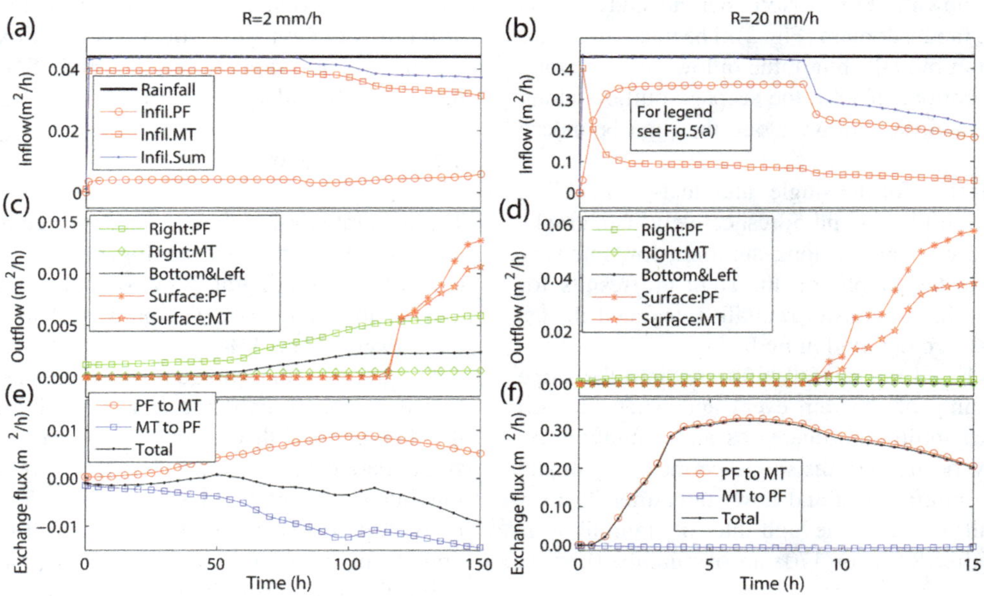

Figure 5. Integrated fluxes for dual-permeability model and $2\,\mathrm{mm\,h^{-1}}$(left panels) and $20\,\mathrm{mm\,h^{-1}}$ (right panels) rainfall. Rainfall and infiltration **(a, b)**; outflow at the right, outflow at the left and bottom, and outflow at the surface **(c, d)**; exchange between matrix domain (MT) and preferential flow domain (PF) **(e, f)**, positive for flow from PF to MT and negative for flow from MT to PF.

model is shown in Fig. 8. The infiltration process of the dual-permeability model differs significantly from that of the single-permeability model.

The initial water content distribution in the matrix, as well as preferential flow domains, is similar for both models. During the rainfall events, the wetting front in the single-permeability model develops parallel to the surface and propagates downward. This holds for both low and high rainfall intensities (Fig. 7 left-hand column). The wetting front generally reaches the groundwater table at the toe of the slope first, after which the infiltrated water continuously enlarges the saturated area.

In the dual-permeability model, the combined effects of the preferential flow and the matrix flow show a more complicated response. For the low-intensity rainfall, infiltration is dominated by matrix flow, as 90 % of the subsurface consists of the matrix. Because the rainfall intensity is lower than the saturated conductivity of the matrix domain, on the surface boundary, 90 % of the rainfall infiltrates into the matrix domain and 10 % of the rainfall infiltrates into the preferential flow domain. The pressure difference between the two domains drives the water exchange at the matrix wetting front (Figs. 5e and 8a). Inside the flow domain of the slope, 10 % of matrix flow transferred to preferential flow due to water

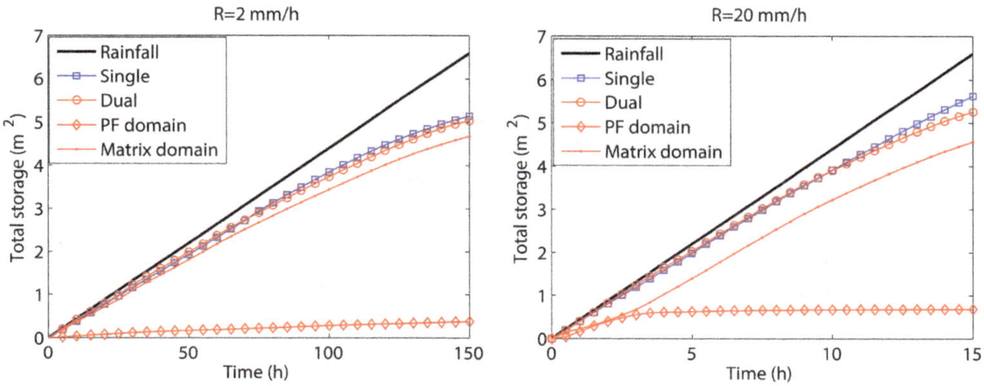

Figure 6. Storage increase of single-permeability model and dual-permeability model.

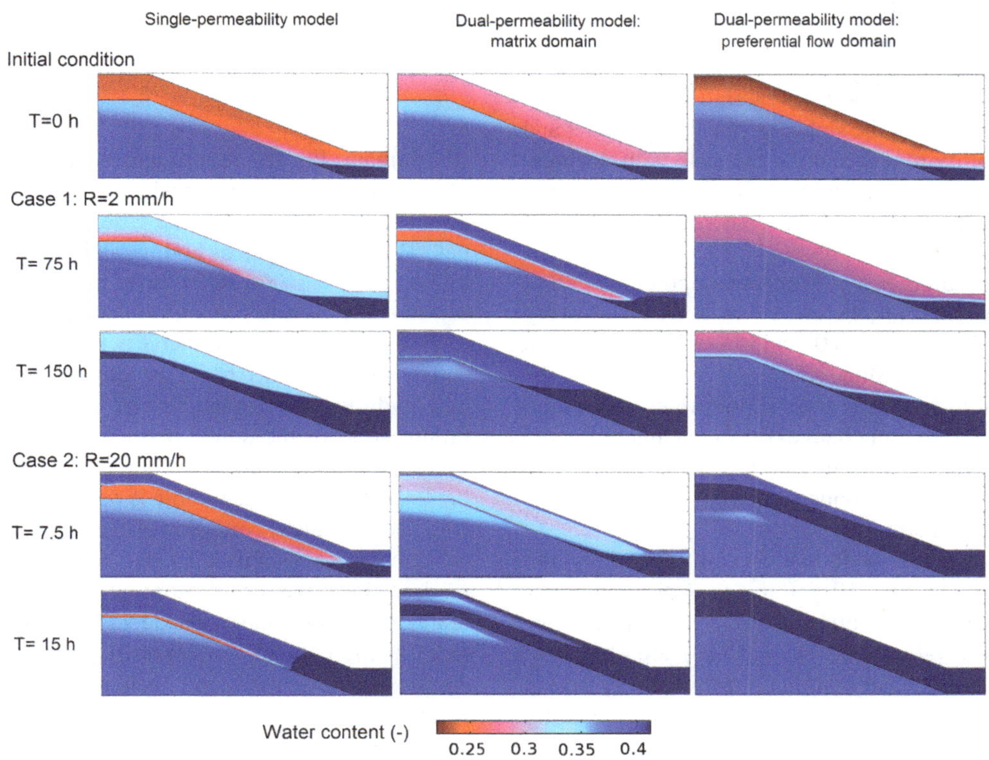

Figure 7. Water content distribution.

exchange in the soil. At first, water quickly reaches the soil layer interface by preferential flow where it transmits to the matrix, although this exchange flux is very small (Figs. 5e and 8a). After sufficient time (70 h), a much stronger matrix flow (taking about 80 % of the infiltrated rainfall) reaches the soil layer interface and generally reverses the water exchange direction (Fig. 5e). Overall, water exchange during low-intensity rainfall in the study area is dominated by flow from the matrix to the preferential flow domain (Fig. 8a and b).

For the high-intensity rainfall, the rainfall intensity is 8.4 times the matrix saturated hydraulic conductivity. The percentage of infiltration into the matrix domain decreases from 90 to 50 % within the first half hour, and continues to decrease to less than 20 % after 1.5 h. In contrast, the percentage of rainfall that infiltrates into the preferential flow domain increases from 10 to over 80 % after 2 h. Water in the preferential flow domain quickly reaches the deeper soil layer and forms a perched groundwater table (Fig. 7), where a significant amount of water infiltrates into the matrix (Fig. 5f).

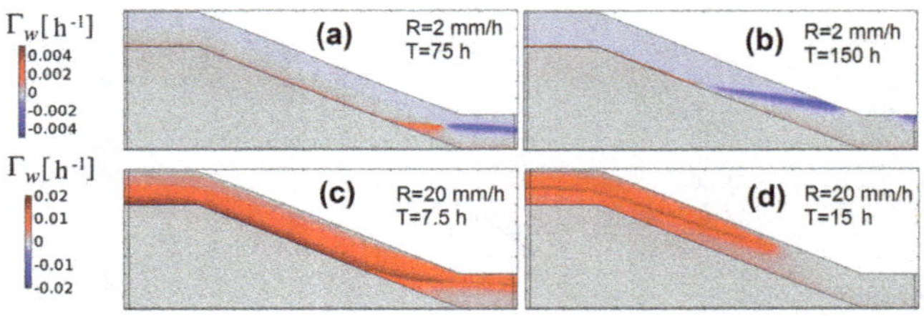

Figure 8. Water exchange rate distribution. Positive values (red): mean water exchange from preferential flow domain to matrix; negative values (blue): mean water exchange from matrix to preferential flow domain.

4.4 Slope stability

The local factor of safety is computed based on the computed water-pressure distribution (Fig. 7). The distribution of the local factor of safety is shown in Fig. 9 for the initial condition and after 150 h (low-intensity rainfall) and 15 h (high-intensity rainfall) for both the single-permeability model and the dual-permeability model and for the case with different cohesion values for the upper and lower layers. The case with equal cohesion values is not shown because the potential failure areas are very small.

A local factor of safety below 1 indicates a potential failure area. The area with a F_{LFS} below 1 was determined every time interval (5 h in case 1, and 0.5 h in case 2) and is shown by the black line in Fig. 9. Slope stability is related to both the specific weight of the wet soil and the porewater pressure in the soil. The specific weight changes due to changes in water storage are relatively small, but changes in water pressure have a significant effect on slope stability, especially in the area of the perched water table.

The size of the potential failure area is plotted vs. the cumulative rainfall in Fig. 10 for the two different rainfall events and two sets of cohesion values. The results for the same cohesion values ($c'_1 = c'_2 = 5\,\mathrm{kPa}$) are shown in Fig. 10a. For the low-intensity rainfall, the failure area is very small and is approximately the same for both permeability models. For the high-intensity rainfall, the failure area in the single-permeability model is larger than for the low-intensity rainfall, but the trend is similar. The failure area in the dual-permeability model is significantly larger. Failure starts after 60 mm of rainfall, and the failure area continues to grow during the rainfall infiltration process.

The results for different cohesion values ($c'_1 = 3\,\mathrm{kPa}$, $c'_2 = 6\,\mathrm{kPa}$) are shown in Fig. 10b. For the low-intensity rainfall, the failure area is $0.7\,\mathrm{m}^2$ in the single-permeability model after 20 mm of cumulative rainfall. The size of this area shows almost no increase until approximately 220 mm of cumulative rainfall, when the groundwater table starts to rise (Fig. 7). The failure area of the dual-permeability model is 40 % smaller than that of the single-permeability model as the preferential flow domain drains more water into the matrix domain. For the high-intensity rainfall, the failure area of the dual-permeability model is larger than of the single-permeability model, as is the case with equal cohesion values. The failure areas of both models increase fairly quickly to $2\,\mathrm{m}^2$, or 5 % of the upper layer in the study area. The failure area increases to $5\,\mathrm{m}^2$ in the dual-permeability model and to $3\,\mathrm{m}^2$ in the single-permeability model.

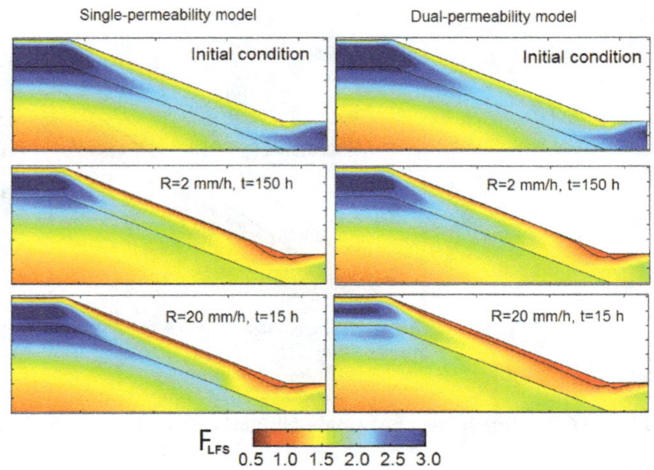

Figure 9. Final slope stability after the rainfall event ($c'_1 = 3\,\mathrm{kPa}$, $c'_2 = 6\,\mathrm{kPa}$). The black line delineates potential unstable area ($F_{\mathrm{LFS}} < 1$).

The slope stability result are directly related with subsurface hydrological results. For the low-intensity rainfall, the failure area for the single-permeability model is very similar in size and location to the dual-permeability model as the location of the water table is very similar in both models (Fig. 7). The initial condition of the dual-permeability model is slightly more stable than that of the single-permeability model, since the preferential flow domain has a higher drainage capacity and, consequently, a lower porewater pressure. In the case of low-intensity rainfall, the matrix flow dominates the groundwater recharge and, consequently, the slope instability. Furthermore, the porewater pressure in the preferential flow domain is very low due to its strong drainage capacity. As a result, the failure area calculated by the dual-permeability model under low-intensity rainfall is slightly smaller than that calculated by the single-permeability model (Fig. 10a). The location of the failure area is similar in the single- and the dual-permeability domain (Fig. 9).

For the high-intensity rainfall, the failure area is significantly larger for the dual-permeability model than for the single-permeability model as the perched water table in the preferential flow domain is much more extensive in the dual-permeability model as compared to the single-permeability model (Fig. 7). The regular wetting front of the single-permeability model does not reach the interface between soil layers, and the failure area is limited to the toe of the slope.

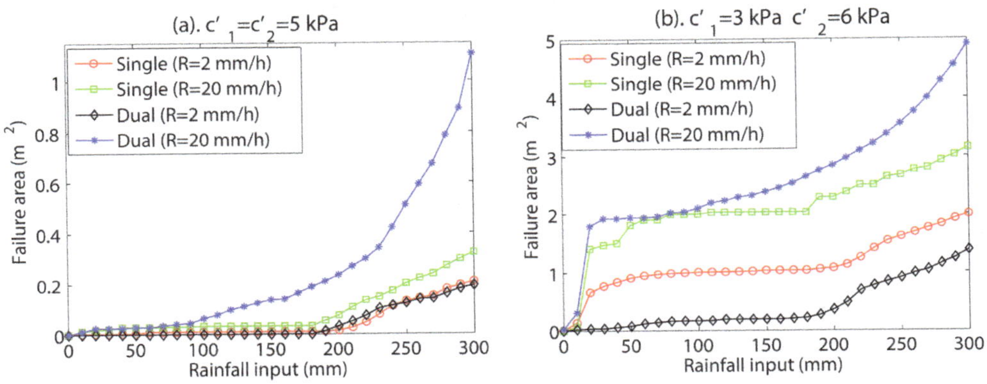

Figure 10. Development of the failure area under different rainfall intensities and soil cohesion.

For the dual-permeability model, the high-intensity rainfall results in a rapid infiltration through preferential flow, which quickly reaches the interface between soil layers, and increases the degree of saturation and pressure head of the deeper soil. Positive porewater pressure occurs in the preferential flow domain before the entire slope is fully saturated, and produces a larger failure area than in the equivalent single-permeability model.

5 Discussion

The role of preferential flow in hydrology focuses mainly on the rapid vertical infiltration of water and contaminant (Christiansen et al., 2004; Kodešová et al., 2005; Laine-Kaulio et al., 2014), or the rapid discharge in hillslope and catchment hydrological studies on discharge generation (Zhang et al., 2006; Mulungu et al., 2005). Preferential flow can cause rapid water-pressure build-up and may trigger landslides, as suggested by many researchers, but it is rarely quantified. Prior to this study, a detailed evaluation of the influence of preferential flow on slope stability using a coupled dual-permeability and slope stability model had not been carried out. In this study, a physically based numerical model was used to investigate the hydrologic response under predefined conditions. Our starting point was an idealised, but representative, 2-D slope with fixed parameters and boundary conditions (except rainfall forcing). A linear-elastic model was used to calculate the stress field to quantify the failure area and timing, but plastic deformation after failure was not considered; the model is not able to quantify landslide displacements. In this section, the underlying approximations of the numerical model are explored and the influence of the chosen parameter sets on the model outcome is discussed. The numerical experimental results are compared with field studies and other published numerical experiments.

5.1 Continuum model

Soil heterogeneity is one of the most difficult problems in both hydrology and soil mechanics studies. As an alternative to the continuum approach used here, preferential flow may be simulated by explicitly including fissures, pipes, or fracture networks in discrete (or discontinuous) models. Several field studies (Hencher, 2010; Uchida, 2004; Verachtert et al., 2013) and numerical experiments (Tsutsumi and Fujita, 2008; Chang et al., 2014) have focused on the investigation and simulation of pipe flow (in soil) and fracture flow (in rock). In order to accurately describe the geometry of the preferential flow paths, the high-resolution macropore image reconstruction approach (Hu et al., 2014) or the statistical approach (Köhne et al., 2009) may be applied. Numerical simulations of these natural macropore networks require large amounts of geometry information (Nieber and Sidle, 2010) and computational time and are consequently limited to small-scale studies with a limited number of pipes (Tsutsumi and Fujita, 2008) or cracks (Moonen et al., 2008).

The dual-permeability model is a useful tool to simulate subsurface stormflow and solute transport in a hillslope when the parameterisation is able to capture the hydraulic characteristics of each domain (Laine-Kaulio, 2011; Laine-Kaulio et al., 2014). As the dual-permeability model describes the subsurface as a continuum of two linked domains, it is suitable for heterogeneous slopes with a high density of preferential flow paths and not for slopes with only a few large fissures or cracks (van der Spek et al., 2013).

In this paper, flow in both domains is described with the Darcy–Richards equation, which is valid when the macropores have a relatively small size, and the macropore flow is still viscous (Köhne and Mohanty, 2005; Laine-Kaulio et al., 2014). When fluid velocities are high and flow becomes turbulent, Darcy's equation is not valid (Beven and Germann, 2013) as may be the case in large cracks or fissures under near-saturated or ponded infiltration (Beven and Germann, 1982). The existence of pore necks and dead ends in pref-

Table 2. Parameters setting of water exchange coefficients in different literature.

Parameter	Vogel et al. (2000) 2-D virtual numerical study	Gerke and Köhne (2004) Bokhorst clay soil	Köhne and Mohanty (2005) macropore: coarse sand matrix: sandy loam	Arora et al. (2011) multiple-macropore in sandy loam soil	This study
w_f	0.05	0.05	0.009675	0.00033	0.1
K_{sf}/K_{sm}	1000	100	3878	63.6	100
K_{sa}/K_{sm}	0.01	0.001	1	32.1	50.5
K_{sm} (cm h^{-1})	0.012	0.1	0.056	0.13	0.238
d (cm)	1	1	11	1.89	–
γ_w	0.4	0.4	0.4	0.001	–
β	3	15	1.0685	0.67	–
α_w (cm^{-2})	1.2	6	2.5×10^{-3}	1.8×10^{-4}	2.0×10^{-5}
$\alpha_w K_{sa}$ (cm^{-1} h^{-1})	1.4×10^{-4}	6.0×10^{-4}	2.0×10^{-4}	7.8×10^{-4}	2.4×10^{-4}

erential flow paths reduce the occurrence of turbulent flow (Jarvis, 2007).

5.2 Coupling term in the dual-permeability model

In the dual-permeability model, the two domains are in general not at equilibrium. The parameterisation of the water exchange between the two domains governs whether a dual-permeability model behaves as a dual-permeability model or mimics the behaviour of a single-permeability model. The water exchange is governed by the pressure difference between the two domains and two parameters: the water exchange coefficient and the average hydraulic conductivity between the two domains (Eq. 6). The average hydraulic conductivity in turn is a function of the hydraulic conductivities of the two domains, which are a function of the pressure head. The larger the product, the quicker the two domains equilibrate. Specifically, the sensitivity analysis of Ray et al. (1997) shows that a very large water exchange coefficient results in almost instantaneous equilibrium between the two domains. Consequently, the dual-permeability model behaves like a single-permeability model, which means that the preferential flow has no influence on the hydrological and soil mechanics results. This behaviour does not agree with experimental studies of natural soils, which often show a non-equilibrium phenomenon between the two domains (Šimůnek et al., 2003; Jarvis, 2007).

The parameterisation of the water exchange used in this study was compared with previous studies (see Table 2). Estimation of the water exchange coefficient from physical measurements is very difficult. The most widely used equation is (Gerke and van Genuchten, 1993b)

$$\alpha_w = \frac{\beta}{d^2}\gamma_w, \tag{18}$$

where β is a scaling factor, d is half the representative distance between two macropores, and γ_w is a geometry-dependent shape factor that equals 3 for rectangular slabs and 15 for spheres (Ray et al., 1997). Parameter values for the water exchange term used in several studies are summarised in Table 2. Vogel et al. (2000) and Gerke and Köhne (2004) conceptualise the preferential flow domain as rectangular matrix blocks arranged as parallel slabs. A reduction factor of 0.01 or 0.001 was used to significantly reduce the water exchange between the two domains, because the hydraulic conductivity at the matrix/fracture interface was conceptualised to be controlled by relatively impermeable coatings that are composed of minerals and organic matter (Ray et al., 2004; Gerke and Köhne, 2002). Köhne and Mohanty (2005) conceptualise the dual domain as a hollow cylindrical matrix that is filled with coarse sand in the middle to mimic the macropore domain. Arora et al. (2011) based their parameters on a high density of macropore columns, and they calculated K_a by averaging the hydraulic conductivities of the two pore domains (as adopted in this paper; see Eq. 7). Arora et al. (2011) and Köhne and Mohanty (2005) did not consider the influence of coatings on the permeability, nor was this done in this study.

It may be seen from Table 2 that the magnitude of the product $\alpha_w K_{sa}$ is similar for all studies, even though some of the other values (notably the ratios K_{sa}/K_{sm} and the values of α_w) differ by several orders of magnitude. As such, the water exchange between all these models is likely similar.

5.3 Computation of effective stress

In the dual-permeability model, the porewater pressure of the matrix and the preferential flow domains are different and water flows from the domain with a higher pressure to the domain with a lower pressure. Van der Spek et al. (2013) showed that in the case of varved clays with a low hydraulic conductivity of the soil matrix and a low density of fissures, the time delay between water entering the fissure network and an increase in pressure in the matrix is relatively large. This study concerns a system with a very high density of macropores and consequently the numerical simulations show only a small time delay for the pressure propagation from the preferential flow domain to the matrix domain. The porewater pressure of the preferential flow domain is used for the effective stress calculation in the slope stability analysis, but failure time and area are only slightly different when the matrix porewater pressure is used for the slope stability analysis. Field evidence (Uchida et al., 2001; Hencher, 2010; Wienhöfer et al., 2011) and numerical experiments (Nieber and Sidle, 2010; Laine-Kaulio et al., 2014) suggest that individual preferential flow networks are hydraulically connected, and that the high porewater pressure build-up in the preferential flow paths is directly correlated with slope stability.

5.4 Implications of preferential flow for slope stability

This study is not the first to address the influence of preferential flow on subsurface flow and slope stability. Preferential flow has an effect on infiltration and drainage fluxes and as such influences the triggering factors for rainfall-induced landslides. Moreover, storage capacity relates to the pore distribution in a soil and controls the antecedent condition or the cause of landslide occurrence (Hamdhan and Schweiger, 2011). The complexity hides in the combination of rainfall characteristics and soil hydraulic properties, together with the physiographic properties like slope, soil thickness, bedrock topography and so on, which determine the resultant porewater pressure response. The model runs and analyses show that rainfall intensity needs to be related to both the soil infiltration rate of the matrix domain and the preferential flow domain. Natural hillslopes show a different response from the matrix as compared to the preferential flow domain depending on the rainfall intensity. The dual-permeability model can simulate the preferential flow which will rapidly reach deep soil and increase porewater pressure. However, a single-permeability model cannot reproduce the phenomenon of preferential flow dominated fast pressure response in deep soil.

Parameterisation of a dual-permeability model is difficult in practice (Laine-Kaulio et al., 2014). Therefore, the use of single-permeability models with effective soil hydraulic parameters prevails in regional hazard assessment (Hamdhan and Schweiger, 2011; Zhou et al., 2014). Rainfall-intensity duration plots for regional hazard assessment are well established and abundantly used but do not include soil and hydrological information (Guzzetti et al., 2007, 2008). They empirically relate precipitation intensity and duration to observed landslides. The inclusion of more detailed hydrometeorological information in these analyses is ongoing. Recently, von Ruette et al. (2014) showed the importance of spatially and temporally heterogeneous rainfall on the initiation of landslides. In a synthetic study they showed that spatially distributed rainfall resulted in an increase of the number of shallow landslides as compared to uniform or intermittent rainfall (short periods with higher rainfall intensities but spatially homogeneous). They concluded that "low-rainfall intensities (below soil infiltration capacity) and long durations resulted in more infiltration, lower stream discharge, and more saturations and thus failure". This is in full agreement with the results for low rainfall intensities in this study. Generally speaking, this holds for every case where infiltration capacity of the matrix remains higher than the rainfall intensity even in the presence of preferential flow paths. For low-intensity rainfall, the water-pressure increase simulated with a single-permeability model is generally larger than with a dual-permeability model as drainage by the preferential flow paths is underestimated. Soil drainage is a typical threshold process of the soil to get rid of its high porewater pressure and in this way stabilises the slope. Consequently, the stability is slightly underestimated with a single-permeability model for low-intensity rainfall.

The reverse is true, however, for high rainfall intensities, when the matrix reaches infiltration capacity early on. In these cases the preferential flow system dominates because water that cannot infiltrate into the matrix domain infiltrates into the preferential flow domain instead, resulting in a large pressure increase with a negative effect on slope stability. A much smaller pressure increase is simulated with a single-permeability model for the same high-intensity rainfall. Consequently, the stability is overestimated with a single-permeability model even when equivalent parameters are used.

6 Conclusions

A coupled dual-permeability and slope stability model was developed to simulate the influence of preferential flow on subsurface hydrology and consequent slope failure area. The dual-permeability model is able to simulate both preferential flow and matrix flow. The slope failure area was determined with a local factor of safety analysis. Numerical experiments were carried out to study the effect of rainfall events on slope stability with both a single-permeability (no preferential flow) and a dual-permeability model. A 23° slope consisting of two soil layers was used in the study. The upper layer is sandy loam and the bottom layer is clay. Both the case where the cohesion of the two layers are equal, and the case where the cohesion of the upper layer is smaller than the lower layer were simulated. Two types of rainfall events were considered: low-intensity, long-duration rainfall and high-intensity short-duration rainfall. The total amount of water of both rainfall events was equal. The effect of preferential flow on slope stability was studied by comparing the failure area obtained with a single-permeability model and a dual-permeability model for the same rainfall event.

For low-intensity rainfall, the failure area of both models is similar when the cohesion of the upper and lower layers is equal, but the failure area is significantly larger in the single-permeability model as compared to the dual-permeability model when the cohesion of the upper layer is lower than the cohesion of the lower layer. During low-intensity rainfall, preferential flow has a positive effect on slope stability as it drains water from the matrix domain and decreases the water pressure.

For high-intensity rainfall, the failure area of the dual-permeability model is significantly larger than the single-permeability model whether the cohesion values of the two layers are equal or not. During high-intensity rainfall, the rainfall intensity is larger than the infiltration capacity of the matrix domain so that most of the rainfall infiltrates into the preferential flow domain. As a result, the water pressure increases very quickly in the preferential flow domain resulting in a much larger failure area than is the case for the single-permeability model.

In summary, two different effects of preferential flow on slope stability were identified with a coupled dual-permeability and soil mechanics model. Preferential flow has a positive effect on slope stability during low-intensity rainfall and a negative effect on slope stability during high-intensity rainfall. The magnitude of the effect is a function of the soil hydraulic properties and soil mechanical properties of a specific slope. Identification of parameter ranges for which this behaviour is significant requires further investigation.

Acknowledgements. The authors are grateful to both Jan Wienhöfer and the anonymous reviewer, and Editor Uwe Ehret for their extensive, profound and constructive comments and suggestions that substantially improved the quality of paper. The first author was financially supported by the China Scholarship Council for his PhD research with the project reference number of 2011671055.

Edited by: U. Ehret

References

Abramson, L. W.: Slope Stability and Stabilization Methods, John Wiley and Sons Incorporated, Hoboken, New Jersey, 2002.

Aleotti, P. and Chowdhury, R.: Landslide hazard assessment: summary review and new perspectives, B. Eng. Geol. Environ., 58, 21–44, 1999.

Allaire, S. E., Roulier, S., and Cessna, A. J.: Quantifying preferential flow in soils: a review of different techniques, J. Hydrol., 378, 179–204, 2009.

Armstrong, A. C., Matthews, A. M., Portwood, A. M., Leeds-Harrison, P. B., and Jarvis, N. J.: CRACK-NP: a pesticide leaching model for cracking clay soils, Agr. Water Manage., 44, 183–199, 2000.

Arnone, E., Noto, L. V., Lepore, C., and Bras, R. L.: Physically-based and distributed approach to analyze rainfall-triggered landslides at watershed scale, Geomorphology, 133, 121–131, 2011.

Arora, B., Mohanty, B. P., and McGuire, J. T.: Inverse estimation of parameters for multidomain flow models in soil columns with different macropore densities, Water Resour. Res., 47, W04512, doi:10.1029/2010WR009451, 2011.

Baum, R. L., Godt, J. W., and Savage, W. Z.: Estimating the timing and location of shallow rainfall-induced landslides using a model for transient, unsaturated infiltration, J. Geophys. Res.-Earth, 115, F03013, doi:10.1029/2009jf001321, 2010.

Berti, M. and Simoni, A.: Observation and analysis of near-surface pore-pressure measurements in clay-shales slopes, Hydrol. Process., 26, 2187–2205, 2012.

Beven, K.: Kinematic subsurface stormflow, Water Resour. Res., 17, 1419–1424, 1981.

Beven, K. and Germann, P.: Macropores and water flow in soils, Water Resour. Res., 18, 1311–1325, 1982.

Beven, K. and Germann, P.: Macropores and water flow in soils revisited, Water Resour. Res., 49, 3071–3092, 2013.

Bogaard, T.: Analysis of Hydrological Processes in Unstable Clayey Slopes, Ph.D. thesis, Universiteit Utrecht, Utrecht, 2002.

Bogner, C., Trancón y Widemann, B., and Lange, H.: Characterising flow patterns in soils by feature extraction and multiple consensus clustering, Ecol. Inform., 15, 44–52, 2013.

Borga, M., Dalla Fontana, G., and Cazorzi, F.: Analysis of topographic and climatic control on rainfall-triggered shallow landsliding using a quasi-dynamic wetness index, J. Hydrol., 268, 56–71, 2002a.

Borga, M., Dalla Fontana, G., Gregoretti, C., and Marchi, L.: Assessment of shallow landsliding by using a physically based model of hillslope stability, Hydrol. Process., 16, 2833–2851, 2002b.

Brinkgreve, R., Engin, E., Swolfs, W., Waterman, D., Chesaru, A., Bonnier, P., and Galavi, V.: PLAXIS 2D 2010, Tech. rep., PLAXIS B. V., The Netherlands, 2010.

Brooks, R. and Corey, A.: Hydraulic Properties of Porous Media, Colorado State University, Colorado, 1964.

Chang, X., Hu, C., Zhou, W., Ma, G., and Zhang, C.: A combined continuous-discontinuous approach for failure process of quasi-brittle materials, Sci. China Ser. E, 57, 550–559, 2014.

Christiansen, J. S., Thorsen, M., Clausen, T., Hansen, S., and Christian Refsgaard, J.: Modelling of macropore flow and transport processes at catchment scale, J. Hydrol., 299, 136–158, 2004.

Chui, T. and Freyberg, D.: Implementing hydrologic boundary conditions in a multiphysics model, J. Hydrol. Eng., 14, 1374–1377, 2009.

Crosta, G. B. and Frattini, P.: Rainfall-induced landslides and debris flows, Hydrol. Process., 22, 473–477, 2008.

Dai, F. C., Lee, C. F., and Ngai, Y. Y.: Landslide risk assessment and management: an overview, Eng. Geol., 64, 65–87, 2002.

Debieche, T. H., Bogaard, T. A., Marc, V., Emblanch, C., Krzeminska, D. M., and Malet, J. P.: Hydrological and hydrochemical processes observed during a large-scale infiltration experiment at the Super-Sauze mudslide (France), Hydrol. Process., 26, 2157–2170, 2012.

Dusek, J., Gerke, H. H., and Vogel, T.: Surface boundary conditions in two-dimensional dual-permeability modeling of tile drain bromide leaching, Vadose Zone J., 7, 1287–1301, 2008.

Flury, M., Flühler, H., Jury, W. A., and Leuenberger, J.: Susceptibility of soils to preferential flow of water: a field study, Water Resour. Res., 30, 1945–1954, 1994.

Gerke, H. H.: Preferential flow descriptions for structured soils, J. Plant Nutr. Soil Sci., 169, 382–400, 2006.

Gerke, H. H. and Köhne, J. M.: Estimating hydraulic properties of soil aggregate skins from sorptivity and water retention, Soil Sci. Soc. Am. J., 66, 26–36, 2002.

Gerke, H. H. and Köhne, J. M.: Dual-permeability modeling of preferential bromide leaching from a tile-drained glacial till agricultural field, J. Hydrol., 289, 239–257, 2004.

Gerke, H. H. and van Genuchten, M.: A dual porosity model for simulating the preferential movement of water and solutes in structured porous media, Water Resour. Res., 29, 305–319, 1993a.

Gerke, H. H. and van Genuchten, M.: Evaluation of a first-order water transfer term for variably saturated dual-porosity flow models, Water Resour. Res., 29, 1225–1238, 1993b.

Godt, J. W., Baum, R. L., Savage, W. Z., Salciarini, D., Schulz, W. H., and Harp, E. L.: Transient deterministic shallow landslide modeling: requirements for susceptibility and hazard assessments in a GIS framework, Eng. Geol., 102, 214–226, 2008.

Greco, R.: Preferential flow in macroporous swelling soil with internal catchment: model development and applications, J. Hydrol., 269, 150–168, 2002.

Greco, R., Comegna, L., Damiano, E., Guida, A., Olivares, L., and Picarelli, L.: Hydrological modelling of a slope covered with shallow pyroclastic deposits from field monitoring data, Hydrol. Earth Syst. Sci., 17, 4001–4013, 2013, http://www.hydrol-earth-syst-sci.net/17/4001/2013/.

Griffiths, D. and Lane, P.: Slope stability analysis by finite elements, Geotechnique, 49, 387–403, 1999.

Griffiths, D. and Lu, N.: Unsaturated slope stability analysis with steady infiltration or evaporation using elasto-plastic finite elements, Int. J. Numer. Anal. Met., 29, 249–267, 2005.

Griffiths, D., Huang, J., and Fenton, G. A.: Probabilistic infinite slope analysis, Comput. Geotech., 38, 577–584, 2011.

Guzzetti, F., Peruccacci, S., Rossi, M., and Stark, C. P.: Rainfall thresholds for the initiation of landslides in central and southern Europe, Meteorol. Atmos. Phys., 98, 239–267, 2007.

Guzzetti, F., Peruccacci, S., Rossi, M., and Stark, C.: The rainfall intensity–duration control of shallow landslides and debris flows: an update, Landslides, 5, 3–17, 2008.

Gwo, J., Jardine, P., Wilson, G., and Yeh, G.: A multiple-pore-region concept to modeling mass transfer in subsurface media, J. Hydrol., 164, 217–237, 1995.

Hamdhan, I. N. and Schweiger, H. F.: Slope Stability Analysis of Unsaturated Soil with Fully Coupled Flow-Deformation Analysis,doi:10.5242/iamg.2011.0063 IMAG 2011, SALZBURG, 2011.

Hammouri, N., Malkawi, A. H., and Yamin, M. A.: Stability analysis of slopes using the finite element method and limiting equilibrium approach, B. Eng. Geol. Environ., 67, 471–478, 2008.

Hencher, S. R.: Preferential flow paths through soil and rock and their association with landslides, Hydrol. Process., 24, 1610–1630, 2010.

Hendrickx, J. M. and Flury, M.: Uniform and Preferential Flow Mechanisms in the Vadose Zone, National Academy Press, Washington, DC, 149–188, 2001.

Hu, Y., Feng, J., Yang, T., and Wang, C.: A new method to characterize the spatial structure of soil macropore networks in effects of cultivation using computed tomography, Hydrol. Process., 28, 3419–3431, 2014.

Huang, M. and Jia, C.-Q.: Strength reduction FEM in stability analysis of soil slopes subjected to transient unsaturated seepage, Comput. Geotech., 36, 93–101, 2009.

Itasca, F.: Fast Lagrangian Analysis of Continua, Version 4.0 User's Guide, Itasca Consulting Group Inc., Thrasher Square East, Minneapolis, MN, 708 pp., 2002.

Jarvis, N. J.: A review of non-equilibrium water flow and solute transport in soil macropores: principles, controlling factors and consequences for water quality, Eur. J. Soil Sci., 58, 523–546, 2007.

Jarvis, N. J., Jansson, P. E., Dik, P. E., and Messing, I.: Modelling water and solute transport in macroporous soil. I. Model description and sensitivity analysis, J. Soil Sci., 42, 59–70, 1991.

Jing, L.: A review of techniques, advances and outstanding issues in numerical modelling for rock mechanics and rock engineering, Int. J. Rock Mech. Min., 40, 283–353, 2003.

Kim, J., Salgado, R., and Yu, H.: Limit analysis of soil slopes subjected to pore-water pressures, J. Geotech. Geoenviron., 125, 49–58, 1999.

Kodešová, R., Kozák, J., Šimůnek, J., and Vacek, O.: Single and dual-permeability model of chlorotoluron transport in the soil profile, Plant Soil Environ., 51, 310–315, 2005.

Köhne, J. M. and Mohanty, B. P.: Water flow processes in a soil column with a cylindrical macropore: experiment and hierarchical modeling, Water Resour. Res., 41, W03010, doi:10.1029/2004wr003303, 2005.

Köhne, J. M., Köhne, S., and Gerke, H. H.: Estimating the hydraulic functions of dual-permeability models from bulk soil data, Water Resour. Res., 38, 26-1–26-11, 2002.

Köhne, J. M., Mohanty, B. P., and Šimůnek, J.: Inverse dual-permeability modeling of preferential water flow in a soil column and implications for field-scale solute transport, Vadose Zone J., 5, 59–76, 2006.

Köhne, J. M., Köhne, S., and Šimůnek, J.: A review of model applications for structured soils: a) Water flow and tracer transport, J. Contam. Hydrol., 104, 4–35, 2009.

Krzeminska, D. M., Bogaard, T. A., van Asch, Th. W. J., and van Beek, L. P. H.: A conceptual model of the hydrological influence of fissures on landslide activity, Hydrol. Earth Syst. Sci., 16, 1561–1576, doi:10.5194/hess-16-1561-2012, 2012.

Laine-Kaulio, H.: Development and Analysis of a Dual-Permeability Model for Subsurface Stormflow and Solute Transport in a Forested Hillslope, Ph.D. thesis, Aalto University, Aalto, 2011.

Laine-Kaulio, H., Backnäs, S., Karvonen, T., Koivusalo, H., and McDonnell, J. J.: Lateral subsurface stormflow and solute transport in a forested hillslope: a combined measurement and modeling approach, Water Resour. Res., 50, 8159–8178, 2014.

Lanni, C., McDonnell, J., Hopp, L., and Rigon, R.: Simulated effect of soil depth and bedrock topography on near-surface hydrologic response and slope stability, Earth Surf. Proc. Land., 38, 146–159, 2013.

Larsbo, M. and Jarvis, N.: MACRO 5.0: aModel of Water Flow and Solute Transport in Macroporous Soil: Technical Description, Tech. Rep. 9157665923, Department of Soil Sciencess, Swedish University of Agricultural Sciences, Uppsala, Sweden, 2003.

Leij, F. J.: The UNSODA Unsaturated Soil Hydraulic Database: User's Manual, University of Michigan, 1996.

Lu, N., Godt, J. W., and Wu, D. T.: A closed-form equation for effective stress in unsaturated soil, Water Resour. Res., 46, W05515, doi:10.1029/2009wr008646, 2010.

Lu, N., Şener Kaya, B., Wayllace, A., and Godt, J. W.: Analysis of rainfall-induced slope instability using a field of local factor of safety, Water Resour. Res., 48, W09524, doi:10.1029/2012wr011830, 2012.

McDonnell, J.: The influence of macropores on debris flow initiation, Q. J. Eng. Geol. Hydroge., 23, 325–331, 1990.

Montrasio, L. and Valentino, R.: A model for triggering mechanisms of shallow landslides, Nat. Hazards Earth Syst. Sci., 8, 1149–1159, doi:10.5194/nhess-8-1149-2008, 2008.

Moonen, P., Carmeliet, J., and Sluys, L.: A continuous–discontinuous approach to simulate fracture processes in quasi-brittle materials, Philos. Mag., 88, 3281–3298, 2008.

Mukhlisin, M., Taha, M., and Kosugi, K.: Numerical analysis of effective soil porosity and soil thickness effects on slope stability at a hillslope of weathered granitic soil formation, Geosci. J., 12, 401–410, 2008.

Mulungu, D. M. M., Ichikawa, Y., and Shiiba, M.: A physically based distributed subsurface-surface flow dynamics model for forested mountainous catchments, Hydrol. Process., 19, 3999–4022, 2005.

Nemes, A., Schaap, M., Leij, F., and Wösten, J.: Description of the unsaturated soil hydraulic database UNSODA version 2.0, J. Hydrol., 251, 151–162, 2001.

Ng, C. W. W. and Shi, Q.: A numerical investigation of the stability of unsaturated soil slopes subjected to transient seepage, Comput. Geotech., 22, 1–28, 1998.

Nieber, J. L. and Sidle, R. C.: How do disconnected macropores in sloping soils facilitate preferential flow?, Hydrol. Process., 24, 1582–1594, 2010.

Pastor, M., Fernández Merodo, J. A., Herreros, M. I., Mira, P., González, E., Haddad, B., Quecedo, M., Tonni, L., and Drempetic, V.: Mathematical, constitutive and numerical modelling of catastrophic landslides and related phenomena, Rock Mech. Rock Eng., 41, 85–132, 2008.

Picarelli, L., Urciuoli, G., Mandolini, A., and Ramondini, M.: Softening and instability of natural slopes in highly fissured plastic clay shales, Nat. Hazards Earth Syst. Sci., 6, 529–539, doi:10.5194/nhess-6-529-2006, 2006.

Pirastru, M. and Niedda, M.: Field monitoring and dual permeability modelling of water flow through unsaturated calcareous rocks, J. Hydrol., 392, 40–53, 2010.

Qiu, C., Esaki, T., Xie, M., Mitani, Y., and Wang, C.: Spatio-temporal estimation of shallow landslide hazard triggered by rainfall using a three-dimensional model, Environ. Geol., 52, 1569–1579, 2007.

Ray, C., Ellsworth, T. R., Valocchi, A. J., and Boast, C. W.: An improved dual porosity model for chemical transport in macroporous soils, J. Hydrol., 193, 270–292, 1997.

Ray, C., Vogel, T., and Dusek, J.: Modeling depth-variant and domain-specific sorption and biodegradation in dual-permeability media, J. Contam. Hydrol., 70, 63–87, 2004.

Roulier, S. and Jarvis, N.: Analysis of inverse procedures for estimating parameters controlling macropore flow and solute transport in the dual-permeability model MACRO, Vadose Zone J., 2, 349–357, 2003.

Shao, W., Boaggrd, T.A., and Bakker, M.: How to Use COMSOL Multiphysics for Coupled Dual-permeability Hydrological and Slope Stability Modeling, Proc. Earth Planet. Sci., 9, 83–90, 2014.

Sharma, R. and Nakagawa, H.: Numerical model and flume experiments of single- and two-layered hillslope flow related to slope failure, Landslides, 7, 425–432, 2010.

Shuin, Y., Hotta, N., Suzuki, M., and Ogawa, K.-I.: Estimating the effects of heavy rainfall conditions on shallow landslides using distributed landslide conceptual model, Phys. Chem. Earth, 49, 44–51, 2012.

Simoni, S., Zanotti, F., Bertoldi, G., and Rigon, R.: Modelling the probability of occurrence of shallow landslides and channelized debris flows using GEOtop-FS, Hydrol. Process., 22, 532–545, 2008.

Šimůnek, J., Jarvis, N. J., van Genuchten, M. T., and Gärdenäs, A.: Review and comparison of models for describing non-equilibrium and preferential flow and transport in the vadose zone, J. Hydrol., 272, 14–35, 2003.

Šimůnek, J., van Genuchten, M. T., and Šejna, M.: Development and applications of the HYDRUS and STANMOD software packages and related codes, Vadose Zone J., 7, 587–600, 2008.

Stead, D., Eberhardt, E., Coggan, J., and Benko, B.: Advanced numerical techniques in rock slope stability analysis – applications and limitations, International Conference on Landslides – Causes, Impacts and Countermeasures, Davos, Switzerland, 615–624, 2001.

Talebi, A., Uijlenhoet, R., and Troch, P. A.: A low-dimensional physically based model of hydrologic control of shallow landsliding on complex hillslopes, Earth Surf. Proc. Land., 33, 1964–1976, 2008.

Therrien, R., McLaren, R., and Sudicky, E.: HydroGeoSphere: a Three-Dimensional Numerical Model Describing Fully-Integrated Subsurface and Surface Flow and Solute Transport, Tech. rep., edited by: Panday, S. M., Hydrogeologic Inc./University of Waterloo, 2005.

Tsai, T.-L. and Yang, J.-C.: Modeling of rainfall-triggered shallow landslide, Environ. Geol., 50, 525–534, 2006.

Tsutsumi, D. and Fujita, M.: Relative importance of slope material properties and timing of rainfall for the occurrence of landslides, Int. J. Erosion Control Eng., 1, 79–89, 2008.

Uchida, T., Kosugi, K., and Mizuyama, T.: Effects of pipeflow on hydrological process and its relation to landslide: a review of pipeflow studies in forested headwater catchments, Hydrol. Process., 15, 2151–2174, 2001.

Uchida, T.: Clarifying the role of pipe flow on shallow landslide initiation, Hydrol. Process., 18, 375–378, 2004.

Uchida, T., Asano, Y., Mizuyama, T., and McDonnell, J. J.: Role of upslope soil pore pressure on lateral subsurface storm flow dynamics, Water Resour. Res., 40, W12401, doi:10.1029/2003WR002139, 2004.

van der Spek, J. E., Bogaard, T. A., and Bakker, M.: Characterization of groundwater dynamics in landslides in varved clays, Hydrol. Earth Syst. Sci., 17, 2171–2183, doi:10.5194/hess-17-2171-2013, 2013.

Verachtert, E., Van Den Eeckhaut, M., Poesen, J., and Deckers, J.: Spatial interaction between collapsed pipes and landslides in hilly regions with loess-derived soils, Earth Surf. Proc. Land., 38, 826–835, 2013.

Vogel, T., Gerke, H. H., Zhang, R., and Van Genuchten, M. T.: Modeling flow and transport in a two-dimensional dual-permeability system with spatially variable hydraulic properties, J. Hydrol., 238, 78–89, 2000.

von Ruette, J., Lehmann, P., and Or, D.: Effects of rainfall spatial variability and intermittency on shallow landslide triggering patterns at a catchment scale, Water Resour. Res., 50, 7780–7799, 2014.

Weiler, M.: An infiltration model based on flow variability in macropores: development, sensitivity analysis and applications, J. Hydrol., 310, 294–315, 2005.

Westen, C. J., Asch, T. W. J., and Soeters, R.: Landslide hazard and risk zonation – why is it still so difficult?, B. Eng. Geol. Environ., 65, 167–184, 2006.

Wienhöfer, J., Lindenmaier, F., and Zehe, E.: Challenges in Understanding the Hydrologic Controls on the Mobility of Slow-Moving Landslides, Vadose Zone J., 10, 496–511, 2011.

Wilkinson, P. L., Anderson, M. G., and Lloyd, D. M.: An integrated hydrological model for rain-induced landslide prediction, Earth Surf. Proc. Land., 27, 1285–1297, 2002.

Wu, Y.-S., Liu, H. H., and Bodvarsson, G. S.: A triple-continuum approach for modeling flow and transport processes in fractured rock, J. Contam. Hydrol., 73, 145–179, 2004.

Zehe, E., Maurer, T., Ihringer, J., and Plate, E.: Modeling water flow and mass transport in a loess catchment, Phys. Chem. Earth Pt. B, 26, 487–507, 2001.

Zhang, G. P., Savenije, H. H. G., Fenicia, F., and Pfister, L.: Modelling subsurface storm flow with the Representative Elementary Watershed (REW) approach: application to the Alzette River Basin, Hydrol. Earth Syst. Sci., 10, 937–955, doi:10.5194/hess-10-937-2006, 2006.

Zhou, C., Shao, W., and van Westen, C. J.: Comparing two methods to estimate lateral force acting on stabilizing piles for a landslide in the Three Gorges Reservoir, China, Eng. Geol., 173, 41–53, 2014.

8

Multiple runoff processes and multiple thresholds control agricultural runoff generation

Shabnam Saffarpour[1], Andrew W. Western[1], Russell Adams[1], and Jeffrey J. McDonnell[2,3]

[1]Department of Infrastructure Engineering, The University of Melbourne, Parkville, 3010, Australia
[2]Global Institute for Water Security, University of Saskatchewan, Saskatoon, SK S7N 3H5, Canada
[3]School of Geosciences, University of Aberdeen, Aberdeen, UK

Correspondence to: Andrew Western (a.western@unimelb.edu.au)

Abstract. Thresholds and hydrologic connectivity associated with runoff processes are a critical concept for understanding catchment hydrologic response at the event timescale. To date, most attention has focused on single runoff response types, and the role of multiple thresholds and flow path connectivities has not been made explicit. Here we first summarise existing knowledge on the interplay between thresholds, connectivity and runoff processes at the hillslope–small catchment scale into a single figure and use it in examining how runoff response and the catchment threshold response to rainfall affect a suite of runoff generation mechanisms in a small agricultural catchment. A 1.37 ha catchment in the Lang Lang River catchment, Victoria, Australia, was instrumented and hourly data of rainfall, runoff, shallow groundwater level and isotope water samples were collected. The rainfall, runoff and antecedent soil moisture data together with water levels at several shallow piezometers are used to identify runoff processes in the study site. We use isotope and major ion results to further support the findings of the hydrometric data. We analyse 60 rainfall events that produced 38 runoff events over two runoff seasons. Our results show that the catchment hydrologic response was typically controlled by the Antecedent Soil Moisture Index and rainfall characteristics. There was a strong seasonal effect in the antecedent moisture conditions that led to marked seasonal-scale changes in runoff response. Analysis of shallow well data revealed that streamflows early in the runoff season were dominated primarily by saturation excess overland flow from the riparian area. As the runoff season progressed, the catchment soil water storage increased and the hillslopes connected to the riparian area. The hillslopes transferred a significant amount of water to the riparian zone during and following events. Then, during a particularly wet period, this connectivity to the riparian zone, and ultimately to the stream, persisted between events for a period of 1 month. These findings are supported by isotope results which showed the dominance of pre-event water, together with significant contributions of event water early (rising limb and peak) in the event hydrograph. Based on a combination of various hydrometric analyses and some isotope and major ion data, we conclude that event runoff at this site is typically a combination of subsurface event flow and saturation excess overland flow. However, during high intensity rainfall events, flashy catchment flow was observed even though the soil moisture threshold for activation of subsurface flow was not exceeded. We hypothesise that this was due to the activation of infiltration excess overland flow and/or fast lateral flow through preferential pathways on the hillslope and saturation overland flow from the riparian zone.

1 Introduction

Thresholds have been an integral part of overland flow theory since the early infiltration excess work of Horton (1933) and saturation excess studies of Dunne and Black (1970a, b). Thresholds in runoff response have also been observed in subsurface stormflow dominated systems (Hewlett and Hibbert, 1967). More recent work has shown these to be a function of catchment wetness status for saturation excess overland flow (Western and Grayson, 1998; Western et al., 2005) and subsurface stormflow (Freer et al., 2002;

Tromp-van Meerveld et al., 2007). Hydrological connectivity is now a useful generic concept that links reservoirs to their downstream conduits (Tetzlaff et al., 2010) and a connectivity framework can provide a powerful explanator of catchment flow and transport response (Ali et al., 2013; Detty and McGuire, 2010; Lehmann et al., 2007; McGuire and McDonnell, 2010; Western et al., 1998, 2001).

In this paper we are interested in connectivity in terms of the movement of water from hillslopes to streams at the timescale of events and longer. We say there is connectivity along a flow pathway when water is moving along that pathway and contributing to stream flow from the catchment. Connectivity and thresholds are intimately related; typically a threshold in some catchment state controls the transition between connected and disconnected states, for example, the observation that subsurface flow becomes connected above some soil water storage and rainfall threshold (Detty and McGuire, 2010; Tromp-van Meerveld and McDonnell, 2006a; Fu et al., 2013; Penna et al., 2015). In this study we use the threshold concept to examine runoff generation mechanisms and to discuss how various mechanisms produce runoff and change in importance during the runoff season.

Despite significant progress in understanding the non-linear behaviour of catchments related to soil moisture thresholds, watertable dynamics, connectivity of surface and subsurface pathways and their influence on runoff generation mechanisms, it is not explicitly understood how the non-linear properties of catchments (connectivity and thresholds) work to convert rainfall to runoff or how such behaviours vary between different types of catchments. It has been argued that interactions between the various processes and thresholds lead to complex non-linear rainfall–runoff behaviour in catchments (Hopp and McDonnell, 2009; Kirchner, 2006; Tetzlaff et al., 2010; Uchida et al., 2005), including thresholds for initiation of hillslope-to-stream connectivity (Ali et al., 2013; Detty and McGuire, 2010; Fujimoto et al., 2008; Lehmann et al., 2007; McGuire and McDonnell, 2010; Tromp van Meerveld and McDonnell, 2005, 2006a), variable flow hysteresis patterns depending on rainfall amount and antecedent soil moisture conditions (Bowes et al., 2009; Holz, 2010; McGuire and McDonnell, 2010), and flushing of nutrients in agricultural catchments (Bracken and Croke, 2007; Ocampo et al., 2006; Tockner et al., 1999; Withers and Lord, 2002). Moreover, the explicit linkage of runoff mechanisms and flow pathways has received less attention in agricultural landscapes compared with forested basins. An exception is the study of Ocampo et al. (2006). Ocampo et al. (2006) investigated hydrologic connectivity, the threshold dependency of connectivity and its influence on seasonal and event based runoff mechanisms in an agricultural catchment in Western Australia.

While the concept of connectivity has been useful in many of these studies, most studies have concentrated on individual mechanisms. It is less clear how catchments behave when subject to a mixture of runoff mechanisms including infiltration excess and saturation excess overland flow, and subsurface stormflow. Only a small number of studies have tried to tease apart the influence of multiple processes in catchments where infiltration excess runoff, saturation excess runoff and subsurface stormflow are all important (e.g. Lana-Renault et al., 2007, 2014; Latron and Gallart, 2008). These catchments are in Mediterranean environments and have typically been studied by field inspection of the catchment surface to identify infiltration excess runoff generation areas, field inspection of saturated areas to identify saturation excess source areas and also by examination of hydrographs in combination with groundwater wells. These studies associate certain controls with specific processes, such as past cultivation being associated with infiltration excess runoff (Lana-Renault et al., 2014). These studies have generally not considered rainfall intensity information at scales finer than daily. Despite the considerable improvement in our understanding of basic catchment functional mechanisms, today it is still challenging to apply our understanding of these specific case studies to other areas due to the problems of catchment heterogeneity, complexity, non-linearity of behaviour, scale, etc. (Beven, 2002).

To aid systematic consideration of the variation in runoff processes between catchments and over time, we develop a summary of the status quo in terms of the combined effects of thresholds and connectivity on runoff processes at the hillslope to zero-order catchment scale (Fig. 1). Often we think of dominant runoff processes and, as Fig. 1 is easiest to interpret in that context, the following discussion takes that view initially. However, there is often a mix of runoff processes either spatially due to heterogeneity or for different events, and Fig. 1 can also be used to interpret such a mixture, which we will return to later. This paper aims to tease apart the influence of different processes by considering 60 rainfall events (resulting in 38 runoff events) in a small agricultural catchment. We show the shifting importance of different processes over time associated with changes in catchment wetness and rainfall intensity and apply the runoff process framework. We consider the role of thresholds in different catchment states and fluxes as well as the role of thresholds in certain timescales in controlling different modes of hydrologic connectivity and associated rainfall–runoff responses. The variety of potential thresholds leads to a variety of runoff processes, as illustrated in the following discussion.

Figure 1 shows the importance of various timescales, durations, fluxes and states, and how these relate to variation in rainfall–runoff processes over time (and between catchments with different physiographic characteristics). Of course, questions of instantaneous flux and also of the relative timescales of various processes are often important in determining the existence of connectivity (Tromp van Meerveld and McDonnell, 2005, 2006b; Western et al., 2005). It would be attractive to think of the problem of runoff response purely in terms of timescales of competing processes following Oldham et al. (2013), who used a generalised Damköhler number

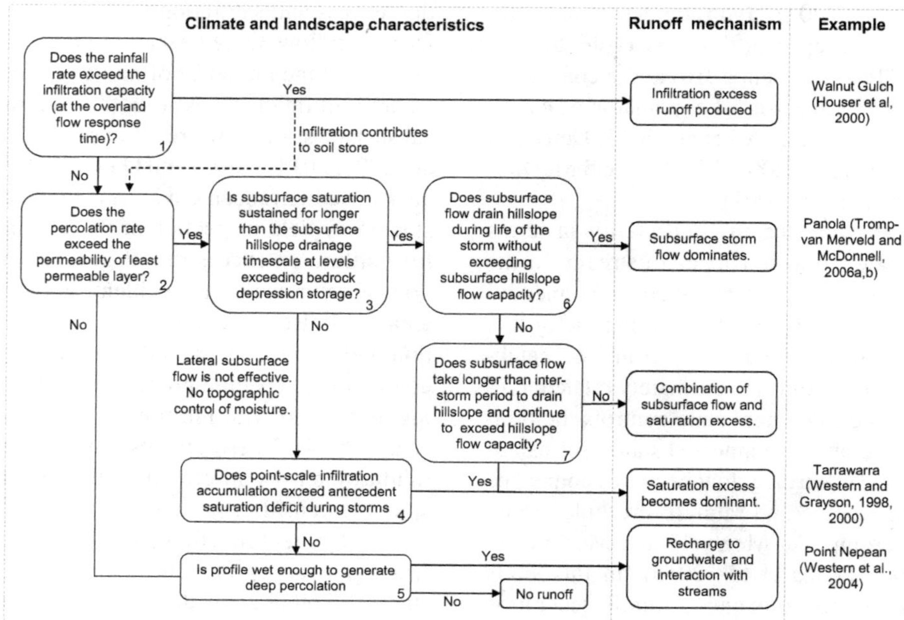

Figure 1. The role of flux and timescale thresholds in determining runoff processes. The red lines indicate cases where there is surface or subsurface hillslope connectivity to the stream.

to represent the competing effects of transport and reaction timescales on the loss of material along a flow path. When the reaction timescale is small compared with the transport timescale, the reactant is consumed before reaching the exit of the flow path it is moving along. A complication here is that both flux and time thresholds are important. This arises because there is finite capacity for flow in various parts of the catchment system.

Figure 1 is divided into three parts; the left-hand area provides a series of catchment thresholds that depend on hydroclimatic and landscape characteristics and influence the type of runoff process and the connectivity between the hillslope and stream, depending on whether they are exceeded or not. The middle area points to the outcome in terms of dominant runoff generation processes and the right-hand area provides example catchments from the literature that exhibit those processes. In Fig. 1 we define the specific runoff processes as follows.

Infiltration excess runoff: runoff that occurs due to the rainfall (or throughfall) rate exceeding the infiltration capacity of the surface and that results in flows of water to the catchment outlet by surface flow pathways.

Subsurface stormflow: runoff due to infiltration that generates rapid lateral subsurface flow (i.e. a quickflow response) that flows to the catchment outlet through subsurface flow paths for at least part of the distance to the outlet. This water may exfiltrate in low convergent parts of the catchment or directly into the stream before reaching the catchment outlet. Often impeding layers and preferential flow paths are implicated.

Saturation excess runoff: runoff due to rainfall on saturated areas that flows to the catchment outlet by surface flow pathways. The saturated area may be generated by either lateral flow in excess of the capacity of the hillslope to transmit the lateral flow, a drainage impediment at depth coupled with a sufficient excess of infiltration over evapotranspiration and drainage, or a combination of these.

Some of the thresholds are posed in terms of flux rate compared with a flow capacity (e.g. box 2) and some in terms of a state threshold (box 5). The flux and state thresholds are considered in the context of process timescales and durations. This is because the threshold needs to be exceeded for a sufficient time for the action of the process to lead to a significant impact. That impact typically involves lateral flow either on the surface or through the subsurface and hence also interactions upslope and downslope. Specific examples are given below.

Consider box 1. Rainfall rates vary across a very wide range of timescales. If the rainfall (or throughfall) intensity exceeds the infiltration threshold for only a very short time, the water that ponds on the surface will continue to infiltrate as it flows down the hillslope toward the stream (run-on infiltration) after the intensity reduces and very little or no runoff will reach the stream (surface connectivity did not become established). However, if average intensities exceed the infiltration capacity for long enough for ponded water to flow to the catchment outlet, the hillslope will connect to the catchment outlet via surface pathways and produce runoff. Thus the duration of high intensity rainfall compared with the overland flow timescale (flow distance divided by wave celerity) is important. The remaining boxes consider thresholds in the context of subsurface flow times. Box 3 considers situations where subsurface saturation exists, allowing water to flow along lateral subsurface flow paths. If any one of deep infiltration through the impeding layer (Jackson et al., 2014), unfilled bedrock storage (Janzen and McDonnell, 2015) or evapotranspiration (including between events) causes the saturation and/or lateral flow to cease before water can move a significant distance downstream, the water will not be effectively redistributed downslope, and subsurface connection will not be established. If the saturation persists for long enough for lateral subsurface flow to move down the hillslope and into the stream, connectivity between the hillslope and the stream will develop. Thus the ratio of the lateral flow timescale for the hillslope and the evaporative drying (or other loss) timescale is important here. At the other extreme (box 7), if lateral flow is persistently exceeding the subsurface flow capacity, surface saturation will exist, leading to saturation excess runoff because saturated areas will exist antecedent to the event in this case. Provided continuous saturation exists to the stream, this will lead to a surface flow path connecting the hillslope and stream. In this case the lateral flow timescale is long enough to lead to memory in the spatial patterns between events (non-local control of Grayson et al., 1997).

While Fig. 1 suggests catchment rainfall–runoff response is dominated by specific processes (e.g. saturation excess runoff), it needs to be recognised that many catchment conditions vary over time and space. For example, in our study, catchment summer rainfall is often more intense than winter rainfall, and this can lead to differences in runoff processes between events. Soil water conditions vary seasonally in response to both rainfall and potential evapotranspiration, sometimes leading to switching between characteristic spatial patterns of soil moisture and prevailing responses to rainfall (Grayson et al., 1997; Western et al., 1999). Topographic, soil and vegetation conditions can also vary across a catchment. This all suggests that catchments could exhibit a mix of processes.

Having introduced the framework above, we use it to understand the behaviour of a catchment in Australia that does indeed exhibit a mix of runoff processes. We examine how soil water storage and shallow water table response influence surface and subsurface connectivity and rainfall–runoff response at seasonal and event based timescales. We also examine the relative role of saturation excess and subsurface flow in generating peak runoff rates and event volumes. Finally we examine circumstances under which rainfall intensity plays a role in runoff generation responses. The field site is a small agricultural catchment in the Lang Lang River catchment, Victoria, Australia, which we examine through the lens of hydrometric and isotope and geochemistry measurements. In the context of Fig. 1, these results are used to examine various runoff generation mechanisms and flow pathways that are important in the study catchment and to determine how they contribute to produce the catchment runoff as the catchment wetness and rainfall intensity vary. Thus the contributions of this paper revolve around the introduction of the framework in Fig. 1 and demonstration of its application using a catchment where multiple runoff processes are important. The paper also contributes to further understanding of the role of thresholds and connectivity in determining flow pathways and runoff processes in agricultural catchments. Most connectivity studies in the past have focussed on forested catchments.

2 Methods

2.1 Study location

The study site is a 1.37 ha catchment (named RBF) located on a dairy farm at Poowong East, in the Lang Lang River catchment, Victoria, Australia, 130 km south-east of Melbourne (Fig. 2). The study area has a humid climate and rainfall is uniformly distributed across the year, with an annual mean (1961–1990) of 1100 mm (Bureau of Meteorology, 2009). Annual areal potential evapotranspiration (1961–1990) is 1040 mm (Bureau of Meteorology, 2005).

A general description of the study catchment can be found in Adams et al. (2014). The study period was between September 2009 and December 2011. Elevation ranges from 160 to 210 mAHD (Australian Height Datum) and the slope varies from 2 to 50 %. Based on field observations, the topography, range of slopes and the groundwater behaviour, the catchment was divided into four different zones: (1) the riparian area located on the relatively flat convergent lower part of the catchment (outlined in red in Fig. 2) included sites 1, 2, 32 and 3; (2) the lower slope (low slope) area; (3) the mid-slope area with sites 4, 5, 6 and 7; and (4) the upper slope (upslope) area with sites 10, 11 and 15 (Fig. 2).

The catchment geology comprises sandstones and mudstones of the Cretaceous Strezlecki Group (VRO, 2013). Outcrops on the lower stream banks of the catchment (just downstream of the monitored hillslope) show weathered sandstone and mudstone bedrock. Hand augering revealed a soil depth of between 1 and 1.6 m, and the lower parts of the profile included mottled clay and weathered bedrock particles. The soils are acidic and mesotrophic brown dermosols (Isbell, 2002) that grade from a fine sandy clay loam to a medium clay with mottles and weathered bedrock. Soil profile depth decreases, moving downslope. These soils typically have a moderate hydraulic conductivity surface horizon (0–40 cm, $K_s \approx 5\times10^{-6}$ m s^{-1}, about 20 mm h^{-1}). The dominant land use is grazing by dairy cows.

2.2 Site instrumentation and hydrometric data monitoring

Stream discharge was measured at the catchment outlet (Fig. 2) using an RBC flume (Clemens et al., 1984) and an

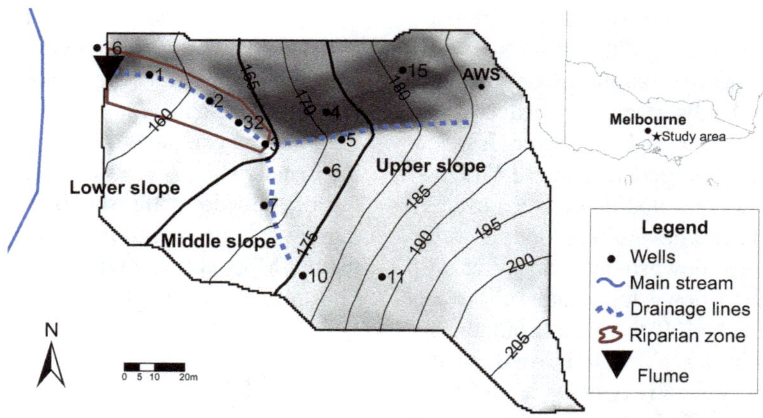

Figure 2. The study site location within Australia and a hill shaded DEM, topography and sampling site locations at RBF. In this figure the black circles show shallow groundwater sites, the red circle shows the soil moisture site and the blue triangle demonstrates the catchment flume.

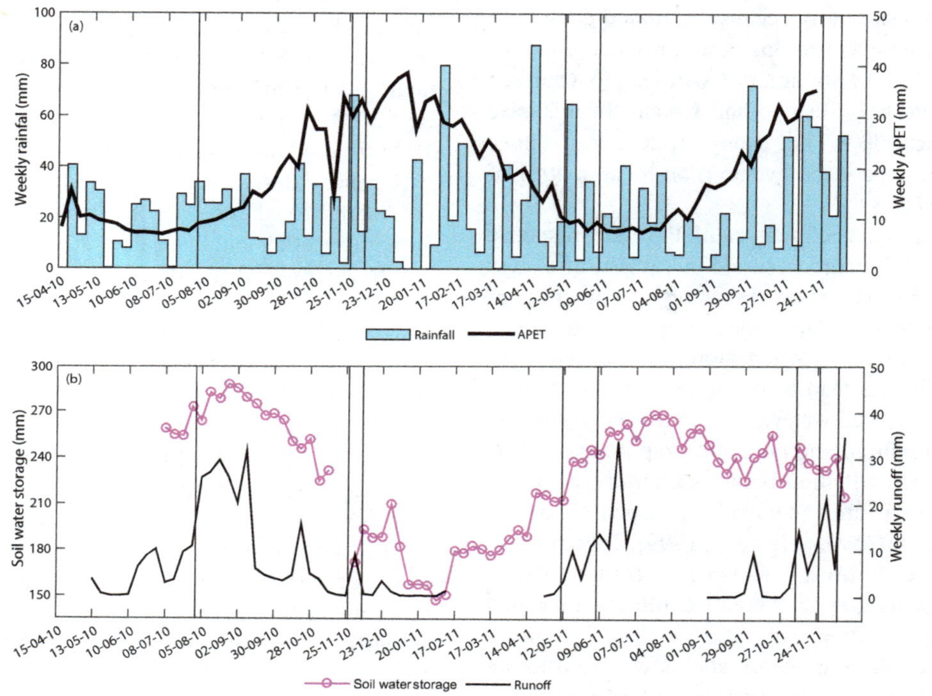

Figure 3. (a) Weekly rainfall and APET (areal potential evapotranspiration) time series, **(b)** soil water storage in the top 60 cm of the soil profile and weekly runoff time series for the flume at the catchment outlet. The rainfall, areal potential evapotranspiration and volumetric soil moisture used to calculate soil moisture storage were all recorded at the AWS (automatic weather station) (Fig. 1). The vertical bars show the timing of events in Fig. 4.

Odyssey (Dataflow Systems inc. Christchurch, NZ) pressure transducer (PT) recorded stream water levels every 10 min, which were used to compute instantaneous discharge rates. After August 2011, the PT was replaced with an ISCO (Teledyne ISCO, Lincoln, NE, USA) model 730 bubbler. The following rating curve was used to calculate Q from water level:

$$Q = 0.001H^2 + 0.0168H, \quad (1)$$

where Q is stream flow discharge and H is water head.

Rainfall data were recorded using a tipping-bucket raingauge at an automatic weather station (AWS) which was installed in 2010 on the upper boundary of RBF. Weather variables (temperature, humidity, wind, rainfall, global radiation) were measured by the AWS. Areal potential evapotranspiration (APET) was also computed using the Morton (1983) wet environment method on a daily basis. APET was strongly seasonal, resulting in strongly seasonal soil moisture contents and intermittent streamflow at RBF. Soil moisture storage was calculated from the volumetric soil water content which was measured for the 0–30 and 30–60 cm layers and recorded hourly by the AWS logger using two vertically installed 30 cm long Campbell Scientific (CS625) soil moisture probes (Campbell Scientific, 2006). The 0–30 cm probe was inserted from the surface. To install the 30–60 cm probe, a 30 cm deep hole was dug, with the excavated soil set aside. The probe was inserted vertically and then the excavated soil was repacked so that the soil was replaced at a similar depth to that from which it had been removed.

The CS625 produces a pulse signal and the pulse period was temperature corrected using measured soil temperature, and the manufacturers recommended temperature correction, as follows (Campbell Scientific, 2006).

$$\tau_{\text{corrected}}(T_{\text{soil}}) = \tau_{\text{uncorrected}} + (20 - T_{\text{soil}}) \cdot \left(0.526 - 0.052 \cdot \tau_{\text{uncorrected}} + 0.00136 \cdot \tau_{\text{uncorrected}}^2\right), \quad (2)$$

where $\tau_{\text{uncorrected}}$ is the probe output period, T_{soil} is the soil temperature and $\tau_{\text{corrected}}$ is the corrected probe output. The VWC was then computed from $\tau_{\text{corrected}}$ using (Campbell Scientific, 2006)

$$\text{VWC} = -0.0663 - 0.0063 \cdot \text{period} + 0.0007 \cdot \text{period}^2. \quad (3)$$

Soil water storage over the top 60 cm soil depth was computed by adding the VWC from the 0–30 and 30–60 cm layers and then multiplying by the 300 mm soil depth. We use this soil water storage at the start of each rainfall event as an index of antecedent soil water (ASI) and assume it represents the catchment wetness condition.

To capture the nature of hydrologic connectivity, runoff mechanisms and flow pathways, shallow (1.5–1.6 m) groundwater wells were installed at 12 sites across the RBF catchment using 40 mm PVC pipes and backfilled with sand, bentonite, the topsoil and grass. Figure 2 shows these sites, of which 1, 2, 3, 16 and 32 were in the riparian zone; 4, 5, 6 and 7 were on the mid-slope; and 10, 11 and 15 were on the upper slope. Sites 4, 5 and 6 were equipped with water level loggers from July 2010 until the end of study period in December 2011. Sites 3, 7, 16 and 32 were logged from winter 2011 until the end of the study period in December 2011. Water level loggers were not installed at sites 1 and 2 since they were nearly always saturated. At sites 1 and 2 groundwater levels were measured manually. Water levels were logged using Odyssey PT loggers.

2.3 Water sampling and analysis

A rainfall sampler (Kennedy et al., 1979) collected up to 10 sequential rainfall samples per event, each being equivalent to 6.6 mm of rainfall. The sampler was initially installed close to the AWS; however, due to instances of damage by animals, it was relocated near to the flume in August 2010 until the end of the study period (December 2011).

An auto sampler (Teledyne ISCO 6712) was installed at the flume and streamflow sampling was triggered based on the rising stage. Following triggering, the sampler was programmed to collect up to 24 samples of streamflow at hourly intervals. Samples were removed from the auto sampler within 48 h. To reduce the laboratory analysis workload, we plotted recorded water levels from the RBF catchment in the field prior to removing the event sample from the auto sampler and selected certain samples for analysis. All samples during the rising limb and the peak were selected and samples were typically selected at an interval of 4 h during the falling limb. Routine grab sampling was undertaken at weekly intervals during the main runoff season when water was flowing through the RBF flume. This was supplemented by additional grab sampling during visits to collect event samples from the auto sampler. Routine grab sampling was also undertaken at weekly intervals during the main runoff season when water was flowing through the RBF flume.

Sub-samples for isotopic analysis were taken of stream water from both manual and auto sampler samples, and from all full rainfall sample bottles; these were collected in glass bottles for isotope analysis. Bottles were completely filled. The samples were refrigerated (+4 °C) until analysis for δ^{18}O and δ^2H. Stable isotope ratios were measured either at Monash University using a Finnigan MAT 252 and ThermoFinnigan Delta Advantage Plus mass spectrometers (2010 samples) or at the University of Melbourne, where a Picarro L2120i cavity ring-down isotope analyser was used to determine isotope ratios (2011 samples). The instrumental precision was $\delta^{18}\text{O} = \pm 0.15$ ‰ and $\delta 2\text{H} = \pm 1$ ‰ for the isotope samples analysed at Monash University, and it was $\delta^{18}\text{O} = 0.1$ ‰ and $\delta^2\text{H} = 0.4$ ‰ for samples analysed at Melbourne University. The results from the isotope analysers were checked for systematic differences, by analysing duplicate samples in both laboratories. We developed and applied a correction between the two laboratories based on these samples. With the exception of determining pre-event end member uncertainty (described later), the analyses presented here only use samples analysed at the Monash laboratory, and hence this difference between laboratories only affects our hydrograph separation uncertainty estimates. For isotope analysis, in total we collected 115 samples of rainfall, 28 samples during low flows and multiple stream water samples of isotopes from five events.

For the measurement of major ions, sub-samples of grab and event streamflow samples were collected using 50 mL plastic bottles. Sub-samples were also taken from each water sample for selected major ion (Na^+, K^+, Ca^{2+}, Mg^{2+} and Cl^-) analyses, which were analysed in the NATA-certified, analytical chemistry laboratory of the Water Studies Centre at Monash University using standard methods. Major ions were determined in non-filtered samples as follows (Cm is ion concentration).

Table 1. Rainfall–runoff event summary at RBF. ASI is the Antecedent Soil moisture Index. RC is the quickflow runoff coefficient. Runoff event grouping is discussed in Sect. 3.2. The events in bold are shown in Fig. 4.

Group no.	Date	Rain duration (h)	Total rainfall depth (mm)	Peak hourly rainfall intensity ($mm\,h^{-1}$)	ASI (mm)	ASI + rainfall depth (mm)	Runoff duration (h)	Total runoff depth (mm)	Event runoff depth (mm)	RC (%)	Saturated area (%)
2	25/06/2010	49	22.4	1.8	237	259	74	7.6	1	4	4.5
2	13/07/2010	33	23.8	5.4	237	261	41	3.5	0.54	2	
2	20/07/2010	15	9.2	4.2	272	281	32	3.1	0.8	9	5.3
2	**01/08/2010**	**30**	**31.4**	**6.2**	**253**	**284**	**67**	**19.5**	**13.7**	**44**	**5.5**
2	05/08/2010	52	23.6	2	275	299	96	21.6	7	30	5.5
2	12/08/2010	34	17.2	5	274	291	94	12.7	4.3	25	5.5
2	16/08/2010	50	19.2	3.4	275	294	73	16.9	7.7	40	5.5
2	18/08/2010	34	14.6	4	286	301	57	13.8	4.6	32	5.5
2	24/08/2010	4	9.6	5.2	273	283	24	4.1	1.6	17	5.5
2	25/08/2010	59	12.6	2	285	298	73	10	1.4	11	5.5
2	31/08/2010	22	12.4	2	271	283	80	9.5	0.8	6	5.5
2	05/09/2010	47	25.2	3.4	272	297	74	23.1	13.5	54	5.5
2	09/09/2010	17	8.0	2.4	264	272	16	1.1	0.2	3	5.3
2	06/10/2010	18	15.2	6.6	231	246	3	0.37	0.27	2	NA
2	15/10/2010	59	34.2	2.8	228	262	86	11.3	3.9	11	NA
2	23/10/2010	11	12.8	5.2	227	240	8	1	0.57	4	NA
2	30/10/2010	29	32.8	8.8	202	235	16	1.2	0.83	3	NA
3	**27/11/2010**	**19**	**54.4**	**30.4**	**161**	**215**	**31**	**7.6**	**6.5**	**12**	**NA**
2	19/12/2010	49	25.6	3.8	191	217	10	0.83	0.18	1	NA
3	04/02/2011	33	71.2	16.4	130	201	10	2.5	2	3	NA
2	**11/05/2011**	**122**	**63.6**	**6.6**	**220**	**284**	**47**	**5.6**	**2.2**	**3**	**NA**
2	22/05/2011	50	21.6	4.4	231	253	3	0.23	0.1	0.46	NA
2	26/05/2011	37	13.6	1.6	244	258	33	2.66	1	7	NA
2	**07/06/2011**	**25**	**17.6**	**2.6**	**254**	**272**	**43**	**6.7**	**2.9**	**16**	**NA**
2	17/06/2011	31	15.6	2.2	248	264	45	3.2	0.2	1	4.6
2	21/06/2011	61	38.8	3.6	255	294	80	21.8	12.1	31	4.6
2	05/07/2011	10	9.8	2.4	253	263	26	2.9	0.8	8	4.7
2	06/07/2011	12	16.6	6.2	267	284	29	8.7	4.8	29	4.7
2	10/07/2011	20	7.4	2.2	262	269	27	3.7	1.4	19	4.7
2	28/09/2011	74	72.0	8.8	204	276	96	20.1	6.1	8	5.4
2	09/10/2011	54	23.2	2.4	232	255	5	0.35	0.14	1	NA
2	28/10/2011	19	30.2	7.8	213	243	6	0.92	0.6	2	NA
3	**08/11/2011**	**12**	**35.2**	**14.8**	**222**	**257**	**31**	**10**	**7.1**	**20**	**5.2**
2	09/11/2011	14	23.8	6.8	239	263	44	15.7	8.9	37	5.2
2	18/11/2011	30	45.5	10.2	219	265	82	17.7	8.9	20	5.5
2	**26/11/2011**	**29**	**36.8**	**8.4**	**216**	**253**	**85**	**27.6**	**16.3**	**44**	**5.5**
2	29/11/2011	47	19.4	4	233	252	70	11.9	0.9	5	5.5
3	**10/12/2011**	**16**	**52.6**	**31**	**192**	**245**	**32**	**41.1**	**35.6**	**68**	**5.5**

Group 2 has ASI + Rain > 250 mm; Group 3 has $I_{peak} > 15\,mm\,h^{-1}$.
"NA" stands for "not available".

- Mg^{2+} was determined by atomic absorption spectrometry (APHA, 2005). The issue of interference with Si, Al, and P was solved using a combination of lanthanum and caesium. The determined uncertainty for Mg^{2+} was $Cm \cdot 0.0596$, with a 95 % confidence interval (CI).

- Ca^{2+} was analysed using atomic absorption spectrometry utilising an air/acetylene flame based on the American Public Health Association procedure (APHA, 2005). The issue of interferences with Si, Al, and P was solved using lanthanum releasing agents. The determined uncertainty for Ca^{2+} was $Cm \cdot 0.0386$ (95 % CI).

- Following the APHA (2005) approach, "flame atomic absorption spectrometry" was used to determine K^+ concentration and caesium was used to solve the issue of ionisation. Uncertainty for K^+ was $Cm \cdot 0.0372$ (95 % CI).

- Na^+ which is "ionized in the air/acetylene flame" was analysed by adding caesium to overcome this problem following the APHA (2005) method. Uncertainty in Na^+ results was $Cm \cdot 0.0432$ (95 % CI).

- Cl^- was analysed by colourimetric method "using flow injection analysis (FIA)" (APHA, 2005). The uncertainty in Cl^- analysis laboratory results was $Cm \cdot 0.05$ (95 % CI).

2.4 Rainfall and runoff events

In order to analyse event behaviour, it was necessary to identify rainfall and runoff events. Based on an examination of the time series of hourly rainfall in the catchment (in the study period which was between April 2010 and December 2011), rain events were defined as having $>= 5$ mm total rainfall, and peak hourly rainfall intensity, $I_{peak} >= 1.5\,mm\,h^{-1}$. Distinct events were separated by > 12 h without rainfall.

The runoff hydrograph was also divided into events. Runoff events began when the stream discharge hydrograph started to rise from its initial low flow value or moved above a threshold of $0.05\,mm\,h^{-1}$ following the commencement of a rainfall event. Events ended either when (1) the discharge returned to its initial value, (2) a new rainfall event started, or (3) 96 h after the end of the rainfall event in unusually wet situations where elevated flow continued. For each event, a number of characteristics were determined as shown in Tables 1 and 2.

The Antecedent Soil moisture Index (ASI) was represented as the amount of the soil water storage in the top 60 cm of the profile at the AWS at the start of each rainfall event. The topography of the catchment was surveyed using a differential Global Positioning System (dGPS) and a 1 m horizontal resolution DEM (digital elevation model) was developed by interpolation methods. Instrument and well locations were also determined using dGPS. Then these data were used to produce maps of the study area, including sampling sites using Arc GIS. The lateral boundary of the saturated area was topographically constrained and field inspections suggested it was stable over time, while the upstream boundary moved up and down the riparian zone. The saturated area was estimated by locating the upper boundary through field inspection and then measuring the distance from either well 2 or 3 (the saturated area extended towards site 3 during very wet conditions). Figure 2 demonstrates the approximate maximum boundary location of the saturated area at site 3 and the main stream located close to the outlet of the catchment. The saturated area was then estimated from this information combined with the mapping of the riparian zone boundary in ArcGIS. These measurements were made between events. The proportion of saturated area was estimated using these data and then used to estimate saturation excess runoff generation for the different events. Consistent with our definitions in Fig. 1, here we conceptually separate return flow and flow resulting from direct precipitation on the saturated area and use saturation excess runoff to refer to the latter. The event runoff depth (mm) and event runoff coefficient (RC %) were calculated by separating the event hydrograph using the method of Hewlett and Hibbert (1967), which has been widely applied (Buttle et al., 2004; Fujimoto et al., 2008; McGuire and McDonnell, 2010). The method assumes that baseflow increases at the rate of $0.55\,L\,s^{-1}\,km^{-2}\,h^{-1}$ ($0.002\,mm\,h^{-1}$) from the start of the rising limb. While this method, like other hydrograph based baseflow separation methods, is arbitrary, the results are insensitive to the assumed rate of rise.

2.5 Isotopic hydrograph separation

Isotope samples are used for hydrograph separation into pre-event and event contributions. We applied the well-known one-tracer, two-component model of hydrograph separation approach (Pinder and Jones, 1969; Sklash and Farvolden, 1979). We determined uncertainties following Genereux (1998). We undertook the analysis using both the ^{2}H and ^{18}O data. We used the mean of the rainfall samples during the event to estimate the event water end member and low flow samples from the few days to the event for the pre-event end member. In the uncertainty analysis, the standard deviations of the event rainfall samples and of all the low flow samples across the study period were used. Half the analytic uncertainty was used to represent the standard deviation of the streamflow sample.

3 Results

The following results first provide an overview of the seasonal behaviour and rainfall–runoff events. They then examine whether thresholds in the antecedent conditions and/or event rainfalls exist. Next, links between the catchment condition and the event runoff are examined using the piezometer and soil moisture data. After that, the recession behaviour of events is examined and linked to catchment wetness conditions. Finally, isotope and major ion data are presented for selected events.

3.1 Overview of runoff behaviour and rainfall–runoff event characteristics

Figure 3 shows time series of weekly rainfall, APET, soil water storage and runoff. The rainfall, although variable from week to week, exhibited seasonality, while there was strong seasonality in PET. This drove a strong seasonality in soil water storage. An examination of the weekly runoff data shows that there was generally no flow from about October to May due to the seasonal nature of this catchment; however, an exception was that persistent low flow occurred from 26 November 2011 to the end of the event on 10 December 2011. During this period the ASI was often relatively low, but there was frequent and substantial rainfall (>200 mm in 30 days). While a strong link between runoff and soil water storage is evident at the seasonal scale in Fig. 3, there are exceptions at the event scale. For example, in February 2011, there was runoff response despite the catchment being near to the lowest soil water storage for the study period.

Moving to the event timescale, Table 1 summarises 38 rainfall–runoff events and Table 2 shows a summary of 22 rainfall events that did not produce a runoff response. A further 16 rainfall events occurred over the study period which

Table 2. Rainfall event summary at RBF with no runoff (Group 1 events). ASI is the Antecedent Soil moisture Index. The events in bold are shown in Fig. 4.

Date	Rain duration (h)	Total rainfall depth (mm)	Peak hourly rainfall intensity ($mm\,h^{-1}$)	ASI (mm)	ASI+ rainfall depth (mm)	Runoff duration (h)	Total runoff depth (mm)
12/10/2010	5	5	2.6	227	232	0	0
12/11/2010	32	28.2	4.0	184	212	0	0
25/11/2010	17	13	3.6	157	170	0	0
02/12/2010	8	7.2	2.0	191	198	0	0
08/12/2010	**11**	**19.2**	**7.8**	**175**	**194**	**0**	**0**
09/12/2010	6	5.2	2.4	193	198	0	0
17/12/2010	21	15	2.4	181	196	0	0
10/01/2011	18	10	3.2	146	156	0	0
11/01/2011	26	10.4	2.4	152	162	0	0
13/01/2011	26	22.2	9.0	151	173	0	0
25/01/2011	9	5.4	1.6	147	152	0	0
05/02/2011	12	8.6	1.6	182	191	0	0
16/02/2011	20	14.8	9.6	167	182	0	0
26/02/2011	13	13.4	4.2	174	187	0	0
21/04/2011	28	8.4	2.6	213	221	0	0
01/05/2011	8	10.8	2.6	207	218	0	0
08/05/2011	14	7.2	4.8	215	222	0	0
05/06/2011	18	9.4	3.0	238	247	0	0
09/09/2011	22	12.8	3.0	221	234	0	0
19/09/2011	18	12.2	4.0	218	230	0	0
24/10/2011	12	15.6	4.2	211	227	0	0

are not included in the analysis due to missing stream discharge data. For the 38 runoff events, total event rainfall varied from 7 to 72 mm, I_{peak} ranged from 2 to 31 $mm\,h^{-1}$, the ASI ranged from 130 to 286 mm and total event runoff varied between 0.23 and 41 mm. For the no-flow events (Table 2), total rainfall varied from 5 to 28 mm, I_{peak} ranged from 2 to 10 $mm\,h^{-1}$ and the ASI ranged from 146 to 238 mm. Figure 4 shows rainfall–runoff responses for selected events at RBF. These graphs are ordered from lowest (27 November 2010) to highest ASI (7 June 2011) for the selected events. All events (except events on 26 November 2011 and 10 December 2011; Fig. 4c) presented in Fig. 4 had zero or very low initial flow. For the events on 26 November 2011 and 10 December 2011, the initial discharge was 0.07 and 0.13 $mm\,h^{-1}$, respectively.

In Fig. 4 most events showed rapid response to rainfall, except for the event on 8 December 2010 (Fig. 4b), which did not produce any significant runoff, and the event on 7 June 2011. The events on 27 November 2010 and 10 December 2011 in particular showed a very flashy response. These events had the highest peak hourly rainfall intensity (30.4 and 31 $mm\,h^{-1}$, respectively) during the study period, and they occurred at the end of the flow season with low ASI (ASI was 161 and 192 mm, respectively). The highest peak runoff rates for the study period were for the events on 27 November 2010 and 10 December 2011, which were 2.4 and 5.6 $mm\,h^{-1}$, respectively. In contrast to most events, the runoff response for the event on 27 November 2010 was transient with very rapid recession. For the event on 10 December 2011, a second peak of moderate rainfall intensity (about 10 $mm\,h^{-1}$) produced a second runoff peak, and there was a more significant recession flow following the rainfall bursts. This was also true for the other events (8 and 26 November 2011, 11 May 2011, 1 August 2010, 7 June 2011) shown in Fig. 4, which were typical of responses to lower intensity rainfall during wetter (in terms of soil water) periods.

For events with $I_{\mathrm{peak}} < 10\,mm\,h^{-1}$ there was a general increase in response as the ASI increased. The event on 12 November 2010 had 184 mm ASI and total rainfall was 28 mm, and it did not produce any runoff. This was a typical example of no-flow events. Coming into the runoff season, as the ASI increased (e.g. 220 mm on 11 May 2011), RBF started to respond gradually, producing small amounts of runoff (e.g. for events on 11 and 14 May 2011). When the ASI was > 250 mm for the event on 7 June 2011, it can be clearly seen that RBF responded to this low intensity, small size rainfall event with a delayed and smooth discharge hydrograph with continued flow following the event. This also occurred for the next event on 1 August 2010.

3.2 Runoff thresholds

In Fig. 1 we set out a number of thresholds that are important in runoff production mechanisms. We now explore the event data from the perspective of thresholds, concentrating on two key ones: catchment wetness and rainfall intensity. Figure 5

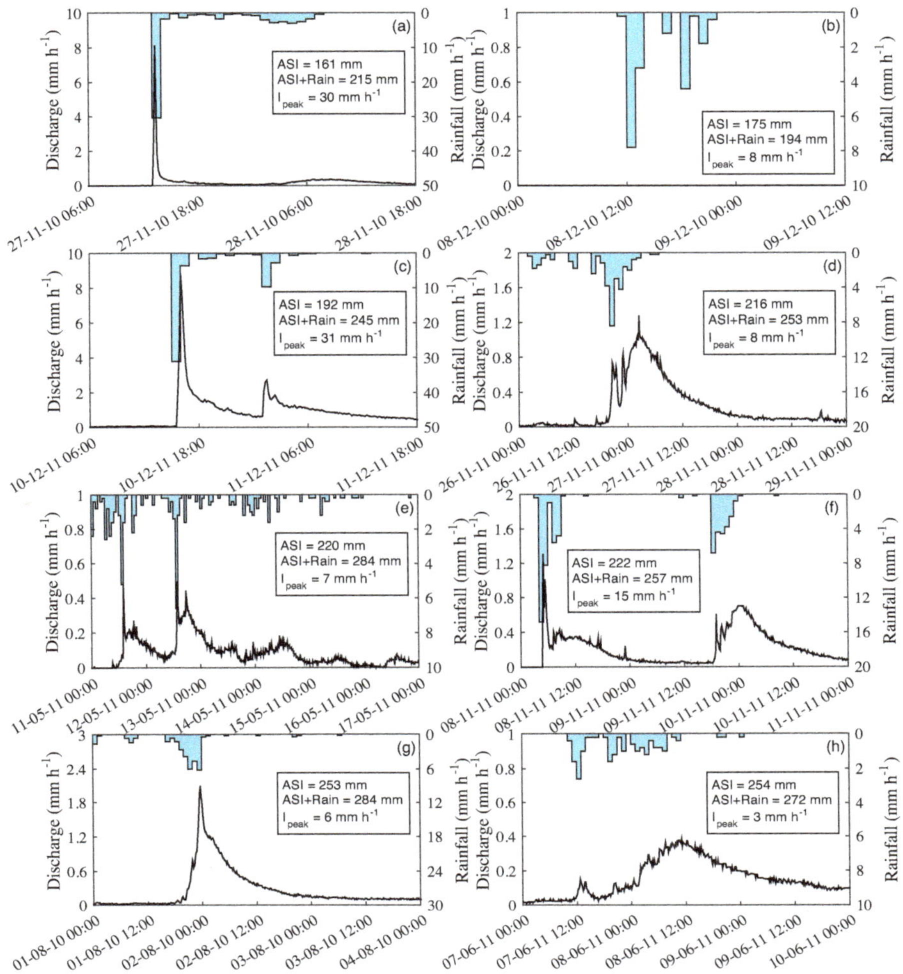

Figure 4. Characteristics of eight selected rainfall–runoff events sorted by their Antecedent Soil moisture Index (ASI). Rainfall is shown in blue. It should be noted that the axis scales vary between events. All events (except those on 26 November and 10 December 2011) had zero or very low initial discharge. For the events on 26 November and 10 December 2011, the initial discharges were 0.07 and 0.13 mm h^{-1}, respectively. Note that the axes vary between events.

builds on approaches by Detty and McGuire (2010), who considered thresholds in ASI and ASI + Rain (i.e. Fig. 5c and d), and Janzen and McDonnell (2015), who considered the impact of event rainfall and rainfall intensity on event runoff (i.e. Fig. 5a). As we move from Fig. 5a to d, the various influences on rainfall–runoff behaviour and thresholds become clearer. Figure 5a shows event runoff as a function of event rainfall, with the highest hourly rainfall intensity (I_{peak}) indicated in colour. This illustrates that there is little relationship between event rainfall and runoff or between rainfall intensity and runoff. For some quite large events (up to ~25mm), zero runoff can occur. There was also a wide variation in runoff coefficients (indicated by the scatter). It is also clear that the events with high peak hourly intensity also had relatively large total rainfall accumulations. Overall, Fig. 5a shows that event rainfall and intensity do not effectively differentiate rainfall–runoff behaviour when considered by themselves.

Figure 5b begins to separate out different effects by showing the impact of five factors together. The cumulative curve shows the distribution of soil water storage as observed throughout the study period. Specific events are shown with the ASI identified (left-hand end of the grey lines) and the ASI + Rain depth (filled markers at the right end of the horizontal grey lines). The length of the lines is the rainfall depth for the event. The colour of each bubble shows the peak hourly rainfall intensity (I_{peak}) and the size of the bubbles shows the event runoff coefficient. Squares indicate events that did not produce any runoff or where the peak runoff rate was less than 0.05 mm h^{-1}. The vertical grey dashed line shows the ASI (left-hand end of the horizontal grey line) or

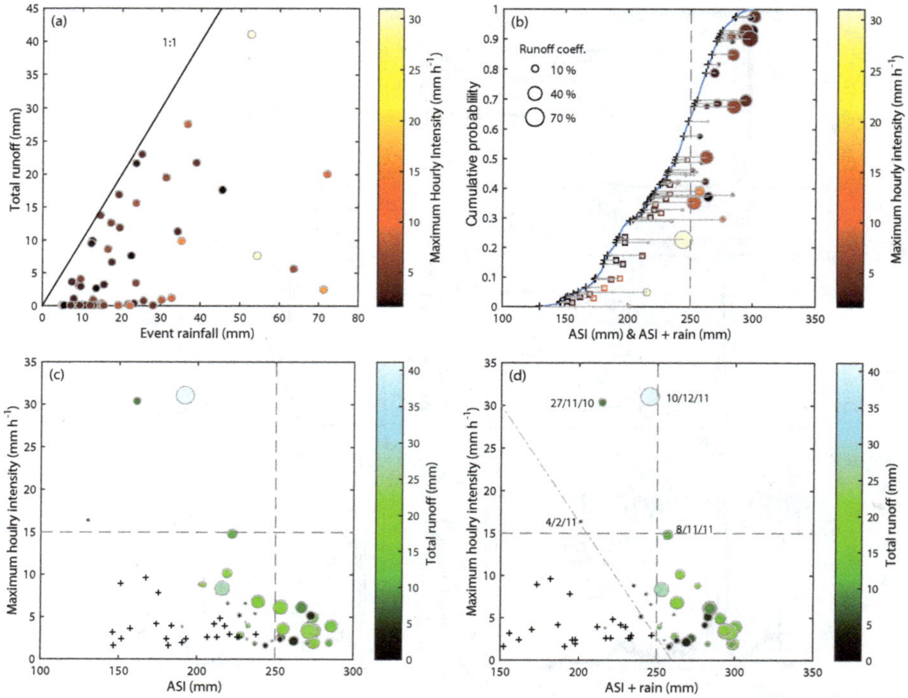

Figure 5. Thresholds of runoff mechanisms at RBF. **(a)** Event rainfall vs. event total runoff, colours indicating the highest hourly rainfall intensity. **(b)** The impact of five factors together, including the cumulative curve of the distribution of soil water storage as observed through the study period, ASI, ASI + Rain; colour shows the peak hourly rainfall intensity (I_{peak}) and the size of the bubbles shows the event runoff coefficient. **(c)** ASI vs. the peak hourly rainfall intensity (I_{peak}), and the size of the bubbles shows the event runoff coefficient and colour shows event total runoff. **(d)** ASI + Rain vs. the peak hourly rainfall intensity (I_{peak}), and the size of the bubbles shows the event runoff coefficient and colour shows event total runoff. Note that ASI is the antecedent soil water storage (mm) in the top 60 cm of the soil at the AWS at the beginning of the event.

ASI + Rain = 250 mm (right-hand end of the horizontal grey line).

There are several trends that can be discerned from Fig. 5b. First the rainfall events analysed here occurred across the full range of catchment wetness and were relatively evenly spread, showing that the rainfall events occur over a representative range of catchment antecedent conditions. The larger rainfall events generally occurred in summer when ASI < 250 mm and a mix of low and high intensity events occurred for these conditions, also in summer. All the events on a wet catchment (ASI > 250 mm) had low I_{peak} ($\leq 6.2\,\mathrm{mm\,h^{-1}}$). Events where the ASI + Rain was less than 250 mm usually did not generate any runoff, although there were some high intensity rainfall events that were exceptions and a small number of events with very low runoff coefficients (1–4 %) where the ASI + Rain was generally between 230 and 250 mm. These low runoff coefficient events were at the end of the runoff season.

Figure 5c and d examine the impact of catchment wetness, quantified as ASI and ASI + Rain, respectively, at the start (Fig. 5c) and end (Fig. 5d) of the event, on event runoff response, combined with the impact of rainfall intensity, I_{peak}. Catchment wetness is plotted on the x axis and rainfall intensity on the y axis. Values ASI = 250 mm (Fig. 5c), ASI + Rain = 250 mm (Fig. 5d) and $I_{peak} = 15\,\mathrm{mm\,h^{-1}}$ are shown by grey dashed lines. The bubble size shows the event runoff coefficient, as before, and crosses indicate rainfall events that did not generate any runoff. Colour indicates the runoff volume. The runoff coefficient behaviour is separated into groups more clearly in Fig. 5d than in Fig. 5c. In Fig. 5d, three different groups of events can be identified. Looking along the x axis, there is a threshold at ASI + Rain = 250 mm that separates most events with a significant runoff response from those without. Looking along the y axis direction, it can be seen that the ASI + Rain threshold is not successful at distinguishing runoff when I_{peak} is high and there is also a threshold in runoff response at $I_{peak} = 15\,\mathrm{mm\,h^{-1}}$. Thus the three groups are (1) events without runoff where ASI + Rain < 250 mm and $I_{peak} < 10\,\mathrm{mm\,h^{-1}}$, (2) events that produce runoff when ASI + Rain > 250 mm and (3) events with ASI + Rain < 250 mm and $I_{peak} > 15\,\mathrm{mm\,h^{-1}}$ that did produce runoff (Tables 1 and 2). Where ASI + Rain exceeded 250 mm (group 2), some runoff was always produced.

Both the first and third groups had ASI + Rain less than 250 mm, but they behaved differently in that some produced runoff and others did not. In the first group low intensity rainfalls mostly happened in drier periods when ASI varied between 146 and 227 mm. It is assumed that rainfall completely infiltrated into the soil and these events did not produce runoff (see Table 2 for event characteristics).

The third group (Table 1) occurred during dry periods at the end of the flow season when the ASI was <200 mm. The runoff coefficients for the four events with peak hourly intensity of 15 mm h^{-1} and higher are 3, 12, 20 and 68 % for peak hourly intensities of 16, 30, 15 and 31 mm h^{-1} and ASI + Rain of 202, 215, 257 and 245 mm, respectively. Note that one of these events exceeds both the wetness and intensity thresholds. These events were distinguished by having maximum hourly rainfall intensities above $\sim$ 15 mm h^{-1}, and they did produce runoff. In particular, two of these events on 27 November 2010 and 10 December 2011 had the highest rainfall intensities observed ($I_{peak} > 30$ mm h^{-1}), and they produced the highest peak runoff rates (8.1 and 9.1 mm h^{-1}) and hourly runoff totals (2.4 and 5.6 mm) observed during the study period (Fig. 3). These runoff peaks happened at the same time as the highest recorded rainfall intensities. Antecedent stream discharge for the events on 27 November 2010 and 4 February 2011 was zero and the hydrograph rose and recessed quickly. For the event on 10 December 2011, the ASI was 192 mm, and the total rainfall was 53 mm, with an initial discharge of 0.13 mm h^{-1} (Table 1). It produced the highest observed peak hourly runoff of 5.6 mm, the runoff duration was 32 h and total runoff was 41 mm. The highest intensity was observed in the first 2 h of the event and the runoff coefficient calculated for the first 2 h of the rainfall event was 18 %. The RC calculated for the duration this event was 68 %. These are large compared with the maximum surface saturated extent throughout the study period of about 6 % of the catchment area, indicating that processes other than saturation excess runoff are important. The event on 10 December 2011 marked the end of a particularly rainy period, with more than 200 mm over 30 days. Note that due to flow measurement equipment being removed after this event, it was the last recorded at RBF.

Table 3 provides a summary of the average conditions for each group of events. In terms of rainfall characteristics, the events that produced runoff (groups 2 and 3) tended to have higher total rainfall, with the highest intensity events also having the largest total rainfall. Of course the grouping criteria mean larger groups are more likely to fall into group 2 compared with group 1. Peak rainfall intensities were both low and almost identical for groups 1 and 2. The runoff behaviour is quite different, with almost no flow and very low runoff coefficients on average for group 1 and average runoff coefficients of 17.9 and 25.8 for groups 2 and 3, respectively. These runoff coefficients are clearly much higher than the observed maximum saturated area proportion (6 %).

We also undertook a bootstrap analysis to test for differences between the group means. This showed that groups 2 and 3 are statistically different from group 1 at the 1 % significant level for total event runoff, quick flow and the quickflow runoff coefficient. Groups 1 and 2 (which are distinguished by ASI + Rain) are statistically similar in terms of I_{peak}, suggesting rainfall intensity does not explain the runoff response differences between these two groups. Groups 1 and 3 (which are distinguished by I_{peak}) are statistically similar in terms of ASI + Rain, suggesting that catchment wetness does not explain the runoff response differences between these two groups.

Overall these results demonstrate a range of different rainfall–runoff responses. The responses depended on both the catchment wetness as quantified by ASI + Rain and on the peak hourly rainfall intensity, I_{peak}, with thresholds of 250 mm and 15 mm h^{-1} being identified. Runoff was produced whenever thresholds in either of these were exceeded.

3.3 Runoff processes and thresholds

The above presentation of results from Fig. 5d identifies a threshold catchment wetness expressed as antecedent soil water storage plus event rainfall depth of 250 mm above which runoff always occurred and another threshold of hourly rainfall depth exceeding 15 mm which also led to runoff production. It should be noted that there is some uncertainty in the I_{peak} threshold as the most intense event (for ASI + Rain < 250 mm) that failed to produce runoff was 10 mm h^{-1}, while the least intense event that did produce runoff was 14.8 mm h^{-1}. Looking at events in the lower right quarter of Fig. 5d also shows that the event runoff coefficient tends to increase as either catchment wetness or peak hourly intensity increases. In fact, runoff and non-runoff producing events are very well separated by the relationship

$$I_{peak} = 3/11 \cdot (\mathrm{ASI} - 260).$$

These results suggest that there are both wetness dependent and intensity dependent runoff production mechanisms operating. This section examines the evidence for different runoff mechanisms contributing to event runoff.

3.3.1 Catchment wetness–flow response relationships

Figure 6 shows the runoff time series together with water level time series at several shallow piezometers: sites 4, 5, 3, 32, 2 and 1 (Fig. 2). All sites except 4 were located in drainage lines. In Fig. 6, the sites are ordered by elevation from highest to lowest in the catchment. Manually read sites are shown with dashed lines. Figure 6 clearly shows that the water table became shallower and that the occurrence of profile saturation (water table at the surface) was restricted to the riparian zone. Within the riparian zone, sites 1 and 2 were saturated much of the time and site 32 quickly became saturated during events. The water table remained below the surface at sites 3 (at the upstream end of the riparian area), 4 and 5 (in mid-slope positions). Comparing the runoff time series with the piezometer record for sites 1, 2 and 32, it is clear that the water table rose to the surface in the upper parts of the ripar-

Table 3. Average characteristics for each event group. ASI is the Antecedent Soil moisture Index.

Group and criteria	Event rainfall (mm)	I_{peak} (mm h^{-1})	ASI (mm)	ASI + Rain (mm)	Total runoff (mm)	Quickflow (mm)	Quickflow runoff coefficient (%)
1: ASI + Rain <= 250 and I_{peak} < 15	14.0	4.3	194.1	208.0	0.2	0.1	0.4
2: ASI + Rain > 250	24.1	4.6	250.0	274.2	10.6	4.5	17.9
3: I_{peak} >= 15	53.4	23.2	176.3	229.5	15.3	12.8	25.8

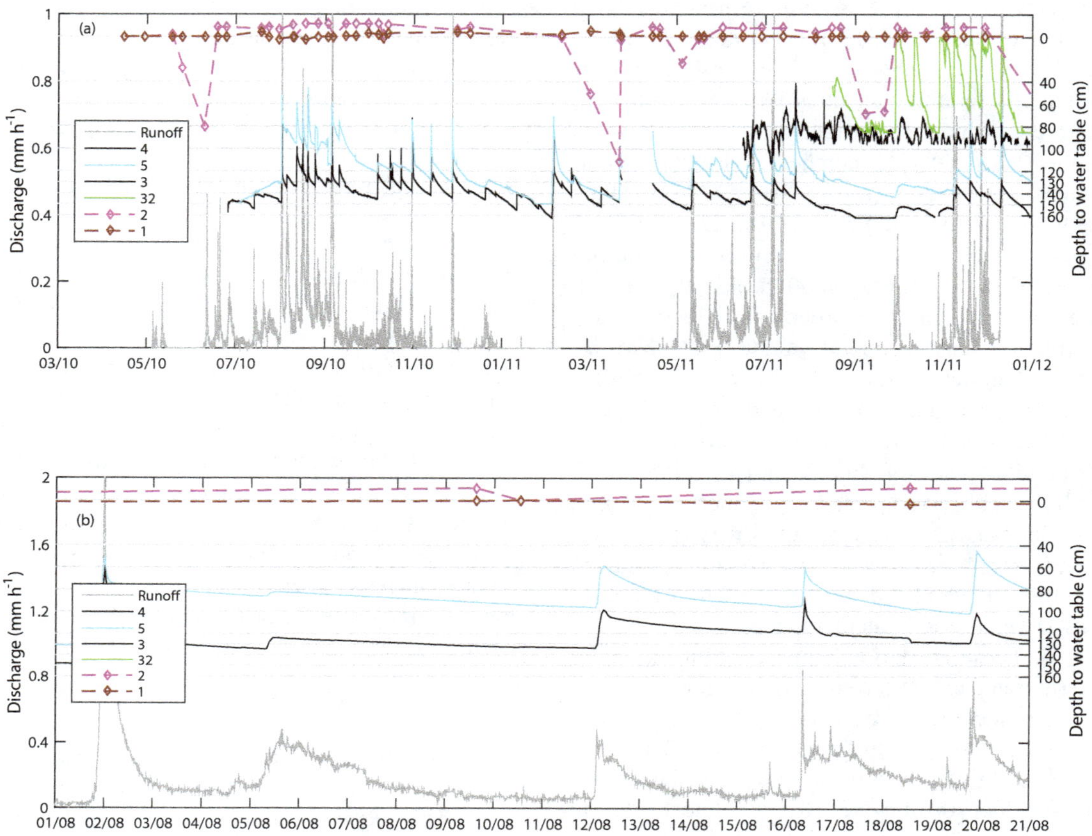

Figure 6. Time series of discharge (grey lines) and groundwater levels at sites 4, 5, 3, 32, 2 and 1; manually read sites are shown with dashed lines and diamonds at measurement points. Levels above zero at sites 1 and 2 occur as the soils are highly pugged in the riparian zone and water pooled on the surface in places, leading to slightly positive water levels being measured, at site 2 in particular.

ian zone during runoff events. Furthermore, the lower half of the riparian zone remained saturated to the surface for long periods during the runoff season. The data recorded at site 3 indicate that the water table at this site did not rise to the surface, even during events.

Looking at the discharge record in Fig. 6a, there were periods where significant baseflow persisted between events. These correspond to periods where the water table at site 5 was above about 120 cm and at site 4 was above about 140 cm deep. Flow became more strongly persistent between rainfall events as the water table at sites 4 and 5 rose further. The water table recessions at sites 4 and 5 correspond to flow recessions when the water table was above 120 and 140 cm at sites 4 and 5, respectively (Fig. 6b).

So far we have examined thresholds in catchment wetness (ASI + Rain) associated with a change in runoff behaviour and the linkage between shallow water tables and catchment response. It is probable that the linkage between catchment wetness and runoff is through the intermediary of shallow subsurface flow controlled by the existence of saturated conditions within a permeable part of the soil horizon. Figure 7 shows the relationship between water table levels in the catchment (sites 4 and 5) and soil water storage at the weather station. These sites represent planar and convergent mid-slope positions, respectively. These sites were

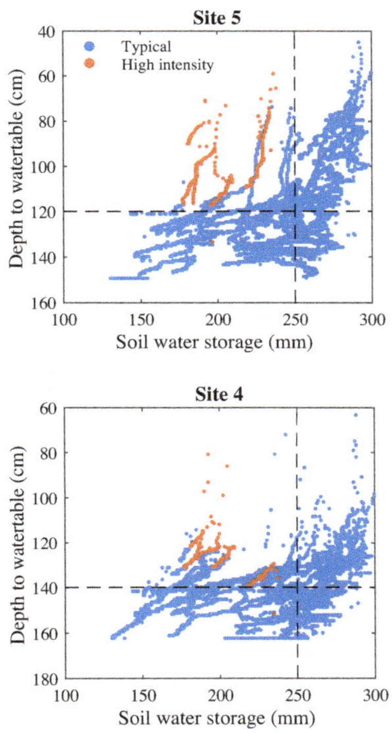

Figure 7. Soil water storage vs. water table level at sites 4 and 5. The red colour distinguishes events with high hourly rainfall intensity, defined as $I_{peak} > 15$ mm, consistent with Fig. 5. The dashed black lines correspond to values of water level and soil water storage discussed in the text. Soil water storage is measured for the top 60 cm at the Automatic Weather Station.

chosen because they show significant dynamics and we had logged records available. Other sites with loggers had limited dynamics or short records. Site 5 in particular shows a rapid change in behaviour for soil water storage around 250 mm, which corresponds to the ASI + Rain threshold identified above. As soil water storage moves above this level, much higher water tables develop and those water tables showed relatively rapid recession when shallower than 120 cm. Similar observations were seen at site 4, but the corresponding depth was 140 cm. Figure 7 thus explains the linkage between the 250 mm ASI + Rain threshold and runoff. When soil water storage exceeded this level, water tables rose and lateral subsurface drainage occurred, as evidenced by the recessions. The recessions in particular suggest that the water table moved into a more permeable zone on these occasions. This is consistent with soil profiles being characterised by mottled clay at depth, which is likely to have lower hydraulic conductivity. This behaviour at these two wells in combination with the soil water and flow data indicate that the hillslopes were becoming connected to the catchment outlet via subsurface flow from the hillslope to the riparian zone. Our analysis of isotope data below will add to the evidence for this. There were a few occasions where the water table responded strongly for soil water storage of less than 250 mm (Fig. 7). As indicated by the red colour, these corresponded to high intensity ($I_{peak} > 15\ \mathrm{mm\,h^{-1}}$) rainfall events.

Flow recessions provide information on the drainage characteristics of catchments. Figure 6 shows that the catchment flow usually ceased between events during the wet period, with no flow during dry periods. However, in August and early September 2010, continuous catchment flow endured for a month (Fig. 6). There was also a marked variation in the recession behaviour during August–September 2010 and at other times during the study period. To explore this, we calculated the recession constant, k (as in $Q = Q_0 e^{-kt}$, where Q_t is discharge at time t during the recession period, Q_0 is the discharge at the beginning of the recession and k is the recession constant). In fitting recessions, we target the period after the recession of the event flow, which typically ceased around 24 h after the end of the rainfall event. This was judged visually by a marked change in slope of the recession hydrograph plotted in semi-log space. Recessions were only fitted where a long enough period of reliable flow data ($>= 24$ h) was available. Data availability was sometimes limited by the commencement of another event or flow falling to very low ($< 0.05\ \mathrm{mm\,h^{-1}}$). k is plotted against soil water storage at the start of the catchment flow recession for individual events (Fig. 8). In total we could estimate values of k for 20 events. The three events in Fig. 8 coloured blue were from November 2011 from a relatively dry period in late spring when there was a drying trend but where flow from the riparian area was just persisting. These were excluded from the fitting because, in our judgement, they appear to represent different behaviour. k for each event and the soil water storage at the start of the recession period were negatively correlated ($R^2 = 0.64$) (Fig. 8). k decreased as soil water storage increased at a rate of about $0.1\ \mathrm{d^{-1}}$ for each 10 mm increase in soil water storage. Considering this and the transient nature of flow during dry periods, it is clear that the wetter the catchment is, the slower the recessions are. By inference, this suggests that greater (perhaps more spatially extensive) subsurface connectivity is providing flows from the hillslope and maintaining catchment flow during wetter conditions.

3.3.2 Isotope and major ion results

The hydrometric results presented in Figs. 6, 7 and 8 suggest that subsurface flow is important in this catchment. Given this, we would expect the hydrograph to be dominated by "old" or pre-event water; however, the saturated area in the lowest parts of the catchment would also be expected to produce direct flows of "new" or event water. This is addressed here using isotope data. We faced some constraints in obtaining meaningful hydrograph separation results. Of the five events sampled, two events were missing rainfall samples, which prevented their analysis. For another two events the rainfall isotope signature was quite close to the typical low flow signature. Uncertainty estimates showed standard deviations for pre-event and event fractions of over 50 % for

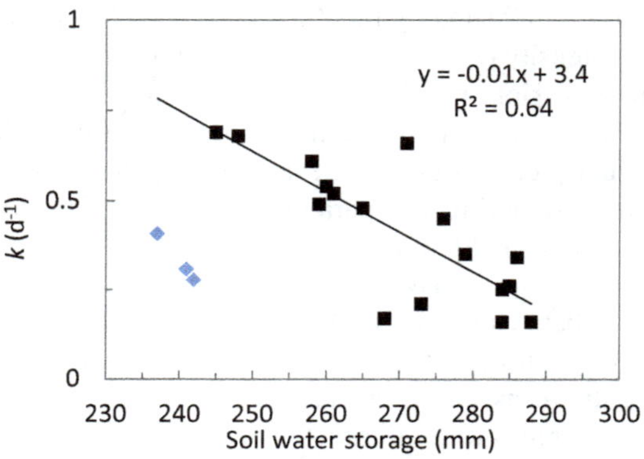

Figure 8. Recession constant (k) and soil water storage at the start of the recession period for individual events. The choice of events is explained in the text, as are the three blue points.

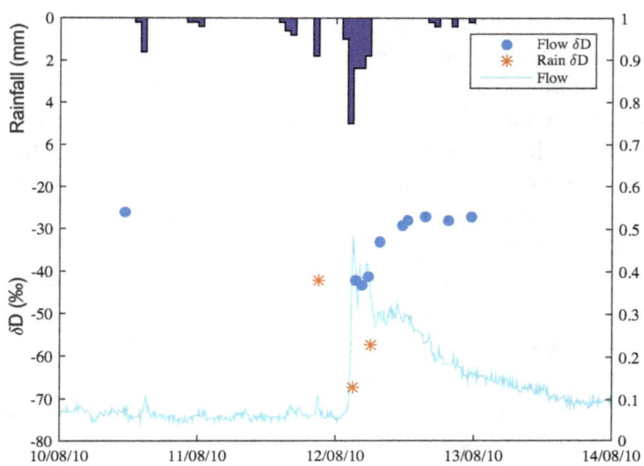

Figure 9. Time series of total rainfall and discharge, ^{2}H of rainfall and runoff for the event on 12 August 2010.

these two events; therefore, these events were also excluded. For the event on 12 August 2010, the standard deviations were approximately 12 and 9 % for ^{18}O and ^{2}H, respectively, which we judged to be acceptable.

Figure 9 shows the event from 12 August 2010 during the wettest part of the study period. The antecedent soil water storage at the beginning of this event was 274 mm and total rainfall was 17 mm. We used a pre-event low-flow sample from $\sim$2 days before the analysed event as the pre-event end member. We estimated the contribution at the time of each stream water sample. Over the study period δ^2H (δ^{18}O) for rainfall varied between −7 and −83 ‰ (−1.4 and 12.6 ‰) and isotopic concentrations for low flows were highly damped. Low flow samples from the RBF flume before and after the event showed a δ^2H (δ^{18}O) of −27 ‰ (−4.1 ‰) and rainfall for this event was strongly depleted (three samples prior to and during the event $\delta^2\text{H} = -42$, −67 and −57 ‰, $\delta^{18}\text{O} = -6.5$, −9.8 and −8.2 ‰), compared with low flow. The runoff samples showed a very different isotopic concentration during the rising limb and the peak of the hydrograph ($\delta^2\text{H} = -43$ ‰, $\delta^{18}\text{O} = -6.8$ ‰) in comparison to antecedent low flow. This shows that the isotopic concentration moves significantly towards the rainfall sample concentrations. To estimate the overall event water contribution we first separated the hydrograph at each sampling time and then interpolated the fraction of event water between stream water sampling times and combined this with the discharge hydrograph to calculate the overall volume of event water. This analysis suggests that the percentage of rain becoming runoff based on δ^2H and δ^{18}O is 4.4 % (3.4–5.4 %) and 3.6 % (3.0–4.2 %), respectively. The figures in the brackets are 95 % confidence intervals for uncertainty based on the hydrograph separation uncertainty only. These results suggest that precipitation on the saturated area generates direct runoff in amounts that are close to what would be expected (i.e. 100 % runoff) given that the saturated area is around 5–6 % of the catchment area.

Another interesting event is the higher intensity ($I_{peak} = 15\,\text{mm}\,\text{h}^{-1}$) event on 8 November 2011. Major ion geochemistry data were available for this event. Figure 10a shows the typical relationship between discharge and chloride concentration, with samples from this event identified by red. Figure 10b shows the time series of chloride concentration along with the hydrograph. The first and second chloride samples, respectively, plot above and within the typical scatter of data in Fig. 10a, while the remaining samples plot well below the typical variation in chloride concentration with flow (Fig. 10a). Similar plots are shown for sodium, magnesium, calcium and potassium in Supplement Fig. S1. Sodium and magnesium show similar behaviour to chloride. Potassium also shows somewhat anomalous behaviour but with anomalously high concentrations. The relationship between discharge and K^+ is also different to Cl^-, Na^+ and Mg^{2+} in that concentrations increase slightly as discharge increases to above about 0.2 mm mm^{-1}, rather than decline. Calcium does not show anomalous behaviour.

Given the late spring timing of this event, the first sample probably reflects some evapoconcentration of solutes in the riparian area. The flow shows a rapid peak in response to the main rainfall burst followed by a sustained relatively low flow and a recession over the second half of the day suggestive of subsurface flow. Up until the end of the first flow peak (i.e. 06:00), there had been 23.4 mm of rainfall and 1 mm of runoff. This runoff volume is 4.3 % of the rainfall volume, which is similar to the proportion of event water that became runoff for the 12 August 2010 event discussed above.

4 Discussion

4.1 Runoff mechanisms

The hydrometric data enable us to identify the important runoff mechanisms under different circumstances. The iso-

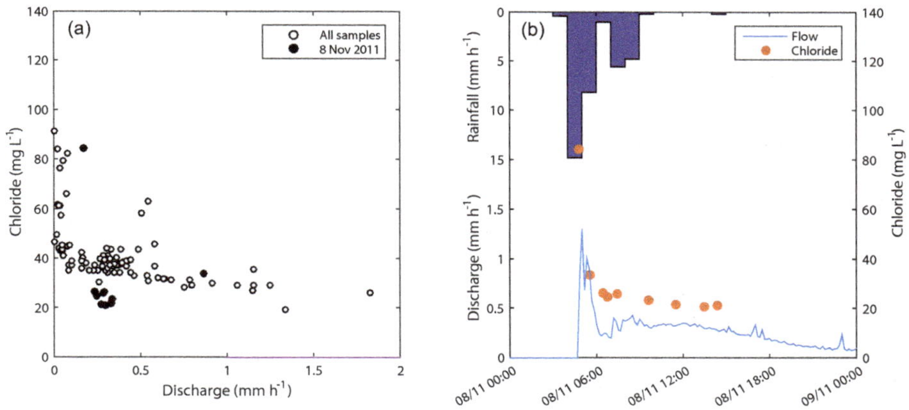

Figure 10. (a) Discharge vs. Cl^- concentration for all events. The red colour identifies samples from the event on 8 November 2011. **(b)** Time series of rainfall, runoff and Cl^- concentration for the event on 8 November 2011.

tope and major ion geochemistry data provide further supporting evidence. The rainfall plus antecedent soil water threshold of 250 mm that needs to be exceeded for runoff in most circumstances shows that wetness dependent runoff processes are important, that is, either saturation excess or subsurface stormflow. Shallow groundwater data combined with field mapping of surface saturated areas show that complete profile saturation is limited to about 5 % of the catchment area and that this saturation is persistent, with only a small variation (see Table 1) over the winter–spring season but reducing over the dry summer–autumn period. The extent of the saturated area at the riparian zone varied seasonally and between events. Field observations of surface saturation extending to the flume and well hydrograph measurements of the water table at the surface showed that the saturated area was highly connected to the catchment outlet and suggest that it would be expected to produce saturation excess runoff. This surface connection disappeared during the dry summer season when the water level in all wells fell below the surface (see Fig. 6), so that the saturated area in the riparian zone disappeared. The isotope results for 12 August 2010 enabled the event runoff to be separated into event water and pre-event water contributions. Four to five percent of the rainfall volume on the catchment appeared in the event runoff, which corresponds well to the proportion of surface saturated area in the catchment (5.5 %, Table 1), supporting the identification of significant saturation excess runoff from this part of the catchment, as observed elsewhere (McGlynn and McDonnell, 2003; Penna et al., 2016).

While saturation excess runoff undoubtedly occurs, many of the event runoff coefficients were well in excess of 5 %, and they approach 100 % under very wet conditions (Fig. 5). The event on 12 August showed a substantial pre-event water contribution; logged shallow wells show that the water table did not reach the surface in the steeper areas of the catchment, even within the convergent drainage lines under very wet conditions (e.g. sites 5 and 7). The recession behaviour of wells in the catchment suggests subsurface flow moves down the catchment under wet conditions and the recession constant analysis shows that this connection becomes stronger as the catchment wets beyond 250 mm of stored water. This is all consistent with a substantial contribution of subsurface flow to event runoff once the catchment is sufficiently wet to establish subsurface connection with the riparian area, as has been inferred in other studies (Buttle et al., 2004; Detty and McGuire, 2010; Hewlett and Hibbert, 1967; Jencso et al., 2009; Penna et al., 2011).

Perhaps more surprisingly there was a group of events that produced runoff under conditions of relatively low soil water storage (ASI + Rain < 250 mm) but high rainfall intensity. This suggests that an intensity dependent runoff process is being triggered when rainfall exceeds some threshold for sufficient time, in this case about 15 mm of rainfall in an hour. It is worth noting that the 15 mm h^{-1} threshold will be dependent on the time step used in calculating the intensity. The choice of time step needs to consider the travel time from hillslope to stream as run-on infiltration processes occurring on the catchment surface will impact the connectivity to the stream. Thus the most appropriate time to use for averaging intensity would be the average travel time to the stream because it is this time period which is available to infiltrate rainfall that has become runoff at the point scale. Here the choice of hourly rainfall was pragmatic, as that was the recording time step of the AWS, but it is also likely that it reasonably approximates the flow times.

It is tempting to assume that this evidence suggested surface runoff occurred due to infiltration excess runoff, but it is also possible that the high rainfall intensities are efficiently activating macropore networks (Beven and Germann, 1982) and that the flow could be following subsurface pathways. For these events there were also rapid responses in water levels in wells on the hillslope and it is not clear exactly how the water moved rapidly into the wells in these cases, but it could be due to preferential flow through macropores. Furthermore, for the high intensity event on 8 November 2011, the flow after the main peak had a surprisingly low concentration of chloride (the cluster of low concentration red points in Fig. 10a), which may suggest that the higher intensity activated either overland or preferential flow paths, limiting soil contact time and leading to this low concentration.

The hydrograph from the event on 8 November 11 (the event that exceeded both thresholds, Fig. 10) shows both a rapid and a delayed runoff response. The concentration of chloride was unusually low during the delayed runoff component compared with all other events (with major ion data available). This may suggest limited contact with the catchment soils which could occur if macropore flow was important, but this explanation is not definitive. Of the two events with peak hourly intensities around 15 mm h^{-1}, one also exceeded the wetness (ASI + Rain) threshold of 250 mm and the other had a very low runoff coefficient (only 3 %), and hence these two events are somewhat equivocal in terms of the importance of intensity. However, the two events with peak hourly rainfall intensities around 30 mm h^{-1} both produced rapid runoff responses without a significant delayed component (Fig. 4) and had runoff coefficients (12 and 68 %) well in excess of the surface saturated area (5 %) in the catchment, showing clear evidence of the role of rainfall intensity of 30 mm h^{-1}. Unfortunately isotope and major ion data were not available for those events to attempt to determine whether surface or subsurface pathways are important. Overall there is clear evidence for intensity dependent runoff mechanisms, especially for the largest 30 mm h^{-1} events.

In summary, the process evidence relating runoff behaviour to catchment wetness thresholds, together with the data from shallow wells, suggests a catchment where subsurface flow leads to a seasonally saturated riparian area that produces saturation excess runoff in immediate response to rainfall. This saturation excess is augmented by subsurface stormflow when the catchment wetness (ASI + Rain) exceeds a 250 mm threshold. This subsurface flow exfiltrates in the riparian area. Volume considerations provide evidence that, for many of the monitored events, water must be coming from outside the riparian area. This area is only about 5 % of the catchment, but quickflow runoff coefficients very regularly exceed this, often (about 1/3 of runoff events) exceeded 20 %, and on occasion exceeded 50 %. The role of both saturation excess runoff and subsurface stormflow is corroborated to some extend by isotopic data from one event where the hydrograph separation shows a volume of event water similar to the volume of rainfall on the saturated area and by the highly damped isotopic signal in the stream in general. When hourly rainfalls exceed a threshold of 15–30 mm h^{-1}, an intensity dependent runoff process is activated that also contributes flow from the hillslope area outside the riparian zone. It is not clear whether this is a purely surface runoff process or not. One of the key contributions of this work is clear hydrometric evidence of the interplay of both wetness and intensity dependent runoff processes in the one catchment. These responses include two wetness dependent processes, saturation excess runoff and subsurface stormflow, and an intensity dependent process for which we cannot distinguish between surface and subsurface flow pathways with our data.

4.2 Thresholds and connectivity in runoff production

We identified two important thresholds in the catchment response. The first is a wetness threshold of ASI + Rain exceeding 250 mm. Under these conditions the water table approaches the surface in the riparian area and water tables rise on the hillslope into what is inferred from relatively rapid hillslope water table recessions to be a more transmissive part of the soil profile (within ~ 120–140 cm of the surface). Based on the volume of event runoff (Table 1), the lack of surface saturation on the hillslope and the behaviour of shallow wells (Fig. 6), we infer that the hillslope becomes connected to the riparian zone under these conditions. The existence of a large volume of pre-event water in the event on 12 August 2010 (Fig. 8) adds further corroboration to this. Similar catchment wetness thresholds for connectivity and runoff generation have been reported elsewhere (Detty and McGuire, 2010; Penna et al., 2011; Tromp-van Meerveld and McDonnell, 2006b). These have been expressed either in terms of rainfall depth, antecedent soil water storage conditions, or a combination of these. Similar to our results, in a study of a forested subcatchment with highly permeable soils and a small riparian area, Detty and McGuire (2010) found a relationship between a threshold of 316 mm for ASI + Rain and the start of the event streamflow response. They showed that the ASI + Rain threshold corresponded to a water table height threshold. Based on this, they suggested that subsurface flow, transmissivity feedback and preferential flow from hillslope to stream could be used to explain runoff mechanisms in their catchment. They did not observe either Hortonian overland flow or saturation overland flow (SOF) even during the largest events.

A second threshold to initiate event streamflow response associated with high rainfall intensities was also evident. A similar role of intensity has also been observed by Janzen and McDonnell (2015), who found that the Panola hillslope can produce significant runoff from dry antecedent conditions when high intensity rainfall occurs. In general, event runoff from Panola is controlled by catchment wetness, similar to our catchment. Mixed wetness and intensity dependent runoff processes have also been reported in Mediterranean catchments (e.g. Lana-Renault et al., 2014). Nevertheless, there are only a few studies we know of that have reported intensity thresholds in catchments where runoff is normally dominated by wetness thresholds. This may be a consequence of such events being relatively rare in any given catchment (roughly 10 % of runoff producing events in our case).

4.3 Runoff processes framework

We now consider the three runoff processes occurring in the catchment in relation to the framework proposed in Fig. 1 and the various flux and timescale thresholds identified therein. Essentially Fig. 1 is posing a series of questions that allow us to systematically think through the runoff processes. Above we have identified three groups of rainfall events: those that do not produce runoff; those that produce runoff by saturation excess and subsurface stormflow from a wet catchment; and those that produce runoff from higher intensity events.

The rainfall events that do not produce runoff are not exceeding infiltration capacity for a sufficient time for runoff to flow from the catchment (box 1, "No"). They may or may not produce significant percolation (box 2), but if any of these events does produce percolation to a perched water table, this only results in an ephemeral water table that dissipates before lateral flow can move water down the hillslope (box 3, "No"), and they do not saturate the full profile (box 4, "No"). As a consequence no event runoff is produced. These conditions correspond to the local control state of Grayson et al. (1997).

Some high intensity rainfall events on a dryer catchment do exceed the infiltration capacity for sufficiently long periods of time (box 1, "Yes"; hourly intensity of 15–30 mm h^{-1}) and these events produce runoff. The increase in runoff coefficient as intensity increases for a given ASI + Rain in Fig. 5 suggests that infiltration thresholds may also be playing a role for wetter conditions (i.e. box 1 "Yes") but that infiltrated water (the dashed link in Fig. 1) also contributes through other mechanisms.

The final group of events are those that produce runoff from a wet catchment at low intensity rainfall. These follow the path box 1 "No", box 2 "Yes" and box 3 "Yes" in Fig. 1. On the upper hillslope segments of the catchment, the subsurface flow capacity is sufficient so that the water table does not reach the surface and water drains either during or shortly after the storm (box 6, "Yes"). Subsurface connectivity develops during the event and subsurface flow dominates. The shallow well data (Fig. 6, wells 1 and 2, to some extent 32) show that the riparian area in the lower part of the catchment drains very slowly. There is a substantial reduction in slopes (and probably lower hydraulic conductivity associated with poorly structured, poorly drained soils) that suggests a substantial reduction in lateral subsurface flow capacity. The surface topography also suggests subsurface flow would converge in this area. From the perspective of Fig. 1, this leads to a situation where water takes longer than the typical time between events in the wet season to drain (box 7, "Yes"), resulting in persistent surface saturation and saturation excess runoff generation from the lower catchment. Hence under wet conditions this catchment produces a mix of saturation excess and subsurface storm flow, but from geographically distinct parts of the catchment.

The above illustrates how the framework in Fig. 1 can be used to understand the role of different thresholds regarding fluxes and timescales in determining runoff mechanisms. Such a framework is likely to be particularly valuable where there is a mix of runoff mechanisms operating for different events or in different parts of the catchment. Our study catchment nicely illustrates such a mixture.

5 Conclusion

This study has examined the role of intensity and wetness thresholds in determining runoff responses for an agricultural catchment in the Lang Lang River catchment, Victoria, Australia. Both intensity dependent and wetness dependent thresholds were identified in the runoff response. During wet conditions, hydrological connectivity has a strong influence on water delivery to the riparian area. Saturation excess runoff from the riparian zone was also important. The results of this study demonstrated the following.

1. Runoff generation in most events is dependent on the catchment connectivity and soil moisture conditions. When the sum of the antecedent soil water storage and event rainfall exceeded 250 mm, runoff was typically produced by a mix of saturation excess and subsurface storm flow. Under these conditions, a water table forms in the soil and a saturated area develops in the riparian zone. When the water level rises to within about 1 m of the surface at mid-slope sites, rapid subsurface flow pathways are activated which connected the mid-slope and riparian area, contributing event flow to the flume at the catchment outlet.

2. When the catchment became very wet, high water levels persisted at the mid-slope sites, which remained hydrologically connected to the riparian area, and baseflow became persistent between events.

3. High rainfall intensity events produced runoff even when the antecedent soil water storage (ASI) plus event rainfall depth was below the 250 mm threshold. This could be due to either Hortonian overland flow or fast subsurface preferential flow paths being activated.

We have also advanced a set of threshold conditions or questions (Fig. 1) that allow a logical examination of which runoff mechanisms are likely to be important in a catchment given thresholds regarding fluxes and timescales. This framework provides a useful way of thinking through the controls on rainfall–runoff response as conditions change either between events or between different parts of the catchment. It is illustrated using the behaviour of this catchment. Our study catchment demonstrates a mix of intensity dependent and wetness dependent processes, something which has been rarely reported for humid catchments.

Acknowledgements. The project was funded by the Australian Research Council (grant DP0987738). Shabnam Saffarpour was supported by an International Postgraduate Research Scholarship (IPRS) from the Australian Government and an International Research Scholarship from the University of Melbourne (MIRS). Peter and Wilma Mackay provided access to their farm for the study. Many people helped in the field, especially Rodger Young, Lucas Dowell, Olaf Klimczak and Jacqui Lloyd. The water samples were analysed for ^{18}O and δ^2H, either by Ian Carwright's Earth Sciences laboratory, Monash University or by Russell Drysdale's isotope laboratory at the University of Melbourne. The chloride data were analysed by Michael Grace's Water Studies Laboratory, Monash University. Anthony Weatherly assisted with soil interpretation at the site.

Edited by: M. Hrachowitz

References

Adams, R., Arafat, Y., Eate, V., Grace, M. R., Saffarpour, S., Weatherley, A. J., and Western, A. W.: A catchment study of sources and sinks of nutrients and sediments in south-east Australia, J. Hydrol., 515, 166–179, doi:10.1016/j.jhydrol.2014.04.034, 2014.

Ali, G., Oswald, C. J., Spence, C., Cammeraat, E. L., McGuire, K. J., Meixner, T., and Reaney, S. M.: Towards a unified threshold based hydrological theory: necessary components and recurring challenges, Hydrol. Process., 27, 313–318, 2013.

APHA: Standard Methods for the Examination of Water and Wastewater, 21st edition, American Public Health Association, 1200 pp., 2005.

Beven, K.: Introduction: Modelling hydrological and nutrient transport processes, in Agriculture, hydrology, and water quality, edited by: Haygarth, P. M. and Jarvis, S. C., North Wyke Research Station, Devon, UK, Institute of Grassland and Environmental Research, CABI, 194–196, 2002.

Beven, K. J. and Germann, P.: Macropores and water flow in soils, Water Resour. Res., 18, 1311–1325, 1982.

Bowes, M. J., Smith, J. T., and Neal, C.: The value of high-resolution nutrient monitoring: A case study of the River Frome, Dorset, UK, J. Hydrol., 378, 82–96, 2009.

Bracken, L. and Croke, J.: The concept of hydrological connectivity and its contribution to understanding runoff-dominated geomorphic systems, Hydrol. Process., 21, 1749–1763, doi:10.1002/hyp.6313, 2007.

Bureau of Meteorology: Australian Climate Averages – Evapotranspiration (Climatology 1961–1990), available at: http://reg.bom.gov.au/jsp/ncc/climate_averages/evapotranspiration/IDCetgrids.jsp (last access: 2 April 2012), 2005.

Bureau of Meteorology: Australian Climate Averages – Rainfall (Climatology 1961–1990), available at: http://reg.bom.gov.au/jsp/ncc/climate_averages/rainfall/index.jsp (last access: 2 April 2012), 2009.

Buttle, J. M., Dillon, P. J., and Eerkes, G. R.: Hydrologic coupling of slopes, riparian zones and streams: an example from the Canadian Shield, J. Hydrol., 287, 161–177, doi:10.1016/j.jhydrol.2003.09.022, 2004.

Campbell Scientific: CS616 and CS625 water content reflectometers instruction manual, Campbell Scientific, Logan, UT, 46 pp., 2006.

Clemmens, A. J., Bos, M. G., and Reploge, J. A.: Portable RBC Flumes for Furrows and Earthen Channels, T. ASAE, 27, 1016–1021, doi:10.13031/2013.32915, 1984

Detty, J. M. and McGuire, K. J.: Threshold changes in storm runoff generation at a till-mantled headwater catchment, Water Resour. Res., 46, W07525, doi:10.1029/2009wr008102, 2010.

Dunne, T. and Black, R. D.: An experimental investigation of runoff production in permeable soils, Water Resour. Res., 6, 478–490, 1970a.

Dunne, T. and Black, R. D.: Partial area contributions to storm runoff in a small New England watershed, Water Resour. Res., 6, 1296–1311, 1970b.

Freer, J., McDonnell, J. J., Beven, K., Peters, N. E., Burns, D., Hooper, R., Aulenbach, B., and Kendall, C.: The role of bedrock topography on subsurface storm flow, Water Resour. Res., 38, W12410, doi:10.1029/2001WR000872, 2002.

Fu, C., Cheng, J., Jiang, H., and Dong, L.: Threshold behavior in a fissured granitic catchment in southern China: (1) analysis of field monitoring results, Water Resour. Res., 49, 1–17, doi:10.1002/wrcr.20191, 2013.

Fujimoto, M., Ohte, N., and Tani, M.: Effects of hillslope topography on hydrological responses in a weathered granite mountain, Japan: comparison of the runoff response between the valley-head and the side slope, Hydrol. Process., 22, 2581–2594, doi:10.1002/hyp.6857, 2008.

Genereux, D.: Quantifying uncertainty in tracer-based hydrograph separations, Water Resour. Res., 34, 915–919, doi:10.1029/98WR00010, 1998.

Grayson, R. B., Western, A. W., Chiew, F. H. S., and Blöschl, G.: Preferred states in spatial soil moisture patterns: Local and non-local controls, Water Resour. Res., 33, 2897–2908, 1997.

Hewlett, J. D. and Hibbert, A. R.: Factors affecting the response of small watersheds to precipitation in humid areas, Forest Hydrol., 275–290, 1967.

Holz, G. K.: Sources and processes of contaminant loss from an intensively grazed catchment inferred from patterns in discharge and concentration of thirteen analytes using high intensity sampling, J. Hydrol., 383, 194–208, 2010.

Hopp, L. and McDonnell, J.: Connectivity at the hillslope scale: Identifying interactions between storm size, bedrock permeability, slope angle and soil depth, J. Hydrol., 376, 378–391, 2009.

Horton, R.: The role of infiltration in the hydrologic cycle, Trans. Am. Geophys.Union, 14, 446–460, 1933.

Houser, P., Goodrich, D., and Syed, K.: Runoff, precipitation, and soil moisture at Walnut Gulch, in: Spatial Patterns in Catchment Hydrology: Observations and Modelling, edited by: Grayson, R. and Blöschl, G., Cambridge University Press, Cambridge, 125, 105–157, 2000.

Isbell, R. F.: The Australian Soil Classification, Australian Soil and Land Survey Handbooks, CSIRO Publishing, Collingwood, Vicoria, Australia, 144 pp., 2002.

Jackson, C. R., Bitew, M., and Du, E.: When interflow also percolates: downslope travel distances and hillslope process zones,

Hydrol. Process., 28, 3195–3200, doi:10.1002/hyp.10158, 2014.

Janzen, D. and McDonnell, J. J.: A stochastic approach to modelling and understanding hillslope runoff connectivity dynamics, Ecol. Model., 298, 64–74, doi:10.1016/j.ecolmodel.2014.06.024, 2015.

Jencso, K. G., McGlynn, B. L., Gooseff, M. N., Wondzell, S. M., Bencala, K. E., and Marshall, L. A.: Hydrologic connectivity between landscapes and streams: Transferring reach-and plot-scale understanding to the catchment scale, Water Resour. Res., 45, W04428, doi:10.1029/2008WR007225, 2009.

Kennedy, V. C., Zellweger, G. W., and Avanzino, R. J.: Variation of rain chemistry during storms at two sites in northern California, Water Resour. Res., 15, 687–702, 1979.

Kirchner, J. W.: Getting the right answers for the right reasons: Linking measurements, analyses, and models to advance the science of hydrology, Water Resour. Res., 42, W03S04, doi:10.1029/2005WR004362, 2006.

Lana-Renault, N., Latron, J., and Regüés, D.: Streamflow response and water-table dynamics in a sub-Mediterranean research catchment (Central Pyrenees), J. Hydrol., 347, 497–507, 2007.

Lana-Renault, N., Regüés, D., Serrano, P., and Latron, J.: Spatial and temporal variability of groundwater dynamics in a sub-Mediterranean mountain catchment, Hydrol. Process., 28, 3288–3299, 2014.

Latron, J. and Gallart, F.: Runoff generation processes in a small Mediterranean research catchment (Vallcebre, Eastern Pyrenees), J. Hydrol., 358, 206–220, 2008.

Lehmann, P., Hinz, C., McGrath, G., Tromp-van Meerveld, H. J., and McDonnell, J. J.: Rainfall threshold for hillslope outflow: an emergent property of flow pathway connectivity, Hydrol. Earth Syst. Sci., 11, 1047–1063, doi:10.5194/hess-11-1047-2007, 2007.

McGlynn, B. L. and McDonnell, J. J.: Quantifying the relative contributions of riparian and hillslope zones to catchment runoff, Water Resour. Res., 39, 1310, doi:10.1029/2003WR002091, 2003.

McGuire, K. J. and McDonnell, J. J.: Hydrological connectivity of hillslopes and streams: Characteristic time scales and nonlinearities, Water Resour. Res., 46, W10543, doi:10.1029/2010WR009341, 2010.

Morton, F. I.: Operational estimates of areal evapotranspiration and their significance to the science and practice of hydrology, J. Hydrol., 66, 1–76, 1983.

Ocampo, C. J., Sivapalan, M., and Oldham, C.: Hydrological connectivity of upland-riparian zones in agricultural catchments: Implications for runoff generation and nitrate transport, J. Hydrol., 331, 643–658, doi:10.1016/j.jhydrol.2006.06.010, 2006.

Oldham, C. E., Farrow, D. E., and Peiffer, S.: A generalized Damköhler number for classifying material processing in hydrological systems, Hydrol. Earth Syst. Sci., 17, 1133–1148, doi:10.5194/hess-17-1133-2013, 2013.

Penna, D., Tromp-van Meerveld, H. J., Gobbi, A., Borga, M., and Dalla Fontana, G.: The influence of soil moisture on threshold runoff generation processes in an alpine headwater catchment, Hydrol. Earth Syst. Sci., 15, 689–702, doi:10.5194/hess-15-689-2011, 2011.

Penna, D., van Meerveld, H. J., Oliviero, O., Zuecco, G., Assendelft, R. S., Dalla Fontana, G., and Borga, M.: Seasonal changes in runoff generation in a small forested mountain catchment, Hydrol. Process., 29, 2027–2042, doi:10.1002/hyp.10347, 2015.

Penna, D., van Meerveld, H. J., Zuecco, G., Dalla Fontana, G., and Borga, M.: Hydrological response of an Alpine catchment to rainfall and snowmelt events, J. Hydrol., 537, 382–397, doi:10.1016/j.jhydrol.2016.03.040, 2016.

Pinder, G. F. and Jones, J. F.: Determination of ground-water component of peak discharge from chemistry of total runoff, Water Resour. Res., 5, 438–445, doi:10.1029/WR005i002p00438, 1969.

Sklash, M. G. and Farvolden, R. N.: Role of groundwater in storm runoff, J. Hydrol., 43, 45–65, doi:10.1016/0022-1694(79)90164-1, 1979.

Tetzlaff, D., Carey, S. K., Laudon, H., and McGuire, K.: Catchment processes and heterogeneity at multiple scales-benchmarking observations, conceptualization and prediction Preface, Hydrol. Process., 24, 2203–2208, doi:10.1002/hyp.7784, 2010.

Tockner, K., Pennetzdorfer, D., Reiner, N., Schiemer, F., and Ward, J. V.: Hydrological connectivity, and the exchange of organic matter and nutrients in a dynamic river-floodplain system (Danube, Austria), Freshw. Biol., 41, 521–535, doi:10.1046/j.1365-2427.1999.00399.x, 1999.

Tromp-van Meerveld, H., Peters, N., and McDonnell, J.: Effect of bedrock permeability on subsurface stormflow and the water balance of a trenched hillslope at the Panola Mountain Research Watershed, Georgia, USA, Hydrol. Process., 21, 750–769, 2007.

Tromp-van Meerveld, H. J. and McDonnell, J. J.: Threshold relations in subsurface stormflow: 2. The fill and spill hypothesis, Water Resour. Res., 42, W02411, doi:10.1029/2004WR003800, 2006a.

Tromp-van Meerveld, H. J. and McDonnell, J. J.: Threshold relations in subsurface stormflow: 1. A 147-storm analysis of the Panola hillslope, Water Resour. Res., 42, W02410, doi:10.1029/2004WR003778, 2006b.

Tromp van Meerveld, I. and McDonnell, J. J.: Comment to "Spatial correlation of soil moisture in small catchments and its relationship to dominant spatial hydrological processes", J. Hydrol., 286, 113–134, J. Hydrol., 303, 307–312, 2005.

Uchida, T., Asano, Y., Onda, Y., and Miyata, S.: Are headwaters just the sum of hillslopes?, Hydrol. Process., 19, 3251–3261, 2005.

VRO (Victorian Resources Online): Department of Economic Development, Jobs, Transport and Resources, West Gippsland Region, Geology of Southern Gippsland, available at: http://vro.dpi.vic.gov.au/dpi/vro/wgregn.nsf/pages/wg_lf_geology, last access: 3 September 2013.

Western, A. W. and Grayson, R. B.: The Tarrawarra data set: Soil moisture patterns, soil characteristics, and hydrological flux measurements, Water Resour. Res., 34, 2765–2768, 1998.

Western, A. W., Blöschl, G., and Grayson, R. B.: How well do indicator variograms capture the spatial connectivity of soil moisture?, Hydrol. Process., 12, 1851–1868, 1998.

Western, A. W., Grayson, R. B., Blöschl, G., Willgoose, G. R., and McMahon, T. A.: Observed spatial organisation of soil moisture and its relation to terrain indices, Water Resour. Res., 35, 797–810, 1999.

Western, A. W. and Grayson, R. B.: Soil moisture and runoff processes at Tarrawarra, in: Spatial patterns in catchment hydrology – Observations and modelling, edited by: Grayson, R. B. and Blöschl, G., Cambridge University Press, 209–246, 2000.

Western, A. W., Blöschl, G., and Grayson, R. B.: Towards capturing hydrologically significant connectivity in spatial patterns, Water Resour. Res., 37, 83–97, 2001.

Western, A. W., Zhou, S.-L., Grayson, R. B., McMahon, T. A., Blöschl, G., and Wilson, D. J.: Spatial correlation of soil moisture in small catchments and its relationship to dominant spatial hydrological processes, J. Hydrol., 286, 113–134, 2004.

Western, A. W., Zhou, S.-L., Grayson, R. B., McMahon, T. A., Blöschl, G., and Wilson, D. J.: Reply to comment by Tromp van Meerveld and McDonnell on Spatial correlation of soil moisture in small catchments and its relationship to dominant spatial hydrological processes, J. Hydrol., 303, 313–315, 2005.

Withers, P. J. and Lord, E. I.: Agricultural nutrient inputs to rivers and groundwaters in the UK: policy, environmental management and research needs, Sci. Total Environ., 282, 9–24, 2002.

9

Negative trade-off between changes in vegetation water use and infiltration recovery after reforesting degraded pasture land in the Nepalese Lesser Himalaya

C. P. Ghimire[1], L. A. Bruijnzeel[2], M. W. Lubczynski[1], and M. Bonell[3,†]

[1]Faculty of Geo-Information Science and Earth Observation (ITC), University of Twente, Enschede, the Netherlands
[2]Critical Zone Hydrology Group, Faculty of Earth and Life Sciences, VU University, Amsterdam, the Netherlands
[3]The Centre for Water Law, Policy and Science under the auspices of Unesco, The Peters Building, University of Dundee, Dundee, DD14HN, Scotland, UK
[†]deceased

Correspondence to: C. P. Ghimire (c_ghimire@yahoo.com)

Abstract. This work investigates the trade-off between increases in vegetation water use and rain water infiltration afforded by soil improvement after reforesting severely degraded grassland in the Lesser Himalaya of central Nepal. The hillslope hydrological functioning (surface and subsurface soil hydraulic conductivities and overland flow generation) and the evapotranspiration (rainfall interception and transpiration) of the following contrasting vegetation types were quantified and examined in detail: (i) a nearly undisturbed, natural broadleaved forest; (ii) a 25-year-old, intensively-used pine plantation; and (iii) a highly degraded pasture. Planting pines increased vegetation water use relative to the pasture and natural forest situation by 355 and 55 mm year^{-1}, respectively. On balance, the limited amount of extra infiltration afforded by the pine plantation relative to the pasture (only 90 mm year^{-1} due to continued soil degradation associated with regular harvesting of litter and understory vegetation in the plantation) proved insufficient to compensate the higher water use of the pines. As such, observed declines in dry season flows in the study area are thought to mainly reflect the higher water use of the pines although the effect could be moderated by better forest and soil management promoting infiltration. In contrast, a comparison of the water use of the natural forest and degraded pasture suggests that replacing the latter by (mature) broadleaved forest would (ultimately) have a near-neutral effect on dry season flows as the approximate gains in infiltration and evaporative losses were very similar (ca. 300 mm year^{-1} each). The results of the present study underscore the need for proper forest management for optimum hydrological functioning as well as the importance of protecting the remaining natural forests in the region.

1 Introduction

The traditional view of forest hydrological functioning, in which wet season rainfall is readily absorbed and stored for subsequent gradual release during the dry season, has often been likened to a "sponge" (Hamilton and King, 1983). Although the forest sponge effect had been the subject of debate before (cf. reviews of older literature by Andréassian, 2004; Bruijnzeel, 2004; Galudra and Sirait, 2009), the concept came under severe scrutiny in the early 1980s when Bosch and Hewlett (1982) reviewed the results from nearly 100 paired catchment experiments around the globe. They concluded that "no experiments in deliberately reducing vegetation cover caused reductions in water yield, nor have any deliberate increases in cover caused increases in yield". As such, the removal of a dense forest cover was seen to lead to higher streamflow totals, and reforestation of open lands to a decline in overall streamflow. These initial results were confirmed by several subsequent reviews, both for the (warm) temperate zone (Stednick, 1996; Brown et al., 2005; Jackson et al., 2005) and the humid tropics (Bruijnzeel, 1990; Grip et al., 2005; Scott et al., 2005). Whilst the majority of experiments concerned small catchments ($< 2\,km^2$), comparable results of increased water yield following forest removal and reduced flows following forestation were also obtained for much larger river basins ($1100–175\,000\,km^2$; Trimble et al., 1987; Madduma Bandara, 1997; Costa et al., 2003; Zhou et al., 2010). The fact that the bulk of the change in streamflow associated with such experiments was observed during conditions of baseflow (Bosch and Hewlett, 1982; Bruijnzeel, 1989; Farley et al., 2005) at first sight contradicted the reality of the forest sponge concept, and its very existence became questioned (Hamilton and King, 1983; Calder, 2005). Indeed, since the early reviews of Bosch and Hewlett (1982) and Hamilton and King (1983), many have emphasised the more "negative" aspects of forests, such as their high water use or inability to prevent extreme flooding (e.g. Calder, 2005; FAO-CIFOR, 2005; Jackson et al., 2005; Kaimowitz, 2005) rather than focus on the positive functions of a good forest cover, including the marked reduction of surface erosion and shallow landslip occurrence (Sidle et al., 2006), improved water quality (Bruijnzeel, 2004; Wohl et al., 2012), moderation of all but the largest peak flows (Roa-García et al., 2011; Ogden et al., 2013) or carbon sequestration (Farley et al., 2005; Malmer et al., 2010).

At the same time, there is ample evidence of severe and widespread soil degradation after tropical forest conversion to unsustainable forms of land use (e.g. overgrazed pastures, subsistence cropping on steep slopes without adequate soil conservation measures, etc.; Oldeman et al., 1991; Bai et al., 2008). In addition, the area occupied by impervious surfaces such as roads, trails, built-up areas and yards is also on the increase (Ziegler and Giambelluca, 1997; Rijsdijk et al., 2007; Sidle and Ziegler, 2012). This is often accompanied by strongly increased stormflow volumes during times of rainfall (Bruijnzeel and Bremmer, 1989; Fritsch, 1993; Chandler and Walter, 1998; Zhou et al., 2002; Ziegler et al., 2004; Ziegler et al., 2007) and shortages of water during extended dry periods (Bartarya, 1989; Madduma Bandara, 1997; Bruijnzeel, 2004; Tiwari et al., 2011). Such changes in streamflow regime effectively reflect the loss of the former forest sponge (cf. Roa-García et al., 2011; Krishnaswamy et al., 2012; Ogden et al., 2013) through critically reduced replenishment of soil water and groundwater reserves due to lost surface infiltration opportunities, despite the fact that the lower water use of the post-forest vegetation should have produced higher streamflows throughout the year (Bruijnzeel, 1986, 1989).

Reforestation of degraded land in the tropics is often conducted in the expectation that disturbed streamflow regimes (commonly referred to as the "too little–too much syndrome": Bartarya, 1989; Schreier et al., 2006) will be restored by the increased rainfall absorption afforded by soil improvement after tree planting (Scott et al., 2005; cf. Ilstedt et al., 2007). At the same time, the water use of fast-growing tree plantations tends to be (much) higher than that of the degraded vegetation they typically replace – particularly where the trees have access to groundwater (Calder, 1992; Kallarackal and Somen, 1997). Furthermore, major improvements in the (near-)surface infiltration capacity of severely degraded soils, after tree planting may well take several decades of undisturbed forest development to fully materialise (Gilmour et al., 1987; Scott et al., 2005; Bonell et al., 2010). As such, reforesting degraded pasture or shrub land may well cause already diminished dry season flows to become reduced even further, depending on the net balance between increases in soil water reserves through improved infiltration versus decreases caused by the higher plant water uptake (the so-called "infiltration trade-off" hypothesis; Bruijnzeel, 1986, 1989).

Although the overwhelming majority of paired catchment experiments have shown major decreases in baseflow volumes, after the establishment of a tree cover on non-forested land (Farley et al., 2005) this does not disprove the possibility of enhanced dry season flows after reforestation at the small catchment scale. As pointed out by Bruijnzeel (2004) and Malmer et al. (2010), only three out of the 26 paired catchment studies of the hydrological impacts of reforestation reviewed by Jackson et al. (2005) and Farley et al. (2005) were located in the tropics (where soil degradation tends to be more acute; Oldeman et al., 1991; Bai et al., 2008) whereas none of the experiments represented degraded soil conditions. In other words, no positive effects of reforestation on soil water intake capacity could manifest and the observed reductions in water yield simply reflected the higher water use of the trees compared to the grasses and scrubs they replaced (cf. Trimble et al., 1987; Waterloo et al., 1999; Scott and Prinsloo, 2008).

Nevertheless, although direct evidence for the infiltration trade-off hypothesis seemed to be lacking until recently (see below), several hillslope-scale or small-basin studies of the hydrological impacts of reforesting severely degraded land (discussed in some detail by Scott et al., 2005) observed reductions in stormflow volumes that were likely to exceed the estimated corresponding increases in vegetation water use (cf. Zhang et al., 2004; Sun et al., 2006). Unfortunately, as the catchments involved in these experiments were either too small or leaky to sustain perennial streamflow, the expected net positive effect of forestation on dry season flows could not be ascertained in these cases (Scott et al., 2005; cf. Chandler, 2006). However, recently published concurrent reductions in stormflow response to rainfall, and positive trends in baseflow over time since reforesting severely degraded land in southern China (Zhou et al., 2010) and South Korea (Choi and Kim, 2013), or along a forest degradation–recovery gradient in the Western Ghats of India (Krishnaswamy et al., 2012, 2013) are highly suggestive of a net positive outcome of the infiltration trade-off mechanism.

A related aspect concerns the vexed question whether or to what extent (or at what scale) reforestation may have a positive influence on the amount of precipitation received. Although it is clear that tree planting at the local scale (< 50 km) cannot be expected to have such an effect (Harding, 1992), claims of a positive effect of *large-scale* forestation on rainfall generation are becoming more frequent (see Ellison et al., 2011, for an in-depth discussion). At the same time, re-establishing forest cover over an area of ca. 67 000 km^2 of wasteland over a period of 50 years in southern China did not produce any change in rainfall (Zhou et al., 2010). Furthermore, in a particularly comprehensive observational study of Amazonian rainfall and its dependence on vegetation (Angelini et al., 2011), isotopic analysis showed the precipitation to originate mostly from large-scale weather systems linking the interior region to the ocean and precipitation to derive only very partly from local evaporation. Nevertheless, there are some indications that large-scale deforestation may negatively affect rainfall during the transition from wet season to dry season in mainland Southeast Asia (Kanae et al., 2001; cf. Kumagai et al., 2013) and therefore it cannot be excluded that the reverse case also applies (Spracklen et al., 2012). All in all, in the absence of strong empirical evidence of a major effect of tree plantation on rainfall, it is difficult to take into account any possibly positive influence at this stage.

The Middle Mountain zone of the Nepalese and Indian Himalayas provides another case in point regarding the need to restore diminished dry season flows. The rivers originating in this densely populated part of the mountain range (Singh et al., 1984; Hrabovszky and Miyan, 1987) are mostly rain-fed and thus do not benefit from increased water yields from melting glaciers under a changing climate scenario (Bookhagen and Burbank, 2010; Andermann et al., 2012; Immerzeel et al., 2013). Land surface degradation in the zone has often progressed to a point where rainfall infiltration has become seriously impaired and overland flow is rampant (Bruijnzeel and Bremmer, 1989; Gerrard and Gardner, 2002; cf. Ghimire et al., 2013), with reduced dry season flows as the result (Bartarya, 1989; Tiwari et al., 2011; Tyagi et al., 2013). Partly in response to the latter problem, some 23 000 ha of severely degraded pastures and shrublands in the Middle Mountain zone of central Nepal were planted to fast-growing coniferous tree species (mainly *Pinus roxburghii* and *P. patula*) between 1980 and 2000 (District Forest Offices, Kabhre and Sindhupalchok, unpublished data, 2010). However, villagers and farmers in central Nepal have expressed serious concerns about diminishing spring discharges and dry season flows following the large-scale planting of the pines (República, 2012). Such considerations assume added importance under the strongly seasonal climatic conditions prevailing in the Middle Mountain zone where ~ 80 % of the annual rainfall falls during the main monsoon (June–September; Merz, 2004; Bookhagen and Burbank, 2010) and where water during the dry season is already at a premium (Merz et al., 2003).

There is a dearth of sound experiments in the Himalayan region to ascertain the hydrological effects of land use change (reviewed by Bruijnzeel and Bremmer, 1989; Negi, 2002) and indeed for the (sub)tropics in general with respect to the extent to which reforestation of degraded land can improve or even restore diminished dry season flows (Scott et al., 2005; Zhou et al., 2010; Krishnaswamy et al., 2013; Choi and Kim, 2013). In view of the difficulty of identifying catchments with a single dominant land cover type (e.g. forest or grassland) as well as being sufficiently large to support perennial flows (required for the evaluation of changes in baseflows during the extended dry season) in this rugged and spatially highly variable terrain (Maharjan, 1991; Merz, 2004; Bookhagen and Burbank, 2006), the present study opted for the "space for time substitution approach" in which experimental plots with contrasting land-cover types were studied in terms of their hydrological processes and taking the infiltration trade-off hypothesis as a starting point. Thus, the primary objective of the present study was to investigate the trade-offs between the changes in water use going from a severely degraded pasture to a mature coniferous tree plantation or well-developed broadleaved forest on the one hand, and the concurrent increases in soil water reserves afforded by improved rainfall infiltration after plantation or forest maturation on the other in a typically Middle Mountain zone setting in central Nepal. Total vegetation water use (evapotranspiration, ET), overland flow production and field-saturated soil hydraulic conductivity profiles with depth were determined in a little disturbed natural broad-leaved forest, a highly degraded pasture, and in a mature 25-year-old planted pine forest near Dhulikhel. The resulting data were used to address the central question: can reforestation of severely degraded hillslopes in the heavily populated Nepalese Middle Mountain zone restore diminished dry season flows upon plantation maturation?

2 Methods

2.1 Site description

The Middle Mountain zone or Lesser Himalaya, which is situated between ~ 800 and ~ 2400 m above mean sea level (a.m.s.l.) and which occupies about 30 % of Nepal, is home to ca. 45 % of the country's population (based on the 2011 population census; http://cbs.gov.np/). The zone is characterised by a complex geology which has resulted in equally complex topography, soil and vegetation patterns (Dobremez, 1976). The geology of the central Nepalese Middle Mountains consists chiefly of phyllites, schists and quartzites (Stocklin and Bhattarai, 1977). Depending on elevation, the climate is humid subtropical to warm temperate and strongly seasonal, with most of the rain falling between June and September. Rainfall varies with elevation and exposure to the prevailing monsoonal air masses (Merz, 2004; cf. Bookhagen and Burbank, 2006). Between 1000 and 2000 m a.m.s.l. the natural vegetation consists of a largely evergreen mixed broadleaved forest dominated by *Schima wallichii* and various chestnuts and oaks (*Castanopsis* spp., *Quercus* spp.), with *Rhododendron arboreum* as their main associate above ca. 1500 m a.m.s.l. Due to the prevailing population pressure, much of this species-rich forest has disappeared (< 10 % remaining), except on slopes that are too steep for agricultural activity (Dobremez, 1976; Merz, 2004).

The present study was conducted in the Jikhu Khola catchment (JKC) (Fig. 1). The 111.4 km^2 JKC (27°35′–27°41′ N, 85°32′–85°41′ E) is situated approximately 45 km east of Kathmandu (the capital of Nepal) along the Araniko Highway in the Kabhre District between 796 and 2019 m a.m.s.l. The general aspect of the catchment is southeast and the topography ranges from flat valleys to steep upland slopes (Maharjan, 1991). The geology consists of sedimentary rocks affected by various degrees of metamorphism and includes phyllites, quartzites, and various schists (Nakarmi, 2000). In general, soils in the upper half of the JKC are Cambisols (FAO, 2007) of loamy texture and moderately well to rapidly drained (Maharjan, 1991).

The climate of the JKC is largely humid subtropical, grading to warm temperate around 2000 m a.m.s.l. Mean (±SD, standard deviation) annual rainfall measured at mid-elevation (1560 m a.m.s.l.) for the period 1993–1998 was 1487 (±155) mm (Merz, 2004). The main seasons are the rainy season (monsoon; June–September), the post-monsoon (October–November), winter (December–February), and the pre-monsoon (March–May). During the monsoon about 80 % of the total annual precipitation falls. In general, July is the wettest month with about 27 % of the annual rainfall. The driest months are November–February, each accounting for about 1 % of the annual rainfall (Merz, 2004). About 40–50 % of monsoon rainfall occurs as small showers (< 5 mm) while larger storms (> 30 mm) are comparatively infrequent (< 8 %). Maximum rainfall intensities are in the order of 75–85 mm h^{-1} (based on 5-min observations) but median values are low at 2–2.5 mm h^{-1} (Ghimire et al., 2012). Average monthly temperatures as measured at 1560 m a.m.s.l. range from 7.7 °C in January to 22.6 °C in June while the average monthly relative humidity varies from 55 % in March to 95 % in September. Strong winds are common during thunderstorms before the onset of the main monsoon, but these are momentary and the average monthly wind speeds are always less than 2 m s^{-1} and show only slight seasonal variation. Annual reference evaporation ET_0 (Allen et al., 1998) for the period 1993–2000 was 1170 mm (Merz, 2004). Vegetation cover in the catchment consists of ~ 30 % forest (both natural and planted), 7 % shrubland and 6 % grassland, with the remaining 57 % largely under agriculture (Merz, 2004). The JKC was subjected to active reforestation until 2004 as part of the Nepal–Australia Forestry Project (NAFP).

We measured vegetation water use (wet- and dry-canopy evaporation), overland flow and saturated soil hydraulic conductivity (K_{fs}) in natural forest, degraded pasture and planted forest. The land use at the respective research sites can be characterised as follows:

Natural Forest: this site (elevation 1500 m a.m.s.l., northwest exposure, average slope angle 24°) consisted of dense, largely evergreen forest facing little anthropogenic pressure. The 14.0 ± 2.2 m high forest had a well-developed understory and litter layer with visual evidence of soil faunal activity in the form of a high carbon content and numerous near-surface macropores. Tree density in the 50 m × 50 m plot as measured in May 2011 was 1869 trees ha^{-1}, whereas the average diameter at breast height (DBH) for trees with DBH ≥ 5 cm was 13.6 ± 4.4 cm and the corresponding basal area 27.1 m^2 ha^{-1}. The forest was floristically diverse. More than half of the trees consisted of *Castanopsis tribuloides* (65 %) followed by *Schima wallichii* (16 %), *Myrica esculenta* (6 %), *Rhododendron arboreum* (5 %), *Quercus lamellosa* (4 %) and various other species (4 %). Although largely evergreen, the natural forest sheds a small proportion of its leaves towards the end of the dry season (February–March) but recovers quickly thereafter. For example, the maximum measured leaf area index (LAI, using a Licor 2000 Plant Canopy Analyzer) was 5.43 ±0.12 (SD) in September 2011 (i.e. at the end of the rainy season), while corresponding values measured during the premonsoon period in March, April and early June were 4.52 ± 0.19, 5.14 ± 0.09, and 5.32 ± 0.10, respectively. The soil was classified as a Cambisol of clay loam texture. Clay content varied little with depth between 5 and 100 cm (26–29 %) as did the sand (24–26 %) and silt (44–48 %) contents. Soil organic carbon (SOC) declined from 4.10 ± 0.25 % at 5–15 cm depths to 0.72 ± 0.13 % between 50 and 100 cm depths. The topsoil had a low bulk density (0.93 ± 0.08 g cm^{-3} at 5–15 cm). During soil profile excavations and in road exposures, roots were observed to penetrate into the underlying (weathered) bedrock. Depth to bedrock within the 50 m × 50 m plot was ca. 2.3 m. Further information on changes in soil characteristics with depth at this and the other two research sites is given by Ghimire et al. (2013).

Degraded Pasture: this 50 m × 120 m site (1620 m a.m.s.l., southeasterly exposure, average slope angle 18°) has been heavily grazed for more than 150 years according to various local sources and is crossed by numerous footpaths. It is located ~ 2700 m from the natural forest plot (Fig. 1). Numerous patches of compacted or bare soil surface were evident. The dominant grass and herb species included *Imperata cylindrica*, *Saccharum spontaneum*, and *Ajuga macrosperma*. Little or no grass cover remained at the peak of the dry season (March–May). The Cambisol underneath the degraded pasture plot had a silty clay texture, with a lower clay content (12–19 %) and a higher sand content (34–44 %) than the soil of the natural forest plot. Because of the intensive grazing and frequent human traffic, topsoil bulk density in the degraded pasture was significantly higher (1.18 ± 0.33 g cm^{-3}) than in the natural forest. Depth to bedrock within the plot was determined at 2.4 m.

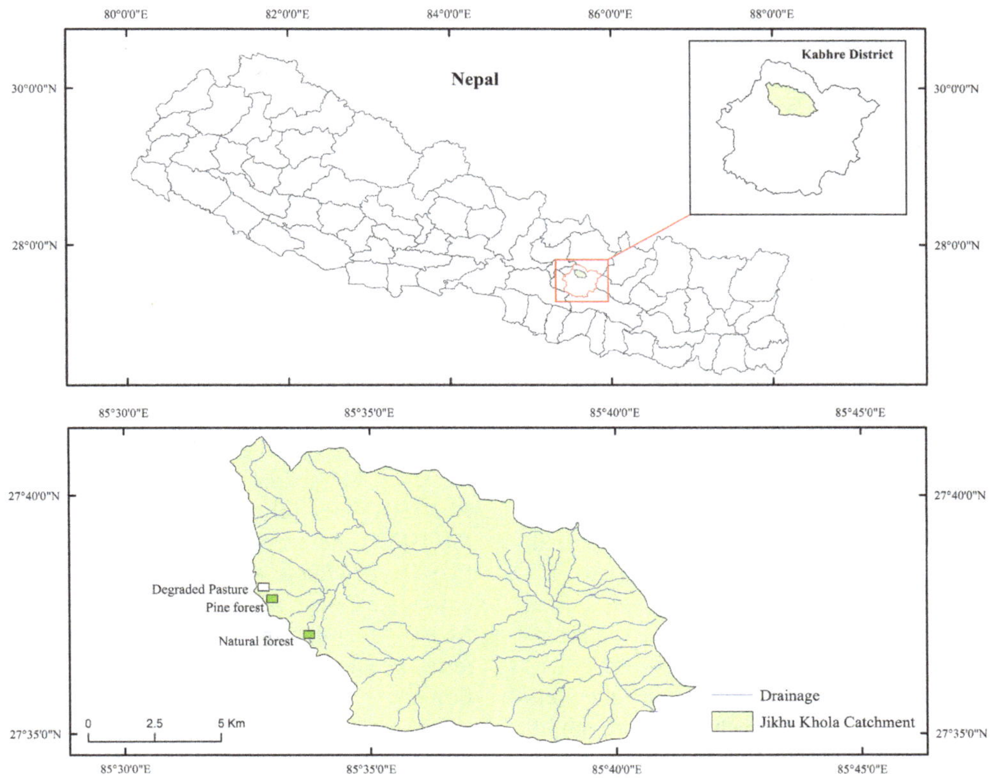

Figure 1. Location of the study sites in the Jikhu Khola catchment in the Middle Mountains of central Nepal.

Pine Forest: this former degraded pasture located approximately 400 m from the studied degraded pasture on a 20° slope of southwesterly exposure at an elevation of 1580 m a.m.s.l. was planted with *P. roxburghii* in 1986. The pine trees in the 60 m × 60 m plot were 25 years old at the start of field work in June 2010. No other tree species were recorded in the plot. An understory was largely absent as grazing by cattle is common. In addition, the local population collects the litter for animal bedding and regularly harvests the grassy herb layer. Pruning of trees for fuelwood, and felling for timber are also common in the pine forests of the JKC (Schreier et al., 2006; cf. Ghimire et al., 2013a) although the research plots themselves remained free from such major disturbances throughout the present investigation. Like the dominant trees in the natural broadleaved forest, the evergreen *P. roxburghii* trees shed a proportion of their needles towards the end of the dry season but also recovered their foliage fairly quickly thereafter. The LAI of the pine forest plot was estimated at 2.21 ± 0.10 in September 2011 whereas corresponding values during the premonsoon period in March, April and early June were 2.05 ± 0.14, 2.15 ± 0.12, and 2.17 ± 0.11, respectively. The stem density, mean DBH and basal area as measured in May 2011 were 853 trees ha^{-1}, 23.65 ± 3.8 cm and 37.6 $m^2\ ha^{-1}$, respectively. The average tree height was estimated at 16.3 ± 3.82 m. The Cambisol underneath the pine forest plot had a silty clay texture, with a lower clay content (11–19 %) and a (much) higher sand content (40–47 %) than the soil of the natural forest or degraded pasture. Because of the regular collection of litter and associated exposure of the soil surface with erosive canopy drip, topsoil organic carbon in the pine forest was much lower (1.7 ± 0.3 %) and bulk density significantly higher (1.24 ± 0.10 g cm^{-3}) than in the natural forest. Roots were observed to penetrate into the underlying (weathered) bedrock which occurred at a depth of ca. 1.5 m.

2.2 Field measurements

2.2.1 Environmental monitoring

Climatic conditions were monitored by an automatic weather station located in the degraded pasture site at a distance of 400 m from the pine forest plot (Fig. 1). In addition, incident rainfall (P, mm) for each plot was recorded using a tipping bucket rain gauge (Rain Collector II, Davis Instruments, USA; 0.2 mm per tip) backed up by a manual gauge (100 cm^2) that was read daily at ca. 08:45 LT (local time). Incoming short-wave radiation (R_S) at the pasture station was measured using an SKS 110 pyranometer (Skye Instruments, UK). The resulting radiation loads were corrected for slope aspect for use at the natural- and pine-forest sites. Air temperature (T, °C) and relative humidity (RH, percentage of saturation) were measured at 2 m height using a Vaisala HMP45C probe protected against direct sunlight and precipitation by a radiation shield. Wind speed and wind direction were measured at 3 m height, using an A100R digital anemometer (Vector Instruments, UK) and W200P potentiometer (Vector Instruments, UK), respectively. All measurements were recorded at 5 min intervals by a Campbell Scientific Ltd. 23X data-logger. Soil water content at each plot was measured at a single location at depths of 10, 25, 50, and 75 cm using TDR (time domain reflectometry) sensors (CS616, Campbell Scientific Ltd.) at 30 min intervals.

2.2.2 Forest hydrological measurements

Wet-canopy evaporation or rainfall interception (E_i) was calculated as the difference between gross rainfall (P) and net rainfall (throughfall + stemflow). Throughfall (Tf, mm) in the natural and pine forest plots was measured daily using 10 (pine plot) or 15 (natural forest) funnel-type collectors (154 cm^2 orifice) that were regularly relocated in a random manner (cf. Holwerda et al., 2006). In addition, Tf was recorded continuously using three tilted stainless steel gutters in each plot (200 cm × 30 cm × 15 cm each). The throughfall measurements were carried out from 20 June to 9 September, 2011 (81 days), thereby covering the bulk of the 2011 rainy season. Stemflow (Sf, mm) was measured on 10 trees which were representative of the dominant species in the natural forest plot and similarly on eight trees in the pine forest plot. Stemflow was collected using 10 L buckets connected to flexible polythene tubing fitted around the circumference of the trunk in a spiral fashion at 1 m from the ground. Stemflow measurements were carried out for 65 days between 28 July and 1 October 2010. Sf was not measured during the 2011 rainy season. Instead, the average values derived for the 2010 rainy season (expressed as a fraction of P) were used when estimating and modelling interception losses for 2011. Annual interception losses were estimated using the revised analytical model of Gash et al. (1995) which was run on a daily basis for the year between 1 June 2010 and 31 May 2011 using the forest structural and average evaporation model parameters established by Ghimire et al. (2012) for the same sites and daily rainfall values as input. A crossed sensitivity analysis was conducted using daily rainfall data from either forest site in the model to examine the effect of potentially contrasting rainfall amounts between the two sites on modelled rainfall interception totals. No significant difference was found, reflecting the fact that the distribution of daily rainfall at the two forest sites was not statistically different ($p < 0.05$). For a more detailed description of the measuring and modelling procedures used in the derivation of annual totals of Tf, Sf and E_i, the reader is referred to Ghimire et al. (2012).

The quantification of tree transpiration (E_t) was accomplished by in situ xylem sap flow rate measurements. Sap flow measurements on individual trees involve the measurement of xylem sap flux density and sap wood cross-sectional area (Granier, 1985; Lu et al., 2004; Lubczynski, 2009). Sap flux density J_p was measured primarily using thermal dissipation probes (TDP; Granier, 1985) because of their low cost, easy installation and overall simplicity, while the heat field deformation method (HFD; Nadezhdhina et al., 2012), and the heat ratio method (HRM; Burgess et al., 2001) were used in addition for the purpose of deriving radial sap flow patterns. Sapwood cross-sectional area was estimated from wood cores using an increment borer at breast height (Grissino-Mayer, 2003). Long-term monitoring of J_p was conducted on nine trees in the natural forest plot that were considered representative in terms of the dominant species present and overall size class distribution, and similarly on six trees in the pine forest plot between June 2010 and May 2011. Radial sap flow patterns were measured using the HFD method on eight trees in the pine forest and on two *C. tribuloides* trees in the natural forest for at least 3 consecutive days per tree. Similarly, radial sap flow patterns for two other dominant species (*S. wallichii*, $n = 2$; *M. esculenta*, $n = 1$) were derived using the HRM. In addition, a sap flow campaign was held from March to May 2011 to capture the sap flow of additional trees that were not covered by the long-term monitoring. A total of 48 additional trees were monitored in the pine forest (distributed over four additional 30 m × 30 m plots) vs. 24 trees in the natural forest (two additional 30 m × 30 m plots) during this campaign using a single TDP sensor per tree. Tree-scale measurements of J_p were scaled up to the plot level using least-squares regressions between total trunk cross-sectional area and corresponding sapwood area, relations between sapwood area and J_p, and information on radial changes in J_p for different tree species before summing the water uptake by all individual trees in a plot to give stand transpiration. For further details on the measurement of J_s, gap filling and the scaling exercise the reader is referred to Ghimire (2014b).

To obtain approximate annual ET totals for the two forests, the respective values of wet- and dry-canopy evaporation (i.e. E_i and E_t) were added and combined with estimates for evaporation from the understory (natural forest only given the absence of understory vegetation in the pine stand; Ghimire et al., 2014b) and from the litter layer based on findings obtained at other sites having comparable forest structural and climatic characteristics. In view of the high LAI of the natural forest (5.4), its northwesterly exposition and the generally low wind speeds, evaporative contributions by the understory were expected to be modest. Motzer et al. (2010) determined the evaporative fraction contributed by the understory at 20 % of that by the overstory in a lower montane forest of comparable stature and LAI in Ecuador. This value was adopted for the Dhulikhel natural forest. Corresponding evaporation from litter layer (E_s) was considered to be low for the reasons given above and in view of the strongly seasonal rainfall regime which causes the forest floor to be moist for four months only. Comparative measurements of E_s in subtropical broadleaved forests are rare but Kelliher et al. (1992) determined a value of 0.3 mm d^{-1} in a well-watered New Zealand forest. Translating this finding to the Dhulikhel situation and assuming the bulk of E_s to take place during the rainy season (4 months) gave an estimated value of 35 mm year^{-1}. Effectively the same result (36 mm year^{-1}) was obtained when applying the ratio of E_s to E_t derived for the same New Zealand forest (0.18; Kelliher et al., 1992) to the Dhulikhel forest. Thus, a value of 35 mm year^{-1} was adopted as a first estimate for E_s in the natural forest. For the more open pine forest (LAI, 2.2), one expects E_s to be somewhat higher than in the nearby natural forest. However, the pine litter is typically harvested by local people after the main leaf-shedding period (Ghimire et al., 2013, 2014a), reducing amounts of litter present and thus its moisture retention capacity. Further, a substantial fraction of the rainfall in the pine forest runs off as overland flow (see Results below), reducing E_s even further. Waterloo et al. (1999) determined E_s in a similarly stocked stand of *Pinus caribaea* in Fiji at 9 % of the Penman open water evaporation, E_0. Applying the same fraction and taking again an effective period of 4 rainy months yielded an estimated E_s for the Dhulikhel pine forest of ca. 35 mm year^{-1}.

2.2.3 Soil hydraulic conductivity and inferred hillslope hydrological pathways

Field-saturated soil hydraulic conductivity (K_{fs}) (Talsma and Hallam, 1980; Reynolds et al., 1985) in the respective plots was measured both at the surface and at depths of 0.05–0.15, 0.15–0.25, 0.25–0.50, and 0.5–1.0 m. The K_{fs} at different depths were subsequently combined with selected percentiles of 5 min maximum rainfall intensities (RI_{5max}) to infer the dominant hillslope hydrological response during intense rainfall following Chappell et al. (2007). A disc permeameter (Perroux and White, 1988; Mckenzie et al., 2002) was used for the measurement of surface K_{fs} in the field and a constant-head well permeameter (CHWP; Talsma and Hallam, 1980) for the measurement of K_{fs} in deeper layers. Use of the CHWP was restricted to the dry season to minimise errors from smearing of the auger hole walls (Chappell and Lancaster, 2007). For surface K_{fs}, 10 (pine forest) to 17 (degraded pasture) replicate measurements were made whereas for subsurface K_{fs}, 45–80 auger holes were used per site. For a detailed description of the measuring procedures and sampling strategy, the reader is referred to Ghimire et al. (2013).

2.2.4 Overland flow

Overland flow (OF) at the natural forest, degraded pasture and pine forest sites was monitored between 20 June and 9 September 2011 (i.e. the bulk of the 2011 rainy season) using a single, large (5 m × 15 m) runoff plot per land-cover type. Runoff was collected in a gutter funneling the water to a first 180 L collector equipped with a seven-slot divider system allowing only one-seventh of the spillover into a second 180 L drum, thereby bringing the total collector capacity to 1440 L (∼ 20 mm). The water levels in the two collectors were measured continuously using a pressure transducer device (Keller, Germany) placed at the bottom. Collectors were emptied and cleaned after measuring the water level manually every day at around 08:45 LT. Event runoff volume was calculated by converting the water levels to volumes using a precalibrated relationship per drum and summing up to obtain total runoff volume. Measured overland flow volumes were corrected for direct rainfall inputs into the runoff collecting system. Overland flow volumes were divided by the projected plot area to give overland flow in millimetres per event.

2.2.5 Grassland evaporation

The HYDRUS-1D model for one-dimensional soil water movement (Šimůnek et al., 2008) was used to estimate evaporation from the degraded pasture site. Like most pastures in the area, the site was heavily overgrazed (Gilmour et al., 1987; Ghimire et al., 2014a, b) and the capacity of the grass to intercept rainfall was considered negligible. Likewise, in view of the very limited live biomass of the grass, transpiration was also considered minimal. Hence, overall, evaporation at the grassland site was set equal to soil evaporation defined by HYDRUS-1D model (Šimůnek et al., 2008). The HYDRUS-1D model is based on the modified Richard's equation. In this study it was assumed that the gaseous phase plays an unimportant role in the overall vapour transport due to thermal gradients.

The modelling was divided into three parts: (i) model calibration against measured soil moisture data (1 September–30 November 2010) using optimisation, (ii) model validation using data collected between 1 February and 31 March 2011, and (iii) complete simulation of soil water dynamics and evaporation for the entire annual period (1 June 2010–31 May 2011). The 1 m-deep soil column was divided into four schematic layers as follows: (i) 0–0.15 m, (ii) 0.15–0.25 m, (iii) 0.25–0.50 m, and (iv) 0.50–1.0 m.

The soil physical parameters employed in HYDRUS-1D include Q_r for the residual water content and Q_s for the saturated water content, as well as the two parameters, α and n, describing the shape and range of the soil water retention curve and the derived relative hydraulic conductivity curve (Van Genuchten, 1980). Other model parameters include the saturated hydraulic conductivity (K_s), and I, a pore-connectivity parameter. All parameters were optimised except for K_s which was measured separately for each layer in the field using well permeametry as described above. The parameters were optimised by fitting observed and modelled soil moisture values using the Marquardt–Levenberg optimisation algorithm. The model was run on an hourly basis. The boundary conditions used in the model were the atmospheric

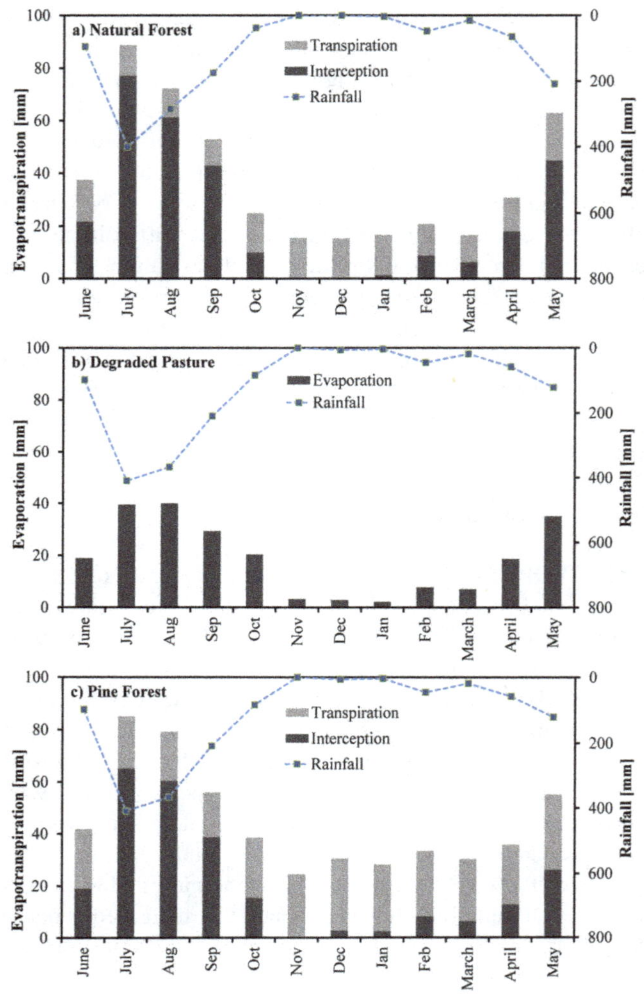

Figure 2. Monthly rainfall, interception and transpiration totals (mm) between 1 June 2010 and 31 May 2011 in **(a)** natural broadleaved forest, **(b)** degraded pasture, and **(c)** planted pine forest near Dhulikhel, central Nepal.

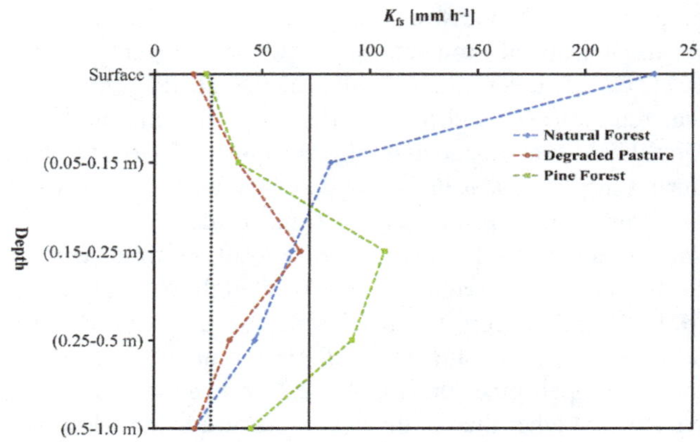

Figure 3. Changes in field-saturated hydraulic conductivity K_{fs} with depth as a function of land use near Dhulikhel, central Nepal. The dotted and solid horizontal lines represent the median and 95 percentile of RI_{5max} rainfall intensity, respectively (modified after Ghimire et al., 2013).

boundary (soil surface) with surface runoff occurrence, and free drainage at the lower boundary.

3 Results

3.1 Evapotranspiration

Figure 2 shows the monthly variation in ET and its two main components, rainfall interception and transpiration, for the three land-cover types studied between 1 June 2010 and 31 May 2011 whereas the respective seasonal and annual evapotranspiration totals are presented in Table 1.

Although the seasonal patterns for ET were similar between vegetation types, monthly ET totals were generally higher for the planted pine forest and lower in the degraded pasture throughout the monitoring period (Fig. 2). All three sites showed higher ET rates and monthly totals during the wet season months (June–September) compared to the dry season months (October–April), with the transitional month of May (marking the first return of the rains; Fig. 2a) showing intermediate values. Such findings can be attributed largely to the (much) higher frequency of wetting and subsequent drying (evaporation) during the wet season (cf. Table 1).

As found earlier for monthly ET totals, the annual ET for the planted forest was the highest for all three vegetation types studied (577 mm), being two and a half times larger than the annual ET in the degraded pasture (225 mm; Table 1). The annual ET for the natural forest was 524 mm, which is 53 mm less than that for the nearby pine forest but 300 mm higher than the ET of the degraded pasture (Table 1). Whilst absolute rainfall interception totals did not differ much between the natural and the planted forest (31 mm year^{-1} higher in the broadleaved forest despite a slightly lower rainfall total), both the seasonal and annual transpiration totals were distinctly higher for the pine forest (Table 1). Wet-season transpiration in the pine forest was 20 mm higher vs. 64 mm during the dry season.

3.2 Soil hydraulic conductivity, overland flow and subsurface flow paths

As expected on the basis of the degree of anthropogenic pressure experienced by the respective sites, the median surface K_{fs} was lowest for the degraded pasture (18 mm h^{-1}) and highest for the natural forest (232 mm h^{-1}), such that the two differed by more than an order of magnitude (Fig. 3). The most striking feature of the K_{fs} data set is that the me-

Table 1. Summary of rainfall and estimated evapotranspiration components for a natural broadleaved forest, a degraded grassland, and a mature planted pine forest near Dhulikhel, Middle Mountains, central Nepal.

	Natural forest			Degraded pasture			Pine forest		
	Wet	Dry	Total	Wet	Dry	Total	Wet	Dry	Total
Rainfall (P, mm)	953	378	1331	1084	338	1423	1084	338	1423
Transpiration (E_t, mm)	48	115	163	–	–	–	78	202	280
Interception (E_i, mm)	203	90	293	–	–	–	184	78	262
Grassland soil evaporation	–	–	–	128	97	225	–	–	–
Understory evaporation (E_{us}, mm)	–	–	33	–	–	–	–	–	–
Litter evaporation (E_s, mm)	35	–	35	–	–	–	35	–	35
Total evapotranspiration (ET, mm)	286	205	524	128	97	225	297	280	577

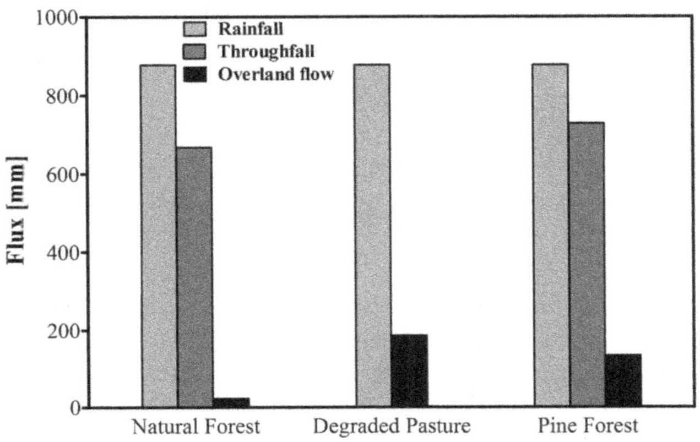

Figure 4. Amounts of rainfall, throughfall and overland flow (mm) during the 2011 monsoon measuring campaign (20 June–9 September) in a natural forest, a degraded pasture, and a mature, intensively used pine plantation near Dhulikhel, central Nepal. Note that values for the natural forest plot were normalised for rainfall amount to allow for more direct comparisons.

dian K_{fs} at the surface and in the shallow soil layer in the 25-year-old pine forest had remained at the same level as the corresponding values for the heavily grazed pasture, suggesting the virtually complete absence of biologically mediated macropores in the pine forest soil down to 0.15 m depth. At 1.0 m depth, however, differences in K_{fs} between the respective land-cover types were mostly non-existent (except for the higher value beneath the pine forest reflecting the much higher sand content listed in the site description), illustrating the lack of influence exerted by cattle grazing and human trampling on the deeper soil layers.

Importantly, the surface K_{fs} in the natural forest exceeded the maximum values of RI_{5max}, suggesting infiltration-excess overland flow (IOF) would never occur at this site. Nevertheless, some overland flow was recorded (Fig. 4) with a monsoonal total of 18 mm (22 mm after normalising for the higher rainfall observed at the other two sites) representing 2.5 % of incident rainfall and 3.3 % of the corresponding amount of throughfall. Given the much lower median K_{fs} derived for the 0.05–0.15 m depth interval in the natural forest (82 mm h^{-1}; Fig. 3), it cannot be excluded that at least some of the recorded overland flow was contributed by the saturation-excess type (SOF) (Bonell, 2005). The median surface K_{fs} values in the degraded pasture and pine forest were below the upper quartile of RI_{5max}, thereby indicating the frequent occurrence of IOF during high-intensity rainfall (Fig. 3). Indeed, overland flow at the degraded pasture site was typically generated after 3–4 mm of rain while the seasonal (monsoonal) overland flow total amounted to 187 mm (21.3 % of incident rainfall; Fig. 4). Corresponding values for the pine forest were comparable at 4.2 mm of rain before overland flow would start and a seasonal total of 136 mm (15.5 % of rainfall and 18.6 % of throughfall; Fig. 4).

With regard to K_{fs} in the 0.05–0.15 m layer, although the upper quartile of RI_{5max} exceeded the median K_{fs} values for this depth interval at both the degraded pasture and pine forest sites (Fig. 3), the actual volumes of water reaching this layer are necessarily restricted by the low value of the surface "throttle" at both sites. Surface K_{fs} at either site did not exceed the K_{fs} at 0.05–0.15 m depth and therefore no perched water table can develop and hence no shallow lateral subsurface stormflow (SSF) is generated at either site. In contrast, the corresponding median K_{fs} at the natural forest is still above (or nearly equal to) most 5 min rainfall intensities, thereby favouring mostly vertical percolation at this site. For the 0.15–0.25 and 0.25–0.50 m depth intervals, the K_{fs} values in the natural forest indicated a similar hydrological response to extreme rainfall, namely mostly lateral SSF because of limited vertical percolation under such conditions (cf. Fig. 3). Finally, at 1.0 m depth, the differences in K_{fs} and inferred hydrological response to rainfall became insignificant between sites (Fig. 3).

3.3 Site water budgets

Combining the above mentioned overland flow percentages for the pine forest and degraded pasture with a long-term mean annual site rainfall of 1500 mm (Merz, 2004), the difference in approximate annual IOF between the two land

Table 2. Summary of changes in annual ET (mm) and OF(mm), and the resultant gains in infiltration (mm) and evaporative losses (mm) when converting degraded pasture to (heavily used) planted pine forest or (little disturbed) natural broadleaved forest near Dhulikhel, central Nepal. Note that overland flow amounts were calculated for a mean annual site rainfall of 1500 mm (Merz, 2004). Note that all values are rounded off to the nearest fifth or tenth.

Experimental plot	ET (mm)	OF (mm)	Gain (mm)	Loss (mm)	Overall net effect (mm)
Degraded pasture	225	320	–	–	–
Pine forest	575	230	90	350	−260
Natural forest	525	35	285	300	−15

covers represents a gain in infiltration of approximately 90 mm year^{-1} under the planted pine forest relative to the degraded grass land (Table 2). However, the relative amounts of surface evaporation/transpiration in the degraded pasture and the pine forest, plus the added rainfall interception losses from the pine forest, suggest a difference in annual evapotranspiration of $\sim$350 mm year^{-1} after reforestation and stand maturation (Table 2). Thus, the added loss through ET is greatly in excess of the estimated gain in infiltration of 90 mm year^{-1} after reforestation causing a net loss of $\sim$260 mm year^{-1}. Repeating the exercise for the natural forest (with an estimated annual ET of $\sim$525 mm and very low overland flow production) suggests the approximate gain in infiltration (285 mm year^{-1}) and the extra evaporative loss (300 mm year^{-1}) are both very similar (Table 2), implying no major change in moisture availability and dry season flow under mature natural forest conditions.

4 Discussion and conclusions

4.1 Human impact on forest hydrological functioning – an understudied dimension

The results obtained for the respective water balance components suggest that soil water replenishment and retention during the monsoon are largely controlled by surface and subsurface soil hydraulic conductivities and the resultant partitioning of rainfall into overland flow, lateral subsurface flow and deep percolation (Table 1, Fig. 3). As long as rainfall intensities remain below the surface K_{fs} threshold for overland flow to occur, soil water reserves are being recharged. However, for intensities above this threshold a major proportion of the rain is redirected laterally over the surface as overland flow and less water is available for soil moisture replenishment. The high surface and near-surface K_{fs} in the natural forest (82–232 mm h^{-1}) ruled out IOF occurrence and favour vertical percolation. In contrast, the corresponding K_{fs} values for the planted forest and degraded pasture were conducive to IOF generation during medium- to high-intensity storms which represents an important net loss of moisture to these hillslopes (Fig. 3, Table 2).

Marked reductions in surface- and near-surface K_{fs} after converting tropical forest to grazed pasture have been observed in many cases (Alegre and Cassel, 1996; Tomasella and Hodnett, 1996; Deuchars et al., 1999; Zimmermann et al., 2006; Molina et al., 2007; Tobón et al., 2010) and the Himalayas are no exception (Patnaik and Virdi, 1962; Gilmour et al., 1987; Gerrard and Gardner, 2002; Ghimire et al., 2013, 2014a). The low surface and near-surface K_{fs} reported for grazing conditions is mostly the result of destroyed macroporosity through trampling by cattle and by the much diminished soil faunal activity after forest clearing and burning with the associated loss of topsoil organic matter and surface exposure with erosive precipitation (McIntyre, 1958a, b; Lal, 1988; Deuchars et al., 1999; Colloff et al., 2010; Bonell et al., 2010). Natural forest regrowth on degraded pasture or planting trees followed by uninterrupted plantation development can be expected to gradually improve the soil water intake capacity again through the steady incorporation of organic matter, soil faunal and insect burrowing activity, and root turnover (Gilmour et al., 1987; Bonell, 2005; Ilstedt et al., 2007; Bonell et al., 2010; Colloff et al., 2010; Hassler et al., 2011; Perkins et al., 2012). Conversely, impervious footpaths, yards and roads will remain runoff-producing features in post-forest landscapes (Ziegler et al., 2004; Rijsdijk et al., 2007) as well as in managed (forest) plantation areas (La Marche and Lettenmaier, 2001; Ziegler et al., 2007; Liu et al., 2009). Likewise, the surface hydraulic conductivity of the intensively used pine forest showed little improvement even 25 years after the trees were planted (Fig. 3). Clearly, the continued human access, grazing and collection of forest products (notably litter from the forest floor to be used for animal bedding and composting; Singh and Sundriyal, 2009; Joshi and Negi, 2011) is having a profound negative effect on the stand's hydrological functioning (cf. Ghimire et al., 2013, 2014a). Thus, the general expectation of restored surface and near-surface K_{fs} with time after reforestation (cf. Gilmour et al., 1987; Ilstedt et al., 2007) is in need of modification under the conditions of high anthropogenic pressure that appear to be the rule rather than the exception in the Middle Mountain zone (Singh et al., 1984; Mahat et al., 1987; Singh and Sundriyal, 2009; Joshi and Negi, 2011) despite claims to the contrary (HURDEC Nepal and Hobley, 2012). Indeed, if the potential benefits of reforestation such as enhanced infiltration, and therefore possibly improved replenishment of soil water and groundwater reserves are to be realised, then a balance will need to be struck between the continued usage of the forests by uplanders whose livelihoods are at stake and sustained forest hydrological functioning (Ghimire et al., 2014a). Naturally, this holds for many other densely populated tropical uplands as well (e.g. Ding et al., 1992; Van Noordwijk et al., 2001; Forsyth and Walker, 2008; Van Noordwijk and Leimona, 2010).

4.2 Trade-off between changes in vegetation water use and surface infiltration after reforestation

The infiltration trade-off hypothesis states that the ultimate hydrological effect of reforestation in terms of site water yield is determined by the net balance between increases in soil water reserves afforded by improved soil infiltration versus decreases caused by the higher plant water uptake (Bruijnzeel, 1986, 1989). In the absence of direct published evidence of improved dry season flows after reforestation (Jackson et al., 2005; Farley et al., 2005) at the time, Scott et al. (2005) discussed a number of tropical studies that had observed marked reductions in stormflow production at the hillslope (e.g. Zhang et al., 2004; Chandler and Walter, 1998) or small-basin scale (e.g. Zhou et al., 2002) after reforesting severely degraded land. They concluded that in a number of cases the increases in water retention should be more than enough to compensate the estimated corresponding increases in forest water use (not measured). It is unfortunate that the catchments involved did not sustain perennial flows and thus the presumed net positive effect of forestation on dry season flows could not be confirmed (Scott et al., 2005; cf. Chandler, 2006). Nevertheless, recently published evidence from southern China (Zhou et al., 2010), South Korea (Choi and Kim, 2013) and southwestern India (Krishnaswamy et al., 2012, 2013) strongly suggests a net positive outcome of the infiltration trade-off mechanism is possible as long as the initial situation is sufficiently degraded and the site receives ample rainfall (cf. Bruijnzeel et al., 2013).

Such experimental catchment studies usually comprise measurement of the change in streamflow following reforestation but generally lack detailed supporting process-based observations within the catchment undergoing the change (Farley et al., 2005; Scott et al., 2005). As such, the present process-based work which integrated the dominant hydrological processes (notably evaporation, infiltration and runoff generation) to quantify the net hydrological impact of reforestation is a first (cf. Chandler, 2006; Bonell et al., 2010; Krishnaswamy et al., 2013). Comparing the hydrological behaviour of the three contrasting land-cover types studied here (degraded pasture, mature near-undisturbed broadleaved forest, and a heavily used mature pine plantation) within the context of the infiltration trade-off hypothesis showed that planting pines increased vegetation water use relative to the pasture situation by $\sim 350\,\text{mm year}^{-1}$ (Table 1). On balance, the limited amount of extra infiltration afforded by the pine trees ($\sim 90\,\text{mm year}^{-1}$) is clearly insufficient to compensate the much higher water use of the pines, giving a net negative balance of $260\,\text{mm year}^{-1}$ (Table 2, Fig. 4). Pertinently, the net effect would still have been negative (by $\sim 120\,\text{mm year}^{-1}$) even if all rainfall would have been accommodated by the pine forest soil through better forest and soil management promoting infiltration (cf. Wiersum, 1985). As such, the observed decline in dry season flows following reforestation in the study area (República, 2012) is likely to

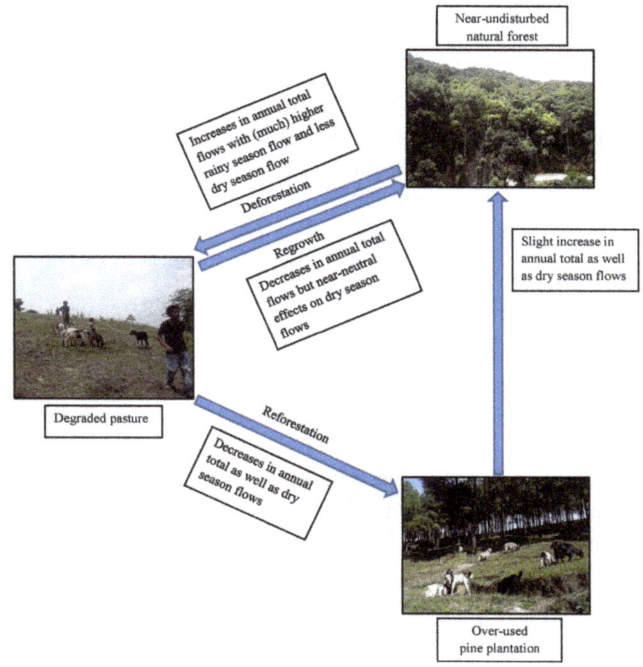

Figure 5. Conceptual diagram of the effect of land-cover transformation on annual total and dry season flows in the study area as well as other comparable regions with similar land-cover transformations.

primarily reflect the higher water use of the pines (Tables 1, 2; Fig. 5).

If the degraded pasture were to revert to natural forest instead (with an estimated annual ET of 525 mm and very limited overland flow production) the ultimate effect on dry season flows would be expected to be near neutral as the approximate gain in infiltration ($\sim 285\,\text{mm year}^{-1}$) and the extra evaporative loss ($\sim 300\,\text{mm year}^{-1}$) are very similar (Table 2). Effects might be more negative in case the water use of the young regenerating broadleaved forest turned out to be enhanced compared to old-growth forest as found for lowland tropical and warm temperate forests (Vertessy et al., 2001; Giambelluca, 2002).

Although the present finding of a slightly higher ET ($\sim 10\,\%$; Table 1) for planted forest compared to natural broadleaved forest is not going to have major hydrological consequences on an annual basis, the much higher water use of the pines during the dry season (Table 1, Fig. 1) is likely to result in a corresponding reduction in dry season water yield upon converting natural broadleaf forest to pine plantations (Fig. 5), especially during the more vigorous early growth stage of the pines (Bruijnzeel, 1997; Scott and Prinsloo, 2008; Alvarado-Barrientos, 2013). The present results further illustrate that the conditions found in the nearly undisturbed natural broadleaved forest and in similarly well-maintained forests elsewhere in the Himalaya (Pathak et al., 1984; Gerrard and Gardner, 2002) will encourage the replenishment of soil water and groundwater reserves through vertical percolation (geology permitting) more than in any other land-cover type studied here (Fig. 5; cf. Choi and Kim, 2013; Krishnaswamy et al., 2013) and thus better sustain baseflows during the long dry season (Hessel et al., 2007; Tiwari et al., 2011). The importance of the latter in the water-scarce Middle Mountain zone can hardly be overemphasised (Merz et al., 2003; Schreier et al., 2006; Bandyopadhyay, 2013).

4.3 Regional implications

The Himalayan river basins are home to about 1.3 billion people and supply water, food and energy to more than 3 billion people in total (Bandyopadhyay, 2013). Thus, large-scale changes in Himalayan land use and hydrology will have important regional consequences. For example, substantial decreases in dry season flows following advanced surface degradation (Bartarya, 1989; Madduma Bandara, 1997) or large-scale reforestation (cf. Trimble et al., 1987; Zhou et al., 2010) would affect the availability of water for millions of people, both those depending directly on agriculture for their livelihoods and downstream city dwellers. Therefore, large-scale reforestation campaigns and the subsequent use of the planted forests must be based on a sound assessment of what is to be expected hydrologically (Peña-Arancibia et al., 2012; Van Dijk et al., 2012) and meteorologically (Ellison et al., 2011). Precipitation fluctuations in Nepal and northern India have been shown to be strongly influenced by surface temperatures in the Indian Ocean and southwestern Pacific, with dry years coinciding mostly with ENSO (El Niño–Southern Oscillation) events and wet years with La Niña events (Shrestha et al., 2000; Yadav, 2009; Varikoden et al., 2014). Therefore, the influence (if any) of increased forest cover on rainfall in this strongly seasonal environment remains to be ascertained (Basistha et al., 2009). Although continued global warming is predicted to ultimately lead to an increase in precipitation in the Himalayas, such predictions are subject to major uncertainty with respect to the inclusion of aerosol effects and large-scale irrigation in the adjacent plains (Craig Collier and Zhang, 2009; Mathison et al., 2012; Kumar et al., 2013).

The presently observed negative hydrological effect of an apparently long-term trend of gradual forest degradation in the Nepalese Middle Mountain zone goes against the optimistic notion regarding the overall improved quality of Lesser Himalayan forests expressed by HURDEC Nepal and Hobley (2012). However, there is reason to believe that the situation of over-intensive use of forests and the correspondingly poor soil hydrological functioning are a rather more widespread phenomenon in Nepal's Lesser Himalaya. For example, Gerrard and Gardner (2002) reported very high overland flow occurrence in degraded (broadleaved) forests in the Likhu Khola catchment north of Kathmandu, whereas Tiwari et al. (2009) and Wester (2013) recently presented similar evidence for community-managed forests further west in Nepal. Although process-based hydrological evidence to this effect from the Indian Himalaya is scarce (Negi, 2002; Tiwari et al., 2011; Tyagi et al., 2013) the continued overexploitation of its forest resources is well-documented (Singh et al., 1984; Singh and Sundriyal, 2009; Joshi and Negi, 2011). Such situations may significantly reduce the recharge of shallow groundwater reserves during the monsoon season, thereby potentially decreasing regional dry season flows (Bartarya, 1989; cf. Andermann et al., 2012).

A key message of the present work thus is the need to protect the remaining natural forests in headwater areas throughout the Middle Mountain zone. The present findings further highlight the need for some form of protection of reforested areas that will enable the forest soils to realise the enhanced infiltration/percolation benefits envisaged at the time of planting. Alternative sources of energy to replace fuelwood (e.g. biogas) may take off some of the pressure on the forests (cf. Schreier et al., 2006). Continued degradation of the remaining old-growth forests and planted forests that are now reaching maturity is likely to cause further increases in overland flow production during the monsoon season due to the corresponding decline in infiltration opportunities (Ghimire et al., 2014a). This may, in turn, have a further negative effect on already declining dry season flows and will cause increased hardship to the rural populace (Merz et al., 2003; Schreier et al., 2006). Finally, and most importantly, the present results point to the need for balancing the societal and hydrological functions of forests (both planted and natural) in densely populated uplands. Like elsewhere in Asia (e.g. Tomich et al., 2004; Hairiah et al., 2006), agroforests appear to represent a viable alternative that is on the increase in Nepal (Schreier et al., 2006; Gilmour and Shah, 2012). Not only do they contain a variety of tree and crop species serving a range of uses as opposed to the monospecific character of most planted forests but agroforests are also likely to consume less water (e.g. Wallace et al., 2005) while minimising surface runoff and erosion, thereby maintaining adequate levels of rainfall infiltration (Wiersum, 1984; cf. Hairiah et al., 2006).

Acknowledgements. We dedicate this work to our co-author, friend and colleague Mike Bonell who passed away during the revision stage of the paper. This study was supported financially by the Faculty of Earth and Life Sciences of the VU University Amsterdam (FALW–VUA) and by the Faculty of Geo-Information Science and Earth Observation (ITC–UT) of the University of Twente, Enschede, the Netherlands. Additional financial support was also received from the Royal Society, UK, through the University of Dundee. Neil Coles, Centre for Ecohydrology at the University of Western Australia, is also thanked for his loan of the disc permeameters and support in the associated logistics in the early stages of the field campaign. We acknowledge the indispensable help in the field received from Paras Poudel, Suraj Karki, Sudish Lal Maskey, Umesh Karki, Laxman Baniya, Khadga Bahadur Kharel, Dawa Lama and Kapil Gyawali. We especially would like to thank Pruna Bahadur Magar and his family for their hospitality and support during the field campaign. Enrico Balugani is thanked for his help with the HYDRUS-1D simulation. Michel Groen (VUA) and Murat Ulcer (ITC-UT) supplied and constructed equipment. Last but not least, we are grateful to D. Tongway, K. Farrick, and M. van Noordwijk for their constructive comments.

Edited by: N. Romano

References

Alegre, J. C. and Cassel, D. K.: Dynamics of soil physical properties under alternate systems to slash-and-burn, Agricultural, Ecosyst. Environ., 58, 39–48, 1996.

Allen, R. G., Pereira, L. S., Raes, D., and Smith, M.: Crop evapotranspiration – Guidelines for computing crop water requirements, United Nations Food and Agriculture Organization, Rome, 1998.

Alvarado-Barrientos, M. S., Hernandez-Santana, V., and Asbjornsen, H.: Variability of the radial profile of sap velocity in *Pinus patula* from contrasting stands within the seasonal cloud forest zone of Veracruz, Mexico, Agr. Forest Meteorol., 168, 108–119, 2013.

Andermann, Ch., Longuevergne, L., Bonnet, S., Crave, A., Davy, Ph., and Gloaguen, R.: Impact of transient groundwater storage on the discharge of Himalayan rivers, Nat. Geosci., 5, 127–132, 2012.

Andréassian, V.: Water and forests: From historical controversy to scientific debat, J. Hydrol., 291, 1–27, 2004.

Angelini, I. M., Garstang, M., Davis, R. E., Hayden, B., Fitzjarrald, D.R., Legates, D. R., Greco, S., Macko, S., and Connors, V.: On the coupling between vegetation and the atmosphere, Theor. Appl. Climatol., 105, 243–261, 2011.

Bai, Z. G., Dent, D. L., Olsson, L., and Schaepman, M. E.: Proxy global assessment of land degradation, Soil Use Manage., 24, 223–234, 2008.

Bandyopadhyay, J.: Sustaining the Himalaya as the water tower of Asia: On the need for innovative policy making from India and China, 19th Pundit Govind Ballabh Pant memorial lecture, 10 September 2013, G.B. Pant Institute of Himalayan Environment and Development, Almora, India, 19 pp., 2013.

Bartarya, S. K.: Hydrogeology, geo-environmental problems and watershed management strategies in a central Himalayan river basin, Kumaun, India, in: Headwater Control, edited by: Kreek, J. and Haigh, M. J., IUFRO/WASWC/CSVIS, Plzen, Czechoslovakia, 308–318, 1989.

Basistha, A., Arya, D. S., and Goel, N. K.: Analysis of historical changes in rainfall in the Indian Himalayas, Int. J. Climatol., 29, 555–572, 2009.

Bonell, M.: Runoff generation in tropical forests, in: Forests, Water and People in the Humid tropics, edited by: Bonell, M. and Bruijnzeel, L. A., Cambridge University Press, Cambridge, UK, 314–406, 2005.

Bonell, M., Purandara, B. K., Venkatesh, B., Krishnaswamy, J., Acharya, H. A. K., Singh, U. V., Jayakumar, R., and Chappell, N.: The impact of forest use and reforestation on soil hydraulic conductivity in the Western Ghats of India: Implications for surface and sub-surface hydrology, J. Hydrol., 391, 47–62, 2010.

Bookhagen, B. and Burbank, D. W.: Topography, relief, and TRMM-derived rainfall variations along the Himalaya, Geophys. Res. Lett., 33, L08405, doi:10.1029/2006GL026037, 2006.

Bookhagen, B. and Burbank, D. W.: Toward a complete Himalayan hydrological budget: spatiotemporal distribution of snowmelt and rainfall and their impact on river discharge, J. Geophys. Res., 115, F03019, doi:10.1029/2009JF001426, 2010.

Bosch, J. M. and Hewlett, J. D.: A review of catchment experiments to determine the effect of vegetation changes on water yield and evapo-transpiration, J. Hydrol., 55, 3–23, 1982.

Brown, A. E., Zhang, L., McMahon, T. A., Western, A. W., and Vertessy, R. A.: A review of paired catchments studies for determining changes in water yield resulting from alterations in vegetation, J. Hydrol., 310, 28–61, 2005.

Bruijnzeel, L. A.: Environmental impacts of (de)forestation in the humid tropics: A watershed perspective, Wallaceana, 46, 3–13, 1986.

Bruijnzeel, L.A.: (De)forestation and streamflow in the tropics: A closer look, J. Tropical Forest Sci., 1, 229–243, 1989.

Bruijnzeel, L. A.: Hydrology of Moist Tropical Forests and Effects of Conversion: A State of Knowledge Review, UNESCO, Paris, and Vrije Universiteit, Amsterdam, 226 pp., 1990.

Bruijnzeel, L. A.: Hydrology of forest plantations in the tropics, in: Management of Soil, Nutrients and Water in Tropical Plantation Forests, edited by: Nambiar, E. K. S. and Brown, A. G., ACIAR Monograph No. 43, ACIAR, Canberra, Australia, 125–167, 1997.

Bruijnzeel, L. A.: Hydrological functions of tropical forests: Not seeing the soil for the trees?, Agr. Ecosyst. Environ., 104, 185–228, 2004.

Bruijnzeel, L. A. and Bremmer, C. N.: Highland-Lowland Interactions in the Ganges-Brahmaputra River Basin: A Review of Published Literature. ICIMOD Occasional Paper no. 11, ICIMOD, Kathmandu, Nepal, 136 pp., 1989.

Bruijnzeel, L. A., Peña-Arancibia, J. L., and Mulligan, M.: Identifying "bright spots" of potentially enhanced stream flow after reforesting degraded land across the tropics, in: Asia-Pacific Workshop on Water and Forests – Beyond Traditional Forest Hydrology, Singh MD (Convener), Forest Research Institute, Dehradun, India, p. 41, 2013.

Burgess, S. S. O., Adams, M., Turner, N. C., Beverly, C. R., Ong, C. K., Khan, A. A. H., and Belby, T. M.: An improved heat pulse method to measure low and reverse rates of sap flow in woody plants, Tree Physiol., 21, 589–598, 2001.

Calder, I. R.: Water use of eucalypts – A review, in: Growth and Water Use of Forest Plantations, edited by: Calder, I. R., Hall, R. L., and Adlard, P. G., J. Wiley, Chichester, UK, 167–175, 1992.

Calder, I. R.: The Blue Revolution. Integrated Land and Water Resource Management (2nd Edn.), Earthscan, London, 353 pp., 2005.

Chandler, D. G.: Reversibility of forest conversion impacts on water budgets in tropical karst terrain, Forest Ecol. Manage., 224, 95–103, 2006.

Chandler, D. G. and Walter, M. F.: Runoff responses among comon land uses in the uplands of Matalom, Leyte, Philippines, Trans. Am. Soc. Agr. Eng., 41, 1635–1641, 1998.

Chappell, N. A. and Lancaster, J. W.: Comparison of methodological uncertainties within permeability measurements, Hydrol. Process., 21, 2504–2514, 2007.

Chappell, N. A., Sherlock, M., Bidin, K., Macdonald, R., Najman, Y., and Davies, G.: Runoff processes in Southeast Asia: Role of soil, regolith, and rock type, in: Forest Environments in the Mekong River Basin, edited by: Swada, H., Araki, M., Chappell, N. A., La Frankie, J. V., and Shimizu, A., Springer-Verlag: Tokyo, 3–23, 2007.

Choi, H. T. and Kim, Y. K.: Impacts of afforestation, reforestation and forest management pratices on the long-term streamflow patterns and water balance at the catchment scale in Korea, in: Asia-Pacific Workshop on Water and Forests – Beyond Traditional Forest Hydrology, edited by: Singh, M. D. (Convener), Forest Research Institute, Dehradun, India, p. 44, 2013.

Colloff, M. J., Pullen, K. R., and Cunningham, S. A.: Restoration of an ecosystem function to revegetaion communities: The role of invertebrate macropores in enhancing soil water infiltration, Restor. Ecol., 18, 65–72, 2010.

Costa, M. H., Botta, A., and Cardille, J. A.: Effects of large-scale changes in land cover on the discharge of the Toscantins River, Southeastern Amazonia, J. Hydrol., 283, 206–217, 2003.

Craig Collier, J. and Zhang, G. J.: Aerosol direct forcing of the summer Indian monsoon as simulated by the NCAR CAM3, Clim. Dynam., 32, 313–332, 2009.

Deuchars, S. A., Townend, J., Aitkenhead, M. J., and Fitzpatrick, E. A.: Change in soil structure and hydraulic properties in regenerating rain forest, Soil Use Manage., 15, 183–187, 1999.

Ding, M. M., Yi, W. M., Liao, L. Y., Martens, R., and Insam, H.: Effect of afforestation on microbial biomass and activity in soils of tropical China, Soil Biol. Biochem., 24, 865–872, 1992.

Dobremez, J. F.:Le Népal – Écologie et Biogéographie. Centre National de la Recherche Scientifique, Paris, France, 356 pp., 1976.

Ellison, D., Futter, M. N., and Bishop, K.: On the forest cover–water yield debate : From demand- to supply-side thinking, Global Change Biol., 18, 806–820, doi:10.1111/j.1365-2486.2011.02589.x, 2011.

FAO: World Reference Base for Soil Resources (2006) First update 2007, World Soil Resources Reports 103; United Nations Food and Agriculture Organization. Rome, 115 pp., 2007.

FAO-CIFOR: Forests and Floods: Drowning in Fiction or Thriving on Facts? RAP Publication 2005/03, Forest Perspectives 2. United Nations Food and Agriculture Organization, Rome, and Center for International Forestry Research, Bogor, 30 pp., 2005.

Farley, K. A., Jobbagy, E. G., and Jackson, R. J.: Effects of afforestation on water yield: A global synthesis with implications for policy, Global Change Biol., 11, 1565–1576, 2005.

Forsyth, T. and Walker, A.: Forest Guardians, Forest Destroyers. The Politics of Environmental Knowledge in Northern Thailand. University of Washington Press, Seattle, WA, USA, 302 pp., 2008.

Fritsch, J. M.: The hydrological effects of clearing tropical rainforest and of implementation of alternative land uses, Int. Assoc. Hydrol. Sci. Public., 216, 53–66, 1993.

Galudra, G. and Sirait, M.: A discourse on Dutch colonial forest policy and science in Indonesia at the beginning of the 20th century, Int. Forestry Rev., 11, 524–533, 2009.

Gash, J. H. C., Lloyd, C. R., and Lachaud, G.: Estimating sparse forest rainfall interception with an analytical model, J. Hydrol., 170, 79–86, 1995.

Gerrard, A. J. and Gardner, R. A. M.: Relationships between runoff and land degradation on non-cultivated land in the Middle Hills of Nepal, Int. J. Sustain. Develop. World Economy, 9, 59–73, 2002.

Ghimire, C. P.: Hydrological impacts of reforesting degraded pasture land in the Middle Mountain Zone, Central Nepal. PhD Dissertation, Vrije Universiteit, Amsterdam (FALW-VUA) and University of Twente (ITC-UT), Enschede, The Netherlands, 169 pp., available at: http://www.itc.nl/library/papers_2014/phd/ghimire.pdf, 2014.

Ghimire, C. P., Bruijnzeel, L. A., Lubczynski, M. W., and Bonell, M.: Rainfall interception by natural and planted forests in the Middle Mountains of Central Nepal, J. Hydrol., 475, 270–280, 2012.

Ghimire, C. P., Bonell, M., Bruijnzeel, L. A., Coles, N. A., and Lubczynski, M. W.: Reforesting severely degraded grassland in the Lesser Himalaya of Nepal: Effects on soil hydraulic conductivity and overland flow production, J. Geophys. Res.-Earth Surf., 118, 2528–2545, 2013.

Ghimire, C. P., Bruijnzeel, L. A., Bonell, M., Coles, N., Lubczynski, M. W., and Gilmour, D.A.: The effects of sustained forest use on hillslope soil hydraulic conductivity in the Middle Mountains of Central Nepal, Ecohydrol., 7, 478–495, doi:10.1002/eco.1367, 2014a.

Ghimire, C. P., Lubczynski, M. W., Bruijnzeel, L. A., and Chavarro-Rincón, D.: Transpiration, canopy conductance and decoupling coefficient of two contrasting forest types in the Lesser Himalaya of Central Nepal, Agr. Forest Meteorol., 197, 76–90, 2014b.

Giambelluca, T. W.: The hydrology of altered tropical forest, Hydrol. Process., 16, 1665–1669, 2002.

Gilmour, D. A., Bonell, M., and Cassells, D. S.: The effects of forestation on soil hydraulic properties in the Middle Hills of Nepal: A preliminary assessment, Mountain Res. Develop., 7, 239–249, 1987.

Gilmour, D. A. and Shah, R.: Enhancing livelihoods and food security from agroforestry and community forestry systems in Nepal: Synthesis paper, IUCN, Kathmandu, Nepal, 2012.

Granier, A.: Une nouvelle méthode pour la mesure du flux de sève brute dans le tronc des arbres, Ann. Sci. Forest., 42, 193–200, 1985.

Grip, H., Fritsch, J. M., and Bruijnzeel, L. A.: Soil and water impacts during forest conversion and stabilisation to new land use, in: Forests, Water and People in the Humid tropics, edited by: Bonell, M. and Bruijnzeel, L. A., Cambridge University Press, Cambridge, UK, 561–589, 2005.

Grissino-Mayer, H. D.: A manual and tutorial for the proper use of an increment borer, Tree-Ring Res., 59, 63–79, 2003.

Hamilton, L. S. and King, P. N.: Tropical Forested Watersheds: Hydrologic and Soils Response to Major Uses and Conversions, Westview Press, Boulder CO, USA, 168 pp., 1983.

Hairiah, K., Sulistyani, H., Suprayogo, D., Widianto Purnomosidhi, P., Harto Widodo, R., and Van Noordwijk. M.: Litter layer residence time in forest and coffee agroforestry systems in Sumberjaya, West Lampung, Forest Ecol. Manage., 224, 45–57, 2006.

Harding, R. J.: The modification of climate by forests, in: Growth and Water Use of Forest Plantations, edited by: Calder, I. R., Hall, R. L., and Adlard, P. G., J. Wiley and Sons, Chichester, UK, 332–346, 1992.

Hassler, S. K., Zimmermann, B., Van Breugel, M., and Hall, J. S., and Elsenbeer, H.: Recovery of saturated hydraulic conductivity under secondary succession on former pasture in the humid tropics, Forest Ecol. Manage., 261, 1634–1642, 2011.

Hessel, R., Gupta, M. K., Singh Dataa, P., van der Elsen, E., and Sharma, S. D.: Rainfall, soil moisture content and runoff in a small catchment in the Indian Himalayas, Int. J. Ecol. Environ. Sci., 33, 115–128, 2007.

Holwerda, F., Scatena, F. N., and Bruijnzeel, L. A.: Throughfall in a Puerto Rican lower montane rain forest: A comparison of sampling strategies, J. Hydrol., 327, 592–602, 2006.

Hrabovszky, J. P. and Miyan, K.: Population growth and land use in Nepal. "The great turnabout", Mountain Res. Develop., 7, 264–270, 1987.

HURDEC Nepal and Hobley, M.: Community forestry impact study: persistence and change. Review of 30 years of community forestry in Nepal, HURDEC, Kathmandu, Nepal, 2012.

Ilstedt, U., Malmer, A., Verbeeten, E., and Murdiyarso, D.: The effect of afforestation on water infiltration in the tropics: A systematic review and meta-analysis, Forest Ecol. Manage., 251, 45–51, 2007.

Immerzeel, W. W., Pellicciotti, F., and Bierkens, M. F. P.: Rising River flows throughout the twenty-first century in two Himalayan glacierized watersheds, Nat. Geosci., 6, 742–745, 2013.

Jackson, R. B., Jobbagy, E. G., Avissar, R., Roy, S. B., Barrett, D. J., Cook, C. W., Farley, K. A., Le Maître, D. C., McCarl, B. A., and Murray, B. C.: Trading water for carbon with biological carbon sequestration, Science, 310, 1944–1947, 2005.

Joshi, G. and Negi, G. C. S.: Quantification and valuation of forest ecosystem services in the western Himalayan region of India, International Journal of Biodiversity Science, Ecosyst. Serv. Manage., 7, 2–11, 2011.

Kaimowitz, D.: Useful myths and intractable truths: The politics of the link between forests and water in Central America, in: Forests, Water and People in the Humid Tropics, edited by: Bonell, M. and Bruijnzeel, L. A., Cambridge University Press, Cambridge, UK, 86–98, 2005.

Kallarackal, J. and Somen, C. K.: Water use by *Eucalyptus tereticornis* stands of differing density in southern India, Tree Physiol., 17, 195–203, 1997.

Kanae, S., Oki, T., and Musiake, K.: Impact of deforestation on regional precipitation over the Indochina Peninsula, J. Hydrometeorol., 2, 51–70, 2001.

Kelliher, F. M., Koestner, B. M. M., Hollinger, D. Y., Byers, J. N., Hunt, J. E., McSeveny, T. M., Meserth, R., Weir, P. L., and Schulze, E. D.: Evaporation, xylem sap flow, and tree transpiration in a New Zealand broad-leaved forest, Agr. Forest Meteorol., 62, 53–73, 1992.

Krishnaswamy, J., Bonell, M., Venkatesh, B., Purandara, B. K., Lele, S., Kiran, M. C., Reddy, V., Badiger, S., and Rakesh, K. N.: The rain-runoff response of tropical humid forest ecosystems to use and reforestation in the Western Ghats of India, J. Hydrol., 472–473, 216–237, 2012.

Krishnaswamy, J., Bonell, M., Venkatesh, B., Purandara, B. K., Rakesh, K. N., Lele, S., Kiran, M. C., Reddy, V., and Badiger, S.: The groundwater recharge response and hydrologic services of tropical forest ecosystems to use and reforestation: Support for the "infiltration-evapotranspiration trade-off hypothesis", J. Hydrol., 498, 191–209, 2013.

Kumagai, T., Kanamori, H., and Yasunari, T.: Deforestation-induced reduction in rainfall, Hydrol. Process., 27, 3811–3814, 2013.

Kumar, P., Wiltshire, A., Mathison, C., Asharaf, S., Ahrens, B., Lucas-Picher, Ph., Christensen, J. H., Gobiet, A., Saeed, F., Hagemann, S., and Jacob, D.: Downscaled climate change projections with uncertainty assessment over India using a high resolution multi-model approach, Sci. Total Environ., 468–469, S18–30, doi:10.1016/j.scitotenv.2013.01.051, 2013.

Lal, R.: Effects of macrofauna on soil properties in tropical ecosystems, Agric. Ecosyst. Environ., 24, 101–116, 1988.

La Marche, J. L. and Lettenmaier, D. P.: Effects of forest roads on flood flows in the Deschutes River, Washington, Earth Surf. Process. Landf., 26, 115–134, 2001.

Liu, W. J., Liu, W. Y., Lu, H. J., Duan, W. P., and Li, H. M.: Runoff generation in small catchments under a native rain forest and a rubber plantation in Xishuangbanna, southwestern China, Water Environ. J., 25, 138–147, 2009.

Lu, P., Urban, L., and Zhao, P.: Granier's thermal dissipation probe method for measuring sap flow in trees: theory and practice, Acta Botanica Sinica, 46, 631–646, 2004.

Lubczynski, M. W., Chavarro-Rincon, D. C., and Roy, J.: Novel, cyclic heat dissipation method for the correction of natural temperature gradients in sap flow measurements. Part I. Theory and application, Tree Physiol., 32, 894–912, 2012.

Madduma Bandara, C. M.: Land-use changes and tropical stream hydrology: Some observations from the upper Mahaweli basin of Sri Lanka, in: Process and Form in Geomorphology, edited by: Stoddart, D. R., Routledge, London, 175–186, 1997.

Maharjan, P. L.: The soil survey of the Jikhu Khola catchment, in: Soil Fertility and Erosion issues in the Middle Mountains of Nepal, edited by: Shah, P. B., Schreier, H., Brown, S. J., and Riley, K. W., Topographical Survey Branch, Kathmandu, and University of British Columbia, Vancouver, 195–200, 1991.

Malmer, A., Murdiyarso, D., Bruijnzeel, L. A., and Ilstedt, U.: Carbon sequestration in tropical forests and water: A critical look at the basis for commonly used generalizations, Global Change Biol., 16, 599–604, 2010.

Mathison, C., Wiltshire, A., Dimri, A. P., Falloon, P., Jacob, D., Kumar, P., Moors, E., Ridley, J., Siderius, Ch., Stoffel, M., and Yasunari, T.: Regional projections of North Indian climate for adaptation studies, Sci. Total Environ., 468–469, S4–S17, doi:10.1016/j.scitotenv.2012.04.066, 2012.

McIntyre, D. S.: Permeability measurements of soil crusts formed by raindrop impact, Soil Sci., 85, 185–189, 1958a.

McIntyre, D. S.: Soil splash and the formation of surface crusts by raindrop impact, Soil Sci., 85, 261–266, 1958b.

McKenzie, N. J., Cresswell, H. P., and Green, T. W.: Field measurement of unsaturated hydraulic conductivity using tension infiltrometers, in: Soil Physical Measurement and Interpretation for Land Evaluation, edited by: Mckenzie, N., Coughlan, K., and Cresswell, H., CSIRO Publishing, Melbourne, 119–130, 2002.

Merz, J.: Water balances, floods and sediment transport in the Hindu Kush-Himalayas. Geographica Bernensia G72, University of Bern, Bern, Switzerland, 339 pp., 2004.

Merz, J., Nakarmi, G., Shrestha, S. K., Dahal, B. M., Dangol, P. M., Dhakal, M. P., Dongol, B. S., Sharma, S., Shah, P. B., and Weingartner, R.: Water: A scarce resource in rural watersheds of

Nepal's Middle Mountains, Mountain Res. Develop., 23, 41–49, 2003.

Molina, A., Govers, G., Vanacker, V., Poesen, J., Zeelmaekers, E., and Cisneros, F.: Runoff generation in a degraded Andean ecosystem: Interaction of vegetation cover and land use, Catena, 71, 357–370, 2007.

Motzer, T., Munz, N., Anhuf, D., and Küpper, M.: Transpiration and microclimate of a tropical montane rain forest, southern Ecuador, in: Tropical Montane Cloud Forests: Science for Conservation and Management, edited by: Bruijnzeel, L. A., Scatena, F. N., and Hamilton, L.S., Cambridge University Press, Cambridge, UK, 447–455, 2010.

Nadezhdina, N., Vandegehuchte, M., and Steppe, K.: Sap flux density measurements based on the heat field deformation method, Trees – Structure and Function, 26, 1439–1448, 2012.

Nakarmi, G.: Geological mapping and its importance for construction material, water chemistry, and terrain stability in the Jikhu and Yarsa Khola catchments, in: The People and Resources Dynamics: The First Three Years (1996–1999), edited by: Allen, R., Schreier, H., Brown, S., and Shah, P. B., ICIMOD, Baoshan, Nepal, 253–261, 2000.

Ogden, F. L., Crouch, T. D., Stallard, R. F., and Hall, J. S.: Effect of land cover and use on dry season river runoff, runoff efficiency, and peak storm runoff in the seasonal tropics of Central Panama, Water Resour. Res., 49, 1–20, 2013.

Oldeman, L. R., Hakkeling, R. T. A., and Sombroek, W. G.: World Map of the Status of Human-Induced Soil Degradation: An Explanatory Note (Second Edition). International Soil Reference and Information Centre, Wageningen, The Netherlands, 1991.

Pathak, P. C., Pandey, A. N., and Singh, J. S.: Overland flow, sediment output and nutrient loss from certain forested sites in the central Himalaya, India, J. Hydrol., 71. 239–251, 1984.

Patnaik, N. and Virdi, S. S.: Field infiltration studies in Doon Valley, Irrig. Power, 19, 1003–1012, 1962.

Peña-Arancibia, J. L., Van Dijk, A. I. J. M., Guerschman, J. P., Mulligan, M., Bruijnzeel, L. A., and McVicar, T.: Using multiple lines of evidence to evaluate the hydrological response to deforestation of large catchments in the dry tropics of Queensland, Australia, J. Hydrol., 416–417, 60–71, 2012.

Perkins, K. S., Nimmo, J. R., and Medeiros, A. C.: Effects of native forest restoration on soil hydraulic properties, Auwahi, Maui, Hawaiian, Islands, Geophys. Res. Lett., 39, L05405, doi:10.1029/2012GL051120, 2012.

Perroux, K. M. and White, I.: Designs for disc permeameters, Soil Sci. Soc. America J., 52, 1205–1215, 1988.

República: Locals blame pine trees for declining water sources. República Vol. IV, No. 202 (19 November 2012), available at: www.myrebulica.com, last access: 21 November 2012.

Reynolds, W. D., Elrick, D. E., and Clothier, B. E.: The constant head well permeameter: effect of unsaturated flow, Soil Sci., 139, 172–180, 1985.

Rijsdijk, A., Bruijnzeel, L. A., and Kukuh Sutoto, C.: Runoff and sediment yield from rural roads, trails and settlements in the upper Konto catchment, East Java, Indonesia, Geomorphology, 87, 28–37, 2007.

Roa-García, M. C., Brown, S., Schreier, H., and Lavkulich, L. M.: The role of land use and soils in regulating water flow in small headwater catchments of the Andes, Water Resour. Res., 47, W05510, doi:10.1029/2010WR009582, 2011.

Schreier, H., Brown, S., and MacDonald, J. R. (Eds.): Too Little and Too Much: Water and Development in a Himalayan Watershed, Institute for Resources and Environment, University of British Columbia, Canada, 258 pp., 2006.

Scott, D. F. and Prinsloo, F. W.: Longer-term effects of pine and eucalypt plantations on streamflow, Water Resour. Res., 44, W00A08, doi:10.1029/2007WR006781, 2008.

Scott, D. F., Bruijnzeel, L. A., and Macksensen, J.: The hydrological and soil impacts of forestation in the tropics, in: Forest, Water and People in the Humid Tropics, edited by: Bonell, M. and Bruijnzeel, L. A., Cambridge University Press, Cambridge, UK, 622–650, 2005.

Shrestha, A. B., Wake, C. P., Dibb, J. E., and Mayewski, P. A.: Precipitation fluctuations in the Nepal Himalaya and its vicinity and relationship with some large scale climatological parameters, Int. J. Climatol., 20, 317–327, 2000.

Sidle, R. C. and Ziegler, A. D.: The dilemma of mountain roads, Nat. Geosci., 5, 437–438, 2012.

Sidle, R. C., Ziegler, A. D., Negishi, J. N., Nik, A. R., Siew, R., and Turkelboom, F.: Erosion processes in steep terrain – Truths, myths, and uncertainties related to forest management in Southeast Asia, Forest Ecol. Manage., 224, 199–225, 2006.

Šimůnek, J., Sejna, M., Saito, H., Sakai, M., and van Genuchten, M. T.: The HYDRUS 1D software package for simulating the one-dimensional movement of water, heat, and multiple solutes in variability-saturated media, University of California Riverside, California, 281 pp., 2008.

Singh, N. and Sundriyal, R. C.: Fuelwood, fodder consumption and deficit pattern in a Central Himalayan village, Nat. Sci., 7, 85–88, 2009.

Singh, J. S., Pandey, U., and Tiwari, A. K.: Man and forests: A Central Himalayan case study, Ambio, 13, 80–87, 1984.

Spracklen, D. V., Arnold, S. R., and Taylor, C. M.: Observations of increased tropical rainfall preceded by air passage over forests, Nature, 489, 282–286, doi:10.1038/nature11390, 2012.

Stednick, J. D.: Monitoring the effects of timber harvest on annual water yield, J. Hydrol., 176, 79–95, 1996.

Stocklin, J. and Bhattarai, K.: Geology of Kathmandu area and Central Mahabharat range, HMG/UNDP Mineral Exploration Project, Kathmandu, Nepal, 1997.

Sun, G., Zhou, G. Y., Zhang, Z. Q., Wei, X. H., McNulty, S. G., and Vose, J. M.: Potential water yield reduction due to forestation across China, J. Hydrol., 328, 548–558, 2006.

Talsma, T. and Hallam, P. M.: Hydraulic conductivity measurement of forest catchments, Aust. J. Soil Res., 18, 139–148, 1980.

Tiwari, K. R., Sitaula, B. K., Bajracharya, R. M., and Børresen, T.: Runoff and soil loss responses to rainfall, land use, terracing and management practices in the middle Mountains of Nepal, Acta Agriculture and Scandinavica, Section B – Soil Plant Sci., 59, 197–207, 2009.

Tiwari, R. K., Rai, S. P., Qazi, N., and Kumar, B.: Stream discharge analysis under different forest covers: a case study from paired microwatersheds of Garhwal Himalaya, India, Indian Forest., 137, 805–813, 2011.

Tobón, C., Bruijnzeel, L. A., Frumau, K. F. A., and Calvo-Alvarado, J. C.: Change in soil physical properties after conversion of tropical montane cloud forest to pasture in northern Costa Rica, in: Tropical Montane Cloud Forests: Science for Conservation and Management, edited by: Bruijnzeel, L. A., Scatena, F. N., and

Hamilton, L. S., Cambridge University Press, Cambridge, UK, 502–515, 2010.

Tomasella, J. and Hodnett, M. G.: Soil hydraulic properties and van Genuchten parrameters for an oxisol under pasture in central Amazonia, in: Amazonian Deforestation and Climate, edited by: Gash, J. H. C., Nobre, C. A., Roberts, J. M., and Victoria, R. L., J. Wiley and Sons, Chichester, UK, 101–124, 1996.

Tomich, T. P., Thomas, D. E., and van Noordwijk, M.: Environmental services and land use change in Southeast Asia: From recognition to regulation or rewards?, Agr. Ecosyst. Environ., 104, 229–244, 2004.

Trimble, S. W., Weirich, F. H., and Hoag, B. L.: Reforestation and the reduction of water yield on the southern Piedmont since circa 1940, Water Resour. Res., 23, 425–437, 1987.

Tyagi, J. V., Qazi, N., Rai, S. P., and Singh, M. P.: Analysis of soil moisture variation by forest cover structure in lower western Himalayas, J. Forest. Res., 24, 1–8, 2013.

Van Dijk, A. I. J. M., Peña-Arancibia, J., and Bruijnzeel, L. A.: Land cover and water yield: Inference problems when comparing catchments with mixed land cover, Hydrol. Earth Syst. Sci., 16, 3461–3473, doi:10.5194/hess-16-3461-2012, 2012.

Van Genuchten, M. T.: A closed form equation for predicting the hydraulic conductivity of unsaturated soils, Soil Sci. Soc. Am. J., 44, 892–898, 1980.

Van Noordwijk, M. and Leimona, B.: Principles for fairness and efficiency in enhancing Environmental Services in Asia: Payments, compensation, or co-Investment?, Ecol. Soc., 15, available at: http://www.ecologyandsociety.org/vol15/iss4/art17/, 2010.

Van Noordwijk, M., Tomich, T. P., and Verbist, B.: Negotiation support models for integrated natural resource management in tropical forest margins, Conserv. Ecol., 5, 21–42, 2001.

Varikoden, H., Revadekar, J. V., Choudhary, Y., and Preethi, B.: Droughts of Indian summer monsoon associated with El Niño and Non-El Niño years, Int. J. Climatol., doi:10.1002/joc.4097, in press, 2014.

Vertessy, R. A., Watson, F. G. R., and O'Sullivan, S. K.: Factors determining relations between stand age and catchment water yield in mountain ash forests, Forest Ecol. Manage., 143, 13–26, 2001.

Wallace, J. S., Young, A., and Ong, C. K.: The potential of agroforestry for sustainable land and water management, in: Forests, Water and People in the Humid tropics, edited by: Bonell, M. and Bruijnzeel, L. A., Cambridge University Press, Cambridge, UK, 652–670, 2005.

Waterloo, M. J., Bruijnzeel, L. A., Vugts, H. F., and Rawaqa, T. T.: Evaporation from *Pinus caribaea* plantations on former grassland soils under maritime tropical conditions, Water Resour. Res., 35, 2133–2144, 1999.

Wester, R. G. C.: Feasibility of assessing water scarcity with 3R techniques, in the Salyan district in the Middle Mountains of western Nepal, M.Sc. Thesis, VU University, Amsterdam, The Netherlands, 148 pp., 2013.

Wiersum, K. F.: Surface erosion under various tropical agroforestry systems, in: Symposium on Effects ofForest Land Use on Erosion and Slope Stability, edited by: O'Loughlin, C. L. and Pearce, A. J., East-West Center, Honolulu, Hawai'i, 231–239, 1984.

Wiersum, K. F.: Effects of various vegetation layers in an *Acacia auriculiformis* forest plantation on surface erosion in Java, Indonesia, in: Soil Erosion and Conservation, edited by: El-Swaify, S., Moldenhauer, W. C., and Lo, A., Soil Conservation Society of America: Ankeny, Wisconsin, 79–89, 1985.

Wohl, E., Barros, A., Brunsell, N., Chappell, N. A., Coe, M., Giambelluca, T. W., Goldsmith, S., Harmon, R., Hendrickx, J. M. H., Juvik, J. O., McDonnell, J. J., and Ogden, F. L.: The hydrology of the humid tropics, Nat. Clim. Change, 2, 655–662, 2012.

Yadav, R. K.: Role of equatorial central Pacific and northwest of North Atlantic 2-metre surface temperatures in modulating Indian summer monsoon variability, Clim. Dynam., 32, 549–563, 2009.

Zhang, B., Yang, Y. S., and Zepp, H.: Effect of vegetation restoration on soil and water erosion and nutrient losses of a severely eroded clayey Plinthudult in southeastern China, Catena, 57, 77–90, 2004.

Zhang, L., Hickel, K., Dawes, W. R., Chiew, F. H. S., Western, A. W., and Briggs, P. R.: A rational function approach for estimating mean annual evapotranspiration, Water Resour. Res., 40, W02502, doi:10.1029/2003WR002710, 2004.

Zhou, G. Y., Morris, J. D., Yan, J. H., Yu, Z. Y., and Peng, S. L.: Hydrological impacts of reafforestation with eucalypts and indigenous species: A case study in southern China, Forest Ecol. Manage., 16, 209–222, 2002.

Zhou, G. Y., Wei, X., Luo, Y., Zhang, M., Li, Y., Qiao, Y., Liu, H., and Wang, C.: Forest recovery and river discharge at the regional scale of Guangdong Province, China, Water Resour. Res., 46, W09503, doi:10.1029/2009WR008829, 2010.

Ziegler, A. D. and Giambelluca, T. W.: Importance of rural roads as source areas for runoff in mountainous areas of northern Thailand, J. Hydrol., 196, 204–229, 1997.

Ziegler, A. D., Ziegler, A. D., Giambelluca, T. W., Sutherland, R. A., Nullet, M. A., Yarnasarn, S., Pinthong, J., Preechapanya, P., and Jaiaree, S.: Toward understanding the cumulative impacts of roads in upland agricultural watersheds of Northern Thailand, Agr. Ecosyst. Environ., 104, 145–158, 2004.

Ziegler, A. D., Negishi, J. N., Sidle, R. C., Gorni, T., Noguchi, S., and Rahim Nik, A.: Persistence of road runoff generation in a logged catchment in Peninsular Malaysia, Earth Surf. Process. Landf., 32, 1947–1970, 2007.

Zimmermann, B., Elsenbeer, H., and De Moraes, J. M.: The influence of land-use changes on soil hydraulic properties: Implications for runoff generation, Forest Ecol. Manage., 222, 29–38, 2006.

10

Analysis of landslide triggering conditions in the Sarno area using a physically based model

G. Capparelli and P. Versace

Dipartimento di Elettronica, Informatica e Sistemistica, Università della Calabria, Via P. Bucci, 41C, Arcavacata di Rende, CS, Italy

Correspondence to: G. Capparelli (giovanna.capparelli@unical.it)

Abstract. Rainfall is recognized as a major precursor of many types of slope movements. The technical literature reports both study cases and models of landslides induced by rainfall. Subsurface hydrology has a dominant role since changes in the soil water content significantly affect the soil shear strength. The analytical approaches used are very different, ranging from statistical models to distributed and complete models. The latter take several components into account, including specific site conditions, mechanical, hydraulic and physical soil properties, local seepage conditions, and the contribution of these to soil strength.

This paper reports a study using a complete model, named SUSHI (Saturated Unsaturated Simulation for Hillslope Instability), to simulate the role of subsurface hydrology in rain-induced landslides, on a case of great interest both in terms of its complexity and its severity.

The landslide-prone area in question is located in Campania (southern Italy), where disastrous mudflows occurred in May 1998. The region has long been affected by rainfall-induced slope instabilities, which often involve large areas and affect many people. The application allows a better understanding of the role of rainfall infiltration and suction changes in the triggering mechanism of the phenomena. These changes must be carefully considered when assessing hazard levels and planning mitigation interventions regarding slope stability.

1 Introduction

The problems and the damage caused by landslides have become increasingly complex, accounting each year for huge property damage in terms of both direct and indirect costs. Social and economic losses due to landslides can be reduced by effective planning and management. This includes limiting of development in landslide-prone areas, following appropriate construction regulations, using physical measures to prevent or control landslides, and setting up early warning systems. To tackle landslide problems, it is necessary to develop a better understanding of trigger mechanisms, propagation and impact.

As a result of rainfall events and subsequent infiltration, the soil moisture can be significantly changed with a decrease in suction in unsaturated soil layers and/or an increase in pore-water pressure in saturated layers. As a consequence, in these cases, the shear strength can be reduced enough to trigger a failure.

The severity of events is also influenced by the heterogeneity of hydraulic and geotechnical properties. The complex hydrological responses of natural slopes are strongly influenced by infiltration into unsaturated soil, surface runoff, slope-parallel flow through perched aquifers, subsurface flows from upstream areas, the effect of vegetation and flows through fractured bedrock. All these issues affect the predictive ability of the models and sometimes impact on the difficulty of interpreting the results.

In addition, the shear strength contribution from soil suction above the groundwater table is usually ignored if the major portion of the slip surface is below the groundwater table. However, negative pore-water pressures can no longer be ignored in situations characterized by a deep groundwater table and shallow failure surface (Lu and Godt, 2013).

Analytical approaches differ in terms of the spatial scale range adopted, the available data and the description of the processes occurring in the slope.

Empirical models that directly analyze the rainfall by identifying threshold values are very popular. These values are assumed on the basis of historical data and are shown in an intensity–duration plot as proposed by Caine (1980); they provide a lower limit of rainfall associated with the occurrence of landslides, shallow landslides and debris flows (Guzzetti, 2008).

Other types of rainfall thresholds (Glade, 2000) consider the effect of previous rainfall events as more important than the rainfall recorded on the day the landslide occurred. Usually this type of approach is related to the study of deep-seated landslides.

Combined with rainfall forecasting models, both stochastic and meteorological, empirical models are an essential tool to support the prediction of landslides in early warning systems (Sirangelo et al., 2003; Versace et al., 2003; Capparelli and Tiranti, 2010).

However, these models do not provide any information on the hydrological processes involved in a landslide area and do not improve our understanding of landslide dynamics.

Complete models can help in understanding triggering mechanisms since they attempt to reproduce the physical behavior of the processes involved at hillslope scale, using detailed hydrological, hydraulic and geotechnical information (Montgomery and Dietrich, 1994; Wu and Sidle, 1995; Iverson, 2000; Baum et al., 2002; Rigon et al., 2006; Arnone et al., 2011). They develop an analysis over wide areas and usually produce a susceptibility map characterizing the landslide-prone zones according to a stability index. They are generally composed of a hydrological and a geotechnical module.

Hydrological modules differ considerably from each other. The geotechnical module in most cases makes use of the limit equilibrium method under the assumption of an infinite slope. Only in a few cases is more detailed analysis carried out, as in Lu and Godt (2013), where the authors provide a quantitative treatment of rainfall infiltration, effective stress and their coupling and roles in hillslope stability by introducing a unified effective stress modeling framework, linking soil suction to effective stress.

The hydrological approach proposed in SHALSTAB (SHALlow Landslides STABility) (Montgomery and Dietrich, 1994) assumes a constant infiltration rate, neglects soil moisture above the water table, does not take into account the transient response to rainfall and considers the groundwater flow parallel to the slope. These assumptions are too restrictive, for example, when pore-water pressure responds very quickly to transient rainfall and its redistribution has a large component that is normal to the slope.

Wu and Sidle (1995) also combined the infinite slope limit equilibrium equation with a subsurface flow model based on the kinematic wave approximation, also taking into account the vegetation root strength. An enhanced version of this model is proposed by Dhakal and Sidle (2004), who investigate the influence of different rainfall characteristics on slope stability.

Iverson (2000) developed a flexible framework by modeling a one-dimensional linear diffusion process in saturated soil, using an analytical solution of Richards equation. The model is valid for hydrological modeling in nearly saturated soil. According to this hypothesis, which is used to find an analytical solution to pressure heads, the infiltration capacity is assumed to be equivalent to the saturated hydraulic conductivity, instead of considering it as variable with time during the rainfall event. In addition, Iverson considers the ground surface of the hillslope subject to uniform rainfall.

In order to take into account the variability of rainfall intensity and duration, dynamic or quasi-dynamic models have been introduced. These include TRIGRS (Transient Rainfall Infiltration and Grid-based Regional Slope-stability model) (Baum et al., 2002), which gives a more precise description of slope hydrology but requires a large number of parameters. The model performs transient seepage analysis, using the linearized solution to the Richards equation proposed by Iverson (2000) and extended to the case of impermeable bedrock located at a finite depth.

A more recent version – TRIGRS-unsaturated (Baum et al., 2008) – predicts the pore-water pressure regime in unsaturated/saturated conditions, coupling the simple analytic solution for transient unsaturated infiltration proposed by Srivastava and Yeh (1991) with the original TRIGRS equation.

Most of the aforementioned approaches rely on the restrictive assumption of a steady-state subsurface flow, which can affect the predictive capability of the models both in terms of accuracy and timing of the prediction. All models must always be validated by checking, for example, the accuracy of the simulation with the experimental data available from real cases. The results, in fact, can sometimes be very different when different models are applied to the same event. This is described in Sorbino et al. (2010), who illustrate how, by applying three different physically based models (SHALSTAB, TRIGRS and TRIGRS-unsaturated) to the same set of geo-environmental cases, different results were obtained. The results reveal the advantages and limitations of each model in landslide forecasting.

These types of spatial, distributed modeling are clearly well-suited for shallow landslides but in deeper landslides their effectiveness is hindered by the higher complexity of the phenomena (van Westen et al., 2003).

In this work a complete model, named SUSHI (Simulation for Saturated Unsaturated Hillslope Instability; Capparelli and Versace, 2011), is applied to a very complex case in order to improve the understanding of the slope failure mechanism during rainfall infiltration.

SUSHI takes into account several components, such as specific site conditions, mechanical, hydraulic and physical soil properties, local seepage conditions and their contribution to soil strength.

It is composed of a hydraulic module to analyze the subsoil water circulation due to the rainfall infiltration under transient conditions and a geotechnical module, which provides the slope stability using the limit equilibrium equation.

The hydraulic process is highlighted by the implementation of a finite difference scheme, which solves the Richards equation used to represent saturated/unsaturated flows within a hillslope. The temporal and spatial distributions of moisture content in the subsurface are calculated, in order to evaluate the different contributions such as the downslope and vertical components of flow in the hillslope caused by unsteady rainfall.

The model was also developed for cases with strongly heterogeneous soils, irregular domains and variable boundary conditions in space and time. After a brief introduction of the model, the paper describes the analysis and results obtained for the volcaniclastic covers of Sarno, where catastrophic mudslides occurred in May 1998.

Mudslides have been analyzed in several papers focusing on the most significant, hydrological and geotechnical features and on the models for triggering mechanisms and landslide propagation.

There is still no commonly accepted interpretation of the triggering mechanism of landslides such as ones occurring in similar geomorphological contexts in Campania.

To summarize, the objectives of the presented application are:

- to formulate a likely interpretation of the mudslide triggering mechanism, highlighting the main factors affecting slope stability;
- study the rainfall infiltration dynamics, checking whether pore-water pressure, just before the landslide events, reached levels never attained before;
- verify the model's ability to analyze a real case, characterized by very marked spatial heterogeneity.

Using a physically based model designed and created ad hoc is mandatory for such purposes, as unlike many commercial models, it allows interaction with the source code in such a way as to adapt it to the specific case by modifying the spatial and temporal resolution as well as the parametrization of the various processes and variables analyzed. It is thus possible to make replacements, to verify the accuracy of the results and to improve its performance by adapting it to specific cases.

Table 1. Features of some recent flowslides in the Campania Region (Versace et al., 2009).

Site	Year	Length (m)	Volume (m^3)
Ischia	2006	450	3×10^4
Cervinara	1999	2×10^3	4×10^4
Avella	1998	15×10^2	2×10^4
San Felice a Cancello	1998	8×10^2	3×10^4
Sarno	1998	$2\text{–}4 \times 10^3$	5×10^5
Bracigliano	1998	$1\text{–}2 \times 10^3$	15×10^4
Siano	1998	14×10^2	4×10^4
Quindici	1998	$1\text{–}4 \times 10^3$	5×10^5
Maiori	1954	10^3	5×10^4
Avellino	2005	4×10^2	2×10^4
Montoro Inf.	1997	2×10^3	3×10^4

2 Study area

2.1 Description of the study site

The case study proposed in the paper is located in Campania (Italy), where catastrophic flowslides and debris flows in pyroclastic soils are frequent. A brief list of some recent events is reported in Table 1, which also includes information on the size of the landslide. Pizzo d'Alvano is a NW–SE oriented morphological structure, consisting of a sequence of limestone, dolomitic limestone and, subordinately, marly limestone dating from the Lower to Upper Cretaceous age. The slopes are mantled by very loose pyroclastic soils which are the result of the explosive activity of the Somma–Vesuvius volcanic complex, both as primary airfall deposits and volcaniclastic deposits, according to the mode of transport and deposition (Rolandi, 1997). Airfall deposits were dispersed along directions ranging from N–NE to S–SE, according to the prevailing wind direction, and covered a wide area reaching distances of up to 50 km.

Pumiceous and ashy deposits belonging to at least five different eruptions have been identified. From the oldest to the youngest, they include: Ottaviano (8000 years BC; E–NE dispersion direction), Avellino (3800 years BC; E–NE dispersion direction), AD 79 (E–SE dispersion direction), AD 472 (N–NE dispersion direction) and AD 1631 (N–NE dispersion direction). The deposits are affected by pedogenetic processes determining paleosoil horizons during resting phases of the volcanic activity. The total thickness of the pyroclastic covers in these areas ranges between a few decimeters, along the steepest slopes, to 10 m, near the uppermost flat areas. The general structure of the soil progressively adapts itself to the morphology of the calcareous substratum thus showing complex and variable geometries (Rolandi, 1997).

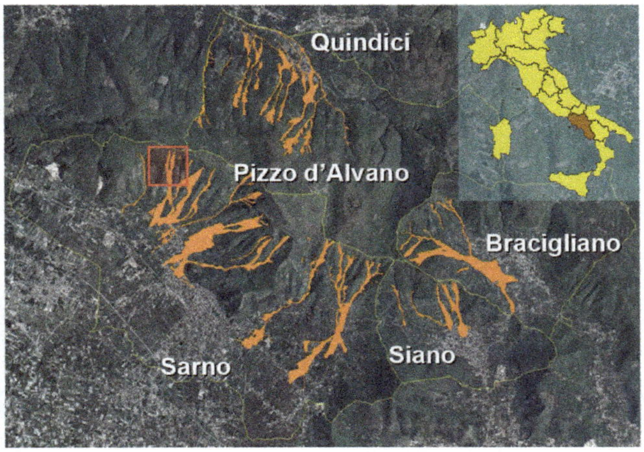

Figure 1. Overview of the Pizzo d'Alvano Massif and the area affected by the May 1998 mudflows. The red square delimits the area where the SUSHI model has been applied.

2.2 Description of the landslide events

On 5 May 1998, a huge number of mudflows were triggered on the slopes of the Pizzo d'Alvano Massif (Fig. 1), involving an extension area of around 60 km^2, a volume of 2 000 000 m^3 (40 % derived from the eroded materials along the channels) and leading to 159 victims and huge damage to the towns of Sarno, Quindici, Siano and Bracigliano.

These landslides were classified as very rapid to extremely rapid soil slip/debris flows (Ellen and Fleming, 1987), which traveled downslope and then propagated in highly urbanized areas.

A characteristic element is represented by the run-out distances, which ranged from a few hundred meters up to distances of over 2 km (Revellino et al., 2004) and speeds which, at the toe of channels, were estimated to be in the range of about 5–20 $m\,s^{-1}$.

Many similar phenomena have afflicted various other parts of the world, (Japan in 1985, California in 1973 and 2005, Brazil in 1967, Venezuela in 1999), in some cases involving similar pyroclastic soils.

Although the triggering mechanisms are different and sometimes the soils involved are not always similar, the common feature seems to be the presence of particles with a high porosity and a very low degree of cementation; these change suddenly due to the action of an external agent (such as an earthquake or more often a rainfall event), which produces a rapid increase in pore-water pressure. What made the May 1998 landslides unique and what made the events even more tragic, was that they all occurred at the same time over a large area, and an extremely high amount of material was involved.

The flowslides of Campania have been analyzed in several papers which indicate the most significant geomorphological, hydrological and geotechnical features of the slopes involved, and propose models for the triggering mechanisms and propagation of landslides.

Cascini et al. (2008) argue that the instabilities in Sarno were caused by a combined effect of water infiltrated into the surface layers plus water from a temporary spring in the bedrock. Calcaterra et al. (2000) discuss the role played by groundwater circulation inside both the pyroclastic deposits and the karst cavities of the underlying limestone bedrock.

Soil water circulation is important due to the typical stratification of the pyroclastic covers involved, where one or more layers of pumice, with high permeability at saturation, and layers of paleosoils, with lower saturated permeability, are present.

When persistent rainfall events occur, it encourages subsurface runoff, which may predispose the slope to instability in limited areas.

The role played by the possible interaction of the unsaturated cover with the underlying groundwater, through an impervious soil–bedrock interface, has also been analyzed, by Greco et al. (2013), for a slope similar to those in Sarno.

Vegetation and plant roots and their possible relations to slope instability have also been analyzed. Some authors emphasize the highly dynamic nature of the soil-vegetation as a whole, where the hydrological processes can be greatly affected by the dynamics of the vegetation. Changes in vegetation cover can produce equally rapid effects on the soil and water regime (Mazzoleni et al., 1998).

2.3 Description of field surveys and rainfall events

In the years following the landslide events, many field surveys have been carried out in order to assess hydraulic and geotechnical soil characteristics of the site.

In order to assess the influence of soil suction on the triggering mechanism, suction measurements were performed along the Tuostolo Basin (Sarno area), very close to areas that collapsed in May 1998 (Cascini and Sorbino, 2004), using quick-draw portable tensiometers and jet fill tensiometers. These measurements were taken at three sites (Fig. 2), at different depths from the ground surface. Site 1 was located in an area not affected by the landslides in 1998; sites 2 and 3 were located in landslide source areas.

A significant spatial and temporal variability of soil suction can be noted (Fig. 3), essentially related to the differences between the sites, the depths at which the measurements were carried out and also to local factors which induce changes at the end of the dry season when the acquired data show very high suction levels (up to 65 kPa).

The data in Fig. 3 also show that slope equilibrium is guaranteed by cohesion rather than by friction. In fact, in situ and laboratory investigations show that the pyroclastic materials involved have friction angles of between 32 and 38° and effective cohesion ranging from 0 (excluding reworked pumices) to 4–5 kPa (ashes). The pyroclastic deposits covering the affected areas often rest on slopes with an inclination

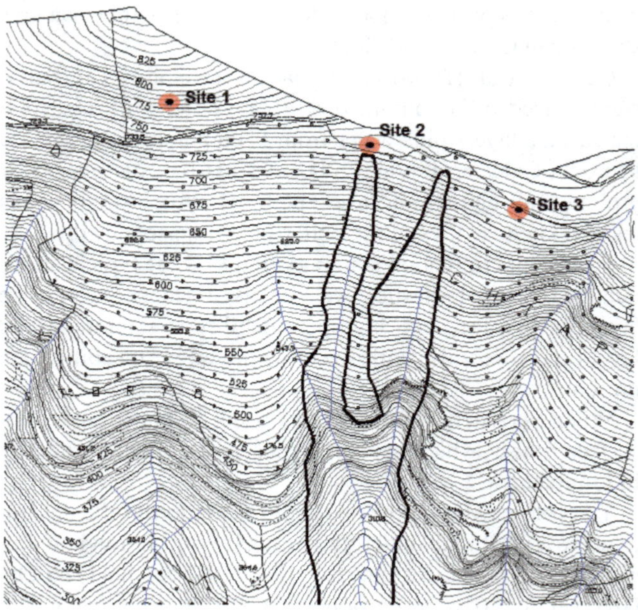

Figure 2. Topographic map and sites where suction measurements were performed.

greater than 40°, much larger than their friction angle. In such geomorphological conditions, soil suction, which increases soil shear strength, is a major contributor to slope stability.

These considerations support the interest in the suction assessment in these pyroclastic soils as possibly the main predisposing cause. The availability of models able to simulate the circulation of water in these complex terrains can provide useful tools to better understand these phenomena.

The rainfall data were collected by the rain gauge of Santa Maria La Foce, around 2 km from the triggering landslide area, which in 1998 recorded a total rainfall value of 900 mm (Fig. 4).

This station is located at 192 m a.s.l., lower than the landslide source areas (700 m a.s.l.). It is therefore likely that the rainfall along the landslide source areas could be greater than that measured by the rain gauge located at a lower altitude. Available data reveal that, the rainfall occurring in May 1998 was not extreme, in fact it was characterized by a return period lower than five years, but the period when it occurred was unusual. In fact, the monthly and maximum daily values of rainfall height observed in April and May 1998 are significantly higher than the mean rainfall and than the maximum daily rainfall values over the period 1967–1997 (Table 2).

3 SUSHI model framework

The model combines two modules: HydroSUSHI, which studies subsoil water circulation, and GeoSUSHI, which assesses slope stability.

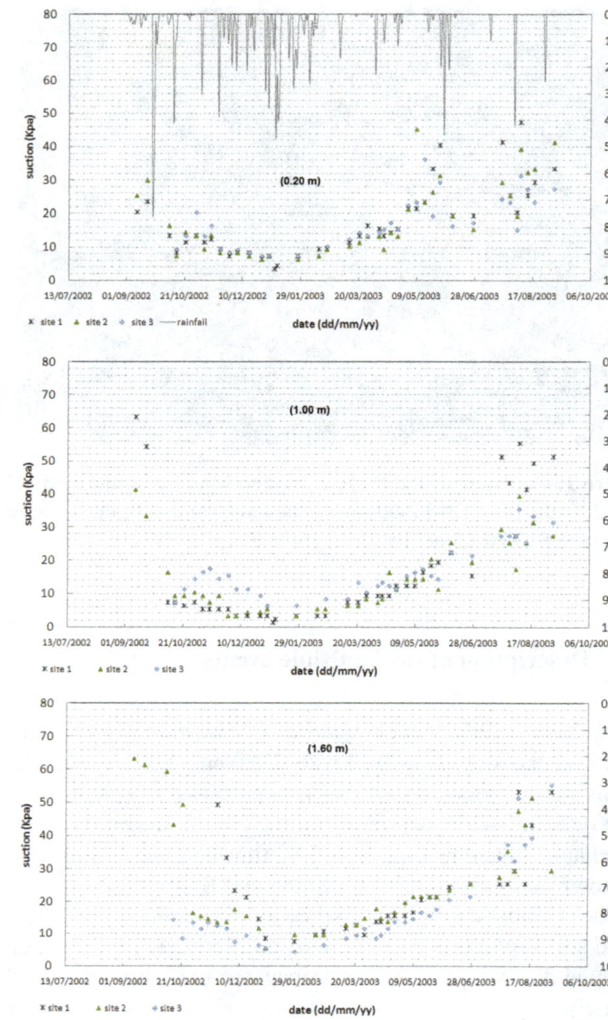

Figure 3. Suction trends recorded and daily rainfall measured by Santa Maria la Foce rain gauge at 0.20, 1.00 and 1.60 m from topsoil.

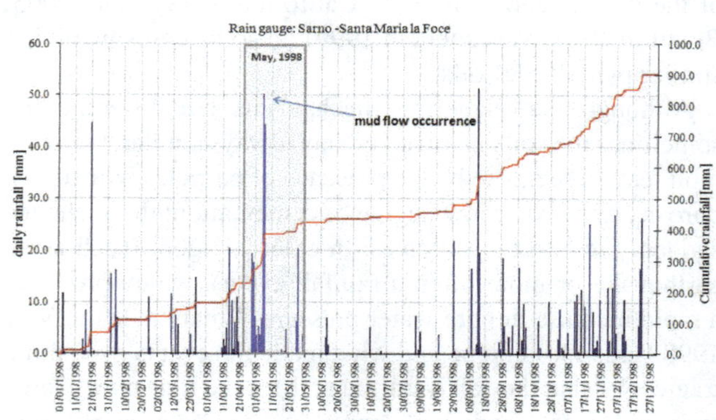

Figure 4. Comparison of daily and cumulative rainfall data recorded at Sarno rain gauge in 1998.

HydroSUSHI analyzes subsoil water circulation in a spatial two-dimensional domain which can be characterized by irregular soil stratigraphy.

Concerning the GeoSUSHI module, a stability analysis is performed to better understand how negative pore-water pressures (or matric suction) increase the shear strength of the soil.

The infiltration analysis is carried out using the Richards equation (1931), expressed in terms of water potential to facilitate applications on layered soils and transient flow regimes for both saturated and unsaturated conditions.

By adopting a Cartesian orthogonal reference system Oxz, with z axis positive downwards, the governing differential equation is

$$\nabla \cdot [K(\psi)\nabla(\psi - z)] = [C(\psi) + S_e(\psi)S_s]\frac{\partial \psi}{\partial t}, \tag{1}$$

where $K(\psi)$ [L/T] is the hydraulic conductivity which depends on pressure head ψ [L] for unsaturated soils (neglecting soil anisotropy). The right-hand side of Eq. (1) is written to simulate water flows in both unsaturated and saturated zones, thus avoiding the use of different algorithms for the resolution of parabolic and elliptic equations. $C(\psi) = \mathrm{d}\vartheta/\mathrm{d}\psi$ [L^{-1}] is the specific soil water capacity in the unsaturated zone, which represents the rate at which soil absorbs or releases water when there is a change in pressure head; S_S [L^{-1}] is the specific volumetric storage, accounting for soil deformation, which is a feature that not many models have; effective saturation $S_e[\psi] = (\vartheta - \vartheta_r)/(\vartheta_s - \vartheta_r)$, where θ is the water content, θ_s is the porosity and θ_r is the residual water content, which can be computed using the soil water retention curve. The saturated flow equation is simply a special case of the Richards equation, in which the conductivity and storage terms are not functions of pressure head.

This module was upgraded by incorporating a method to describe the evapotranspiration process although this component usually produces only secondary effects when slope mobilizations occur during very rainy periods (Capparelli and Versace, 2011). Since the case study here proposed took place during a very rainy spring, the effects related to evapotranspiration were neglected.

3.1 Model structure

In the HydroSUSHI module the finite differences method (FDM) with a fully implicit method is adopted. It is well known that the Richards equation only allows analytical solutions where simplifying hypotheses and/or particular boundary conditions are introduced (Iverson, 2000; Srivastava and Yeh, 1991).

Finite difference algorithms which deal with either variably saturated or fully unsaturated conditions have been proposed by Freeze (1978) and Vauclin et al. (1979). In this approach, the continuous problem domain is discretized so that the dependent variables are considered to exist only at discrete points.

Figure 5 shows an example of the spatial discretization created by a regular mesh (Δx; Δz). The size to be assigned to Δx and Δz should be selected on the basis on the complexity of the stratigraphy. The layers need to be faithfully reproduced, so as to guarantee a realistic representation of water flow exchanges.

With reference to the generic node with coordinates $x = x_0 + i\Delta x$, $z = z_0 + j\Delta z$ according to the finite difference method, Eq. (1) can be written as

$$\begin{aligned}
&\frac{1}{\Delta x}\left[K\left(\psi_{i+1/2,j}^{(k+1)}\right)\left(\frac{\psi_{i+1,j}^{(k+1)} - \psi_{i,j}^{(k+1)}}{\Delta x}\right)\right.\\
&\left.-K\left(\psi_{i-1/2,j}^{(k+1)}\right)\left(\frac{\psi_{i,j}^{(k+1)} - \psi_{i-1,j}^{(k+1)}}{\Delta x}\right)\right]\\
&+\frac{1}{\Delta z}\left[K\left(\psi_{i,j+1/2}^{(k+1)}\right)\left(\frac{\psi_{i,j+1}^{(k+1)} - \psi_{i,j}^{(k+1)}}{\Delta z} - 1\right)\right.\\
&\left.-K\left(\psi_{i,j-1/2}^{(k+1)}\right)\left(\frac{\psi_{i,j}^{(k+1)} - \psi_{i,j-1}^{(k+1)}}{\Delta x} - 1\right)\right]\\
&= C_{SU}\left(\psi_{i,j}^{(k+1)}\right)\left(\frac{\psi_{i,j}^{(k+1)} - \psi_{i,j}^{(k)}}{\Delta t}\right),
\end{aligned} \tag{2}$$

where $C_{SU} = [C(\psi) + S_e(\psi)S_s]$, the subscripts $i \pm 1/2, j$ and $i, j \pm 1/2$ indicate quantities evaluated at the spatial coordinates $(x_0 + (i \pm 1/2)\Delta x, z_0 + j\Delta z)$ and $(x_0 + i\Delta x, z_0 + (j \pm 1/2)\Delta z)$, Δt is the time step, and the superscripts (k) and $(k+1)$ indicate quantities referring to time instants $t = t_0 + k\Delta t$ and $t = t_0 + (k+1)\Delta t$.

To solve Eq. (1), boundary conditions along the edges of the integration domain must be specified. A general form of the boundary conditions for the considered PDE (partial differential equation) can be written as (McCord, 1991)

$$\alpha(\zeta)\psi + \beta(\zeta)\left.\frac{\partial \psi}{\partial n}\right|_{\partial G} = B(\zeta, t), \tag{3}$$

where $\alpha(\zeta)$, $\beta(\zeta)$ and $B(\zeta, t)$ are given functions evaluated on the boundary ∂G, the expression $\partial/\partial n$ is the normal derivative operator and $\boldsymbol{\zeta}$ is the spatial local vector.

By applying this general formulation for water flow modeling, the boundary conditions are as in Fig. 5: along the basal impermeable boundary BC and vertical boundary AB a Neumann condition is considered, with flux equal to zero, then $\alpha(\zeta) = 0$, $\beta(\zeta) = K(\zeta)$ and $B(\zeta, t) = q(\zeta, t) = 0$. In terms of total hydraulic head $h = \psi - z$:

Table 2. Comparison between mean monthly rainfall computed for the period 1964–1997 and the year 1998 (rows a–b); and between maximum daily rainfall for the period 1964–1997 and the year 1998 (rows c–d). Some values are in bold because they reflect the situation of the specific period analyzed. This allows to differentiate the values from other reported data.

		Jan	Feb	Mar	Apr	May	Jun	Jul	Aug	Sep	Oct	Nov	Dec
(a)	Monthly expected value (mm) 1964–1997	87	85	73	73	38	26	18	27	60	104	129	111
(b)	1998	77	46	44	**109**	**150**	12	6	39	122	52	124	94
(c)	Daily maximum (mm)	29	24	23	22	15	13	9	16	27	32	38	34
(d)	1998	42	21	13	**37**	**74**	6	5	20	47	17	25	37

$$\left.\frac{\partial h}{\partial z}\right|_{\mathrm{BC}} = 0, \tag{4}$$

$$\left.\frac{\partial h}{\partial x}\right|_{\mathrm{AB}} = 0. \tag{5}$$

For the vertical downslope side DC we take into account the influence of the increasing subsurface flow, by considering both unsaturated and saturated layers. This is computed by adopting boundary conditions moving from the Neumann to Dirichlet conditions, with specified flux or pressure head, respectively, $\alpha(\zeta) = 1, \beta(\zeta) = 0$ and $B(\zeta, t) = h(\zeta, t)$. Then

$$\left.\frac{\partial h}{\partial x}\right|_{\mathrm{DC}} = q(x,t) = 0 \tag{6}$$

or

$$\psi|_{\mathrm{DC}} = 0. \tag{7}$$

Along the upper boundary AD a time-dependent rainfall intensity r [L/T] is applied. The boundary condition can be stated by considering the infiltration rate $I(x,t)$ as

$$K(x,t)\left.\frac{\partial h(x,t)}{\partial z}\right|_{\mathrm{AD}} = B(\zeta,t) = I(x,t). \tag{8}$$

In particular

$$\begin{cases} B(\zeta,t) = I(x,t) = r(x,t) & \text{if } r(x,t) \le K(x,t)\frac{\partial h(x,t)}{\partial z} \\ B(\zeta,t) = I(x,t) = K(x,t)\frac{\partial h(x,t)}{\partial z} & \text{if } r(x,t) > K(x,t)\frac{\partial h(x,t)}{\partial z}. \end{cases} \tag{9}$$

The value $K(x,t)$ will depend on the values of $\psi(x,t)$ at the point x at time t and on the nature of the $K(\psi)$ curve for the surface soil at x.

As previously mentioned, in the GeoSUSHI module the stability analysis is carried out by evaluating the influence of negative pore-water pressures as well.

It may be a reasonable assumption to ignore negative pore-water pressures for many situations where most of the slip surface is below the groundwater table. However, for situations where the groundwater table is deep or where there is concern over the possibility of shallow failure surface, negative pore-water pressures cannot be ignored.

The procedure adopted here is an extension of conventional limit equilibrium methods adapted to unsaturated soils as suggested by Fredlund and Rahardjo (1993).

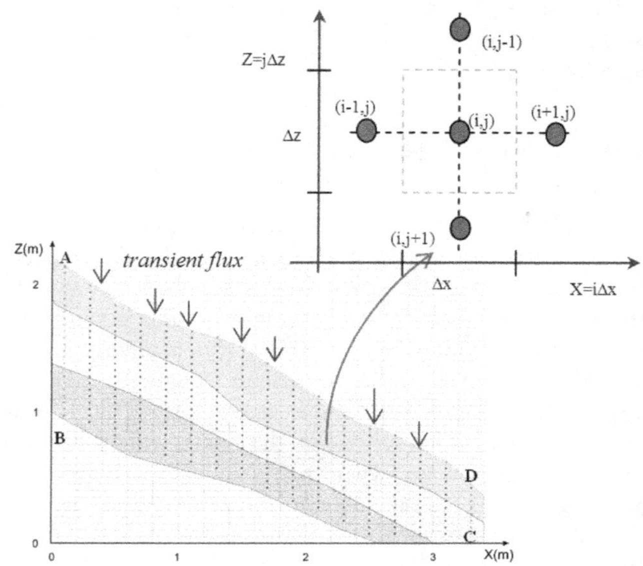

Figure 5. Nodal network implemented for development of FDM equation.

The shear strength of an unsaturated soil can be formulated in terms of independent stress state variables $(\sigma - u_a)$ and $(u_a - u_w)$ as follows:

$$\tau_{ff} = c' + (\sigma_n - u_a)_f \tan\varphi' + (u_a - u_w)_f \tan\varphi^b, \tag{10}$$

where the subscripts f indicate quantities evaluated on the failure plane at failure, and τ_{ff} is shear stress, c' is effective cohesion, $(\sigma_n - u_a)_f$ net normal stress state, u_a pore-air pressure, φ' effective friction angle, $(u_a - u_w)$ matric suction and φ^b angle indicating rate of increase in shear strength relative to the matric suction. This last term is evaluated using the expression proposed by Vanapalli et al. (1996).

Equation (10) is an extension of the shear strength equation for a saturated soil. As the soil approaches saturation, the pore-water pressure, u_w, approaches the pore-air pressure u_a and matric suction $(u_a - u_w)$ goes to zero. The general limit equilibrium method (GLE) provides a general theory whereas other methods can be viewed as special cases. It is well known that the elements used in the GLE method to derive the safety factor (FS) are the summation of forces in two directions and of the moments about a generic point

Figure 6. Picture of the case study area with some topographic details.

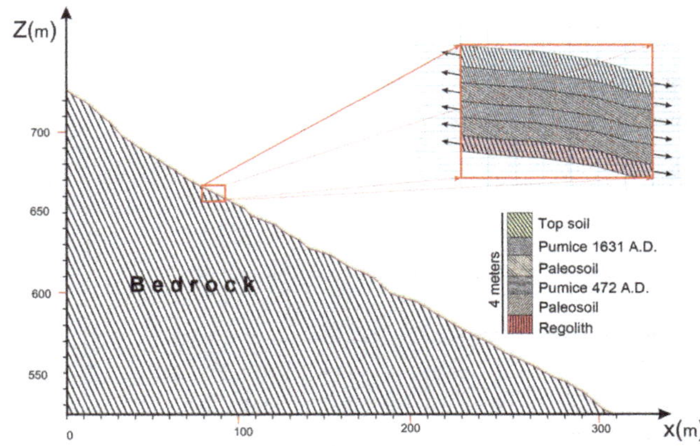

Figure 7. Geometric and stratigraphic characterization of the investigated slope.

(Fredlund and Rahardjo, 1993). FS is defined as the factor by which the shear strength parameters must be reduced in order to bring the soil mass into a state of limiting equilibrium along the assumed slip surface.

Here the calculations for evaluating the stability of a slope are performed by dividing the soil mass above the slip surface into vertical slices. The mobilized shear force at the base of a slice can be written using the shear strength for an unsaturated soil:

$$S_{\mathrm{m}} = \frac{\beta \tau_{\mathrm{ff}}}{\mathrm{FS}}, \tag{11}$$

where S_{m} is the shear force mobilized at the base of the slice; and β is the sloping distance across the base of a slice.

3.2 Input data and slope scheme

The SUSHI model was applied to the mudflow occurring in the Tuostolo Basin, highlighted by the red square in Fig. 1, which destroyed the village of Sarno.

The actual geometry of the mudflow is the result of the coalescence of further landslides that took place over a period of 6–8 h.

The landslide mobilized a volume of about 92 000 m^3 of volcaniclastic materials resting over carbonate bedrock, including the eroded material within the channel. It developed from an altitude of about 725 m to a subvertical limestone wall (a morphological frame) situated at an altitude of about 500 m (Fig. 6).

Most of the landslide occurred in May 1998. It started from the point where morphological discontinuities are located, also represented by topographic variations or anthropogenic discontinuities such as tracks.

To define the dynamics of the water circulation in the subsoil, the solution requires a description of the investigated domain, the soil water characteristic curves, the permeability functions, the mechanical properties of the involved soils, as well as the boundary and initial conditions.

Surveys and studies carried out by also using information available in the literature indicated the presence of alternating layers of pumice with a composition and thickness related to the characteristics of the eruptions and to the distance from the eruptive centers.

This sequence comprises both primary airfall and volcaniclastic deposits. The primary deposits are composed of alternating layers of pumice, with interbedded paleosoils. At the base of this sequence, above the bedrock, there is a layer of dark-red clayey ashy soil (regolith) with rare limestone fragments.

By using the available topographic maps showing the top surface of the ground before the events, the soil cover thickness and their distributions were obtained.

At the main scarp, the average thickness of the pyroclastic cover was about 4 m. From top to bottom under a topsoil formed by humified ashes including roots and organic matter (about 90 cm thick), the following layers were identified: (A) an upper layer (60 cm) of coarse pumice; (B) a layer (70 cm) of paleosoil; (C) a horizon (60 cm) of finer pumices; (D) a layer (80 cm) of paleosoil; (E) a bottom layer (40 cm) of weathered dark-red clayey ash in contact with the fractured limestone bedrock (Fig. 7).

To determine the mechanical and hydraulic properties of the cover involved, undisturbed specimens were collected, both from the investigated area and from other triggering areas on the Pizzo d'Alvano slopes. Table 3 reports the mean values of the physical properties of the various materials.

The hydraulic properties of the ashy soils in saturated conditions were investigated by conventional permeameter tests. In the unsaturated conditions, suction-controlled oedometer was used.

The experimental data were fitted by the expression proposed by van Genuchten and Nielsen (1985). The values of the parameters estimated for the various layers are given in Table 4.

As shown in Fig. 8, the SWRC (soil water retention curve) are extremely variable, butt in all cases they are typical of coarse soils with low air-entry pressure, low residual water content and steep slope within the transition zone.

Due to a lack of direct measurements, representative values have been considered as the specific storage S_S, ranging from 3×10^{-3} (m^{-1}) for plastic clay to 1.3×10^{-4} (m^{-1}) for dense sand (Freeze and Cherry, 1979).

The variable boundary conditions were provided using both Dirichlet and Neumann conditions. At the top (i.e., on the ground surface) a flux boundary condition equal to the rainfall infiltration capacity was applied; the runs enabled the infiltration rate to be defined step by step for each node of the domain. At the bottom (i.e., at the contact between the pyroclastic cover and the bedrock) no flux was imposed, since the bedrock was assumed to be impervious.

Similarly, for the upslope, left side, a Neumann condition of no water flow was fixed, since the morphology of the analyzed area reasonably leads to the hypothesis of coincidence between the surface and underground watershed. These contributions of fluxes from upstream may be assumed to be equal to zero. For the right-hand side of the downslope, along the morphological frames, two different boundary conditions were imposed using a Neumann or a Dirichlet condition depending on whether or not saturation occurred.

For the slope section shown in Fig. 7 the mesh was made up of 130 000 nodes, using regular quadrilaterals with lengths and heights, respectively, equal to $\Delta x = 0.20$ m and $\Delta z = 0.05$ m.

The initial conditions were defined in a non-arbitrary way, thanks to the data provided by the tensiometers, which were located, as previously mentioned, very close to the selected study area. Such information was very useful for setting the initial conditions. Constant distribution suction throughout the domain was initially hypothesized by selecting the following values:

$$\psi(x, z; t = 0) = 3; 4; 5; 8; 10; 14 \quad \text{[kPa]}. \tag{12}$$

By starting a simulation with no rain, a warm-up was performed for each of these values, in order to redistribute the water content all over the domain. The warm-up was stopped when the standard deviation values for each spatial node were less than 10^{-5} [m]. The distribution thus obtained was considered as being representative of an equilibrium condition.

The profiles obtained were compared with the available in situ evidence recorded by tensiometers at the end of summer, which were in considerable agreement with the warm-up results. By comparing these profiles, a strong similarity was evident with the distribution performed with $\Psi(x, z; t = 0) = 6$ kPA. This pore-water pressure distribution was set as the initial condition for simulating the evolution between 1 October 1997 and 5 May 1998.

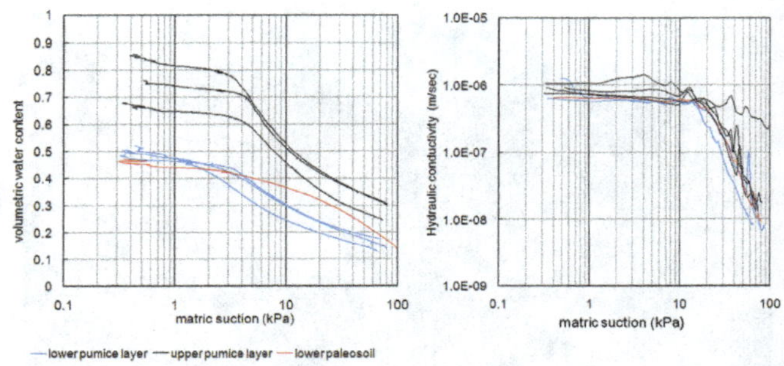

Figure 8. Soil water characteristic curves.

4 Results and discussion

The period analyzed (from 1 October 1997 to 5 May 1998) was characterized by a total rainfall of 891 mm, with greater values of rainfall intensity having occurred between the end of October and December 1997.

Given the large extension of the domain investigated, the results presented refer to the conditions reached in two zones considered as representative of the domain: one in the upslope part, at $Z = 720$ m a.s.l. (hereafter referred to as "section A"), the other at the toe of the slope, at $Z = 520$ m a.s.l., ("section B").

The temporal variations in suction profiles are given in Figs. 9 and 10 for sections A and B, respectively.

For the sake of clarity, in the two figures, the various computed suction profiles are not all plotted on the same graph. The upper panel shows the first three computed soil profiles, for around the end of October, November and December 1997. The middle panel shows another two profiles, evaluated around the end of January and February 1998. To let the reader compare the new profiles with the old ones, the latter are also plotted in the same panel in light gray. Finally, the bottom panel shows the suction profile computed on 3, 4 and 5 May 1998, during the rainfall event which led to a catastrophic flowslide at the investigated slope. Also in this case, the previous suction profiles are plotted in light gray to help the reader follow the evolution of soil suction with time.

Figure 9 shows that no water table is ever computed along the upslope area (section A), since the values of the pressure head are always negative. This situation is consistent with the morphological characteristics of the investigated site, where the steepness of the slope prevents water from accumulating and lets it move through the layers towards the foot of the slope.

The situation is very different at the toe of the slope (section B, Fig. 10), where the lower layers soon reached saturated conditions, already in November 1997, much before May 1998. From the former date onwards, the lower layers remained saturated for the entire simulation period.

The results of the analysis do not lead to a clear interpretation of the triggering phenomenon, but suggest that the

Table 3. Average values of pyroclastic soil properties.

Soil properties	Topsoil	Pumice (A)	Paleosoil (B)	Pumice (C)	Paleosoil (D)	Regolite (E)
Dry unit weight [$\mathrm{kN\,m^{-3}}$]	10.99	6	7	6	9	10.75
Saturated unit weight [$\mathrm{kN\,m^{-3}}$]	17.2	13	13	13	15	15.3
Saturated soil water content θ_s	0.55	0.82	0.61	0.68	0.61	0.60
Residual soil water content θ_r	0.14	0.23	0.18	0.05	0.18	0.10
Saturated hydraulic conductivity K_s [$\mathrm{m\,s^{-1}}$]	3.2×10^{-5}	1.0×10^{-3}	1.0×10^{-6}	1.0×10^{-2}	4.0×10^{-6}	7.6×10^{-7}
Effective cohesion c' [kPa]	2	0	4.5	0	4.7	15
Friction angle φ' [°]	15	30	24	32	28	21

Table 4. Parameters of van Genuchten and Nielsen model. The values of the bubbling pressure, or air-entry tension, ψ_b, were determined through the graphic method proposed by Fredlund and Xing (1994).

Topsoil	$n = 1.6; m = 0.38$	$\psi_b = 1.65$ (kPa
Layer A	$n = 1.71; m = 0.42$	$\psi_b = 0.2$ (kPa)
Layer B	$n = 1.66; m = 0.40$	$\psi_b = 2.5$ (kPa)
Layer C	$n = 1.8; m = 0.44$	$\psi_b = 0.3$ (kPa)
Layer D	$n = 1.9; m = 0.47$	$\psi_b = 2.5$ (kPa)
Layer E	$n = 2; m = 0.50$	$\psi_b = 2.7$ (kPa)

saturation of the underlying layers at the toe of the slope was clearly not the cause of the instability of the slope, although it contributed to it.

The suction values achieved on 5 May 1998 in the lower layers of the soil profile had been already predicted by the model for previous periods.

However, a big difference lies in the fact that, on 5 May 1998, the calculated vertical suction profiles of water content show high values, not far from saturation, in the upper layers as well.

The wetting front generated in early May by the rainfall infiltration did not reach the lower layers and did not contribute to a further increase in pore-water pressure at these depths but led to the shallowest layers becoming close to saturation.

This result is even clearer when analyzing the evolution of the pore pressure distribution along the slope throughout the considered period. Figure 11 shows the pore-water pressures evaluated at 3 and 0.7 m below the ground surface. The first depth is representative of the behavior of a relatively deep layer, which reaches saturation during the first months of the rainy season, while the second represents the conditions of the upper layers well.

In the lower layers (Fig. 11a), the pressure levels remain approximately the same with the rainfall in late April and early May, while it increases sharply in the upper layers (Fig. 11b) from 4 to 5 May.

Regarding the computed pore pressure, a slope stability analysis was carried out to simulate failure conditions and their relation to increasing soil water content.

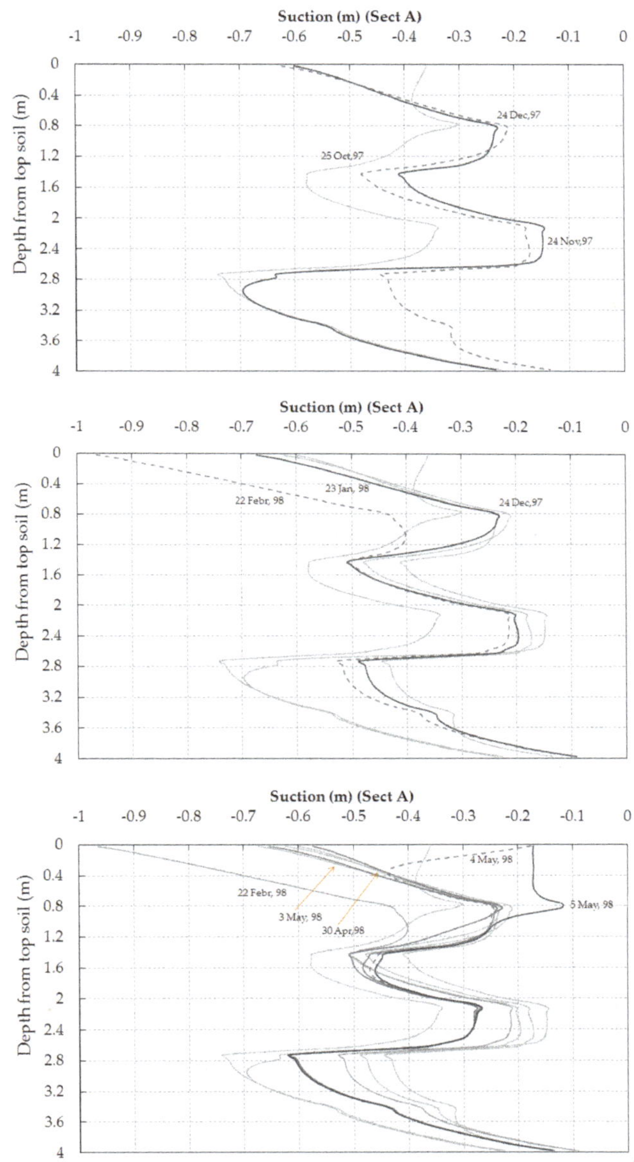

Figure 9. Computed suction profile for the upslope section (Section A).

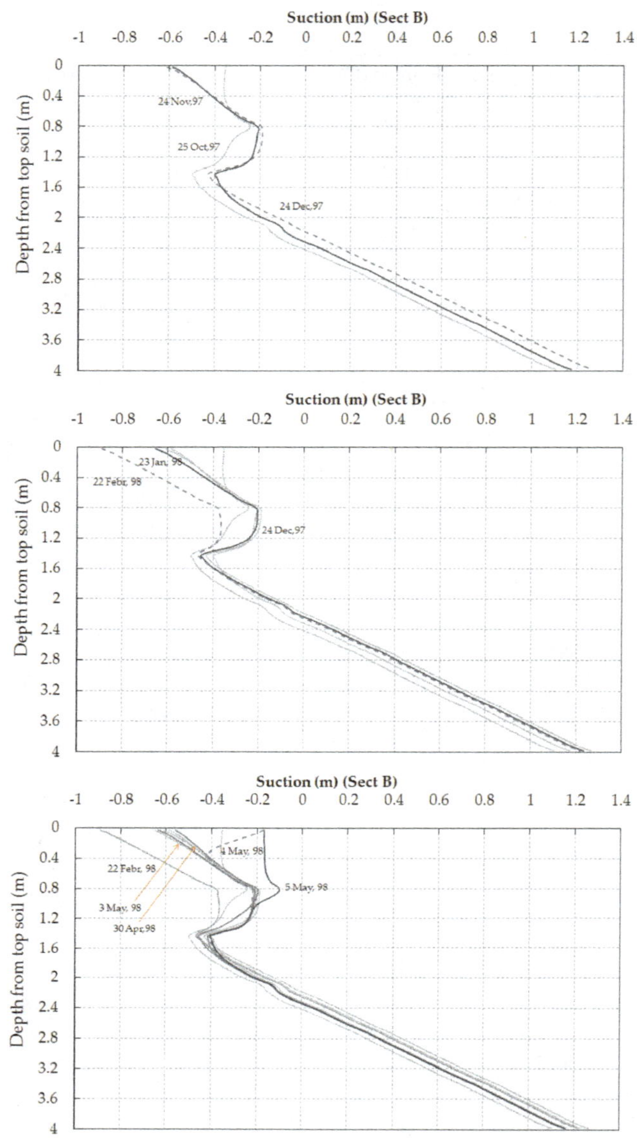

Figure 10. Computed suction profile for downslope section (Section B).

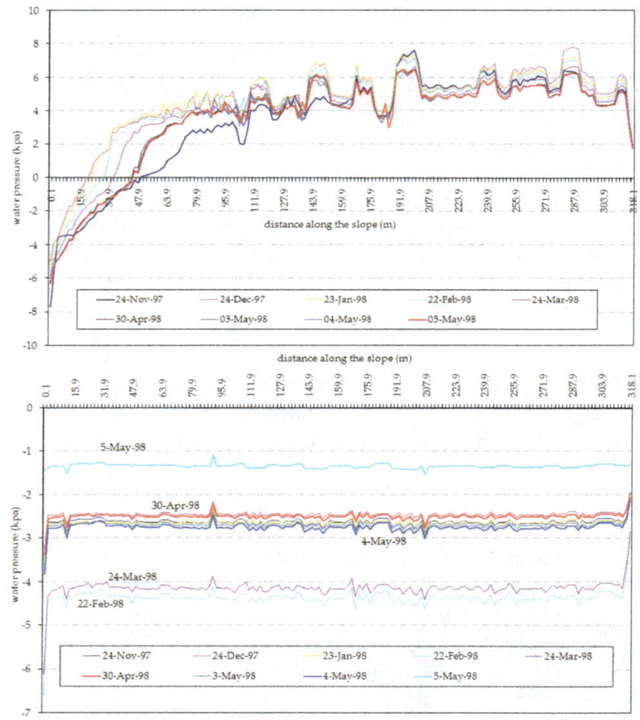

Figure 11. Pore-water pressures performed at **(a)** 3 m and **(b)** 0.7 m below the ground surface.

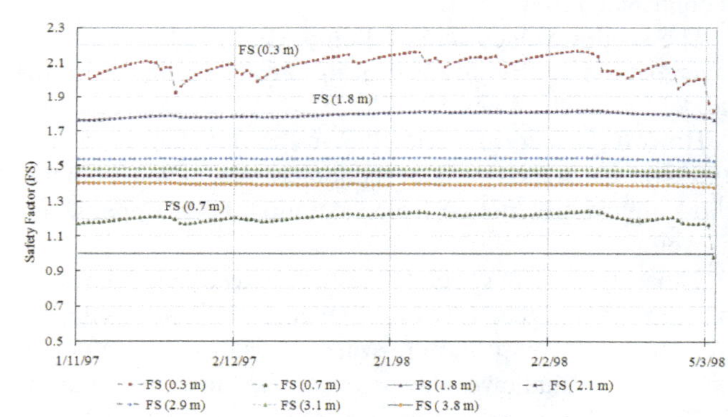

Figure 12. Slope safety factor depending on pore-water pressures and soil mechanism properties at different depths from the ground surface.

For several depths below the ground surface (0.3, 0.7, 1.8, 2.1, 2.9, 3.1 and 3.8 m), the average pore pressure was calculated, and the corresponding FS was estimated under the hypothesis of an infinite slope. This approach is the simplest limit equilibrium method and gives reliable results for slides where the longitudinal dimension prevails over the depth of the landslide, as with the landslide analyzed here.

The plots in Fig. 12 provide the time history of the simulated FS values, which help in understanding the evolution of the slope stability conditions.

The values in the lower layers are always indicative of stability; lower values, but nevertheless above 1, are due to the greater thickness of the soil cover and to higher values of pore-water pressure. By contrast, in the more superficial layers the trends are more variable and show, at a depth of 0.7 m, a decrease in FS to 0.98, corresponding to instability conditions on 5 May, 1998.

5 Conclusions

Landslide prediction is a problem that has concerned the general population for a long time and more recently also the scientific community. The ability to predict landslide onset is closely related to the ability to understand the dynamics of the underlying processes affecting the event.

The application reported here confirms how it is possible to approach the interpretation of a natural but complex and difficult event such as the triggering of a landslide, via simulation with physically based mathematical models.

The analysis outlined a possible interpretation of the Sarno landslides that could explain the triggering mechanisms occurring in May 1998. The model's output suggests that the saturation of deep layers was not the only reason for slope instability, and that the reduction in suction across the shallowest layers may have been the actual cause of the triggering mechanism.

Monthly rainfalls occurring in April and May 1998, though not exceptional, were quite unusual for that period of the year. The rain occurring at the end of April caused the superficial layers to reach a high water content and, consequently, an increase in hydraulic conductivity. Thus, the heavy rainfall in the early days of May produced a steep wetting front which did not reach the deeper layers and considerably reduced the shear strength of the upper soil layers.

This interpretation is partly confirmed by the slope stability analysis, which reveals high values of the safety factor in the lower layers and values close to one in the upper layers with regard to the landslide date.

New applications to other cases occurring after May 1998 would certainly better define the critical conditions and could provide useful information for a possible early warning system.

Further analyses should be carried out in order to better evaluate the influence of the bedrock, of the road cuts located in the upper zone of the triggering areas and other factors that could have helped trigger the landslide.

By pursuing the overall objectives, the SUSHI model confirmed its capacity to (i) simulate, with sufficient detail, the phenomena induced by rainfall in soils characterized by complex stratigraphy and hydraulic properties; (ii) achieve reliable numerical solutions even when running very long simulations, thus enabling antecedent rainfall to be taken into account, and (iii) provide a convincing interpretation of the phenomenon, without the need to introduce other hydrological forcing.

Edited by: R. Greco

References

Arnone, E., Noto, L. V., Lepore, C., and Bras, R. L.: Physically-based and distributed approach to analyze rainfall-triggered landslides at watershed scale, Geomorphology, 133, 121–131, doi:10.1016/j.geomorph.2011.03.019, 2011.

Baum, R. L., Savage, W. Z., and Godt, J. W.: TRIGRS – a FORTRAN program for transient rainfall infiltration and grid-based regional slope stability analysis, US Geological Survey Open-File Report, 02-0424, 2002.

Baum, R. L., Savage, W. Z., and Godt, J. W: TRIGRS-A Fortran program for transient rainfall infiltration and grid-based regional slope-stability analysis, version 2.0, US Geological Survey Open-File Report 2008–1159, available at: http://pubs.usgs.gov/of/2008/1159/, 2008.

Caine, N.: The rainfall intensity-duration control of shallow landslides and debris flows, Geogr. Ann. A, 62, 23–27, 1980.

Calcaterra, D., Parise, M., Palma, B., and Pelella, L.: Multiple debris flows in volcanoclastic materials mantling carbonate slopes, in: Proceedings of the 2nd International Conference on Debris-Flow Hazard Mitigation, Taipei, Taiwan, 16–18 July 2000, 99–107, 2000.

Capparelli, G. and Tiranti, D.: Application of the MoniFLaIR early warning system for rainfall-induced landslides in the Piedmont region (Italy), Landslides, 7, 401–410, doi:10.1007/s10346-009-0189-9, 2010.

Capparelli, G. and Versace, P.: FLaIR and SUSHI: Two mathematical models for Early Warning Systems for rainfall induced landslides, Landslides, 8, 67–79, 2011.

Cascini, L. and Sorbino, G.: The contribution of soil suction measurements to the analysis of flowslide triggering, Invited Lecture, Proc. of the Int. Workshop "Flows 2003 – Occurrence and Mechanisms of Flows in Natural Slopes and Earthfill", Patron Editore, Sorrento, 77–86, 2004.

Cascini, L., Cuomo, S., and Guida, D.:Typical source areas of May 1998 flow-like mass movements in the Campania region, Southern Italy, Eng. Geol., 96, 107–125, 2008.

Dhakal, A. S. and Sidle, R. C.: Pore water pressure assessment in a forest watershed: Simulations and distributed field measurements related to forest practices, Water Resour. Res., 40, W02405, doi:10.1029/2003WR002017, 2004.

Ellen, S. D. and Fleming, R. W.: Mobilization of debris flow from soil slips, San Francisco Bay region, California, Eng. Geol., 7, 31–40, 1987.

Fredlund, D. G. and Rahardjo, H.: Soil Mechanics for Unsaturated Soils, John Wiley & Sons, 1993.

Fredlund, D. G. and Xing, A.: Equations for the soil-water characteristic curve, Can. Geotech. J., 31, 521–532, 1994.

Freeze, R. A.: Mathematical models of hillslope hydrology, in: Hillslope Hydrology, edited by: Kirkby, M. J., John Wiley & Son, 177–226, 1978.

Freeze, R. A. and Cherry, J. A.: Groundwater, Prentice Hall, Englewood Cliffs, New Jersey, 604 pp., 1979.

Glade, T.: Modelling landslide-triggering rainfalls in different regions in New Zealand – the soil water status model, Z. Geomorphol., 122, 63–84, 2000.

Greco, R., Comegna, L., Damiano, E., Guida, A., Olivares, L., and Picarelli, L.: Hydrological modelling of a slope covered with shallow pyroclastic deposits from field monitoring data, Hydrol. Earth Syst. Sci., 17, 4001–4013, doi:10.5194/hess-17-4001-2013, 2013.

Guzzetti, F.: Book Review of "Measuring Vulnerability to Natural Hazards", Nat. Hazards Earth Syst. Sci., 8, p. 521, 2008.

Iverson, R. M.: Landslide triggering by rain infiltration, Water Resour. Res., 36, 1897–1910, 2000.

Lu, N. and Godt, J.: Hillslope Hydrology and Stability, Cambridge University Press, New York, USA, 2013.

Mazzoleni, S., Strumia, S., Di Pasquale, G., Migliozzi, A., Amato, M., and Di Martino, P.: Il ruolo della vegetazione nelle frane di Quindici, in: L'instabilità delle coltri piroclastiche delle dorsali carbonatiche in Campania: primi risultati di uno studio interdisciplinare, 20 Rapporto informativo dell'Unità Operativa 4/21N del CNR/GNDCI, 1998 (in Italian).

McCord, J. T.: Application of second-type boundaries in unsaturated flow modeling, Water Resour. Res., 27, 3257–3260, 1991.

Montgomery, D. R. and Dietrich, W. E.: A physically based model topographic control on shallow landsliding, Water Resour. Res., 30, 1153–1171, 1994.

Revellino, P., Hungr, O., Guadagno, F. M., and Evans, S. G.: Velocity and runout simulation of destructive debris flows and debris avalanches in pyroclastic deposits, Campania region, Italy, Environ. Geol., 45, 295–311, 2004.

Rigon, R., Bertoldi, G., and Over, T. M.: GEOtop: A Distributed Hydrological Model with Coupled Water and Energy Budgets, J. Hydrometeorol., 7, 371–388, 2006.

Rolandi, G.: The eruptive history of Somma-Vesuvius, in: Volcanism and archeology in Mediterranean area, edited by: Cortini, M. and De Vivo, B., Research Signpost (Trivandrum), 77–88, 1997.

Sirangelo, B. , Versace, P., and Capparelli, G.: Forwarning Model for Landslides Triggered by Rainfall Based on the Analysis of Historical Data File, IAHS Red Book, 278, 298–304, 2003.

Sorbino, G., Sica, C., and Cascini, L.: Susceptibility analysis of shallow landslides source areas using physically based models, Nat. Hazards, 53, 313–332, 2010.

Srivastava, R. and Yeh, T.-C. J.: Analytical Solution for One-Dimensional, Transient Infiltration towards the water table in homogeneous and layered soils, Water Resour. Res., 27, 753–762, 1991.

Vanapalli, S. K., Fredlund, D. G., Pufahl, D. E., and Clifton, A. W.: Model for the prediction of shear strength with respect to soil suction, Can. Geotech. J., 33, 379–392, 1996.

van Genuchten, M. T. and Nielsen, D. R.: On describing and predicting the hydraulic properties of unsaturated soils, Ann. Geophys., 3, 615–628, 1985, http://www.ann-geophys.net/3/615/1985/.

van Westen, C. J., Rengers, N., and Soeters, R.: Use of geomorphological information in indirect landslide susceptibility assessment, Nat. Hazards, 30, 399–419, 2003.

Vauclin, M., Khanji, D., and Vachaud, G.: Experimental and numerical study of a transient, two-dimensional unsaturated-saturated water table recharge problem, Water Resour. Res., 15, 1089–1101, 1979.

Versace, P., Sirangelo, B., and Capparelli, G.: Forewarning model of landslides triggered by rainfall, in: Proceedings of Debris-flow Hazards mitigation: Mechanics, Prediction, and Assessment, Davos, Switzerland, 2, 1233–1244, 2003.

Wu, W. and Sidle, R. C.: A distributed slope stability model for steep forested basin, Water Resour. Res., 31, 2097–2110, 1995.

11

Identification of runoff formation with two dyes in a mid-latitude mountain headwater

Lukáš Vlček[1], Kristýna Falátková[1], and Philipp Schneider[2]

[1]Department of Physical Geography and Geoecology, Charles University, Prague, Czech Republic
[2]Department of Geography, University of Zurich, Zurich, Switzerland

Correspondence to: Philipp Schneider (philipp.schneider@geo.uzh.ch)

Abstract. Subsurface flow in peat bog areas and its role in the hydrologic cycle has garnered increased attention as water scarcity and floods have increased due to a changing climate. In order to further probe the mechanisms in peat bog areas and contextualize them at the catchment scale, this experimental study identifies runoff formation at two opposite hillslopes in a peaty mountain headwater; a slope with organic peat soils and a shallow phreatic zone (0.5 m below surface), and a slope with mineral Podzol soils and no detectable groundwater (> 2 m below surface). Similarities and differences in infiltration, percolation and preferential flow paths between both hillslopes could be identified by sprinkling experiments with Brilliant Blue and Fluorescein sodium. To our knowledge, this is the first time these two dyes have been compared in their ability to stain preferential flow paths in soils. Dye-stained soil profiles within and downstream of the sprinkling areas were excavated parallel (lateral profiles) and perpendicular (frontal profiles) to the slopes' gradients. That way preferential flow patterns in the soil could be clearly identified. The results show that biomat flow, shallow subsurface flow in the organic topsoil layer, occurred at both hillslopes; however, at the peat bog hillslope it was significantly more prominent. The dye solutions infiltrated into the soil and continued either as lateral subsurface pipe flow in the case of the peat bog, or percolated vertically towards the bedrock in the case of the Podzol. This study provides evidence that subsurface pipe flow, lateral preferential flow along decomposed tree roots or logs in the unsaturated zone, is a major runoff formation process at the peat bog hillslope and in the adjacent riparian zone.

1 Introduction

Hydrologic extremes in central Europe during recent decades have stimulated debates over sustainable solutions and suitable, cost-efficient strategies to prevent or mitigate the impacts of droughts and floods. Floods on the Vltava and Elbe River are documented back to Middle Ages (Brázdil et al., 2015; Faist, 1967) as having had devastating impacts on infrastructure and the economy. However, a system of dams and reservoirs, which was built in the Vltava catchment – the main drainage of the Šumava Mts. in the 1960s – failed to prevent major floods in 2002 (Hladný et al., 2005; Hladný, 2009) or 2013. These floods were mainly caused by heavy rainfall in summer or by rain on snow events. As the Šumava Mts. and the Vltava catchment are promising water resources, research efforts have focused on this area.

Peat bogs (PB) are prominent in the Šumava Mts., which affects water quality and storm discharge (Ferda et al., 1971; Janský and Kocum, 2008; Vlček et al., 2016). Streamflow in peaty catchments is characterized by its quick rise and fall, and huge volatility: very low baseflows during dry periods and spiky storm hydrographs caused by heavy rainfall events (Evans et al., 1999; Holden et al., 2001; Holden, 2005). However, most of these studies have focused on pure peat areas only. In the Šumava Mts., peat bog covers approx. 35 % of the catchments in this region, but the larger proportion of these catchments is covered by mineral soils. Although peat bogs are not dominating the catchment area, their hydrology determines the runoff processes (Vlček et al., 2012). In order to establish the relative roles of peat bog and Podzol in this type of catchment, this study focused on a comparison of runoff formation on contrasting soil types – namely a mountain Histosol/peat bog and a Podzol (PZ) on opposite hillslopes of the headwater – representing organic and mineral soil properties, respectively.

Sprinkling experiments with sorptive dye tracers have

been successfully applied at the plot and hillslope scale in many different environments (Bachmair et al., 2009, 2012; Hümann et al., 2011; Flury and Flühler, 1995; Schneider et al., 2014; Weiler and Flühler, 2004; Weiler, 2011; Wienhöfer et al., 2009) and are an established method to identify interactions between infiltration, preferential flow, matrix flow and percolation. The sorptive dye Brilliant Blue (BB) FCF, is probably the most widely used dye tracer in such sprinkling experiments (Flury and Flühler, 1995). Successful experiments using BB at mineral soil test sites identified significant processes during the formation of storm runoff such as infiltration and initiation of vertical preferential flow in macropores (Weiler and Flühler, 2004), infiltration and vertical preferential flow dependence on soil structures, soil type and land use (Bachmair et al., 2009; Weiler and Flühler, 2004), lateral preferential flow in a soil pipe network (Anderson et al., 2009; Wienhöfer et al., 2009) and lateral preferential flow in organic topsoil layer/biomat flow (Schneider et al., 2014).

Almost all headwaters of the Šumava Mts., including the Rokytka catchment, have been affected by bark beetle calamity and by storm activity; e.g., the storm Kyrill in 2007 uprooted many trees. These disturbances have been shown to have a long-term impact on runoff formation in headwater systems (Langhammer et al., 2015). However, it is unclear how these disturbances modify specific runoff formation processes like infiltration, lateral drainage (soil piping) or percolation in riparian soils.

However, runoff formation at hillslopes, particularly subsurface stormflow (SSF), is highly variable and complex (Bachmair and Weiler, 2012). To evaluate both vertical and lateral preferential subsurface flow in the Rokytka headwater, we conducted two separate sprinkling experiments on each soil type with two different dye tracers with different properties. The experiments in this study were conducted in the vicinity of a headwater stream where the riparian zone connects to the two hillslopes. The test site was selected for several reasons:

1. Runoff formation control – the riparian zone represents a potential buffer in hydrologic connectivity between hillslopes and a stream (Von Freyberg et al., 2014; Seibert et al., 2009).

2. Representativeness – the two slopes are covered by prevailing soil types in the Šumava Mts. (Podzol, Histosol).

3. Tracer detection in a stream – hydrologic connectivity of a hillslope and a stream can be detected and potentially quantitatively described by conservative tracers such as Fluorescein sodium (FLC, also Uranine). Therefore, proximity of a stream is favorable in terms of time-saving and lesser dilution of a tracer.

Table 1. Selected dye-sprinkling experiments sorted by land use and scale of dye application.

	Land use	Plot studies	Hillslope studies
Brilliant Blue	Forest studies	Bachmair et al. (2009); Wienhöfer et al. (2009)	Anderson et al. (2009); Wienhöfer et al. (2009)
	Grassland studies	Weiler and Flühler (2004); Bachmair et al. (2009); Schneider et al. (2014)	–
Fluorescein sodium	Forest studies	Gerke et al. (2008, 2013, 2015)	Wienhöfer et al. (2009); Weiler and Naef (2003)
	Grassland studies	–	Weiler and Naef (2003); Schneider (2007); Schneider et al. (2014)

The sorptive tracer BB was used mainly to detect vertical flow, and the conservative tracer FLC was used to detect vertical and lateral flow at two opposite hillslopes with different soil types (peat bog and Podzol). Furthermore, based on other studies in dark-colored organic soils (Markus Weiler, personal communication, 2014), we considered that BB-stained soil patterns may be difficult or impossible to identify in the peat bog. Thus, we adapted a successful dye tracer soil staining experiment applied in organic forest soils using FLC (Gerke et al., 2008). FLC can be used to identify preferential flow patterns by staining soil particles (Gerke et al., 2015), and as a tracer detecting lateral subsurface flow and thus hydrologic connectivity and potentially tracer breakthrough into a stream or a spring equipped with automated water samplers or field fluorometers. Thus, FLC provides two functions in parallel: vertical and lateral soil staining of preferential flow structures and lateral connectivity between the sprinkling plots (lower, convex part of the hillslopes) and the stream or springs. In theory, such a combined FLC experiment could link plot and hillslope experiments and thus provide additional insights into the mechanical understanding of the entire hillslope-riparian zone-stream system and thus providing an estimate where (flow paths), when (timing, delay) and possibly how much hillslopes with different soils contribute to discharge in headwaters. Examples of applications of the dyes BB and FLC are shown in Table 1. To summarize, the objectives of this study are to identify runoff formation at both prevailing soil types (Histosol/Peat, entic Podzol) with emphasis on the following aspects:

- Identify infiltration and vertical preferential flow in the unsaturated zone in soil profiles – as well as possible infiltration barriers causing lateral flow on the plot scale.

- Identify soil horizons and/or structures, where vertical flow translates into lateral preferential flow, e.g., whether lateral preferential flow occurs in partially saturated parts in the *unsaturated* zone, or expanding *fully saturated* horizons promote transmissivity feedback.

- Identify vertical percolation in the saturated zone; determine whether infiltrated water percolates deep into the bedrock and thus a "secondary" (strongly delayed) drainage system is recharged during storms – plot scale.
- Identify lateral preferential flow paths and possibly estimate lateral flow velocities in the soil – hillslope scale.

2 Materials and methods

2.1 Study site

The 3.8 km^2 Rokytka headwater is a tributary to the Otava River, located in the central part of Šumava Mts. The second-order stream drains 0.6 km^2 of the Rokytka headwater covering an altitude range between 1100 and 1260 m a.s.l. The test site can be divided into two parts – two opposite hillslopes – with different soils and vegetation cover (Fig. 1). The mineral soil hillslope consists of a Podzol (PZ hillslope) and is covered by beech stands at the upper hillslope zone PZ1 and dead spruce stands with healthy seedlings at the lower hillslope zone PZ2. The soil profiles are similar throughout the slope without a clear gradient towards the stream. The soil type has been identified as an entic Podzol with a shallow organic top layer (< 5 cm) and similar soil texture (Table 2) to a depth of 1 m. Some small parts of the mineral soil hillslope PZ are covered by haplic Podzol, but these areas are hardly identifiable without excavation. Neither there was a sharp transition between the mineral soil and the bedrock (well-weathered gneiss or granite) perceptible with ERT measurements, nor could a persistent groundwater level be detected. The organic soil hillslope is covered by a well-developed mountain peat bog (PB hillslope). The entire PB area consists of a mixture of various stages of decomposed peat; however, acrotelm and lower catotelm can be distinguished at depths ranging from 8 to 25 cm. Vegetation and soil depth vary according to the position along the hillslope forming a catena. The upper organic soil hillslope zone is covered mostly by cotton grass (*Eriophorum* sp. L.) or moss (*Sphagnum* sp. L.; PB1 in Fig. 1). This zone has the highest water table fluctuations and the depth of the PB is 4–5 m. The vegetation cover at the lower hillslope zone (PB2 in Fig. 1) consists of pine (*Pinus mungo*), blueberry (*Vaccinium myrtillus*) and moss. The riparian zone (PB3 in Fig. 1) forms the bottom of the valley, which is covered by a waterlogged spruce forest with blueberry and moss. The depth of the PB varies from 1 m in the riparian zone PB3 up to 5 m in the upper hillslope zone PB1 (Fig. 1).

Despite the differences between an organic and a mineral soil at the two hillslopes, the basic soil properties – which have a strong impact on infiltration and subsurface stormflow – are rather similar (Table 1). Vertical hydraulic conductivity (HC_V) was measured on-site with a single-ring infiltrometer (Flow-Group Comp.). Low values of HC_V are contrasted by the high effective porosity. This contrast is caused by relatively few macropores compared to other soils and a high percentage of small pores, which are mostly not active during the infiltration process. A low HC_V in the topsoil is supposed to generate surface flow – likely saturation overland flow (SOF) and possibly Hortonian overland flow (HOF) to a minor extent – or near-surface biomat flow (BMF; Sidle et al., 2007) during high intensity storms. However, at the mineral soil hillslope PZ no surface flow has been observed even during large storms with daily precipitation sums of up to 80 mm. At the organic soil hillslope PB, surface flow can be observed at times when the peat bog is saturated.

In general, the dominant runoff formation process in most forested mountain headwater catchments can be described as SSF (Weiler et al., 2006). However, site-specific soil types and their properties, such as those at our test site, peat bog (PB3) and entic Podzol (PZ2), may result in a characteristic and possibly unique combination of runoff formation processes. Based on our field surveys and soil mapping using the *Hydrology of Soil Types* classification (HOST, Boorman et al., 1995), the Podzol at the mineral soil hillslope PZ2 was classified as hydrological soil class 4, meaning it is a "mineral soil, aquifer > 2 m depths, no impermeable layer, consolidated substrate" and conceptual runoff formation model A. The latter model implies "The dominant water movement is downwards through the vadose zone to an aquifer at least two meters below the surface. Lateral movement is largely confined to the saturated zone, with the hydrological response being controlled by the flow mechanisms of the substrate. Where the rock is more coherent but deeply weathered or fissured, the dominant flow is via the fissures as the bulk of the rock is only slightly porous at best. Aquifers or groundwater are more rarely found in this group" (Boorman et al., 1995). The peat bog at the organic soil hillslope PB3 was classified as hydrological soil class 12 meaning organic soils, no significant aquifer, raw peaty topsoil, substrate raw peat, upper soil layers remain saturated for much of the year and conceptual model K, which implies "Where there is deep peat, the flow is dominated by surface and immediate subsurface flow, with the underlying substrate having little influence on the hydrological response" (Boorman et al., 1995).

2.2 Hydrologic conditions of the Rokytka headwater

In these catchments, the retention ability depends mainly on the shallow depth of the phreatic zone in the peat bog, whereas the deep phreatic zone in the Podzol plays a minor role (Vlček et al., 2012). Peat bog areas are hypothesized to control storm runoff formation in these headwaters. Peat bogs can significantly contribute to stormflow when the peat is fully saturated, i.e., storm events exceeding a threshold of 10–15 mm (Vlček et al., 2016). According to a geochemical study based on 2 years of monthly stream water sampling (Kocum et al., 2016), peat bogs contribute only 10 % to baseflow at the outlet of the entire Rokytka catchment (3.8 km^2).

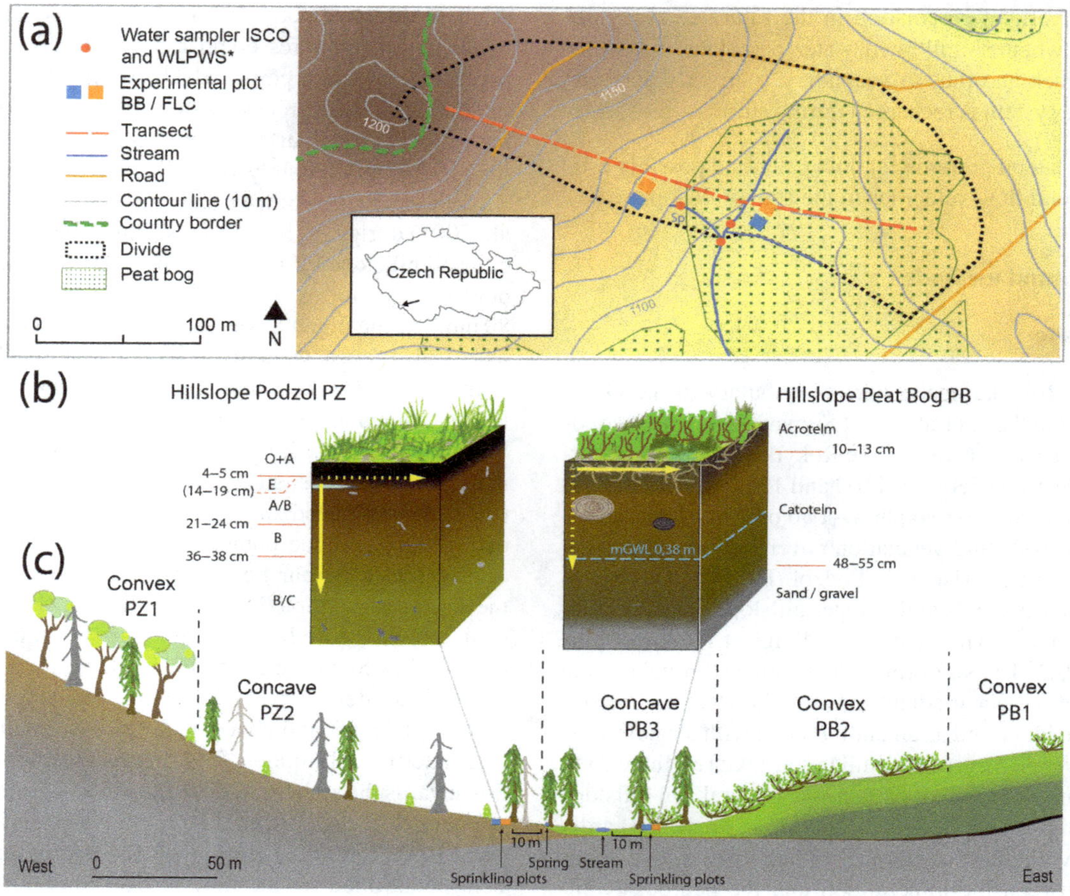

Figure 1. (a) Overview of the Rokytka headwater test site (0.6 km^2) in the Šumava Mts. in southwestern (SW) Czech Republic; Sp = spring; * Water-level proportional water sampler (Schneider et al., 2013). **(b)** The yellow arrows at the experimental Podzol plot PZ2 and the peat bog plot PB3 represent simplified conceptual models of assumed runoff formation during stormflow according to the *Hydrology of Soil Types* classification (HOST; Boorman et al., 1995): The soil type entic Podzol at plot PZ2 matches HOST soil class 4 with conceptual model A; the soil type peat bog at plot PB3 matches HOST soil class 12 with conceptual model K (Fig. 8). **(c)** Cross section of the Rokytka headwater and its two opposite hillslopes. The mineral soil hillslope Podzol (PZ) consists of entic or at least haplic Podzol and is covered by beech and spruce stands. The organic soil hillslope (PB1-PB3) represents a typical peat bog (PB) of the Šumava Mts. The upper hillslope zone PB1 (cotton grass), the lower hillslope zone PB2 (pine) and the riparian zone PB3 (waterlogged spruce forest) represent zones of the PB catena with different vegetation cover, groundwater and peat soil depths.

However, some zones of a peat bog area, such as springs or soil pipe systems connected to the stream, exhibit high fluctuations in discharge (Holden and Burt, 2002). This could explain the observed spiky storm hydrographs at the entire Rokytka catchment outlet (3.8 km^2) and at the Rokytka headwater test site (0.6 km^2). Pronounced discharge fluctuations from Peat areas could be caused by surface flow (field observation at the Rokytka peat bog), near-surface (Holden et al., 2001; Holden, 2005) or subsurface stormflow in soil pipes (Rizzuti et al., 2004; Holden, 2005; Gerke et al., 2015). Results of Holden and Burt (2003) at a blanket Peat site showed that near-surface flow (Biomat flow, BMF) up to the depth of 10 cm can contribute more than 90 % to the plot's outflow. BMF is defined as lateral stormflow in the organic litter layer with high porosity and high hydraulic conductivity in the topsoil (Gerke et al., 2015).

Storm hydrographs at the Rokytka headwater are highly volatile and are characterized by quick and steep rising and falling limbs. The hydrologic response to rainfall events is fast and the recession to antecedent baseflow occurs rather quickly (Fig. 2). The average annual mean flow Mq at the Rokytka headwater outlet is about 0.098 mm h^{-1} (860 mm a^{-1}); yet at 330 days of the year (> 90%) the discharge is lower. Compared to the average, the hydrologic year 2015 (1 November–31 October) was a rather dry year with total annual precipitation of 840 mm (long-term average 1220 mm) and total runoff of 580 mm or 0.07 mm h^{-1} (long-term average 860 mm). Mean annual maximum flow MHq is 2.24 mm h^{-1} and mean annual minimum flow MNq

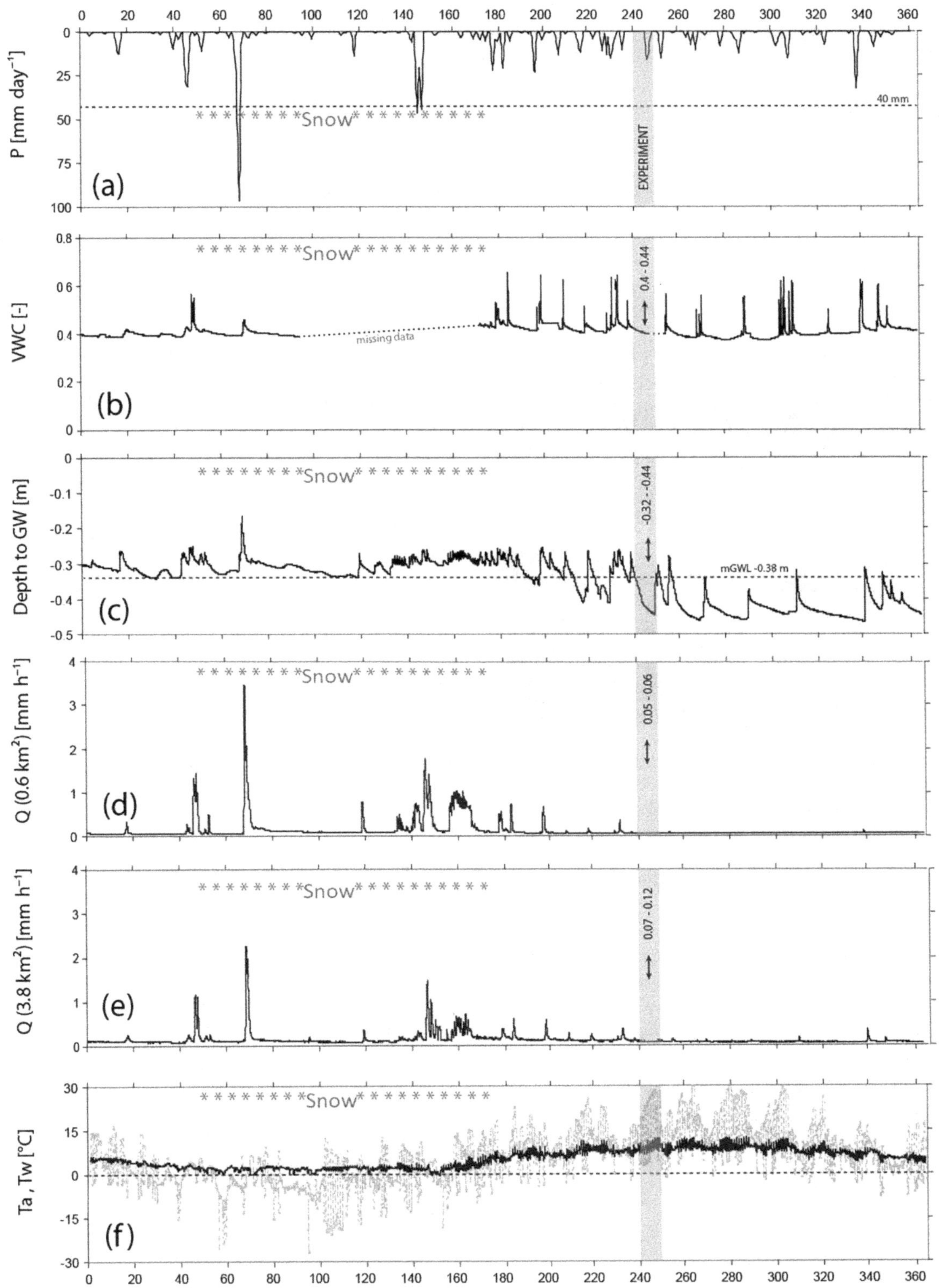

Figure 2. Hydro-meteorological data of the water year 2015 (1 November 2014–31 October 2015). **(a)** P: precipitation [mm d^{-1}]. **(b)** VWC: volumetric water content at hillslope PZ in 0.2 m depth [–]. **(c)** Depth to groundwater level at slope PB3. **(d)** Q (0.6 km^2): discharge [mm h^{-1}] of Rokytka headwater. **(e)** Q (3.8 km^2): discharge [mm h^{-1}] of Rokytka catchment (3.8 km^2). **(f)** T_a: air temperature [°C] (dashed gray line), T_w: stream water temperature [°C] (solid black line). The gray bars mark a 10-day period starting with dye tracer sprinkling in the evening of 29 June (day 240 of the water year) and 8 July (day 249 of the water year).

Table 2. Soil characteristics at the two experimental plots PZ and PB in the Rokytka headwater. PZ is Podzol (mineral soil), PB is peat bog (organic soil); RH is coverage of a selected soil type at the Rokytka Headwater; OR is coverage of a selected soil type at the Otava River Headwater; * peat in general is Histosol according to IUSS Working Group WRB (2006); ** depth at hillslope PB3; SL is sandy loam; L is loam.

Slope	Soil type	Soil area [%] RH/OR	Soil depth [m] RH/OR	Depth to [m] groundwater	Slope [%]	Soil layer	Eff. [%] porosity	Hydrol. [mm h^{-1}] cond.	Soil texture
PZ	entic Podzol	51/45	0.7/0.4–0.7	>2.0	4–6	A/B	80.0	0.50	SL-L
						B	53.5	0.38	SL
PB	Histosol*	44/20	2.1/1.2**	0.35	3–5	T	89.2	0.30	–

Figure 3. Pictures of the application plots of the dye-sprinkling experiments at the Rokytka headwater, Šumava Mts., Czech Republic. **(a, b)** Sprinkling plots at the mineral soil hillslope Podzol (PZ2). **(a)** Brilliant Blue plot (BB); **(b)** Fluorescein sodium plot (FLC). **(c, d)** Sprinkling plots at the organic soil hillslope peat bog (PB3). **(c)** Brilliant Blue plot (BB); **(d)** Fluorescein sodium plot (FLC).

is 0.04 mm h^{-1}. The peak discharge Hq_p in 2015 reached 3.5 mm h^{-1} (Fig. 2); thus, the ratio Hq : Mq $\approx$ 50 is relatively high. From June to October 2015 daily precipitation rarely exceeded 10 mm d^{-1} and thus storm-flow events were unfrequent and small. Yet, the Rokytka creek did not fall dry due to two persistent springs at the mineral soil hillslope PZ. In contrast, springs at the organic soil hillslope PB are susceptible to desiccation.

2.3 Dye tracer experiments

The dye tracer experiments were carried out at both hillslopes (mineral soil slope PZ2 and organic soil slope PB3) of the Rokytka headwater during baseflow conditions in late June 2015. At each hillslope two 1.5 m × 1.5 m plots were sprinkled with both dyes (Brilliant Blue, CAS no. 3844-45-9, concentration 5 g L^{-1}; Fluorescein sodium, CAS no. 518-47-8, concentration 2 g L^{-1}). All sprinkling plots are located at the transition between the concave, lower part of the hillslope and riparian zone in the vicinity of the stream (distance to stream $\approx$ 10 m; Fig. 3).

First, all plots were pre-sprinkled with 45 L ($\approx$ 20 mm) of local stream water to raise soil moisture and connect the pathways for water percolation, and then the plots were sprinkled with 45 L ($\approx$ 20 mm) of dye solution. The overall sprinkling time at each plot was $\sim$ 2 h in order to simulate a rainfall intensity of 20 mm h^{-1}. These amounts and intensities represent a heavy rainfall storm in the Šumava Mts. Due to previous rainfall events, the soil moisture ranged between 0.40 and 0.45 VWC; 40 mm rainfall usually causes significant stormflow and also represents frequently occurring amounts of daily precipitation in central Europe low mountain ranges (Hümann et al., 2011). Storms of this magnitude occur 2–3 times in an average year in the Šumava Mts. (Fig. 2). The groundwater level in the peat bog was initially about 0.35 m below the terrain surface (Fig. 2), which represents average peat bogs summer groundwater levels at the near-riparian organic soil hillslope.

pH is an important parameter when using FLC as a soil staining dye for preferential flow identification (Gerke et al., 2008, 2013). The soils in the Šumava Mts. are characterized by low pH values ranging from extremely acidic pH 3.8 for peat bogs to moderately acidic pH 6.0 for Cambisols. In the Rokytka headwater soil pH is very strongly acidic pH 4.8 at the peat bog plot PB3 and strongly acidic pH 5.4 at the mineral soil plot PZ2. Therefore, a NaOH solution was added during the initial pre-sprinkling with dye-free water (pH 12) to raise and buffer the soil pH to reduce fluorescence suppression of FLC caused by a very strongly acidic environment.

The experiment continued with excavation of the FLC-sprinkling plots. The excavation of soil profiles and the photography of FLC-stained soil structures were performed under short-time ultraviolet (UV) illumination (410 nm) at night, approx. 4 h after sprinkling, as FLC is strongly light sensitive (Käss, 1998).

In the following we describe the soil profile excavation and photo documentation procedure performed after the sprinkling step by step. First, we visually surveyed the terrain surface and all micro-depressions downslope of the sprinkling plots along the primary topographic gradient (thalweg) towards the stream to identify potential preferential flow pat-

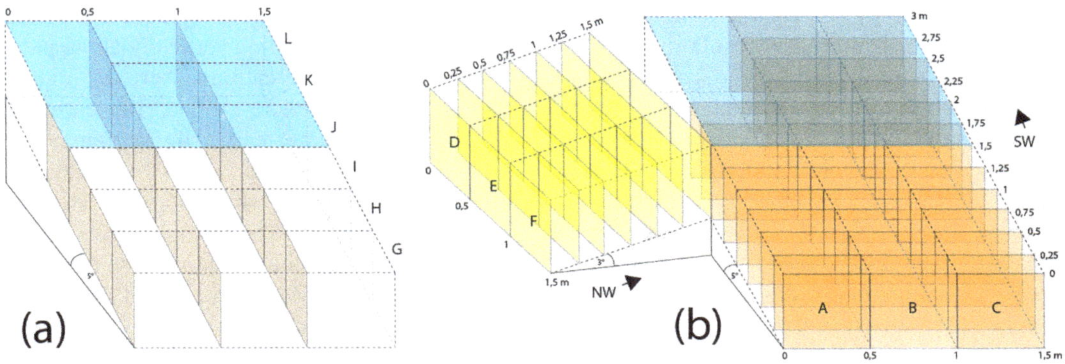

Figure 4. Sketch of the dye tracer experiment excavation directed along the flow lines. **(a)** Scheme of excavation and photography of lateral soil profiles (gray) parallel to the lateral–horizontal preferential flow paths (rows G–L). Blue areas show the dye-sprinkling plots, where "platform" images where taken of horizontal layers of the soil (these layers have the same orientation as the soil layer boundaries). **(b)** Scheme of excavation and photography of frontal soil profiles (orange and yellow sections) up the slope ("columns" A–F). The yellow inclined part (columns D–F) was only excavated and photographed at plot PB3 to capture the dominant lateral–horizontal preferential flow paths.

terns near or at the surface. In case of visible dye patterns at the surface, exploratory frontal soil pits, perpendicular to the horizontal/lateral flow, would start at the stained surfaces nearest to the stream. Second, the systematic frontal profile excavation started 1.5 m downslope from the sprinkling plots outside of the sprinkling area along the projected flow direction along the thalweg (Fig. 4). Third, the preplanned systematic excavation was extended along possible secondary gradients using additional exploratory frontal profiles at local terrain depressions. These were excavated up to 10 m down the slope to find potential secondary flow directions and to identify the maximum distance of dye-stained flow structures. Fourth, lateral profiles, oriented parallel to the flow, were excavated in a systematic way at all places where the dyes were detected outside of the sprinkling area. Fifth, the sprinkling area was excavated in a predefined systematic way (Fig. 4). The pictures of soil profiles were taken from two sides in frontal and lateral orientation. Frontal images were taken towards the slope (perpendicular to the direction of horizontal–lateral flow), whereas lateral images were taken along the slope (parallel to the direction of horizontal–lateral flow). The size of each image (photograph) was 50 cm × 50 cm.; lateral pictures (Fig. 4a) were taken in soil pit profile lines every 0.5 m (G–L), frontal pictures in soil pit profile rows at every 0.25 m (Fig. 4b). A similar system of excavation was used by Schneider et al. (2014) and Gerke et al. (2015).

Pictures of the soil profiles were taken during the excavation with a digital Micro Four Third camera with a crop factor of 2.0 (Panasonic Lumix DMC-G1 with a 12 MP MOS sensor, 13 mm × 17.3 mm sensor area and a 14–45 mm zoom lens, mostly at a focal length of 14 mm, equivalent to 28 mm in full format/35 mm film) under daylight conditions beneath a shading tarp to avoid direct sunlight and shadow effects in case of the BB plots. White balance, white point and black point reference were calibrated using a Datacolor Spider Cube, which was placed in every image. Pictures at the FLC plot were taken at night with the same camera. Each FLC soil profile was illuminated separately with two light sources:

1. a 500 W Halogen lamp (approx. light temperature 5500 K in the visible spectra with maximum 550 nm) to document the natural soil profile color with its horizons;

2. a 27 W UV LED lamp (9 × 3 W UV LEDs, Highlite International BV Comp.) producing UV light 410 nm with a 120° beam angle to visualize fluorescent FLC-stained soil structures similar to Gerke et al. (2013).

The dye-stained flow patterns for both dyes BB and FLC at all soil profiles were analyzed according to a method and with an analytical tool described by Weiler and Flühler (2004). This method was originally developed for analyzing BB. Therefore, the color space of the photographs is converted from the red–green–blue (RGB) color space taken by the camera sensor into the hue–saturation–value (HSV) color space and then classified and spatially analyzed with an algorithm written in IDL code (Weiler and Naef, 2003). For the Rokytka experiments, this procedure was applied for both dyes, BB and FLC, thus for two different groups of photographs. To detect and analyze FLC in the soil profile photographs similarly to the BB photographs, the dye detection routine in the original IDL code was adapted for optimal FLC identification.

2.4 Headwater stream and spring sensing and sampling

All sprinkling plots were located approx. 10 m away from the headwater stream in the lower concave part of the hillslope.

Rainfall runoff data at the Rokytka catchment indicate that 10–15 mm of precipitation produce a noticeable response in the stream when the peat bog is in moist, near-saturated conditions (Vlček et al., 2016). Hence, the amount of sprinkling water (40 mm per plot) simulated a rainfall that connects the hillslopes to the riparian zone and the headwater stream. Thus, the tracer FLC could potentially appear in the stream. During early summer conditions, similar to the conditions of the experiments with comparable antecedent soil moisture and groundwater levels, 40 mm of daily rainfall (on the entire catchment) rises the discharge from baseflow to peak flows on the order of 1.5 mm h^{-1}. To detect whether and when the hillslopes connect to the stream, we installed two automated water samplers (ISCO 6700) in the vicinity of the sprinkling area in the stream and one at the catchment outflow to sample stream water for future analysis (Fig. 1). Each sampler detected one hillslope. A field fluorometer (Albilia GGUN-FL30, detection limit 2×10^{-11} g mL^{-1}) was installed in the stream at the Rokytka headwater gauge in order to detect hydrologic connectivity, to capture the travel time of FLC from the sprinkling plots to the stream, and to possibly monitor the FLC break-through curve. The sampling interval was 15 min, allowing for continuous operation of the field fluorometer for 2 weeks. Moreover, water-level proportional water samplers WLPWS (Schneider et al., 2013) were installed at the same places as the ISCO samplers (two springs and at the gauge).

3 Results

3.1 Mineral soil hillslope, Podzol (PZ2)

Visible BB dye-stained patterns of lateral preferential flow in the soil profiles were observed up to a distance of 0.5 m outside the BB-sprinkling plot in the downslope direction (± along the thalweg). Additional exploratory trenches were excavated 2 and 3 m downslope from the plot to detect further dye-stained patterns of lateral subsurface flow in the soil, but no traces of BB were found. Thus, the systematic excavation started 1.5 m downslope of the BB-sprinkling plot (Fig. 4). Figure 5a shows a selected lateral soil pit profile (IL1), which is mostly within the BB-sprinkling plot at the mineral soil hillslope. BB infiltrated rather homogeneously into the upper soil horizon (O + A) and percolated rather heterogeneously deeper into the soil. Thus, BB-stained patterns are placed irregularly at lower soil horizons, some of which reached down to the B/C horizon without continuous connection to the topsoil in the excavation plain. The frontal soil profile (Fig. 5b) located outside the sprinkling plot confirms the prominence of BB-stained lateral–horizontal flow paths in the shallow subsurface, namely in the uppermost soil horizons (O + A, A/B). Some BB-stained patterns were observed at deeper soil horizons of the profile indicating preferential infiltration independent of specific soil horizon or depth. Stones and roots occur rarely and thus do not significantly modify lateral or vertical dye-stained flow structures. BB was transported vertically along patches in the soil matrix, typically at locations where the higher soil horizons O + A and/or A/B were stained. These vertical preferential flows were rather created by slight differences in texture and porosity of the soil matrix than by vertical macropore structures such as root networks or burrows. There were no visible vertical macropores. The root systems of the spruce stands are mostly limited to a depth of 0.2 m (O + A and A/B horizons) and traces of edaphone are very limited. The local soil at profile AC0.5 (Fig. 5b) was not pre-sprinkled and thus likely less saturated when compared with the lateral profile located within the sprinkling plot (Fig. 5a). This may explain the more pronounced dye-stained vertical flow structures in the AC0.5 frontal profile compared to the pre-sprinkled IL1 lateral profile.

The dye FLC was not visible at the soil surface outside of the sprinkling plot. First exploratory trenches were excavated at a distance of 2 m downslope of the sprinkling plot; none of them showed any dye-stained patterns in the soil profiles. The systematic excavation started 1.5 m downslope from the sprinkling plot similar to the BB plot at hillslope PZ2 (Fig. 1). FLC dye-stained patterns are located without any visible link to the soil horizons, roots or stones (Fig. 6a and b). FLC dye was sparsely distributed in the topsoil horizons (O + A, A/B). Almost no dye-stained patterns were found in the lateral soil profiles, especially in the lower soil horizons, within the sprinkling plot area (e.g., Fig. 6a). The largest occurrence of FLC dye-stained soil patterns were found at frontal profile AC0.5 in the Bh soil horizon - horizon enriched with humus substances (Fig. 6b).

The smallest stained spots (< 1 cm^2) were likely caused by UV light reflected from small grains of quartz from the weathered gneiss. The adopted version of the FLC dye classification algorithm could not distinguish these pixel-scale patches from truly FLC dye-stained features. However, this analytical bias has a negligible impact on the main findings concerning the FLC dye-staining in the mineral soil profiles (PZ2).

3.2 Organic soil hillslope, peat bog (PB3)

The visual survey of the soil surface in the vicinity of the BB-sprinkling plot revealed near-surface flow in the northwestern (NW) direction towards the stream. BB was detected in a small, water-filled depression 10.5 m downslope from the sprinkling plot. A few excavations were performed from this point uphill towards the dye-sprinkling plot; systematic excavation and profile photographic documentation started 1.5 m downslope of the plot with a spacing of 0.25 m (Fig. 4a, yellow section). The BB-stained flow paths did not strictly follow the terrain gradient but went from the NW side of the sprinkling plot and followed mostly lateral preferential flow structures formed by decomposed trees or roots. This lateral preferential flow path was later identified as the main

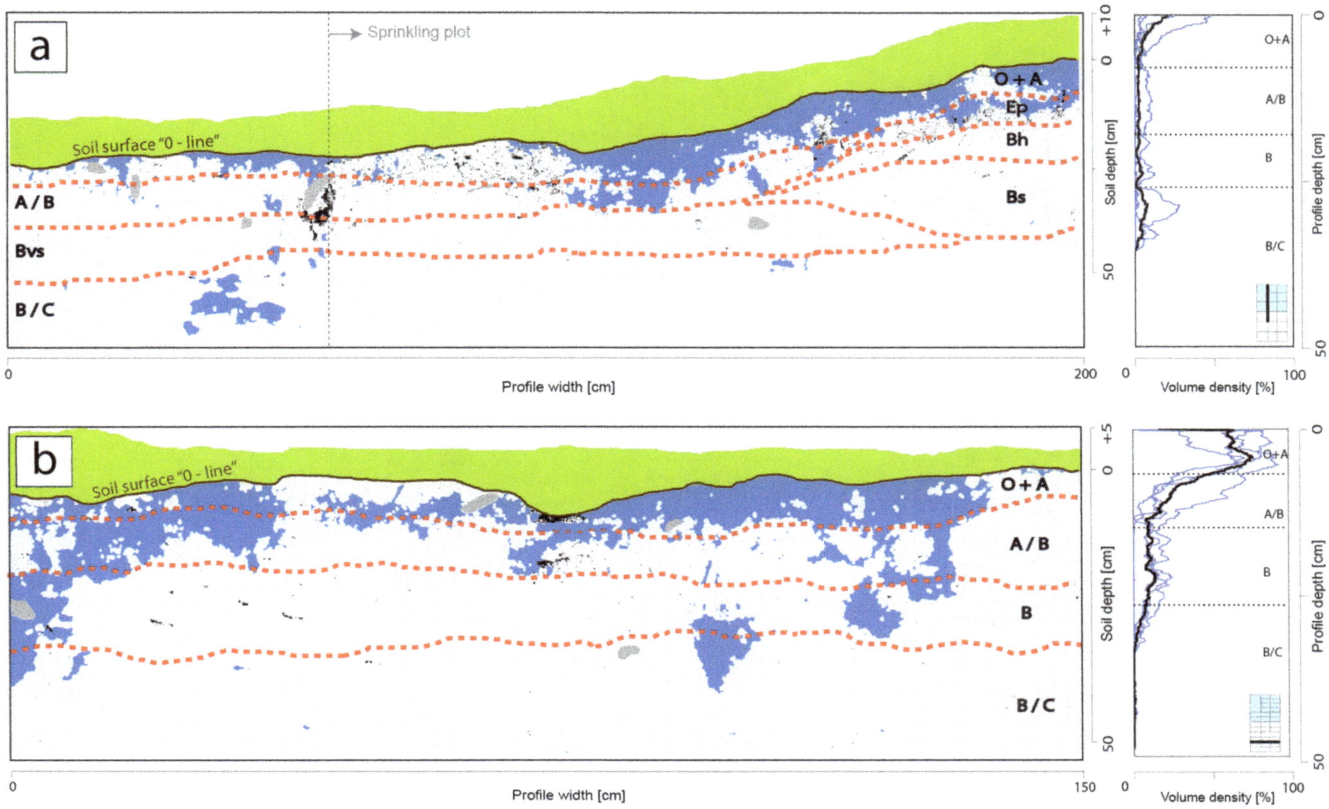

Figure 5. **(a)** Lateral profile IL1 and **(b)** frontal profile AC0.5 at the Brilliant Blue (BB)-sprinkling plot PZ2 at the mineral soil hillslope (Podzol). The position of the profile is indicated in bottom right corner. Blue: BB dye; gray: stones, roots; green: vegetation; black: unclassified shadows, roots; red-dotted line: soil horizon divide. The charts on the right represent the distribution of the volume density of BB in different soil depths. Ep is elusive podzol horizon; Bh is podzol horizon enriched with humus substances; Bs is podzol horizon with significant sesquioxide content.

direction of the subsurface flow. Relatively smaller and less stained flow paths were observed downslope from the sprinkling plot (Fig. 4a, orange section), with BB disappearing 2 m from the sprinkling plot. The BB excavation started the day after the sprinkling, yet dye stored in large macropore pockets started flowing down the trench walls when these soil structures were truncated. BB followed lateral soil pipes that were formed by decomposed roots or fallen trees. Healthy trees and undecomposed timber did not create such effective lateral preferential flow paths; therefore, they had no significant impact on dye-stained patterns (Fig. 7). BB created clearly detectable dye-stained patterns on the dark peat particles as well, so the major flow paths of BB could be detected even several days after the dye application. The excavation of BB-stained soil patterns at the organic soil hillslope PB3 proceeded from two directions (NW and SW, Fig. 4a) following the stained flow paths in the soil. Near the sprinkling plot, most of the dye was detected at the surface and in near-surface soil horizons, which correlates with acrotelm (Fig. 7). About 2.0 m downslope from the BB-sprinkling plot at hillslope PB3 (Fig. 1), the dye-stained patterns diminished in the acrotelm and were observed mainly in and around macropores in the catotelm. The excavation caused problems at location FD0.25/profile D 0 – 1 as dye-filled macroporous pockets in the soil drained BB when disturbed during excavation and thus secondarily stained these soil profiles. Such secondary patterns were cleaned to minimize falsely detected BB along the excavation front.

The excavation of the FLC-sprinkling plot at PB3 at the organic soil hillslope (peat bog) started 3 h after sprinkling after dark. Repeated visual surveys of the terrain surface and of the excavated soil trenches downslope of the sprinkling plot using the UV lamp (410 nm) and UV torches (385 nm) did not detect any traces of FLC at the surface or in the soil pits. Even in the soil horizons within the sprinkling plot, no FLC dye-stained flow patterns were identified. Only parts of the vegetation at the surface of the sprinkling plot itself were visibly stained with FLC. FLC was later detected in a small water-filled depression about 2.5 m downslope from the sprinkling plot PB3.

Traces of FLC were not detected in the headwater stream, neither in any water samples taken by automated water sam-

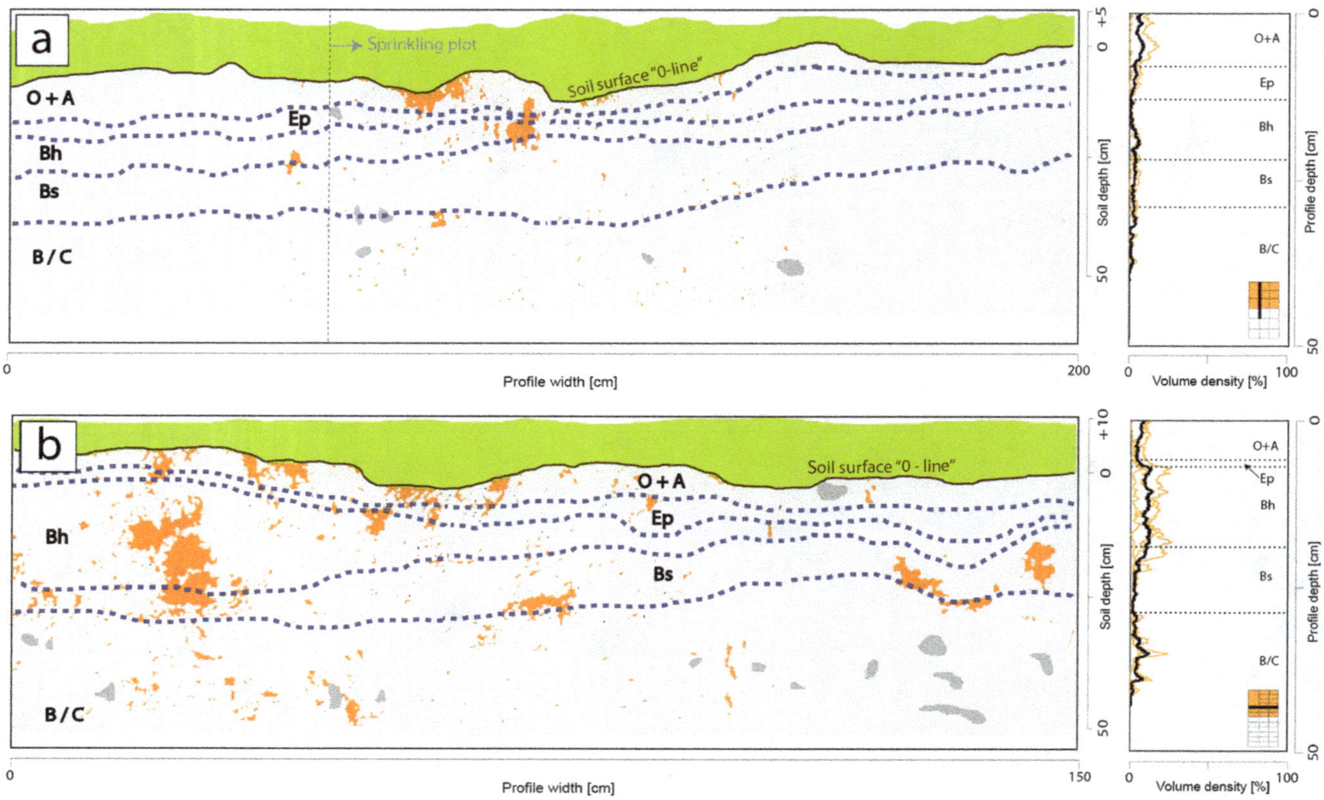

Figure 6. **(a)** Lateral soil profile IL1 and **(b)** frontal soil profile AC0.5 at the Fluorescein sodium (FLC)-sprinkling plot PZ2 at the mineral soil hillslope (Podzol). The position of the profile is indicated in bottom right corner. Orange: FLC dye; gray: stones, roots; green: vegetation; black: unclassified shadows; dark-blue dotted lines: soil horizon divide. The charts on the right represent the dye volume in different depths below the soil surface.

plers (ISCO 6700) or water-level proportional water samplers WLPWS, nor with the field fluorometer during the 14-day monitoring period following the dye application.

4 Discussion

Hillslope hydrology is concerned with the partition of precipitation as it passes through the vegetation and soil between overland flow and subsurface flow (Kirkby, 1988). Runoff formation at hillslopes (zero-order basin response) and in riparian zones (first- and second-order stream response) are the main controls defining the hydrologic response of mountainous headwaters in humid climates. Vertical processes and lateral connectivities are strongly dependent on soil properties, thus soil types are a major regulator of the interplay of these processes. Soil types can be classified in hydrologically meaningful terms, e.g., according to HOST (Boorman et al., 1995).

Hillslope processes define how small catchments respond to rainfall (Anderson and Burt, 1990). Specifically, hillslope processes control how long water is stored in soil or bedrock, which determines how quickly small catchments respond to rainfall (Uhlenbrook et al., 2008). Our experiments in the Šumava Mts. showed that the peat bog hillslope connected much more quickly to the stream and contributed considerably more to the runoff response of the headwater than the Podzol hillslope. The larger Rokytka catchment ($3.8\,km^2$, third-order stream) showed similar hydrologic behavior (Fig. 2) – low baseflow and flashy storm hydrographs – to the smaller Rokytka headwater ($0.6\,km^2$, second-order stream). This is noteworthy since the proportion of the peat bog ranges from 60 % at the second-order stream headwater to less than 30 % at the third-order stream catchment; the remaining areas are covered by Podzol. This illustrates that the hydrologic response of the catchment is dominated by the runoff formation at the peat bog, whereas we speculate that a deep groundwater system is fed by percolation in the Podzol that is rather disconnected from the second- and third-order streams.

4.1 Mineral soil hillslope, Podzol (PZ2)

Based on the properties of Podzol, surface flow (SOF or HOF) at the mineral soil hillslope PZ was not expected. This hypothesis was supported by the results of the BB staining

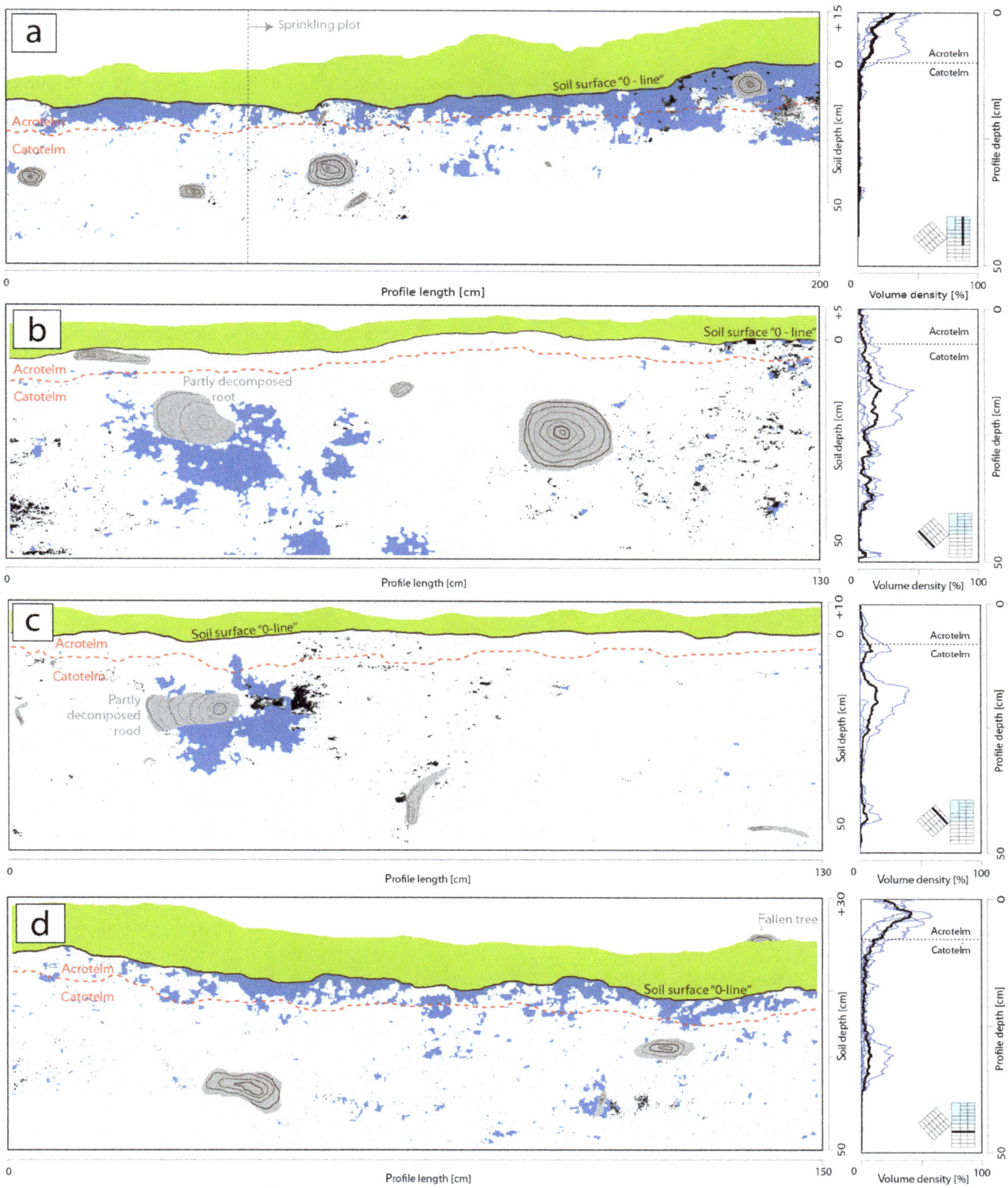

Figure 7. **(a)** Lateral soil profile IL0.5, and **(b)** frontal soil profile FD0.25 at the Brilliant Blue (BB)-sprinkling plot PB3 at the organic soil hillslope (peat bog). The position of the profile is indicated in bottom right corner. Blue: BB dye; gray: roots; green: vegetation; black: unclassified shadows; red-dotted line: soil horizon divide. The charts on the right represent the vertical distribution of the volume density of the BB. The reduced width of the soil profiles shown in Fig. 7b and c are due to the presence of a tree stump on the right side of these profiles. **(c)** Profile FD1.25 and **(d)** frontal profile AC1 at the BB-sprinkling plot PB3 at the organic soil hillslope (peat bog). The position of the profile is indicated in bottom right corner. Blue: BB dye; gray: roots; green: vegetation; black: unclassified shadows; red-dotted line: soil horizon divide. The charts on the right represent the vertical distribution of the volume density of the BB. The reduced width of the soil profiles shown in Fig. 7b and c is due to the presence of a tree stump on the right side of these profiles.

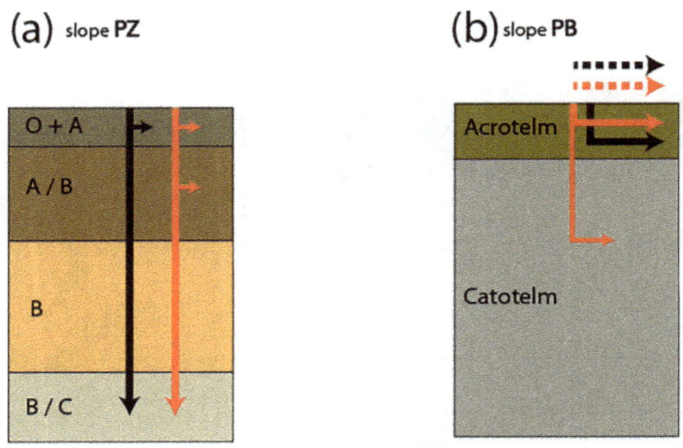

Figure 8. Conceptual models of runoff formation and (subsurface) stormflow for the organic soil hillslope (peat bog PB3) and the mineral soil hillslope PZ (Podzol) derived by parallel plot sprinkling experiments with the conservative dye Fluorescein sodium (FLC) and the sportive dye Brilliant Blue (BB) at the Rokytka headwater. Detected subsurface-flow components at the PZ hillslope **(a)** biomat flow, shallow lateral subsurface flow and mostly deep percolation (vertical), and at the organic soil hillslope PB (peat bog) **(b)** biomat flow at short distances and mostly lateral pipe flow following decayed tree-root systems with long lateral subsurface-flow distances. The dashed arrows represent surface flow (saturation overland flow, SOF), which could not be detected during the experiments but has been observed on-site during natural storm events. Black arrows: hypothesized runoff formation processes according to the *Hydrology of Soil Types* classification (HOST; Boorman et al., 1995); Red arrows: results of the Šumava experiments.

patterns in the soil profiles at PZ2. However, the dye did not identify a hydrologic active soil horizon as clearly as with the Gleysol hillslope study (Schneider et al., 2014). The most abundantly stained soil structures (volume density of up to 85 %, Fig. 5b) were found within the uppermost soil horizon O + A, up to a depth of 0.1 m below the surface. The shape of the BB depth distribution was similar at the plot; however, the volume density of max. 50 % is significantly lower (Fig. 5a). Based on the soil properties at the mineral soil hillslope PZ2, such as porosity and hydraulic conductivity, infiltration was expected to be rather stable and homogeneous. However, BB infiltrated heterogeneously. Parts of the topsoil created conditions for the occurrence of fingering (DiCarlo et al., 2013) or similar types of matrix preferential flow (Weiler and Flühler, 2004; Anderson et al., 2009; Wienhöfer et al., 2009). Similar to previous work, we attribute the heterogeneous infiltration to the corrugated transition between the dark organic topsoil horizons (O, A) and the lower mineral horizons (A/B, B). However, Weiler and Flühler (2004), Anderson et al. (2009) and Wienhöfer et al. (2009) conducted their studies at steeper slopes and on different soil types compared to the Šumava experiments. A sharp interface between an upper organic and a lower "organo-mineral" layer like in the Šumava sprinkling experiments can initiate significant biomat flow (Gerke et al., 2015) that can be attributed to water repellency (Doerr et al., 2000). Water repellency of the soil surface was observed in the organic topsoils at the Rokytka headwater during dry periods. However, due to rainfall events prior to our experiments, the antecedent soils moisture conditions (0.45–0.5) likely did not support water repellency in the topsoil (Fig. 2). At PZ2, BB was clearly visible at deeper soil horizons, where it created seemingly detached stained patterns. These stained patterns represent vertical preferential flow paths (Nobles et al., 2010; Gerke et al., 2015; Uchida et al., 2005), but the excavation spacing (0.25–0.5 m) was probably too coarse to detect connected stained flow paths in full detail. Horizontal–lateral preferential flow paths dominated at the topsoil layers (O + A), whereas vertical flow directions (percolation) dominated at the lower soil horizons. These differences in the direction of stained pathways are consistent with the results of Schneider et al. (2014) or Gerke et al. (2015).

FLC stained significantly fewer soil structures or pathways compared to BB at both hillslopes. FLC was almost absent with few exceptions in the near-surface organic topsoil layers. This could be because of the relatively small amount of dyed water (20 mm dyed sprinkling water and 20 mm undyed pre-sprinkling water). Previous FLC soil staining experiments used simulated rainfalls of 50–100 mm (Gerke et al., 2015). The size of the irrigation plots (1.5 m × 1.5 m) does not appear to be a factor, as previous work used irrigation plots that were 1 m × 1 m, and successfully detected FLC (Gerke et al., 2015). The FLC dye solution could have bypassed the topsoil horizons via macropores and soil pipes without visibly staining these preferential flow structures due to various causes, e.g., local hydrophobicity (Doerr et al., 2000) or strong acidity. On the other hand, FLC-stained patterns in the lower soil horizons at the Podzol plots are similar to the BB-stained patterns. This indicates, that the organic topsoils at the Šumava test sites may suppress the fluorescence of FLC in addition to the well-known pH induced fluorescence elimination. At the Podzol hillslope PZ2 the FLC-stained soil patterns suggest rather a subsurface lateral pipe-flow network as described in Uchida et al. (2005) than biomat flow as identified by Gerke et al. (2015).

4.2 Organic soil hillslope, peat bog (PB3)

BB at the peat bog hillslope PB3 did not detect surface flow SOF or significant vertical deep percolation. BB patterns in the soil profiles supported the hypothesis of near-surface biomat flow, which can be attributed to lateral preferential flow in the acrotelm. Differences in porosity and hydraulic conductivity between acrotelm and catotelm create similar stained patterns when compared to the Gleysol hillslope studied in the experiment of Schneider et al. (2014). The infiltration process at the hillslope PB3 matches fairly well the definition of lateral subsurface stormflow (Wienhöfer et al., 2009). Lateral preferential flow was detected with BB at PB3; however, it was limited mostly to few, but well-connected pipe networks with high drainage capacity. These soil pipe networks are created by decomposed dead trees or dead roots in the Acro- and catotelm. Such lateral soil pipe networks are an important runoff formation process at peaty catchments (Jones, 1997; Holden and Burt, 2002). Holden and Burt (2003) identified the dominant lateral stormflow as shallow subsurface flow (SSF) down to the depth of 0.1 m. However, BB at the organic soil hillslope PB3 showed that lateral soil pipes were connected both in Acro- and catotelm (depth 0.1–0.4 m) to jointly form the major preferential flow paths through the peat bog. Hence, during the BB excavation soil pipes were only observed in the catotelm. BB dye-stained flow paths in the soil appeared prominently further downslope (outside of the dye-sprinkling plot) and connected mostly laterally via soil pipes rather than vertically penetrating the acrotelm. However, most soil profiles – both frontal and lateral – document that the BB dye-stained flow paths are rather limited to macropore structures and rarely to matrix flow. Deep percolation at the organic soil hillslope PB3 was not detected. The assumption, that BB may be difficult to optically detected in such dark soils as peat bog, was not confirmed. It could be shown that BB can be successfully applied in peat bogs to stain vertical flow structures in soil profiles and has the potential to trace rather long-distance lateral preferential flow paths (distances > 10 m) in waterlogged areas with shallow groundwater (~ 0.5 m below surface).

The critical role of pH affecting Fluorescein's fluorescence and thus its on-site optical detection is well known (Gerke et al., 2013; Käss, 1998). Therefore, the FLC solution was buffered with NaOH to compensate for the strong acidity of the organic soil, similar to the approach successfully applied in FLC staining experiments in organic topsoils in Japan (Gerke et al., 2015). The rarely detected FLC at the peat bog hillslope does not necessarily indicate that no dye infiltrated into the upper organic layer as the dye's fluorescence could not be optically detected in situ. The FLC solution could have bypassed the organic topsoil horizons via macropores and soil pipes at the peat bog plot PB3 as well as at the Podzol plot PZ2. The very low soil pH in the peat bog might be the reason why FLC fluorescence was not observed. Although we attempted to increase the soil water pH by pre-sprinkling the plots with NaOH-enriched water, and similarly buffered the FLC solution, these countermeasures at the sprinkling plots were probably not enough to significantly change the pH conditions in the soil at both hillslopes, but especially at PB3. FLC was mostly visible on organic surfaces such as plants (moss, grass) on the surface but not in the peat itself. This might be attributed to the tendency of FLC to attach to organic matter, which significantly counteracts its conservative behavior in mineral soils, especially in thick organic soils such as peat bogs. Furthermore, the "very strong acidity" in the peat bog soil, the phreatic zone and the capillary fringe together may have diluted any buffering effect of the NaOH-enriched sprinkling water. However, FLC dye was visible the next day during daylight in a small, water-filled depression approximately 3.0 m downslope from the FLC-sprinkling plot. This indicates that the dye had been transferred via preferential flow in lateral soil pipes in the vadose zone and had not been completely (ab)sorbed by organic soil particles.

4.3 Plots–stream hydrologic connectivity with Fluorescein

Limitation and possible causes of failure of FLC staining and tracing experiments have been well described for organic and mineral soils by Gerke et al. (2008) and for groundwater and surface-water applications by Käss (1998). The most common problems were addressed for the Šumava experiments by

- pre-sprinkling of the plot with NaOH buffered water (pH 12);
- addition of NaOH to the FLC solution to create a secondary pH buffer;
- avoiding bright sunlight as FLC is light sensitive and decays quickly, by performing the FLC sprinkling and excavation work at night with controlled short-time exposure to UV and visual spectra light.

Despite our advanced and dense FLC monitoring network in all tributaries and springs draining the hillslopes field fluorometer, automated water samplers and water-level proportional water samplers (Schneider et al., 2013), FLC was not detected in the stream or at the springs. As a result, we could neither measure a tracer breakthrough nor could we delineate a transit time or prove hydrologic connectivity from the sprinkling plots to the drainage system. The following explanations may explain why FLC did not reach the drainage system:

- small sprinkling water volume and limited area of the sprinkling plots (1.5 m × 1.5 m) compared to the distance from the plots to the stream or springs (10 m);
- masking of the FLC fluorescence by very low background pH values (Käss, 1998);
- fluorescence reduction and masking of FLC by organic substances (Käss, 1998);
- sorption of FLC to vegetation and topsoil organic matter.

As the sprinkling solution mixed with soil and shallow groundwater, its pH probably dropped back to low background pH values. Another reason for null detection of FLC might be the relatively small amount of water sprinkled at the plot compared to the soil and groundwater volume. The influence of the capillary fringe and the soil matrix to dilute the pH-enriched sprinkling water might be significant, but the presented data cannot confirm or disprove this possibility.

4.4 Dominant runoff formation

In extension to our hypothesis, which was based on the conceptual runoff formation model, HOST model A, (Boorman et al., 1995; Figs. 1 and 8) we found that additional lateral subsurface flow in deeper soil horizons occurred at the mineral hillslope PZ2 (Podzol). As predicted, a near-surface runoff formation process occurred in parallel to deep percolation (Fig. 8). Biomat flow, a shallow subsurface flow in the topmost soil horizon O + A (Sidle et al., 2007; Gerke et al., 2015) is thus a relevant runoff formation process at the mineral hillslope PZ2 (Podzol). The absence of a shallow groundwater body and percolation-restricting soil layers buffers the lateral stormflow at the hillslope PZ2, as a large portion of precipitation infiltrates into the soil and percolates into the underlying bedrock. These findings are supported by the lack of temporary or fully saturated zones, perched aquifers or a groundwater body rising into the soil (transmissivity feedback), which may connect the hillslope to the stream and thus effectively drained the Podzol (Fig. 8). According to the runoff formation decision scheme by Scherrer and Nae (2003), the dominant runoff formation process at the Podzol hillslope can be classified as a combination of delayed HOF and delayed subsurface stormflow (SSF2). The sprinkling experiments with the dyes (BB and FLC) showed the existence of vertical and lateral preferential pathways in the topsoil and indicated that delayed subsurface-flow SSF occurs also.

At the organic soil hillslope PB3, the hypothesized HOST model K (Boorman et al., 1995) was supported by our results. Biomat flow and lateral preferential flow in soil pipe networks formed by decaying fallen trees and roots in the acrotelm and – in addition to our prediction also in the catotelm – are the primary runoff formation processes at the peat bog hillslope (Fig. 8). Our hypothesis of HOF was confirmed for the peat bog hillslope PB3 by the results of the sprinkling experiments with Brilliant Blue. It is necessary to notice that the results should be interpreted with respect to the relatively small size of the sprinkling plots (1.5 m × 1.5 m) compared to the entire hillslopes, which were not sprinkled.

5 Conclusions

In this study runoff formation during stormflow was investigated at two opposite hillslopes with different soil types (Histosol/Peat, Podzol) in a second-order mountain headwater catchment in the Czech Republic. Two dye tracers with different attributes – the sorptive dye Brilliant Blue (BB), and the conservative dye Fluorescein sodium (FLC) – were applied to stain preferential flow paths in the soil and at its surface.

At the peat bog hillslope, BB staining identified a quickly activated and effective shallow lateral subsurface drainage system in the acrotelm. Preferential flow structures connected the hillslope with the not-sprinkled riparian zone via lateral pipe flow along decayed roots and fallen trees in the acrotelm, and the upper catotelm. Healthy roots did not create similar drainage-effective, well-connected lateral preferential flow structures as decomposed roots or dead trees at the peat bog site. In contrast, subsurface flow at the Podzol hillslope was created only near surface in the organic topsoil (biomat flow). The lateral subsurface transport in the unsaturated zone at the organic soil hillslope (peat bog) was about an order of magnitude higher (10 m lateral flow) than at the mineral soil hillslope (Podzol, 1 m lateral flow). At the mineral soil hillslope (Podzol) both dye tracers worked well in the subsoil and delivered similar results. Percolation in the soil and deep percolation into the bedrock dominated, as expected; lateral preferential drainage was rather limited. Compared to the organic soil hillslope (peat bog), the lateral subsurface-flow distance is reduced by an order of magnitude (1 m vs 10 m). The findings at the Podzol hillslope with prevailing vertical flow agree with the facts that groundwater influenced soil horizons could not be detected in soil profiles.

We were more easily able to detect hydrologic connectivity from the hillslope to the riparian zone with BB than with FLC at the peat bog hillslope. This is surprising, as BB is considered to be more sorptive (less conservative) than FLC. Moreover, the dark-blue BB is often difficult to optically detect in dark organic soils like Peat. We attribute this finding to the fact that BB is less affected than FLC by the very strongly acidic soil and groundwater ($pH < 5$) found in peaty environments such as the Šumava headwaters.

Author contributions. LV, KF, and PS participated in the experiment and contributed to the manuscript.

Competing interests. The authors declare that they have no conflict of interest.

Acknowledgements. The authors would like to thank Jan Seibert from the University of Zürich for providing equipment and tools, Bohumír Janský from Charles University, Prague, for help and financial support, and Markus Weiler from the University of Freiburg for providing his IDL code for the Brilliant Blue image analysis. We are indebted to Václav Královec, Miroslav Šobr, Michal Jeníček, Zdeněk Kliment, Luděk Šefrna, Jan Kocum, Julius Česák, Jitka Vlčková and Jiří Vlček for support and assisting with fieldwork, to Sarah Kliegman, Keck Science Department of Claremont McKenna, Pitzer, and Scripps Colleges, for helpful comments, suggestions and proofreading, and to the National Park Šumava for permission to conduct our research on its land. This work was supported by project GAČR 13-32133S "Headwaters retention potential with respect to hydrological extremes", and by project COST ES1306, LD 15130 "Impact of landscape disturbance on stream and basin connectivity".

Edited by: Markus Weiler

References

Anderson, A. E., Weiler, M., and Alila, Y., and Hudson, R. O.: Dye staining and excavation of a lateral preferential flow network, Hydrol. Earth Syst. Sci., 13, 935–944, https://doi.org/10.5194/hess-13-935-2009, 2009.

Anderson, M. G. and Burt, T. P.: Process Studies in Hillslope Hydrology, Wiley, Winchester, UK, 1990.

Bachmair, S. and Weiler, M.: Hillslope characteristics as controls of subsurface flow variability, Hydrol. Earth Syst. Sci., 16, 3699–3715, https://doi.org/10.5194/hess-16-3699-2012, 2012.

Bachmair, S., Weiler, M., and Nützmann, G.: Controls of land use and soil structure on water movement: Lessons for pollutant transfer through the unsaturated zone, J. Hydrol., 369, 241–252, ISSN 0022-1694, 2009.

Boorman, D.B., Hollis, J.M., and Lilly, A.: Hydrology of soil types: a hydrologically-based classification of the soils of the United Kingdom, Institute of Hydrology report no. 126, 1995.

Brázdil, R., Trnka, M., Mikšovský, J., Řezníčková, L., and Dobrovolný, P.: Spring-summer droughts in the Czech Land in 1805–2012 and their forcings, Int. J. Climatol., 35, 1405–1421, https://doi.org/10.1002/joc.4065, 2015.

Dicarlo, D. A.: Stability of gravity-driven multiphase flow in porous media: 40 Years of advancements, Water Resour. Res., 49, 4531–4544, https://doi.org/10.1002/wrcr.20359, 2013.

Doerr, S. H., Shakesby, R. A., and Walsh, R. P. D.: Soil water repellency: Its causes, characteristics and hydrogeomorphological significance, Earth Sci. Rev., 51, 33–65, https://doi.org/10.1016/S0012-8252(00)00011-8, 2000.

Evans, M. G., Burt, T. P., Holden, J., and Adamson, J. K.: Runoff generation and water table fluctuations in blanket peat: evidence from UK data spanning the dry summer of 1995, J. Hydrol., 221, 141–160, https://doi.org/10.1016/S0022-1694(99)00085-2, 1999.

Faist, H.: Die jährlichen Hochwasserabflüsse der Elbe bei Magdeburg im Zeitraum von 1870 bis 1960, Dokumentation der Wasserwirtschaftsdirektion Mittlere Elbe Sude-Elbe, 1967.

Ferda, J., Hladný, J., Bubeníčková, L., and Pešek, L., 1971.Odtokový režim a chemismus vod v povodí Horní Otavy se zaměřením na výskyt rašelinišť, Sborník prací HMÚ, 17, HMÚ, Praha, 22–126, 1971.

Flury, M. and Flühler, H.: Tracer Characteristics of Brilliant Blue FCF, Soil Sci. Soc. Am. J., 59, 22–27, https://doi.org/10.2136/sssaj1995.03615995005900010003x, 1995.

Gerke, K. M., Sidle, R. C., and Tokuda, Y.: Sorption of Uranine on forest soils, Hydrol. Res. Lett., 2, 32–35, https://doi.org/10.3178/hrl.2.32, 2008.

Gerke, K. M., Sidle, R. C., and Mallants, D.: Criteria for selecting fluorescent dye tracers for soil hydrological applications using Uranine as an example, J. Hydrol. Hydromech., 61, 313–325, https://doi.org/10.2478/johh-2013-0040, 2013.

Gerke, K. M., Sidle, R. C., and Mallants, D.: Preferential flow mechanisms identified from staining experiments in forested hillslopes, Hydrol. Process., 29, 4562–4578, https://doi.org/10.1002/hyp.10468, 2015.

Hladný, J.: Jaké následky přineslo sucho v roce 2003? In Němec, J., Kopp, J. (Eds.): Vodstvo a podnebí v České republice, MZE ČR, Consult Praha, 81, 2009.

Hladný, J., Kašpárek, L., Krátká, M., Kněžek, M., and Martínková, M.: Katastrofální povodeň v České republice v srpnu 2002, MŽP ČR, Praha, 68, 2005.

Holden, J.: Piping and woody plants in peatlands: Cause or effect?, Water Resour. Res., 41, 1–10, https://doi.org/10.1029/2004WR003909, 2005.

Holden, J. and Burt, T. P.: Piping and pipeflow in a deep peat catchment, Catena, 48, 163–199, https://doi.org/10.1016/S0341-8162(01)00189-8, 2002.

Holden, J. and Burt, T. P.: Hydrological studies on blanket peat: The significance of the acrotelm-catotelm model, J. Ecol., 91, 86–102, https://doi.org/10.1046/j.1365-2745.2003.00748.x, 2003.

Holden, J., Burt, T. P., and Cox, N. J.: Macroporosity and infiltration in blanket peat: The implicaions of tension disc infiltrometer measurements, Hydrol. Process., 15, 289–303, https://doi.org/10.1002/hyp.93, 2001.

Hümann, M., Schüler, G., Müller, C., Schneider, R., Johst, M., and Caspari T.: Identification of runoff processes – The impact of different forest types and soil properties on runoff formation and floods, J. Hydrol., 409, 3–4, 637–649, ISSN 0022-1694, https://doi.org/10.1016/j.jhydrol.2011.08.067, 2011.

IUSS Working Group WRB: World reference base for soil resources 2006, 2nd edition. World Soil Resources Reports No. 103. FAO, Rome, ISBN 92-5-105511-4, 2006.

Janský, B. and Kocum, J.: Peat Bogs influence on runoff process: Case study of the vydra and křemelná river basins in the šumava mountains, Southwestern Czechia, Geografie, 113, 383–399, 2008.

Jones, J. A. A.: Pipeflow contribution areas and runoff response, Hydrol. Process., 11, 35–41, 1997.

Käss, W.: Tracing technique in geohydrology: Rotterdam, A. A. Balkema, 581 pp., 1998.

Kirkby, M.: Hillslope runoff processes and models, J. Hydrol., 100, 315–339, 1988.

Kocum, J., Oulehle, F., Janský, B., Bůzek, F., Hruška, J., and Vlček, L.: Geochemical evidence for Peat Bog contribution to streamflow generation process: case study of Vltava River headwaters, Czech Republic, Hydrol. Sci. J., 61, 2579–2589, https://doi.org/10.1080/02626667.2016.1140173, 2016.

Langhammer, J., Su, Y., and Bernsteinová, J.: Runoff response to climate warming and forest disturbance in a mid-mountain basin, Water (Switzerland), 7, 3320–3342, https://doi.org/10.3390/w7073320, 2015.

Nobles, M. M., Wilding, L. P., and Lin, H. S.: Flow pathways of bromide and Brilliant Blue FCF tracers in caliche soils, J. Hydrol., 393, 114–122, https://doi.org/10.1016/j.jhydrol.2010.03.014, 2010.

Rizzuti, A. M., Cohen, A. D., and Stack, E. M.: Using hydraulic conductivity and micropetrography to assess water flow through peat-containing wetlands, Int. J. Coal Geol., 60, 1–16, https://doi.org/10.1016/j.coal.2004.03.003, 2004.

Schneider, P.: Hydrologische Vernetzung und ihre Bedeutung für diffuse Nährstoffeinträge im Hotzenwald/Südschwarzwald, Physiogeographica – Basler Beiträge zur Physiogeographie, 177, available at: edoc.unibas.ch/823/1/DissB_8420.pdf (last access: 10 December 2016), 2007.

Schneider, P., Lange A., and Doppler, T.: Development and evaluation of a water level proportional water sampler, American Geophysical Union, Fall Meeting 2013, abstract #H43H-1579, 2013.

Schneider, P., Pool, S., Strouhal, L., and Seibert, J.: True colors – experimental identification of hydrological processes at a hillslope prone to slide, Hydrol. Earth Syst. Sci., 18, 875–892, https://doi.org/10.5194/hess-18-875-2014, 2014.

Scherrer, S. and Naef, F.: A decision scheme to indicate dominant hydrological flow processes on temperate grassland, Hydrol. Process., 17, 391–401, https://doi.org/10.1002/hyp.1131, 2003.

Seibert, J., Grabs, T., Köhler, S., Laudon, H., Winterdahl, M., and Bishop, K.: Linking soil- and stream-water chemistry based on a riparian flow-concentration integration model, Hydrol. Earth Syst. Sci., 13, 2287–2297, https://doi.org/10.5194/hess-13-2287-2009, 2009.

Sidle, R. C., Hirano, T., Gomi, T., and Terajima, T.: Hortonian overland flow from Japanese forest plantations – an aberration, the real thing, or something in between?, Hydrol. Process., 21, 3237–3247, https://doi.org/10.1002/hyp.6876, 2007.

Uchida, T., Tromp-Van Meerveld, I., and McDonnell, J. J.: The role of lateral pipe flow in hillslope runoff response: An intercomparison of non-linear hillslope response, J. Hydrol., 311, 117–133, https://doi.org/10.1016/j.jhydrol.2005.01.012, 2005.

Uhlenbrook, S., Didszun, J., and Wenninger, J.: Source areas and mixing of runoff components at the hillslope scale – a multi-technical approach, Hydrological Sciences Journal – Journal des Sciences Hydrologiques, 53, 741–753, https://doi.org/10.1623/hysj.53.4.741, 2008.

Vlček, L., Kocum, J., Janský, B., Šefrna, L., and Kučerová, A.: Retenční potenciál a hydrologická bilance horského vrchoviště: případová studie Rokytecké slatě, povodí horní Otavy, JZ Cesko, Geografie, 4, 2012.

Vlček, L., Kocum, J., Janský, B., Šefrna, L., and Blažková, Š.: Influence of peat soils on runoff process: case study of Vydra River headwaters, Czechia, Geografie, 121, 235–253, 2016.

Von Freyberg, J., Radny, D., Gall, H. E., and Schirmer, M.: Implications of hydrologic connectivity between hillslopes and riparian zones on streamflow composition, J. Contam. Hydrol., 169, 62–74, https://doi.org/10.1016/j.jconhyd.2014.07.005, 2014.

Weiler, M.: Field-Based Observation of Hydrological Processes, in: Treatise on Water Science Academic Press, 339–350, 2011.

Weiler, M. and Flühler, H.: Inferring flow types from dye patterns in macroporous soils, Geoderma, 120, 137–153, https://doi.org/10.1016/j.geoderma.2003.08.014, 2004.

Weiler, M. and McDonnell, J. J., Tromp-van Meerveld, I., and Uchida, T.: Subsurface stormflow, in: Encyclopedia of Hydrological Sciences, edited by: Anderson, M. G. and McDonnell, J. J., John Wiley & Sons, New York, 1719–1732, 2006.

Weiler, M. and Naef, F.: An experimetnal sudy of the role of macropores in infiltration in grassland soils, Hydrol. Process., 17, 477–493, 2003.

Wienhöfer, J., Germer, K., Lindenmaier, F., Färber, A., and Zehe, E.: Applied tracers for the observation of subsurface stormflow at the hillslope scale, Hydrol. Earth Syst. Sci., 13, 1145–1161, https://doi.org/10.5194/hess-13-1145-2009, 2009.

12

The influence of grid resolution on the prediction of natural and road-related shallow landslides

D. Penna[1,*], M. Borga[1], G. T. Aronica[2], G. Brigandì[2], and P. Tarolli[1]

[1]Department of Land, Environment, Agriculture and Forestry, University of Padova, Agripolis, viale dell'Università 16, 35020 Legnaro (PD), Italy
[2]Department of Civil Engineering, University of Messina, Contrada di Dio, Villaggio S. Agata, 98166 Messina, Italy
[*]now at: Dept. of Environmental Systems Science, Swiss Federal Institute of Technology (ETH), Universitätstrasse 16, 8092, Zurich, Switzerland

Correspondence to: P. Tarolli (paolo.tarolli@unipd.it)

Abstract. This work evaluates the predictive power of the quasi-dynamic shallow landslide model QD-SLaM to simulate shallow landslide locations in a small-scale Mediterranean landscape, namely, the lower portion (2.6 km^2) of the Giampilieri catchment, located in Sicily (Italy). The catchment was impacted by a sequence of high-intensity storms over the years 2007–2009, resulting in widespread landsliding, with a total landslide initiation area amounting to 2.6 % of the basin area. The effect of high-resolution digital terrain models (DTMs) on the quality of model predictions is tested by considering four DTM resolutions: 2, 4, 10 and 20 m. Moreover, the impact of the dense forest road network on the model performance is evaluated by separately considering road-related landslides and natural landslides. The landslide model does not incorporate the description of road-related failures and is applied without calibration of the model parameters. The model predictive power is shown to be DTM-resolution dependent. Use of coarser resolution has a smoothing effect on terrain attributes, with local slope angles decreasing and contributing areas becoming larger. The percentage of watershed area represented by the model as unconditionally unstable (i.e. failing even without the addition of water from precipitation) ranges between 6.3 % at 20 m DTM and 13.8 % at 2 m DTM, showing an overestimation of the mapped landslide area. We consider this prediction as an indication for likely failing sites in future storms rather than areas proved stable during previous storms. When assessed over the sample of mapped non-road-related landslides, better model performances are reported for 4 and 10 m DTM resolution, thus highlighting the fact that higher DTM resolution does not necessarily mean better model performances. Model performances over road-related failures are lower than for the natural cases, and slightly increase with decreasing DTM resolution. These findings indicate that to realize the full potential of high-resolution topography, more extensive work is needed aiming more specifically to identify the extent of the artificial structures and their impact on shallow landsliding processes.

1 Introduction

The occurrence of precipitation-triggered shallow landslides is of concern in hydro-geomorphic and natural hazards science due to the high ranking of such events among natural disasters in terms of both the number of people affected globally and the proportion of fatalities on the affected population. Shallow landsliding can evolve into debris flows, resulting in high risk where vulnerable targets are involved (Petley, 2012).

Landslide susceptibility maps represent one of the key elements for landslide risk management. Landslide susceptibility is the probability that a landslide will occur in a specific area (van Westen, 2000). Quantitative susceptibility assessment is typically accomplished either through the use of empirical models (e.g. Baeza and Corominas, 2001; Lee et al., 2003; Fell et al., 2008), or spatially distributed process-based models of slope stability and hydrology (e.g. Montgomery and Dietrich, 1994; Casadei et al., 2003; Frattini et al., 2009). A number of recent process-based susceptibility models combine steady-state or quasi-steady-state hydrology concepts with the infinite-slope stability model to calculate the critical rainfall – i.e. the rainfall required to initiate a landslide (e.g. Montgomery et al., 1998; Burton and Bathurst, 1998; Borga et al., 2002; Claessens et al., 2005; Rosso et al., 2006). One of the most important factors controlling the landslide location is surface topography through slope and concentration of shallow subsurface flow (Borga et al., 2002; Penna et al., 2011). Other relevant factors are rainfall intensity and soil and vegetation properties.

Digital terrain model (DTM) resolution affects the calculation of critical rainfall (Zhang and Montgomery, 1994; Claessens et al., 2005; Tarolli and Tarboton, 2006). The last few years have been characterized by the development of new remotely sensed technologies for mapping and analysis of the surface topography (e.g. LiDAR) (Tarolli, 2014). Through high-resolution topography it is possible to recognize in detail local variations in hillslope and valley morphology, and better detect the landslide locations (Lin et al., 2013). However, Zhang and Montgomery (1994) suggested that a resolution of 10 m would be enough for DTM-based geomorphic and hydrological modelling. Similarly, Tarolli and Tarboton (2006) noticed that a very high-resolution DTM might decrease the performance of a shallow landslide model. These findings suggest that such processes are better represented by a smoother topography, which may provide a more realistic approximation of water table distribution than very detailed ones (Freer et al., 2002; Lanni et al., 2013). However, De Sy et al. (2013) noted that for a small-scale catchment in New Zealand, landslide locations were better distinguished from stable areas by using the 1 m resolution topography than coarser resolutions. Contrary to earlier approaches, these authors calibrated the model for each of the DTM resolutions they tested.

Given the advances in high-quality digital elevation data and the lack of guidance, it is desirable to investigate the effect of DTM resolution on the performances of landslide models. Tarolli and Tarboton (2006) reported that one of the reasons for the reduced discriminating capability at fine (< 10 m) DTM resolution was the excessive detail in the surface features, leading to unrepresentative values of terrain slopes. Additionally, there might be other error sources. For instance, a specific error source that arises when using LiDAR elevation data is related to the presence of small-scale artificial structures, such as roads. Road networks and their associated earthworks and drainage structures often result in significant changes to hydrologic and geomorphic responses (Luce, 2002; Dutton et al., 2005). In steep terrain prone to landsliding, roads and the associated drainage structures may significantly impact on the surface/subsurface flow pathways, thus influencing the landsliding potential (Montgomery, 1994; Wemple et al., 2001; Borga et al., 2004; Tarolli et al., 2013). For example, Miller and Burnett (2007) documented a density of road-related landslides double than that of natural surface failures. Specific models have been developed to account for the potential of forest roads to modify landsliding susceptibility (Borga et al., 2004; Dutton et al., 2005). However, we consider here generic models, which are not extended to include road-related processes. This is often the case in practical situations of shallow landslide susceptibility assessment, given the data-intensive characteristics of the road-related model applications (Borga et al., 2004).

The presence of a forest road network may affect the analysis of the predictive power of a shallow landslide model, depending on the DTM resolution. Firstly, a number of observed landslides, used for model assessment, may reflect road-related influences (Montgomery et al., 1998). An error may occur when the model predicts that these landslides are not likely to occur. Caution should be used when considering these errors to assess the model predictive power for road-free areas. Secondly, the high-resolution digital topography reflecting the existing road features may alter model predictions of landslide likelihood. An error may occur when these predictions do not match the mapped failures. This error is mostly influenced by DTM resolution, since road-geometry features cannot be recognized in a coarse-resolution DTM. Also in this case, this kind of error should not be used to assess model performances in road-free areas.

The objective of this work is to investigate the predictive power of a quasi-dynamic process-based landslide model (Borga et al., 2002) in a high-resolution landscape, as a function of the DTM resolution and by considering the specific impact of forest roads. Examination of the DTM resolution sensitivity of process-based shallow landsliding models has received considerable attention in the literature and the distribution of the landsliding susceptibility proved to be to some degree dependent on DTM resolution (Claessens et al., 2005; Tarolli and Tarboton, 2006; de Sy et al., 2013). However, to the best of our knowledge, this is the first time that the assessment is focused on a model which uses the quasi-dynamic wetness index to simulate the groundwater dynamics. The landslide susceptibility model is applied and validated by using lidar-derived DTM data at four different resolutions, ranging from 2 to 20 m, on a small size catchment in Sicily (Italy) characterized by a dense forest road network. For this area repeated field surveys provided an accurate inventory of landslide scars that were distinguished between road-related and natural slides. The model performances are quantified separately for the two types of landslides, to clarify the interplay existing between digital topography and process representation.

2 The shallow landslide model

The QD-SLaM model (Borga et al., 2002; Tarolli et al., 2008) is used in this work to predict the spatial distribution of shallow landslide susceptibility. The QD-SLaM model is based on the coupling of a topography-driven model of subsurface flow with an infinite slope, Mohr–Coulomb failure model to describe the shallow landsliding process. The model may predict duration and intensity of the rainfall required for landslide initiation. The model uses a quasi-dynamic wetness index (QDWI) to predict the spatial distribution of soil saturation in response to a rainfall of specified duration. The QDWI is the ratio between the effective contributing area and the local slope (Borga et al., 2002). The effective contributing area $a(d)$ is the fraction of the total specific contributing area which contributes subsurface flow to the contour segment within a specified drainage period d corresponding to a rainfall duration. This is based on the hypothesis that all precipitation infiltrates and that vertically infiltrating volumes quickly redistribute and produce lateral subsurface flow. The assumption of instantaneous infiltration, which leads to disregarding the effects of infiltration on near-surface pore-pressure distributions and consequent slope stability, is motivated by the observation of the importance of macropores, pipeflow and preferential flow in a large number of landscapes susceptible to shallow landsliding, including the study area. These processes may have a considerable impact on the response timescale of vertical infiltration and lateral groundwater. Decades of work (see the recent review by Beven and Germann (2013), and reference therein) have shown the significance of vertical and lateral flow through soils containing macropores under unsaturated soil conditions. The influence of these processes on the triggering of shallow landslides has been analysed in detail by Krzeminska (2013), as well as by Uchida et al. (2001) and Hencher (2010), among others.

The effective contributing area is computed under the hypothesis of kinematic lateral flow routing (Barling et al., 1994) as follows:

$$c = \frac{K_s \sin\theta}{\varepsilon}, \tag{1}$$

where c is the celerity of subsurface flow, K_s is the saturated hydraulic conductivity, θ is the local slope angle, and ε is the drainable porosity. We assume ε and K_s vertically uniform, such that the celerity of the subsurface flow is independent of time and of the local wetness conditions. Under the assumptions that there is an impeding layer at depth z (corresponding to the soil bedrock interface) and the layer is parallel to the slope, the height h of the water table above the impeding layer, measured perpendicular to the slope, is computed as follows:

$$h(d) = \min\left[\frac{r}{K_s \sin\theta} a(d), z\right], \tag{2}$$

where r is the constant rainfall rate. The subsurface flow discharge per unit contour width $q(d)$ may be computed by using:

$$q(d) = r a(d). \tag{3}$$

This yields a simple model capable of incorporating the combined effect of storm duration and intensity in the dynamics of the groundwater flow. The methodology used for the computation of the quasi-dynamic contributing area is based on a D8 single-flow direction algorithm. The local slope is estimated by a cell-to-cell computation, along the steepest-descended flow path according to the D8 flow direction method (O'Callaghan and Mark, 1984).

Coupling the groundwater model and the infinite slope stability model under the assumption that the failure plane is on the impeding layer (Taylor, 1948; Haefeli, 1948) provides a relationship for the critical rainfall rate $r_c(d)$, in other words the rainfall rate required to trigger slope failure for the specific topographic element. The relationship reads as follows:

$$r_c(d) = \frac{T \sin\theta}{a(d)} \left[\frac{C}{\rho_w g z \cos\theta \tan\varphi} + \left(\frac{\rho_s}{\rho_w} + \frac{W}{\rho_w g z}\right)\left(1 - \frac{\tan\theta}{\tan\varphi}\right)\right], \tag{4}$$

where C combines soil and root cohesion, W is the vegetation surcharge, φ is the internal friction angle of the soil, ρ_s is the wet soil density, ρ_w is the density of water, g is the gravitational acceleration, and T is the soil transmissivity, defined as the product of the saturated lateral hydraulic conductivity K_s and soil thickness. Since root strength produces significant reinforcement in vegetation-covered slopes, the formulation of QD-SLaM includes the effective soil cohesion due to vegetation. However to avoid making assumptions about landslide size, we considered only basal cohesion and not cohesion around the perimeter of the slide. For a predefined storm duration d, Eq. (4) allows the determination of the minimum uniform rainfall r_c needed to cause instability, which is the meaning of the critical rainfall. Equation (4) extends the definition of critical rainfall provided by Montgomery and Dietrich (1994) by removing the assumption of steady rainfall of infinite duration. In the Mediterranean climate, many slides are actually triggered by the transient response of pore pressures to bursts of intense rainfall, which may occur on short timescales of less than one day. By introducing a dynamic drainage area, the QD-SLaM may offer an efficient way to model the subsurface flow response at short temporal scales (Barling et al., 1994). Moreover, the model provides a framework to relate the characteristics of the critical rainfall (rate and duration) to their probability of exceedance. Application of Eq. (4) allows the definition of three slope stability conditions: unconditionally stable, unconditionally unstable and conditionally unstable. A slope is defined as (a) unconditionally stable if it is stable even when it is saturated, (b) unconditionally unstable if it is unstable even when the soil is dry, and (c) conditionally unstable if the slope instability depends on rainfall conditions. One should note that the model, as applied here, does not include the description of the effects of road drainage on surface/subsurface flow dynamics.

2.1 Coupling the slope stability model with the generalized extreme value (GEV) simple scaling model

Equation (4) provides the critical rainfall rate for a precipitation of a given duration, thus offering a way to quantify the return time of the critical rainfall. The variability of storm intensity with duration for a specified frequency level is often represented by the intensity-duration-frequency (IDF) relation. A power function is often used to model the IDF relation (Koutsoyiannis et al., 1998):

$$r_F(d) = \varsigma_F d^{m_F - 1}, \tag{5}$$

where r_F is the rainfall rate which can be exceeded with a probability of (1-F), and ς_F and m_F are model parameters. The scaling properties of the statistical moments of rainfall depths of different durations are used in this work to derive the IDF relationship (Ceresetti et al., 2010). Aronica et al. (2012) showed that a generalized extreme value (GEV) simple scaling model described well the distribution of annual maximum series of rainfall for the study region. The GEV simple scaling distribution of the rainfall rate $r_F(d)$ can be determined as

$$r_F(d) = \zeta_1 \left\{ u + \frac{\alpha}{k}\left[1 - \exp\left(y_{T_r} k\right)\right]\right\} d^{m-1}, \tag{6}$$

where u, α and k are the parameters of the GEV distribution, ζ_1 and m are scaling parameters, y_{T_r} is derived by the following relation:

$$y_{T_r} = \ln\left[\ln\left(\frac{T_r}{T_r - 1}\right)\right] \tag{7}$$

and T_r, recurrence interval, corresponds to the exceedance probability (1-F). The values of the parameters ζ_1 and m can be estimated by linear regression of mean values of annual maxima of precipitation depth against their durations, after log transformation. Combining Eqs. (4) and (6) yields the following equation for the value of y_{T_r}:

$$\exp(y_{T_r} k) = 1 - \frac{k}{\alpha}\left\{\frac{T \sin\theta}{a(d)}\left[\frac{C_r + C_s}{\rho_w g z \cos\theta \tan\varphi} + \left(\frac{\rho_s}{\rho_w} + \frac{W}{\rho_w g z}\right)\left(1 - \frac{\tan\theta}{\tan\varphi}\right)\right]\frac{d^{1-m}}{\varsigma_1} - u\right\}. \tag{8}$$

Based on Eq. (8), the values of exp $(y_{T_r} k)$ can be determined for each topographic element and for a given rainfall duration d. Once the values of exp$(y_{T_r} k)$ (represented here as Γ) are determined, the value of return period T_r may be computed as follows:

$$y_{T_r} = \tfrac{1}{k}\ln\Gamma \qquad T_r = \frac{\exp(\exp y_{T_r})}{\exp(\exp y_{T_r}) - 1}. \tag{9}$$

The critical duration d_{cr} is the rainfall duration, which minimizes the value of y_{T_r} accordingly with Eqs. (8) and (9). The concept of the critical rainfall rate and duration, derived accordingly with Eq. (8), incorporates in a compact way both the topographic control (represented by the parameters of local slope and effective contributing area) and the climate control (represented by parameters $\varsigma_1, u, \alpha, k$ and m) on shallow landsliding.

3 Study area and model application

The work is focused on the lower portion of the Giampilieri catchment (2.6 km^2), located on the Ionic Sea in the northeastern part of Sicily, southeast of the city of Messina (Fig. 1). The topography is very rugged: elevation ranges from 0 to 596 m a.s.l. with an average value of 236 m a.s.l., and an average slope of 28.5°. The geology of the area is characterized by a meta-sedimentary terrain belonging to the Peloritani Belt that represents the westernmost part of the Calabria–Peloritani Arc and by alluvial deposits and Pleistocenic conglomerates. Phyllites and metarenites develop a soil cover especially at medium/low elevations a.s.l. as the result of weathering; the thickness of colluvium is in the range of 0.7–3.0 m (Messina et al., 1996). The climate is typically Mediterranean, with rainfall events (mainly convective) characterized by short duration and high intensity during the wet season (October–April) and few events during the dry season (May–September). The mean annual rainfall is about 970 mm with 84 % in the wet season and about 16 % in the dry season.

The catchment is predominantly rural, with grassland and crop cultivation (46 %) and shrubs and sparse forest (42.4 %) in the upper mountainous part. The floodplain is densely urbanized by the municipality of Giampilieri Superiore village (Aronica et al., 2012). A dense network of forest road is reported for the catchment (Fig. 1). The forest road system has a length of 19.2 km, and a density of 7.4 km km^{-2}. Roads are unpaved; road cuts truncate colluvial fills in topographic hollows and weathered bedrock in noses and divert both surface and subsurface flow into an in-board ditch system that drains through culverts into the valley bottom. The road width ranges between 4 and 6 m. Consequently, at 10 and 20 m DTM the features related to the road network are generally not recognized because of coarse resolution, while at higher resolution the digital topography is affected by the road network.

In the last 10 years, the study area has been impacted by several large storm events, triggering landslides and floods. On 25 October 2007 more than 120 mm of rain fell in less than 3 hours, with 50 % of the recorded rain concentrated in about 20 min (Aronica et al., 2008). A more extreme storm occurred on 1 October 2009, with high intensity rainfall (exceeding 230 mm in 8 hours) on an area of about 60 km^2 around Giampilieri. The event caused more than 500 landslides (mainly soil slides and debris flows), widespread inundation associated with massive erosion and deposition of

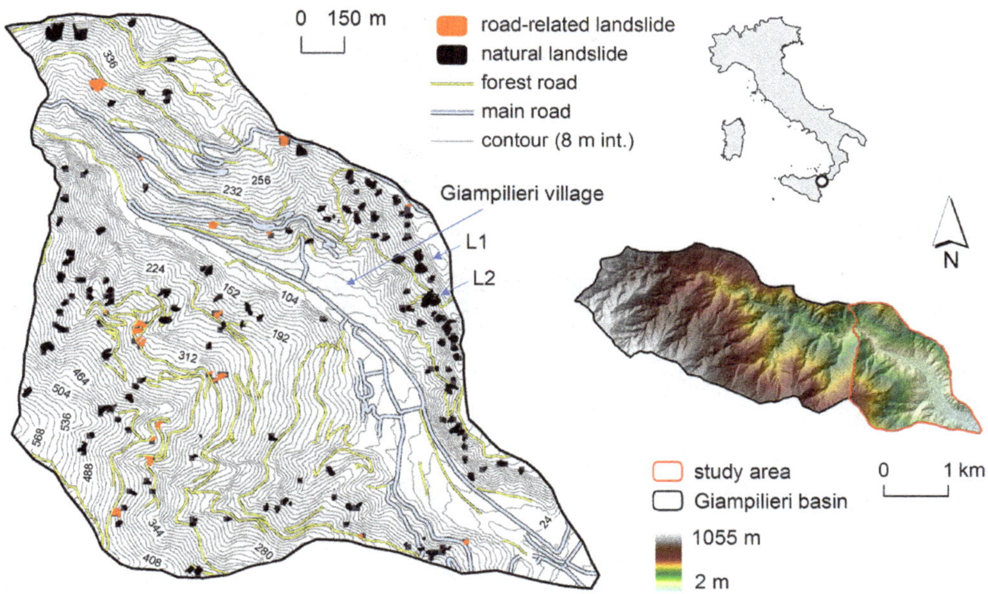

Figure 1. Location map of the study area. Road-related landslides are highlighted in red, while the others are reported in black. The landslides L1 and L2, described as example in Fig. 4, are also reported.

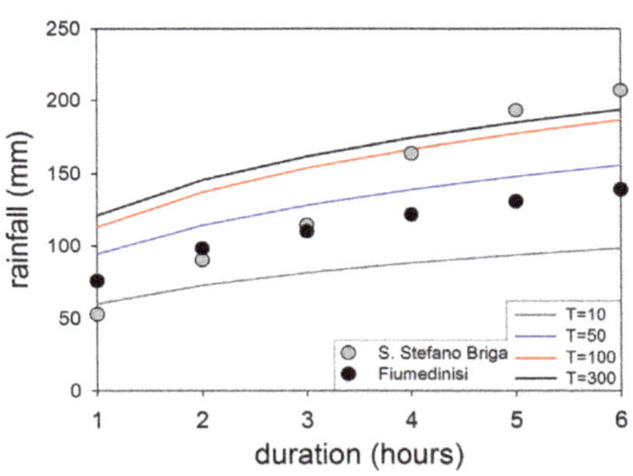

Figure 2. Regional depth-duration-frequency relationship for the study area for return periods ranging from 10 to 300 yr. Observed maxima for rainfall durations ranging from 1 to 6 h measured at the rain gauge stations of S. Stefano and Fiumedinisi during the 2009 event are also reported.

debris along the drainage network. Landslides and inundation caused 31 deaths, 6 missing persons, and an undetermined number of injured people. The evacuees and the homeless people exceeded 2500. Rainfall depths and corresponding durations and return periods for the rainfall data collected at the rain gauges of S. Stefano Briga and Fiumedinisi are reported in Fig. 2, showing that the return period for durations of 5 and 6 hours ranges from 100 to 300 yr. These two rain gauges are located within 6 km of the study basin.

For the study area, a detailed shallow landslide inventory is available (Fig. 1), which includes both landslides triggered by the 2009 event and previous events. The landslides have a rather high length/depth ratio, generally higher than 25. This condition ensures proper application of the infinite length assumption within the infinite slope stability model (Milledge et al., 2012). The landslides were subdivided into road-related and natural slides. Previous investigations (Miller and Burnett, 2007) used a fixed-width buffer on both sides of the road to identify road-related landslides from other slides. In our analysis we examined also the road drainage system features and the coupled hillslope-road topography in order to discriminate road-related landslides. Given the characteristics of the shallow landsliding model used here, which describes only the slope stability processes, only the initiation area is reported for the surveyed landslides. The total landslide initiation area amounts to 67 593 m^2, with 9489 m^2 representing road-related slides. Counter to the conventional wisdom that roads cause the vast majority of shallow slides in steep, soil-mantled catchments (see Gucinski et al., 2001, and references therein), road-related slides account for just 14 % of the mapped slides. This may be due to the catastrophic nature of the event, which caused widespread slope instability in the catchment. Figure 3 provides pictures of two landslides (indicated as L1 and L2 in Fig. 1) just before and after the storm. The figure clearly shows the initiation area as well as the depositional area that impacted the village. The debris flows generated by landslide L1 destroyed or damaged several houses along the creek course, causing several fatalities. Landslide 2 severely impacted the primary school of the village (the school is the large building at the lower boundary of the deposition area). The pictures taken before the storm show that some scars could have been generated by the previous heavy storms that occurred in 2007.

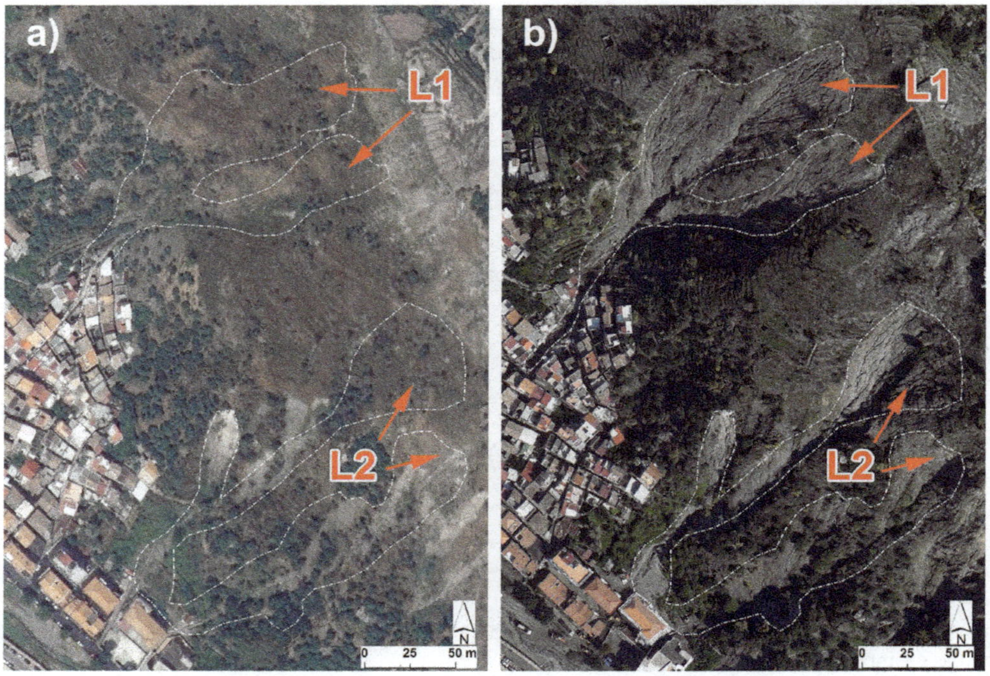

Figure 3. Comparison of aerial photographs taken before (**a**) and after the storm (**b**), showing landslides L1 and L2, upslope Giampilieri village.

Unlike other works on the topic (Keijsers et al., 2011; De Sy et al., 2013), we decided not to calibrate the landslide model for each DTM resolution considered in the analysis. Recalibrating the model parameters would clearly partially compensate for the changing representation of the morphological attributes associated with each resolution. However, this would also cloud the comparison over the road-related and natural failures, making the examination of the results less straightforward. Instead, we decided to base the model parameterization on the geotechnical and hydraulic parameters obtained from in situ analysis (Regional Department of Civil Protection for Sicily Region, personal communication, 2010). These parameters are reported in Table 1. The influence of vegetation surcharge and root strength on slope stability is confined to a limited portion of the area, where shrubs and forests are widespread. The remaining land is covered by grass which provides some root cohesion. Based on these observations, a value of 1000 Pa, averaged over the basin, is used as combined root and soil cohesion. An average value of soil depth is used, equal to 1 m. The vegetation surcharge was not considered in the analysis since the study area is not covered by forest stands, and the dominant component is grassland. Parameters of the GEV distribution (Table 2) were estimated based on a regional analysis by using the 1, 3, 6, 12 and 24 h duration annual maximum rainfall for a total of 388 station years in a homogeneous region defined accordingly with the L-moment procedures (Hosking an Wallis, 1997), as reported by Aronica et al. (2012).

Table 1. Hydraulic and geotechnical parameters for the study site.

Parameter	Value
Density ratio (ρ_s/ρ_w)	1.8
Cohesion C (Pa)	1000
Porosity ε (-)	0.35
Friction angle φ (degrees)	35
Saturated hydraulic conductivity K_s (m s^{-1})	10^{-3}
Transmissivity T (m^2 s^{-1})	10^{-3}
Soil depth z (m)	1

The LiDAR-derived DTM at 2 m was regridded on 4, 10 and 20 m grid-cell resolution by using the mean aggregation function in order to obtain coarser digital terrain models. The mean aggregation function was found to ensure a high degree of consistency between surface flow paths extracted from gridded elevation data having different resolutions.

4 Results and discussion

4.1 Influence of DTM resolution on slope, effective specific contributing area and QDWI

To investigate how different DTM resolutions (2, 4, 10 and 20 m) effect the relative shallow landslide susceptibility distribution, the behaviour of three topographic indices derived from the DTM and used in the susceptibility assessment, that is local slope, effective contributing area and QDWI, was

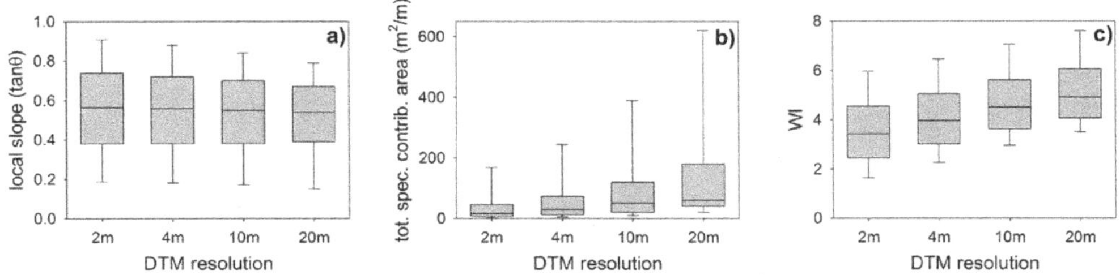

Figure 4. Box plots of topographic attributes for the four different DTM resolutions: **(a)** local slope, **(b)** total specific contributing area, **(c)** WI. The boxes show the 25th and 75th percentile, the whiskers show the 10th and 90th percentile, the horizontal line within the box indicates the median.

Table 2. Parameters estimated for application of GEV model.

Parameter	Value
ζ_1	35.8 mm
α	0.2930
u	0.7550
k	−0.2096
m	0.279

analysed (Figs. 4 and 5). In the figures we show the distribution of both the total (Fig. 4) and the effective (Fig. 5) contributing area, and the relevant wetness indexes. We term here wetness index (WI) as the ratio between the total contributing area and the local slope. Although the total contributing area and WI are not used in QD-SLaM, these topographic attributes represent limiting conditions that help to understand the behaviour of their quasi-dynamic counterparts at very long precipitation durations.

Figure 4a shows the distribution of the slope values for the four DTM resolutions. Although the distributions are similar, it is clear that steep slopes are less represented at the coarser resolutions. The 90th and the 75th percentiles decrease from 0.91 to 0.79 and from 0.74 to 0.67, respectively, with increasing DTM size from 2 to 20 m. On the contrary, the left side of the local slope distribution is almost unaffected by the changes in the DTM resolution.

The distribution of the total specific contributing area (contributing area per unit contour length) for the different resolutions is shown in Fig. 4b. It is possible to observe larger values of specific catchment area for coarser DTM. Similar smoothing effects of coarser resolutions, with contributing areas becoming larger and the local slope angles decreasing, were reported by several researchers for various landscapes (Zhang and Montgomery, 1994; Claessens et al., 2007; Keijsers et al., 2011).

The distribution of the WI for the four resolutions is reported in Fig. 4c. The distribution reflects the behaviour of the specific contributing area and the slope, clearly showing a shift towards higher values with increasing resolution. The 90th and the 75th percentiles increase from 5.98 to 7.60 and from 4.54 to 6.05, respectively, with increasing the DTM size from 2 to 20 m. The distributions of the effective specific contributing area are reported in Fig. 5a–c for the three different durations of 1, 3 and 24 h, respectively, and for the four DTM resolutions. The distributions show that the influence of the resolution increases with the rainfall duration. The distribution of the 24 h-specific contributing area is similar to that reported for the total specific contributing area for the four resolutions. It is interesting to note that, unlike the total case, the minimum specific contributing area values are not directly related to the resolution, since fractions of grid area are computed when the rainfall duration is less than the time taken by the subsurface flow to travel the grid. This is particularly evident for 1 h rainfall duration, where the four distributions exhibit the same behaviour for the right tail and are limited to a range of 1 to 5 m. The distributions of the QDWI are reported in Fig. 5d–f for the three different durations of 1, 3 and 24 h, respectively, and for the four resolutions. The distributions clearly display the smoothing effect related to the decreasing DTM resolution on both slope and effective specific contributing area. Two different patterns emerge. For small rainfall duration (1 and 3 h), the distributions tend be more peaked with decreasing DTM resolution, showing the impact of slope smoothing. For longer rainfall duration (24 h), the distributions are shifted towards larger values with decreasing DTM resolution. Since different rainfall durations are used to generate the critical rainfall return time map, the different sensitivity of the QDWI to the DTM resolution directly translates into a complex pattern of critical rainfall frequency sensitivity to DTM resolution.

4.2 Distribution of the critical rainfall values

The spatial distributions of the critical rainfall frequency are reported in Fig. 6 for the four different DTM resolutions. The Figures show that the main characteristics of the mapped landslides are well represented at the four resolutions, with the observed failures characterized either as unconditionally unstable or with low critical rainfall return period. More specifically, this is the case of the steep valley above the

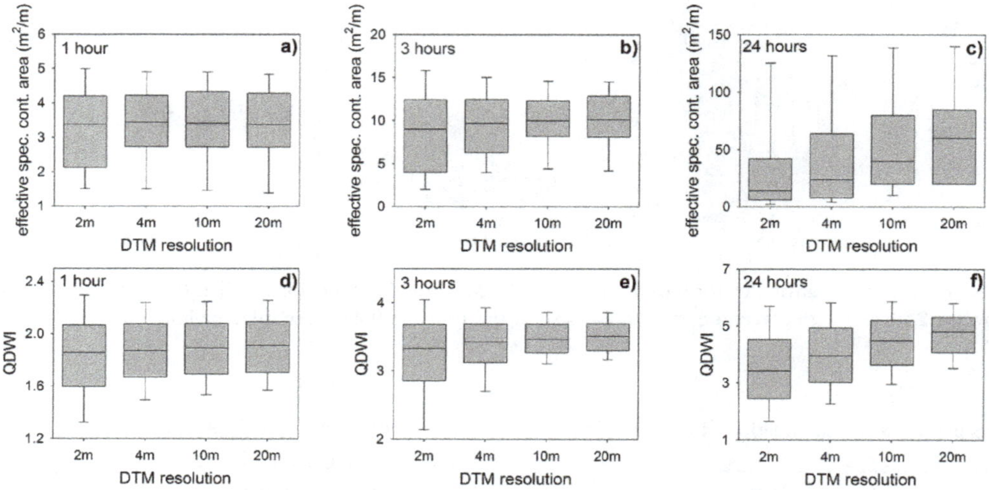

Figure 5. Box plots of topographic attributes for the four different DTM resolutions: **(a)** to **(c)** effective specific contributing area for drainage time ranging from 1 to 24 h, **(d)** to **(f)** QDWI for drainage time ranging from 1 to 24 h. The boxes show the 25th and 75th percentile, the whiskers show the 10th and 90th percentile, the horizontal line within the box indicates the median.

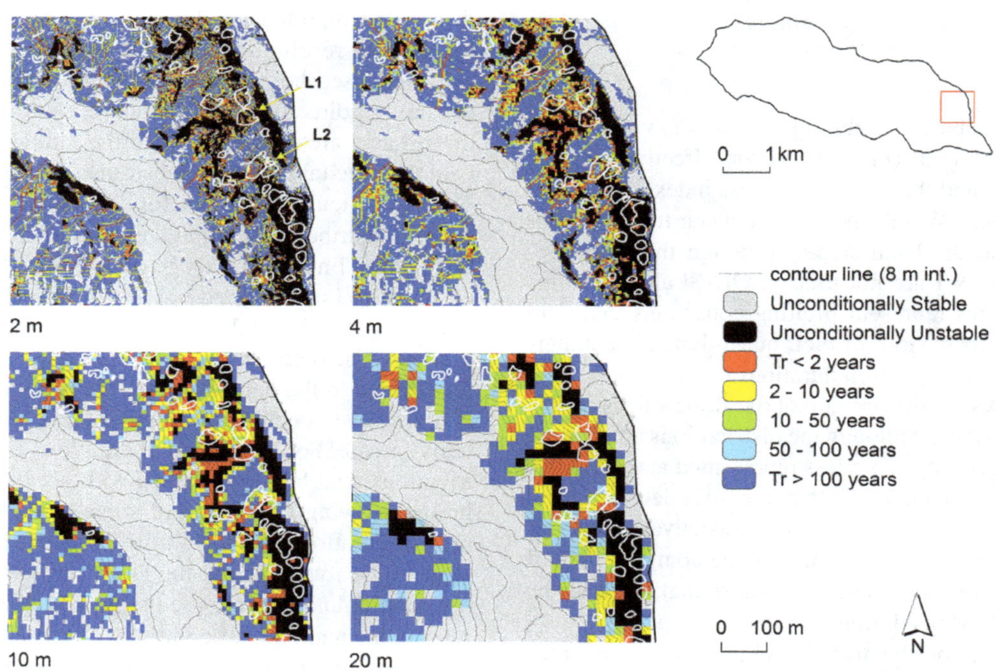

Figure 6. Return period of critical rainfall for four different DTM resolutions: 2, 4, 10, and 20 m (detail of the steep valley above the village of Giampilieri). The landslides are surrounded by white.

village of Giampilieri, on the left side of the creek, where most of the mapped failures are concentrated.

Tables 3 and 4 summarize some key characteristics of the critical rainfall distributions. Table 3 provides the percentages of catchment area and of observed landslide area considered unconditionally unstable and unconditionally stable with the QD-SLaM application for the four DTM resolutions, whereas Table 4 provides the same percentages for two main ranges of critical rainfall frequency. Table 3 shows that more unconditionally unstable cells are reported over the study basin with increasing the DTM resolution, ranging from 6.3 % at 20 m to 13.8 % at 2 m. On the other hand, the percentage of the unconditionally stable cells behaves in a similar way, albeit in a much more reduced range of values, comprised between 30 % at 2 m and 29 % at 20 m. In our application the distribution of unconditionally stable

Table 3. Percentages of catchment area and of observed landslide area considered either unconditionally unstable or unconditionally stable with the QD-SLaM application and for the DTM resolution of 2, 4, 10, and 20 m.

	Catchment area				Landslide area			
	DTM resolution				DTM resolution			
Stability condition	2 m	4 m	10 m	20 m	2 m	4 m	10 m	20 m
Unconditionally unstable	13.75	11.95	9.30	6.34	56.2	52.6	48.0	42.9
Unconditionally stable	30.00	29.77	29.23	29.07	0.	0.	0.	0.

Table 4. Percentages of slope-stability categories in terms of catchment area and of observed landslide area in two main ranges of critical rainfall frequency (return period) for the QD-SLaM application and for the DTM resolution of 2, 4, 10, and 20 m.

	Catchment area				Landslide area			
	DTM resolution				DTM resolution			
Critical rainfall return period	2 m	4 m	10 m	20 m	2 m	4 m	10 m	20 m
< 10 yr	18.3	19.0	18.2	14.1	58.1	70.0	68.7	52.0
≥ 10 yr	81.7	81.0	81.8	85.9	41.9	30.0	31.3	48.0

and unstable states is mainly influenced by local slope. Thus, these effects are related to the distribution of local slope as reported in Fig. 4a, showing a marked impact of the DTM resolution on the higher values of slope, which control the unconditionally unstable condition, and much less on the central values, which control the unconditionally stable condition.

Table 4 shows that the relative frequency of the two main classes of critical rainfall frequency changes in negligible way by varying the DTM resolution from 2 to 10 m. Only for the 20 m DTM do large values of return time increase noticeably. According to Eq. (4), the critical rainfall depends on both the local slope and the QDWI. The results reported in Table 4 imply that the two effects of DTM size on slope and QDWI compensate each other effectively, at least between 2 and 10 m. For a DTM of 20 m the smoothing effect on slope is stronger, thus leading to larger values of critical rain rate and hence of return period.

4.3 Performance assessment methodology

The quality of the model representation of landsliding susceptibility was first assessed by comparing the locations of the observed landslides with those predicted by the model. Better model performance is achieved when a larger difference between fractions of catchment and landslide area corresponding to short recurrence interval of critical rainfall is observed. Tables 3 and 4 report these comparisons for all landslides, without distinction between road-related slides and natural scars.

For a more complete model verification over the three landslide groups (road-related landslides, natural landslides, all landslides), we adopted the assessment methodology developed by Tarolli et al. (2011), which is an extension of the method described above. The methodology can be summarized in a plot where the fraction of basin T_r area $F_B(T_r)$ is reported on the x axis and fraction of landslide T_r area $F_L(T_r)$ on the y axis. The steeper the curve of the plot, the better the performance of the model, because a higher fraction of low T_r area is reported over the observed landslides. The 1 : 1 line represents the "naïve" model, in other words a model without predictive power. The measure of the area below the empirical function is selected here to provide a measure of the model performance. We term this statistic as "efficiency index" (EI). The statistic ranges between 0 and 1, with 1 representing perfect model performances, and 0.5 representing naïve model performances. We examine the model performance for three different set of landslides: (i) road-related slides; (ii) natural slides, and (iii) all landslides, including both road-related and natural slides. Due to the methodology used for the assessment, the comparison is applied only to the conditionally unstable portion of the catchment where the recurrence interval of the critical rainfall is available.

4.4 Evaluation of the model performances at different DTM resolutions

A first assessment is based on results reported in Tables 3 and 4 and concerns all landslides, without distinction between road-related slides and natural failures. Similarly to the results already reported for the basin area (Sect. 4.2), Table 3 shows that more unconditionally unstable cells are predicted over the landslide area with increasing DTM resolution, ranging from 42.9 % at 20 m to 56.2 % at 2 m. This shows again the effect of decreasing slope with coarsening DTM resolution. Interestingly, the percentage of landslide area corresponding to unconditionally stable cells is always

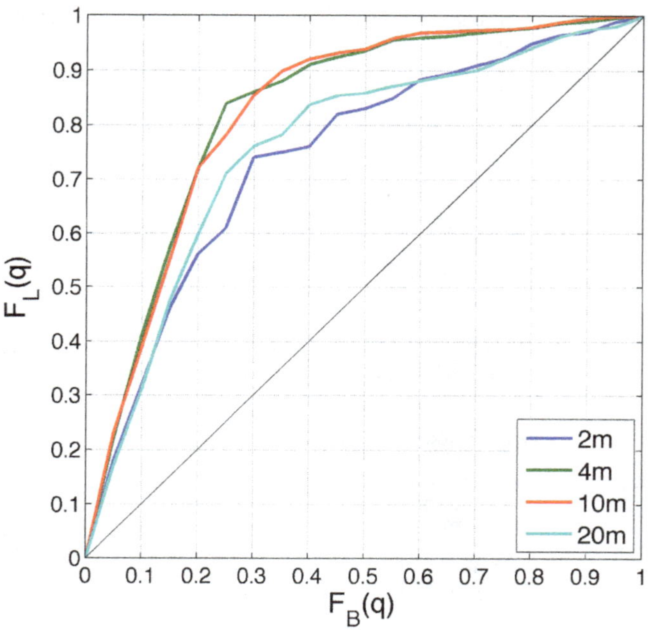

Figure 7. Relationship between cumulative frequencies $F_L(q)$ and $F_B(q)$ considering all landslides and for the DTM resolution of 2, 4, 10, and 20 m. The q variable represents the return period for the QD-SLaM.

zero. The occurrence of mapped landslide in areas predicted to be unconditionally stable is generally considered as an indication about the occurrence of processes external to the model framework (Borga et al., 2002; Tarolli et al., 2011). So, these results show that the processes which are external to the model representation, such as road-related processes, do not emerge over areas considered as unconditionally stable.

Results reported in Table 4, which are weighted on the percentage of area considered as conditionally unstable, offer a first assessment of the DTM resolution impact on the model prediction accuracy. Examination of the results shows that even though the catchment percentages falling into the two main critical rainfall frequencies are roughly similar with coarsening DTM resolution (Table 4), the corresponding spatial organizations of the critical rainfall frequency with respect to the mapped scars are markedly different. Indeed, more landslides are reported in the first range of critical T_r (< 10 yr) for the DTM resolution corresponding to 4 and 10 m, whereas DTMs of 2 and 20 m provide less accurate results, with less landslides reported in the first range of critical T_r. This shows that the 2 and 20 m resolution appear to be of similar poorer quality compared to intermediate resolutions.

Overall, Tables 3 and 4 show that many more cells are predicted to be unconditionally unstable or to fail with moderate rainfall ($T_r < 10$ yr) than are mapped as landslides (which amounts to 2.7 % of the study area). This result is consistent with the general tendency of several process-based models to overestimate the areas predicted as unconditionally unstable or failing with low critical rainfall (Huang and Kao, 2006). Montgomery and Dietrich (1994) reported percentages of unconditionally unstable areas ranging between 1 and 13 %, clearly exceeding the observed landslide area, for the first application of SHALSTAB. In the frame of a more comprehensive validation of SHALSTAB, Dietrich et al. (2001) reported percentages of unconditionally unstable areas ranging between 1 and 23 %, with a mean value of 8 %, for seven well-studied watersheds in northern California. Gorsevski et al. (2006) reported a percentage of 31.2 % for unconditionally unstable areas for a SHALSTAB application in the US. Typically, the assumption of a constant soil depth, usually exceeding the actual one over the steeper portions of the hillslopes, is responsible for this error (Lanni et al., 2012). In our case study, the distribution of soil depth may be affected by a legacy effect due to the presence of terraces. Terraces were built in some steep portions of the basin, by artificially increasing the soil depth distribution. The terraces are now abandoned, hence representing critical conditions for shallow landsliding triggering in unconditionally unstable terrain (Tarolli et al., 2014).

Table 5. Efficiency index for different DTM resolutions and for road-related landslides, natural landslides and for all landslides. Efficiency index equal to 1 represents perfect model performances, and 0.5 represents naïve model performances.

Landslide group	2 m	4 m	10 m	20 m
Road-related	0.54	0.56	0.60	0.61
Natural	0.78	0.87	0.86	0.70
All	0.71	0.82	0.83	0.67

Table 5 reports results in terms of EI for road-related, natural and all landslides and allows for a more complete examination of the interplay between digital topography features and forest roads. Figure 7 provides an example of the comparison when considering all landslides and the four DTM resolutions. Confirming the results shown in Table 4, the statistics reported in Table 5 for all landslides indicate better landslide capability recognition for DTM resolutions corresponding to 4 and 10 m, with relatively high values of EI around 0.83. EI is lower and ranges between 0.67 and 0.71 when considering a DTM of 20 and 2 m, respectively.

The same ranking arises for the natural landslides, with slightly better results. Even in this case, the worst predictions are obtained by using a 20 m DTM (EI = 0.67), whereas predictions with 4 and 10 m DTMs are marked with EI ranging around 0.87.

Model performances over road-related failures are lower than for the natural cases. This is not surprising, given that the QD-SLaM model does not incorporate any module to predict road-related (or other man-related) failures. The model still performs better than the naïve model (characterized by EI = 0.5) because it has a dependence on local slope, which remains important for road-related landslides. The performances over the road-related failures show a well-defined ranking, with EI decreasing and thus increasing the DTM resolution. This result is likely associated with the increasing impact of the road system on the drainage patterns, for the finer digital topography. At 10 and 20 m resolution, the features related to the road network are not recognized because of coarse resolution, whereas at higher resolution the digital topography is affected by the road network with some likely impact on the computation of the contributing area and slope. More insight on the effect of DTM resolution on road-related failures may be obtained by applying a shallow landsliding model that explicitly recognizes the influence of road-related hydrological processes on shallow landsliding, such as the one developed by Borga et al. (2004). These processes include road interception and re-routing of subsurface flow, and the flow concentration effects due to the road drainage system.

Overall, the results reported here show that caution should be used when assessing the impact of DTM resolution on shallow landslide model predictive power, particularly when the terrain attributes may themselves be dependent on the resolution of the digital topography.

5 Conclusions

In this work a quasi-dynamic shallow landslide model has been applied to the 2.6 km^2 Giampilieri catchment, with the main goal of examining the model predictive power as a function of the DTM resolution. The investigation is carried out separately for road-related landslides and natural landslides. Five main remarks arise from our work.

1. Use of coarser resolution has a smoothing effect on terrain attributes, with local slope angles decreasing and contributing areas becoming larger. However, the effects on the effective contributing areas are shown to be dependent on the precipitation duration, with DTM effects which are negligible for 1 h precipitation and which increase with the duration. This pattern of sensitivity is transmitted to the QDWI. Since the critical rainfall spatial pattern is composed by several precipitation durations, depending on local topography and climate a complex pattern of model sensitivity to the DTM resolution results.

2. The DTM resolution has a remarkable impact on the percentage area considered as unconditionally unstable, which decreases with increasing DTM size. An almost negligible impact is reported for the unconditionally stable areas.

3. Comparison with mapped landslides shows that, even though the catchment percentages falling into the main critical rainfall frequencies are similar when considering both fine and coarse resolutions, the corresponding spatial organizations of the critical rainfall frequency over the mapped scars are markedly different. The model predictive power is shown to be DTM-resolution dependent. When assessed over the sample of mapped natural landslides (i.e. landslides that are not road-influenced), better model performances are reported for 4 and 10 m DTM resolution. This agrees with earlier findings (Tarolli and Tarboton, 2006; Keijsers et al., 2011) and highlights the fact that higher DTM resolution does not necessarily mean better model performance.

4. Model results over the mapped road-related failures outline specific prediction and DTM-sensitivity patterns. Model performance for road-related landslides is worse than for natural landslides. This is not surprising, given that road-related processes are not represented in the model structure. Nevertheless, the model performances are still higher than those corresponding to the naïve model, because it has a dependence on local slope, which remains important for road-related landslides.

5. The model predictive power decreases with increasing DTM resolution, when assessed over the road-related landslides. This is due to the interaction of the model structure (which does not describe the road-related processes) and the impact of road geometry on the digital topography. Model structural error is DTM-resolution neutral and as such it is expected to affect model outcomes independently on DTM resolution. The road impact on digital topography increases with DTM resolution and adds to the error, as shown by the pattern of EI.

Overall, these findings indicate that to realize the full potential of high-resolution topography, more extensive work is needed aiming more specifically at identifying the extent of the artificial structures and their impact on shallow landsliding processes.

Acknowledgements. This research is funded by the Era.Net CICRLE Mountain project ARNICA (10-MCGOT-CIRCLE-2-CVS-185 116). The Regional Department for Civil Protection of Sicily Region is acknowledged for supplying data. The Interdepartmental Research Center of Geomatics (CIRGEO) at the University of Padova is thanked for help in the elaboration of topographic data. The modules of the QD-SLaM model are available at the site: http://intra.tesaf.unipd.it/people/tarolli/Software.htm, software section.

Edited by: T. Bogaard

References

Aronica, G. T., Brigandì, G., Marletta, C., and Manfrè, B.: Hydrological and hydraulic analysis of the flash flood event on 25 October 2007 in North-Eastern part of Sicily, Italy, Proceedings of FLOODrisk 2008, 30 September–2 October 2008, Oxford, UK, 2008.

Aronica, G. T., Brigandí, G., and Morey, N.: Flash floods and debris flow in the city area of Messina, north-east part of Sicily, Italy in October 2009: the case of the Giampilieri catchment, Nat. Hazards Earth Syst. Sci., 12, 1295–1309, doi:10.5194/nhess-12-1295-2012, 2012.

Baeza, C. and Corominas, J.: Assessment of shallow landslide susceptibility by means of multivariate statistical techniques, Earth Surf. Process. Landf., 26, 1251–1263, 2001.

Barling, D. B., Moore, I. D., and Grayson, R. B.: A quasi-dynamic wetness index for characterizing the spatial distribution of zones of surface saturation and soil water content, Water Resour. Res., 30, 1029–1044, 1994.

Beven, K. and Germann, P.: Macropores and water flow in soils revisited, Water Resour. Res., 49, 3071–3092, doi:10.1002/wrcr.20156, 2013.

Borga, M., Dalla Fontana, G., and Cazorzi, F.: Analysis of topographic and climatic control on rainfall-triggered shallow landsliding using a quasi-dynamic wetness index, J. Hydrol., 268, 56–71, 2002.

Borga, M., Tonelli, F., and Selleroni, J.: A physically-based model of the effects of forest roads on slope stability, Water Resour. Res., 40, W12202, doi:10.1029/2004WR003238, 2004.

Burton, A. and Bathurst, J. C.: Physically based modeling of shallow landslide sediment yield at a catchment scale, Environ. Geol., 35, 89–99, 1998.

Casadei, M., Dietrich, W. E., and Miller, N. L.: Testing a model for predicting the timing and location of shallow landslide initiation in soil mantled landscapes, Earth Surf. Process. Landf., 28, 925–950, 2003.

Ceresetti, D., Molinié, G., and Creutin, J.-D.: Scaling properties of heavy rainfall at short duration: A regional analysis, Water Resour. Res., 46, W09531, doi:10.1029/2009WR008603, 2010.

Claessens, L., Heuvelink, G. B. M., Schoorl, J. M., and Veldkamp, A.: DEM resolution effects on shallow landslide hazard and soil redistribution modelling, Earth Surf. Process. Landf., 30, 461–477, 2005.

Claessens, L., Schoorl, J. M., and Veldkamp, A.: Modelling the location of shallow landslides and their effects on landscape dynamics in large watersheds: an application for Northern New Zealand, Geomorphology, 87, 16–27, 2007.

De Sy, V., Schoorl, J. M., Keesstra, S. D., Jones, K. E., and Claessens, L.: Landslide model performance in a high resolution small-scale landscape, Geomorphology, 190, 73–81, 2013.

Dietrich, W. E., Bellugi, D., and Real de Asua, R.: Validation of the shallow landslide model, SHALSTAB, for forest management, in: Land Use and Watersheds: Human influence on hydrology and geomorphology in urban and forest areas, edited by: Wigmosta, M. S. and Burges, S. J., Am. Geoph. Union, Water Science and Application, 2, 195–227, 2001.

Dutton, A. L., Loague, K., and Wemple, B. C.: Simulated effect of a forest road on near-surface hydrologic response and slope stability, Earth Surf. Process. Landf., 30, 325–338, 2005.

Fell, R., Corominas, J., Bonnard, C., Cascini, L., Leroi, E., and Savage, W. Z.: Guidelines for landslide susceptibility, hazard and risk zoning for land-use planning, Eng. Geol., 102, 99–111, 2008.

Frattini, P., Crosta, G., and Sosio, R.: Approaches for defining thresholds and return periods for rainfall-triggered shallow landslides, Hydrol. Process., 23, 1444–1460, 2009.

Gorsevski, P. V., Gessler, P. E., Boll, J., Elliot, W. J., and Foltz, R. B.: Spatially and temporally distributed modeling of landslide susceptibility, Geomorphology, 80, 178–198, 2006.

Gucinski, H., Furniss, M. J., Ziemer, R. R., and Brookes, M. H.: Forest roads: A synthesis of scientific information, Gen. Tech. Rep. PNW-GTR-509, US Dept. of Agric. For. Serv. Pac. Northwest Res. Stat., Portland, Oreg., 103 pp., 2001.

Haefeli, R.: The stability of slopes acted upon by parallel seepage, paper presented at International Conference on Soil Mechanics and Foundation Engineering, 1948.

Hencher, S. R.: Preferential flow paths through soil and rock and their association with landslides, Hydrol. Process., 24, 1610–1630, 2010.

Hosking, J. R. M. and Wallis, J. R.: Regional Frequency Analysis, Cambridge University Press, 224 pp., 1997.

Huang, J. C. and Kao, S. J.: Optimal estimator for assessing landslide model performance, Hydrol. Earth Syst. Sci., 10, 957–965, doi:10.5194/hess-10-957-2006, 2006.

Keijsers, J. G. S., Schoorl, J. M., Chang, K.-T., Chiang, S.-H., Claessens, L., and Veldkamp, A.: Calibration and resolution effects on model performance for predicting shallow landslide locations in Taiwan, Geomorphology, 133, 168–177, 2011.

Koutsoyiannis, D., Kozonis, D., and Manetas, A.: A mathematical framework for studying rainfall intensity-duration-frequency relationships, J. Hydrol., 206, 118–135, 1998.

Krzeminska, D.: The influence of fissures on landslide hydrology, PhD Thesis, University of Delft, 2013.

Lanni, C., Borga, M., Rigon, R., and Tarolli, P.: Modelling shallow landslide susceptibility by means of a subsurface flow path connectivity index and estimates of soil depth spatial distribution, Hydrol. Earth Syst. Sci., 16, 3959–3971, doi:10.5194/hess-16-3959-2012, 2012.

Lanni, C., McDonnell, J., Hopp, L., and Rigon, R.: Simulated effect of soil depth and bedrock topography on near-surface hydrologic response and slope stability, Earth Surf. Process. Landf., 38, 146–159, 2013.

Lee, S. J., Ryu, H., Min, K. D., and Won, J. S.: Landslide susceptibility analysis using GIS and artificial neural network, Earth Surf. Process. Landf., 28, 1361–1376, 2003.

Lin, C. W., Tseng, C.-M., Tseng, Y.-H., Fei, L.-Y., Hsieh, Y.-C., and Tarolli, P.: Recognition of large scale deep-seated landslides in forest areas of Taiwan using high resolution topography, J. Asian Earth Sci., 62, 389–400, 2013.

Luce, C. H.: Hydrological processes and pathways affected by forest roads: What do we still need to learn?, Hydrol. Process., 16, 2901–2904, 2002.

Messina, A., Russo, S., and Stagno, F.: The crystalline basements of the Calabrian-Peloritani Arc. 6th field Meeting IGCP Project no 276, "The Calabrian-Peloritani Arc and its correlation with Northern Africa and Southern Europe", Messina, 27 September–2 October 1993, Newsletter 6, 94 pp., 1996

Milledge, D. G., Griffiths, D. V., Lane, S. N., and Warburton, J.: Limits on the validity of infinite length assumptions for modelling shallow landslides, Earth Surf. Process. Landf. 37, 1158–1166, 2012.

Miller, D. J. and Burnett, K. M.: Effects of forest cover, topography, and sampling extent on the measured density of shallow, translational landslides, Water Resour. Res., 43, W03433, doi:10.1029/2005WR004807, 2007.

Montgomery, D. R.: Road surface drainage, channel initiation, and slope instability, Water Resour. Res., 30, 1925–1932, 1994.

Montgomery, D. R. and Dietrich, W. E.: A physically based model for the topographic control on shallow landsliding, Water Resour. Res., 30, 1153–1171, 1994.

Montgomery, D. R., Sullivan, K., and Greenberg, H. M.: Regional test of a model for shallow landsliding, Hydrol. Process., 12, 943–955, 1998.

Montgomery, D. R., Dietrich, W. E., and Heffner, J. T.: Piezometric response in shallow bedrock at CB1: Implications for runoff generation and landsliding, Water Resour. Res., 38, 10-1–10-18, doi:10.1029/2002WR001429, 2002.

O'Callaghan, J. and Mark, D.: The extraction of drainage networks from digital elevation models, Comput. Vis. Graph. Image Process., 28, 328–344, 1984.

Penna, D., Tromp-van Meerveld, H. J., Gobbi, A., Borga, M., and Dalla Fontana, G.: The influence of soil moisture on threshold runoff generation processes in an alpine headwater catchment, Hydrol. Earth Syst. Sci., 15, 689–702, doi:10.5194/hess-15-689-2011, 2011.

Petley, D. N.: Global patterns of loss of life from landslides, Geology, 40, 927–930, 2012.

Rosso, R., Rulli, M. C., and Vannucchi, G.: A physically based model for the hydrologic control on shallow landsliding, Water Resour. Res., 42, W06410, doi:10.1029/2005WR004369, 2006.

Tarolli, P.: High-resolution topography for understanding Earth surface processes: opportunities and challenges, Geomorphology, 216, 295–312, doi:10.1016/j.geomorph.2014.03.008, 2014.

Tarolli, P. and Tarboton, D. G.: A new method for determination of most likely landslide initiation points and the evaluation of digital terrain model scale in terrain stability mapping, Hydrol. Earth Syst. Sci., 10, 663–677, doi:10.5194/hess-10-663-2006, 2006.

Tarolli, P., Borga, M., and Dalla Fontana G.: Analyzing the influence of upslope bedrock outcrops on shallow landsliding, Geomorphology, 93, 186–200, 2008.

Tarolli, P., Borga, M., Chang, K. T., and Chiang, S. H.: Modeling shallow landsliding susceptibility by incorporating heavy rainfall statistical properties, Geomorphology, 133, 199–211, 2011.

Tarolli, P., Calligaro, S., Cazorzi, F., and Dalla Fontana, G.: Recognition of surface flow processes influeced by roads and trails in mountain areas using high-resolution topography, Eur. J. Remote Sens., 46, 176–197, 2013.

Tarolli, P., Preti, F., and Romano, N.: Terraced landscapes: from an old best practice to a potential hazard for soil degradation due to land abandonment, Anthropocene, doi:10.1016/j.ancene.2014.03.002, in press, 2014.

Taylor, D. W.: Fundamentals of Soil Mechanics, Wiley, New York, 1948.

Uchida, T., Kosugi, K., and Mizuyama, T.: Effects of pipeflow on hydrological process and its relation to landslide: A review of pipeflow studies in forested headwater catchments, Hydrol. Process., 15, 2151–2174, 2001.

van Westen, C. J.: The modeling of landslide hazard using GIS, Surv. Geophys., 21, 241–255, 2000.

Wemple, B. C., Swanson, F. J., and Jones, J. A.: Forest roads and geomorphic process interactions, Cascade Range, Oregon, Earth Surf. Process. Landf., 26, 191–204, 2001.

Zhang, W. and Montgomery, D. R.: Digital elevation model grid size, landscape representation, and hydrologic simulations, Water Resour. Res., 30, 1019–1028, 1994.

13

Contrasting transit times of water from peatlands and eucalypt forests in the Australian Alps determined by tritium: implications for vulnerability and the source of water in upland catchments

Ian Cartwright[1,2] **and Uwe Morgenstern**[3]

[1]School of Earth, Atmosphere and Environment, Monash University, Clayton, Victoria 3800, Australia
[2]National Centre for Groundwater Research and Training, G.P.O. box 2100, Flinders University, Adelaide, SA 5001, Australia
[3]Isotope Hydrology & Water Dating Lab, GNS Science, Lower Hutt 5040, New Zealand

Correspondence to: Ian Cartwright (ian.cartwright@monash.edu)

Abstract. Peatlands are a distinctive and important component of many upland regions that commonly contain distinctive flora and fauna which are different from those of adjacent forests and grasslands. Peatlands also represent a significant long-term store of organic carbon. While their environmental importance has long since been recognised, water transit times within peatlands are not well understood. This study uses tritium (^{3}H) to estimate the mean transit times of water from peatlands and from adjacent gullies that contain eucalypt forests in the Victorian Alps (Australia). The ^{3}H activities of the peatland water range from 2.7 to 3.3 tritium units (TUs), which overlap the measured (2.9 to 3.0 TU) and expected (2.8 to 3.2 TU) average ^{3}H activities of rainfall in this region. Even accounting for seasonal recharge by rainfall with higher ^{3}H activities, the mean transit times of the peatland waters are < 6.5 years and may be less than 2 years. Water from adjacent eucalypt forest streams has ^{3}H activities of 1.6 to 2.1 TU, implying much longer mean transit times of 5 to 29 years. Cation / Cl and Si / Cl ratios are higher in the eucalypt forest streams than the peatland waters and both of these water stores have higher cation / Cl and Si / Cl ratios than rainfall. The major ion geochemistry reflects the degree of silicate weathering in these catchments that is controlled by both transit times and aquifer lithology. The short transit times imply that, unlike the eucalypt forests, the peatlands do not represent a long-lived store of water for the local river systems. Additionally, the peatlands are susceptible to drying out during drought, which renders them vulnerable to damage by the periodic bushfires that occur in this region.

1 Introduction

Globally, peatlands occupy parts of the upland areas of many river catchments and provide water to the headwater streams in those catchments. It is estimated that peatlands cover $\sim 4 \times 10^6$ km^2 or ~ 3 % of the world's land area (Dixon et al., 1994). Peatlands have received much attention because they represent a major (1.8 to 4.5×10^{11} t) long-lived store of terrestrial organic carbon (Gorham, 1991; Page et al., 2002). Peatlands commonly contain distinctive flora such as sphagnum moss, sedges, and orchids that may not occur in surrounding forests or grasslands (Grover and Baldock, 2013; Grover et al., 2005; McDougall, 1982). Due to their unique flora, there may be also differences in fauna between the peatlands and the surrounding regions.

Draining of peatlands is a major environmental concern. This can occur directly due to anthropogenic activities such as peat extraction and the conversion of peatland to agricultural land or forest. Additionally, since peatlands form in high rainfall environments, drainage can occur due to a reduction in rainfall. Draining of peatlands destroys the habitats of flora and fauna and also causes oxidation of the organic matter, which in turn releases CO_2 to the atmosphere (Gorham, 1991; Grover and Baldock, 2010). In addition, the

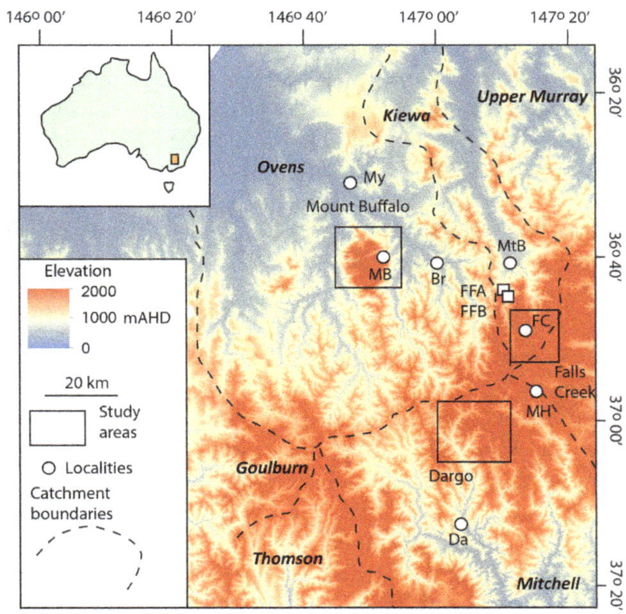

Figure 1. Location of study sites in northeast Victoria superimposed on the digital elevation model. Places: Br indicates Bright; Da indicates Dargo; MB indicates Mount Buffalo; MtB indicates Mount Beauty; MH indicates Mount Hotham; My indicates Myrtleford. FFA and FFB are the Falls Forest sampling sites A and B, other sampling sites are in the indicated study areas (see the Supplement Fig. S1). Catchment boundaries from Department of Environment, Land, Water and Planning (2016); elevations are relative to the Australian Height Datum (AHD).

dried peat is susceptible to bushfires that commonly occur during droughts and which can destroy the peatlands (van der Werf et al., 2010).

Peatlands comprise $\sim 520\,km^2$ or $\sim 2.5\,\%$ of the alpine region of the southeast Australian mainland (Fig. 1) (Costin, 1972; Grover et al., 2005; Western et al., 2009). The peat generally occupies shallowly sloping areas in the upland plains that are poorly drained. While it forms only a minor part of the Australian landscape, understanding the water balance in the Australian peatlands is important for assessing the potential impacts of environmental change. If water transit times are short, then the peatlands may dry significantly during droughts, making them prone to degradation and vulnerable to bushfires. Additionally, the peatlands provide water to numerous small upland streams. Most of the upland peatlands are flanked by eucalypt forests that occupy steeper slopes and gullies. Eucalypts have high transpiration rates, and groundwater recharge rates in eucalypt-dominated areas are low (Allison et al., 1990), which results in relatively long water transit times (on the order of years to decades) in eucalypt-dominated catchments (Cartwright and Morgenstern, 2015).

1.1 Determining mean transit times

The mean transit time represents the average time required for water to flow through the aquifer or soil from where it recharges to where it discharges at the surface (e.g. at a spring, stream, or lake) or is sampled from a bore (Cook and Bohlke, 2000; Małoszewski and Zuber, 1982; McDonnell et al., 2010). Documenting mean transit times is important for understanding and managing catchments. The time required for nutrients or contaminants to be transported from recharge areas to streams is a direct function of the transit time (Morgenstern and Daughney, 2012). Additionally, catchments with long transit times are more likely to be resilient to short-term (years to decades) variations in rainfall but will respond to climate or land use changes that cause longer-term (decades to centuries) changes in groundwater recharge and flow.

Tritium (^3H) has a half-life of 12.32 years which, together with the fact that it is part of the water molecule, makes it an ideal tracer for determining transit times of young shallow groundwater, soil water, or surface water (Clark and Frtiz, 1997; Cook and Bohlke, 2000; Morgenstern et al., 2010). Other potential tracers for determining transit times of young waters, such as ^3He, chlorofluorocarbons, and SF_6, are gases that degas to the atmosphere and are thus difficult to use for water sampled from streams. Because the seasonal variation of stable isotope ratios or major ion concentrations in recharge is attenuated with increasing transit times, time series of stable isotope ratios or Cl concentrations are also commonly used to estimate mean transit times (e.g. McGuire and McDonnell, 2006; Hrachowitz et al., 2010, 2013; Kirchner et al., 2010). However, this is only practicable where mean transit times are shorter than ~ 5 years, as longer transit times attenuate the seasonal variation of these tracers (Stewart et al., 2010). Additionally, the use of these tracers requires a detailed (preferably at least weekly) record of stable isotope and/or major ion geochemistry in rainfall for a period which exceeds that of the transit times of water in the catchment (e.g. Timbe et al., 2015).

Coupled with models that describe the distribution of flow paths within an aquifer, commonly referred to as lumped parameter models (Cook and Bohlke, 2000; Małoszewski, 2000; McGuire and McDonnell, 2006), ^3H may be used to determine transit times of waters that are up to ~ 100 years old. The ^3H activities in rainfall over the last several decades have been measured globally (e.g. International Atomic Energy Association, 2016; Tadros et al., 2014), which allows the ^3H input to catchments over time to be estimated. Rainfall ^3H activities peaked in the 1950s to 1960s due to the production of ^3H in atmospheric thermonuclear tests (commonly termed "bomb-pulse" ^3H).

There are two different approaches in using ^3H to estimate mean transit times. In the Northern Hemisphere, the ^3H activities of remnant bomb-pulse waters are currently above those of modern rainfall. This situation results in individual measurements of ^3H activities in groundwater or surface water yielding non-unique estimates of transit times (Morgenstern et al., 2010). Mean transit times may, however, be determined using time series of ^3H measurements (e.g. Małoszewski and Zuber, 1982; Michel, 1992; McGuire and McDonnell, 2006). The use of time series measurements is potentially advantageous as, in addition to determining a mean transit time, it allows the most appropriate lumped parameter model to be assessed by comparison of the predicted and observed changes to ^3H activities over time (Małoszewski and Zuber, 1982). However, constraining lumped parameter models using sequential ^3H data collected over several years has commonly made the assumption that the data can be described by a single lumped parameter model and that the flow system is time invariant, which may not be the case in reality (e.g. Kirchner, 2016b).

The bomb-pulse ^3H activities in the Southern Hemisphere were several orders of magnitude lower than in the Northern Hemisphere (Clark and Fritz, 1997; Tadros et al., 2014) and the ^3H activities of the remnant bomb-pulse waters have now decayed below those of modern rainfall. This permits ^3H to be used in a similar manner to other radioactive isotope tracers, such as ^{14}C or ^{36}C, whereby a residence or transit time may be estimated from individual ^3H activities (Morgenstern et al., 2010). Since the ^3H activities reflect the net residence time in the catchment, there is no requirement for the flow system to be time invariant. However, this approach to determining mean transit times does not permit the optimal lumped parameter model to be determined, as is the case when time series ^3H measurements at similar flow conditions are available. The attenuation of the bomb-pulse ^3H in the Southern Hemisphere is such that constraining the lumped parameter models using time series measurements of ^3H is probably no longer practicable. For example, the predicted decrease in ^3H activities for a water with a mean transit time of 10 years between 2016 and 2026 calculated using the lumped parameter models and ^3H input employed in this study is only ~ 0.2 TU, and the predicted ^3H vs. time trends is similar for the different models. Instead, a plausible lumped parameter model must be proposed based on factors such as the geometry of the flow system or information from previous time series studies. Given that the different lumped parameter models yield different mean transit times for the same ^3H activity, this introduces uncertainties. However, because the ^3H activities of the remnant bomb-pulse waters are below those of modern rainfall, the relative transit times of Southern Hemisphere waters are largely independent of assumed flow models (i.e. water with a low ^3H activity has a longer mean transit time than water with a high ^3H activity).

Rivers aggregate water from different flow systems and the mean transit times of water contributed by individual tributaries may vary considerably. Kirchner (2016a) discussed the impacts of aggregation for mean transit times calculated from time series major ion and stable isotope data. That study demonstrated that the mean transit time calculated from the aggregated water was generally lower than the actual mean transit time (i.e. that which would be calculated from the weighted average of each of the end-members contributing to the mixture). Aggregation also affects mean transit times calculated from individual ^3H activities. Mean transit times calculated from the ^3H activity of the aggregated water are again generally lower than the actual mean transit times with the error increasing as the difference between the mean transit time of the individual end-members increases (Stewart et al., 2016).

1.2 Transit times in upland peat

Compared with catchments developed on other vegetation types, the transit time of peatland water, especially upland or mountain peat, is less well known. Due to peatlands containing poorly drained organic-rich soils and occupying shallow slopes, it is sometimes assumed that peatland waters have long transit times. However, there has been little assessment of that assumption (e.g. as discussed by Western et al., 2009). Using ^3H activities, transit times of years to decades were proposed for water from metre- to tens-of-metres-thick peat deposits in Quebec, Canada (Dever et al., 1984), Dartmoor, UK (Charman et al., 1999), Lithuania (Mažeika et al., 2009), and Minnesota, USA (Siegel et al., 2001). From the preservation of seasonal $\delta^{18}O$ and δ^2H values, Aravena and Warner (1992) estimated that the water from the upper ~ 0.1 m of peat from Ontario, Canada, was < 1 year old. Mean transit times from an upland peat-bearing catchment in Zhurucay, Ecuador, were estimated at < 1 year using oxygen isotopes (Mosquera et al., 2016). Mean transit times from the extensively studied upland catchments in the United Kingdom, which include extensive regions of peat, were estimated using time series of stable isotopes and Cl as < 5 years (e.g. Soulsby et al., 2006; Kirchner et al., 2010; Hrachowitz et al., 2013; Tetzlaff et al., 2014; Benettin et al., 2015). However, the mean transit times of the water specifically draining the peat in those catchments are generally not known. Morris and Waddington (2011) modelled water transit times in peat and suggested that they vary from < 1 year for the upper 0.1 m to several years in deeper layers, which is broadly consistent with the above estimates. While not estimating transit times, Western et al. (2009) concluded, on the basis of hydrograph analysis and water balance calculations, that peatlands in southeast Australia did not act as long-term water stores for the local streams.

1.3 Objectives

This study was designed to test the hypothesis that peatlands in the Victorian Alps, southeast Australia, represent a relatively long-lived store of water. Specifically, we utilise ^3H to provide the first estimates of the mean transit times of water draining Australian peatlands during summer baseflow conditions and compare these with mean transit times of streams that drain adjacent eucalypt forests. Documenting mean transit times is important for determining the relative importance of the peatlands and eucalypt forests in providing water to the local rivers and also in assessing potential environmental impacts resulting from land use or climate change. In addition, we assess whether the major ion geochemistry of the water may be used as a first-order proxy for mean transit times and discuss the controls on mean transit times in these upland areas.

2 Methods

2.1 Setting

Water from streams draining peatlands and eucalypt forests was sampled in the Mount Buffalo (Ovens catchment: Fig. S1a), Falls Creek (Kiewa and Upper Murray catchments: Fig. S1b), and Dargo (Mitchell catchment: Fig. S1c) regions of the Victorian Alps (Fig. 1). Catchment areas in the peatlands and eucalypt forests are similar, ranging from 0.5 to 6.4 km^2 (Table 1). The Falls Creek and Dargo areas consist of indurated Ordovician metasedimentary rocks, Devonian granites, and minor Tertiary basalts, whereas the Mount Buffalo plateau comprises Devonian granite (van den Berg et al., 2004; Energy and Earth Resources, 2016).

Peatlands are developed in poorly drained low-relief upland areas of the Victorian Alps with eucalyptus forests occupying steeper valleys that dissect the uplands and the higher better-drained areas at the margins of the peatland (Grover and Baldock, 2013; Grover et al., 2005; McDougall, 1982; Western et al., 2009). The peatlands are characterised by sphagnum mosses (*Sphagnum cristatum*), rushes (*Empodisma minus* and *Baloskion australe*), and heaths (*Epacris paludosa* and *Richea continentis*) (McDougall, 1982). Most of the peatlands in the Alpine region are <2 m thick. The upper few centimetres comprise little decomposed plant material (the fibric zone). The fibric zone grades downwards through the hemic zone (typically 50 cm to 1 m thick) where the soils are organic rich and fibrous, with abundant discernible roots into a denser soil (the sapric zone) comprising mainly decomposed organic matter with clays and fragments of rock. The peat typically has a thin (<20 cm thick) layer of weathered rocks at its base but mainly overlies little-weathered basement rocks. The hydrology of the peat comprises the upper acrotelm, which alternates between saturated and unsaturated as the water table rises and falls, and the lower permanently saturated catotelm (e.g. Grover et al., 2005). The boundary between the acrotelm and catotelm is generally in the hemic zone.

In the alpine regions of southeast Australia, the dominant eucalypt species include mountain ash (*Eucalyptus regnans*) at altitudes of <1000 m and alpine ash (*Eucalyptus delegatensis*) and snow gum (*Eucalyptus pauciflora*) at higher altitudes (McDougall, 1982). The soils of the eucalypt forests include sandy lithosols; these are thin, well-drained, and poor in organic matter and largely occur on the upper slopes of the gullies and on the elevated areas surrounding the peatlands. In the centres of the gullies, the soils also include alpine humus loams that contain higher contents of organic matter and which are less well drained. The soils overlie weathered regolith that is a few tens of centimetres to a few metres thick. There are also minor deposits of colluvium and alluvium along the streams (Shugg, 1987). Groundwater flow is restricted to the weathered zones and fractures in the granites and metasediments; the minor alluvial and colluvial sediments are more permeable but represent only a minor part of the landscape.

Average precipitation is 1250 mm yr^{-1} at Falls Creek, 1650 mm yr^{-1} at Mount Hotham (near Dargo), and 1790 mm yr^{-1} at Mount Buffalo (Bureau of Meteorology, 2016) (Fig. 1). Approximately 60 to 70 % of the precipitation occurs in the austral autumn to winter (May to September) with a proportion of the winter precipitation occurring as snow. February and March have the lowest precipitation (4 to 6 % of the annual total in each month). No currently active river gauges exist in these upland areas and previous flow measurements were only from a few streams at Falls Creek (Department of Environment, Land, Water, and Planning, 2016). These limited records show that river flow varies seasonally with the majority of streamflow occurring in the winter months; however, flows persist over summer (Fig. 2a) and the streams generally do not record zero flows (Fig. 2b). Water from some of the peatlands at Falls Creek drains into the Rocky Valley Reservoir ($\sim 2.8 \times 10^7$ m^3), which is used to provide water to the Falls Creek resort and the Kiewa Valley hydroelectric system. Water from some of the peatlands at Mount Buffalo feeds Lake Catani, which is an artificial recreational lake with a volume of $\sim 2 \times 10^6$ m^3.

Mean transit times in the Ovens Valley, which is dominated by eucalypt forests (Fig. 1), were estimated using ^3H as being between 4 and 30 years, with higher mean transit times recorded during summer low flow conditions (Cartwright and Morgenstern, 2015). That study focused on the major tributaries to the Ovens River that had catchment areas between 6 and 435 km^2, and the transit times in the upper reaches of the catchments where streamflow commences remain unknown. Additionally, it was not established whether there are differences between mean transit times for water draining the peatlands and the eucalypt forests in those upper catchments.

2.2 Sampling and analytical techniques

Samples of rainfall representing the total precipitation over periods of several weeks to months (Table 1) were collected from a rainfall collector at Mount Buffalo (Fig. 1). Grab samples of stream water from the peatlands and eucalypt forests were taken from flowing reaches. Water was also sampled from within the peat using a hand-operated siphon pump with a soft PVC inlet hose that was inserted into a rigid 5 cm diam-

Table 1. Geochemistry of water samples from the Australian Alps.

Site[a]	Date	Area[b] km^2	^3H TU	δ^2H ‰ V-SMOW	$\delta^{18}O$ ‰ V-SMOW	EC µS cm^{-1}	Na mg L^{-1}	Mg mg L^{-1}	K mg L^{-1}	Ca mg L^{-1}	Si mg L^{-1}	Cl mg L^{-1}	Br mg L^{-1}	NO_3 mg L^{-1}	SO_4 mg L^{-1}
Peatlands															
Mount Buffalo															
Buffalo Cresta A	26/11/2015	1.5	3.32 ± 0.053	−26	−5.0	19.12	1.12	0.125	0.192	0.621	2.714	0.94	0.0031	0.040	0.072
Buffalo Cresta A	23/02/2015	1.5	2.84 ± 0.058	−41	−7.7	13.72	2.06	0.125	0.168	0.645	2.831	0.95	0.0039	0.048	0.185
Buffalo Cresta B(P)[b]	23/02/2015		2.75 ± 0.055	−41	−7.7	22.33	1.96	0.152	0.273	1.123	3.244	1.08	0.0037	0.285	0.304
Buffalo Cresta C	23/02/2015	2.1	2.75 ± 0.056	−40	−7.5	19.92	1.06	0.121	0.160	0.432	1.632	0.95	0.0036	0.063	0.080
Buffalo Cresta D(P)	26/11/2015		2.70 ± 0.048	−37	−7.3	15.21	1.79	0.131	0.184	0.653	2.601	1.14	0.0038	0.073	0.194
Buffalo Crystal A	23/02/2015	3.5	2.74 ± 0.055	−35	−6.6	12.01	1.97	0.153	0.178	0.684	2.031	1.01	0.0038	0.066	0.207
Buffalo Crystal B	23/02/2015	0.2	2.93 ± 0.058	−30	−5.7	12.65	2.06	0.130	0.250	0.354	3.466	0.99	0.0037	0.320	0.245
Falls Creek															
Falls Cope	24/02/2015	1.9	3.14 ± 0.061	−41	−7.2	9.62	0.92	0.247	0.119	0.519	1.417	0.47	0.0031	0.084	0.076
Falls Langford A	24/02/2015	0.5	3.10 ± 0.060	−39	−7.1	21.42	1.43	0.512	0.203	0.882	2.496	0.64	0.0035	0.250	0.271
Falls Langford B(P)	24/02/2015		2.98 ± 0.059	−43	−8.3	8.52	0.88	0.223	0.160	0.490	0.812	0.70	0.0033	0.165	0.167
Falls Rocky A	24/02/2015	3.5	2.96 ± 0.059	−42	−8.0	7.65	0.79	0.194	0.057	0.463	1.097	0.55	0.0026	0.307	0.111
Falls Rocky A	25/11/2015	3.5	2.90 ± 0.050	−42	−8.1	11.25	0.78	0.149	0.075	0.318	2.103	0.51	0.0018	0.164	0.163
Falls Rocky B(P)	25/11/2015		2.86 ± 0.049	−42	−8.2	7.47	0.62	0.136	0.118	0.255	1.029	0.54	0.0023	0.035	0.130
Falls Rocky C(P)	25/11/2015		3.17 ± 0.055	−35	−6.5	9.54	0.79	0.257	0.245	0.540	1.125	0.51	0.0025	2.097	0.185
Falls Rocky D	25/11/2015	2.6	3.14 ± 0.059	−33	−6.5	16.47	1.04	0.227	0.223	0.403	1.113	0.93	0.0033	0.087	0.079
Dargo															
Dargo A	25/11/2015	5.5	2.84 ± 0.048	−40	−7.7	21.17	1.15	0.930	0.133	1.427	3.259	1.10	0.0040	0.804	0.142
Dargo B	25/11/2015	1.2	2.69 ± 0.043	−43	−8.1	31.9	1.34	1.381	0.334	2.112	2.147	0.91	0.0038	0.493	0.050
Dargo C(P)	25/11/2015		2.47 ± 0.040	−39	−7.0	29.2	1.56	1.667	0.308	1.900	3.012	0.93	0.0038	0.388	0.063
Dargo D	25/11/2015	2.7	3.36 ± 0.053	−35	−6.9	13.9	0.77	0.343	0.097	1.095	2.788	0.83	0.0025	1.845	0.192
Eucalypt forest															
Buffalo Forest A	24/02/2015	3.3	2.21 ± 0.048	−35	−6.3	42.7	3.89	2.056	0.618	1.727	4.433	1.16	0.0045	0.052	0.740
Buffalo Forest A	26/11/2015	3.3	2.35 ± 0.042	−34	−6.4	39.5	3.27	1.878	0.470	1.536	4.844	1.30	0.0048	0.038	0.763
Falls Forest A	24/02/2015	3.3	1.56 ± 0.037	−38	−7.4	65.2	5.70	3.230	0.543	3.686	5.060	2.60	0.0097	0.104	0.499
Falls Forest B	24/02/2015	1.5	2.10 ± 0.047	−37	−7.5	45.7	3.24	2.092	0.445	2.149	4.233	1.09	0.0045	0.038	0.608
Falls Forest C	25/11/2015	6.4	2.26 ± 0.040	−31	−6.2	45.3	3.41	1.855	0.566	1.623	6.236	1.20	0.0048	0.418	0.901
Falls Forest D	25/11/2015	5.8	1.70 ± 0.035	−37	−7.3	56.0	4.78	2.793	0.400	3.092	9.540	1.22	0.0045	0.022	0.581
Dargo Forest A	25/11/2015	6.1	2.00 ± 0.036	−29	−5.7	44.1	4.12	1.898	0.542	1.654	5.328	1.24	0.0045	0.240	0.408
Rainfall															
Buffalo	Feb 2015 (12)[d]		2.85 ± 0.057	−36	−7.0	11.58	0.68	0.011	0.058	0.214	0.081	0.98	0.0039	0.015	0.081
Buffalo	Nov 2015 (9)		2.88 ± 0.051	−35	−6.5	12.20	0.53	0.117	0.108	0.361	0.077	0.80	0.0028	0.106	0.171
Buffalo	Dec 2013 (17)		2.99 ± 0.046	−38	−7.3		0.87					1.10			
Buffalo	Feb 2014 (2)		2.90 ± 0.049	−37	−7.1										
Buffalo	Jul 2014 (< 1)		2.52 ± 0.043	−40	−7.9										
Buffalo	Sep 2014 (3)		2.71 ± 0.044	−34	−6.8										
Melbourne[e]	Jul 2013 (12)		2.72 ± 0.045	−33	−6.3										
Otways[e]	Sep 2014 (6)		2.45 ± 0.041	−22	−4.4										
La Trobe[e]	Aug 2015 (12)		2.76 ± 0.043	−36	−6.8										

Notes: [a] Sampling regions in Fig. 1. [b] Catchment area. [c] P indicates piezometer sample. [d] Month collected and number of months represented by the sample. [e] Other rainfall from southeast Australia. The rainfall samples represent aggregated samples collected over the number of months indicated; all other samples are individual samples.

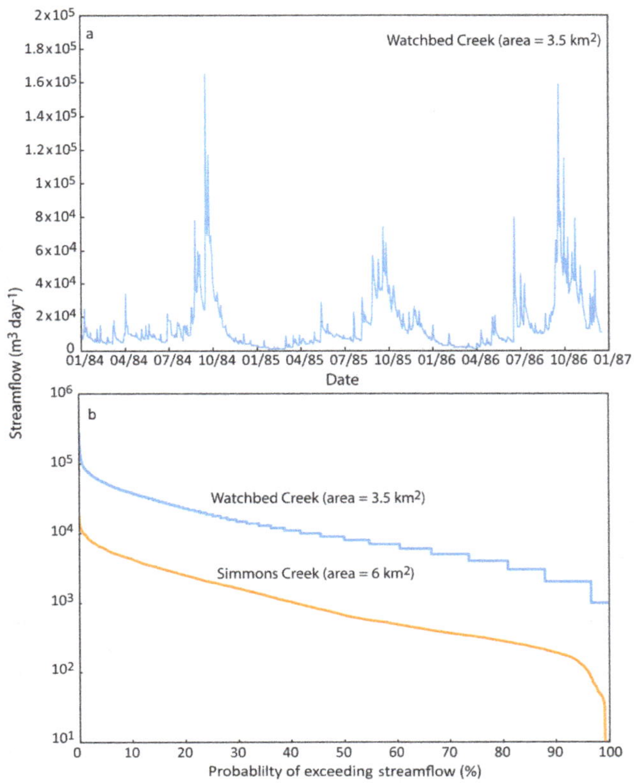

Figure 2. (a) Streamflow at Watchbed Creek, which drains peatlands at Falls Creek (Fig. 1), between 1984 and 1987. **(b)** Flow duration curves for Watchbed Creek and Simmons Creek, which drains the eucalyptus forest in the Upper Ovens (Fig. 1). Data are from the Department of Environment, Land, Water and Planning (2016).

eter PVC piezometer with a ~ 20 cm screen pushed directly into the peat. The piezometer was inserted to the maximum possible extent (i.e. until resistance prevented it from being pushed deeper). Field observations indicated that this was at or close to the base of the peat and within the hemic zone and catotelm. At least three bore volumes of water were extracted prior to sampling.

Tritium activities (Table 1) were measured on water samples that had been vacuum distilled and electrolytically enriched using liquid scintillation spectrometry (Quantulus ultra-low-level counters) at Geological and Nuclear Sciences, New Zealand (Morgenstern and Taylor, 2009). The ^{3}H activities are expressed in tritium units (TUs) where 1 TU represents a ^{3}H / ^{1}H ratio of 1×10^{-18}. Enrichment of ^{3}H was by a factor of 95, which results in a detection limit of 0.02 TU, while the use of deuterium calibration for each sample results in a 1 % reproducibility of the tritium enrichment. Precision (1σ) is ~ 1.8 % at 2 TU.

Major ion concentrations (Table 1) were measured at Monash University using a Thermo Finnigan ICP-OES (cations), Thermo Finnigan ICP-MS (Si), and a Thermo Fisher ion chromatograph (anions). Samples for cation and Si analysis were filtered through 0.45 µm cellulose nitrate filters and acidified to pH <2 using double-distilled 16M HNO_3. Samples for anion analysis were filtered but unacidified. The precision of the major ion concentrations determined by replicate analyses of samples is ±2 % and the accuracy as determined by analysis of certified water standards is ±5 %.

Stable isotope ratios of water (Table 1) were determined using a Thermo Finnigan DeltaPlus Advantage mass spectrometer at Monash University. A Thermo Finnigan Gas Bench was used for the ^{18}O analyses. Waters were equilibrated with He-CO_2 at 32 °C for 24 to 48 h and the gas subsequently analysed by continuous flow. A Thermo Finnigan H-Device was used for the ^{2}H analyses. H was produced from water via reaction with Cr at 850 °C and analysed by dual-inlet measurement. Internal standards that have been calibrated using IAEA SMOW, GISP, and SLAP standards were employed to normalise the δ^{18}O and δ^2H values (following Coplen, 1988). The δ^{18}O and δ^2H values are expressed in ‰ relative to V-SMOW with a precision (1σ) determined from replicate analyses of samples of ±0.1 ‰ (δ^{18}O) and ±1 ‰ (δ^2H). The deuterium (D) excess is defined as $\delta^2\text{H} - 8\,\delta^{18}\text{O}$ (Clark and Fritz, 1997).

Catchment areas (Table 1) were estimated from Google Earth satellite images (Fig. S1a to c) and 1 : 30 000 topographic map sheets on which the streams are clearly distinguished. Attempts to define catchment areas using a digital elevation model (DEM) did not reproduce the drainage pattern in the peatlands due to the low topography relative to the DEM resolution.

2.3 Estimating mean transit times using ^{3}H

Water flowing through aquifers or soils follows flow paths of varying length, and thus the water discharging into streams or sampled from bores has a range of transit times (McGuire and McDonnell, 2006). Mean transit times were calculated using lumped parameter models (Małoszewski, 2000; Małoszewski and Zuber, 1982; Zuber et al., 2005) as implemented in the TracerLPM Excel workbook (Jurgens et al., 2012). For steady-state groundwater flow, the ^{3}H activity in water at the catchment outlet at the time of sampling ($C_o(t)$) may be estimated using the convolution integral:

$$C_o(t) = \int_0^\infty C_i(t-\tau)g(\tau)e^{-\lambda\tau}\,\mathrm{d}\tau. \quad (1)$$

In Eq. (1), C_i is the ^{3}H activity of recharge, τ is the transit time, $t-\tau$ is the time when water entered the catchment, λ is the ^{3}H decay constant (0.0563 yr^{-1}), and $g(\tau)$ is the response function that describes the distribution of transit times in the flow system.

As discussed above, the use of single ^{3}H activities to estimate mean transit times requires that a lumped parameter model be assigned. The following lumped parameter models were considered in this study: the exponential flow model, the exponential-piston flow model, and the dispersion model. Together these represent the most commonly used lumped parameter models in determining mean transit times (e.g. as reviewed by McGuire and McDonnell, 2006). In catchments where time series data are available, these models generally reproduce the predicted variation in ^{3}H activities over time (e.g. Małoszewski and Zuber, 1982; Małoszewski et al., 1983; Zuber et al., 2005; Stewart et al., 2007; Morgenstern et al., 2015).

The exponential flow model predicts transit times in homogeneous unconfined aquifers of constant thickness with uniform recharge where water from the entire aquifer discharges to the stream or is sampled by a bore. Flow through aquifers where flow paths are the same length and no mixing occurs results in all water discharging to the stream at any given time having the same transit time and is described by the piston flow model. Transit times in aquifers that have regions of both piston and exponential flow are described by the exponential-piston flow model, for which the response function is

$$g(\tau) = 0 \qquad \text{for } \tau < \tau_m(1-f) \qquad (2a)$$

$$g(\tau) = (f\tau_m)^{-1} e^{-\tau/f\tau_m + 1/f - 1} \qquad \text{for } \tau > \tau_m(1-f), \qquad (2b)$$

where τ_m is the mean transit time and f is the proportion of the aquifer volume that exhibits exponential flow. At $f = 1$, this model becomes the exponential flow model, while at $f = 0$, it becomes the piston flow model. TracerLPM specifies the ratio of exponential to piston flow as the EPM ratio (equivalent to $1/f - 1$).

The dispersion model is based on the solution to the one-dimensional advection–dispersion transport equation. The response function for the dispersion model is

$$g(\tau) = \frac{1}{\tau\sqrt{4\pi D_P \tau/\tau_m}} e^{-\left(\frac{(1-\tau/\tau_m)^2}{4D_P\tau/\tau_m}\right)}, \qquad (3)$$

where D_P is the dispersion parameter defined as $D_P = D/(vx)$, where D is the dispersion coefficient (m^2 day^{-1}), v is velocity (m day^{-1}), and x is distance (m). This model is generally considered to be a less realistic representation of flow systems; however, it does commonly reproduce the observed distribution of tracers (Małoszewski, 2000).

In this study, mean transit times were calculated by matching the predicted ^{3}H from the lumped parameter model to the measured value at the time of collection using TracerLPM. For these calculations, the form of the lumped parameter model and its parameters (i.e. the EPM ratio or D_P) and the ^{3}H input function were specified.

3 Results

This section presents the streamflow and geochemistry data (Table 1) from the Victorian Alps; interpretation of these data appears in later sections.

3.1 Streamflow

As noted above, very few streamflow measurements exist in these catchments. Watchbed Creek drains the peatlands at Falls Rocky A (Fig. S1b) and there is a near-complete streamflow record for this site between 1940 and 1986. The average discharge for Watchbed Creek over that time period was 5.64×10^6 m^3 yr^{-1} (Department of Environment, Land, Water and Planning, 2016). For the average annual rainfall of 1.25 m yr^{-1} (Bureau of Meteorology, 2016), the runoff coefficient (i.e. the volume of rainfall that is exported by the stream) was 78 %. This is much higher than the runoff coefficients in the eucalypt-dominated catchments of the upper Ovens Valley which range from 6 to 64 % (Cartwright and Morgenstern, 2015). That Watchbed Creek exports more rainfall than streams from the eucalypt forests is also apparent in the flow duration curves (Fig. 2b). Relative discharges from Watchbed Creek are around an order of magnitude higher than those from Simmons Creek in the upper Ovens catchment. Simmons Creek has a slightly larger catchment area (6 km^2 vs. 3.5 km^2) and receives similar rainfall but drains eucalypt forest.

3.2 Stable isotope ratios

Stable isotope ratios are presented in Table 1 and Fig. 3. Rainfall has δ^{18}O values of −7.9 to −6.5 ‰ and δ^2H values of −40 to −34 ‰. The δ^{18}O and δ^2H values of the peatland water range from −8.3 to −5.0 and −43 to −26 ‰, respectively. The eucalyptus forest streams have smaller ranges of δ^{18}O and δ^2H values (−7.4 to −5.7 and −38 to −29 ‰, respectively). As is commonly the case in southeast Australia (Cartwright et al., 2012; Leaney and Herczeg, 1999), the δ^{18}O and δ^2H values of all waters including rainfall lie to the left of the global and Melbourne meteoric water lines. Overall, the waters define an array with a slope of ∼5 and have D excess values between 14 and 24 ‰ with a median of 19 ‰. These D excess values are significantly higher than the mean D excess of the Melbourne meteoric water line of 9.6 ‰ (Hughes and Crawford, 2012).

3.3 Tritium activities

The ^{3}H activities of three multi-month rainfall samples from Mount Buffalo are 2.85 TU (12-month sample collected in February 2015), 2.88 TU (9-month sample collected in November 2015), and 2.99 (17-month sample collected in December 2013) (Table 1). Three samples of rainfall collected over periods of 2 weeks to 3 months in 2014 also from Mount Buffalo have ^{3}H activities of 2.52 to 2.90 TU

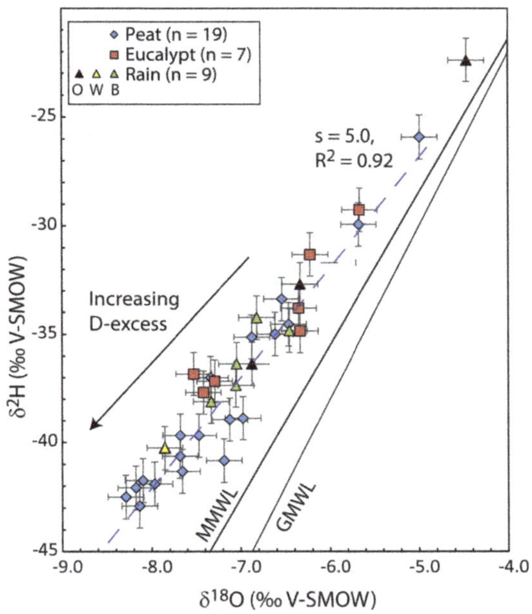

Figure 3. $\delta^{18}O$ vs. δ^2H values of waters from the Australian Alps (data from Table 1) relative to the global meteoric water line (GMWL: Clark and Fritz, 1997) and the Melbourne meteoric water line (MMWL: Hughes and Crawford, 2012). Each data point represents an individual sample collected on the date indicated in Table 1. Error bars are analytical uncertainties (0.2 ‰ for $\delta^{18}O$ and 1 ‰ for δ^2H). Rainfall samples: B indicates Mount Buffalo multi-month samples; W indicates the 2-week winter precipitation sample; O indicates other samples from southeast Australia. Dashed line (slope, $S = 5$) is fit to all of the data by linear regression.

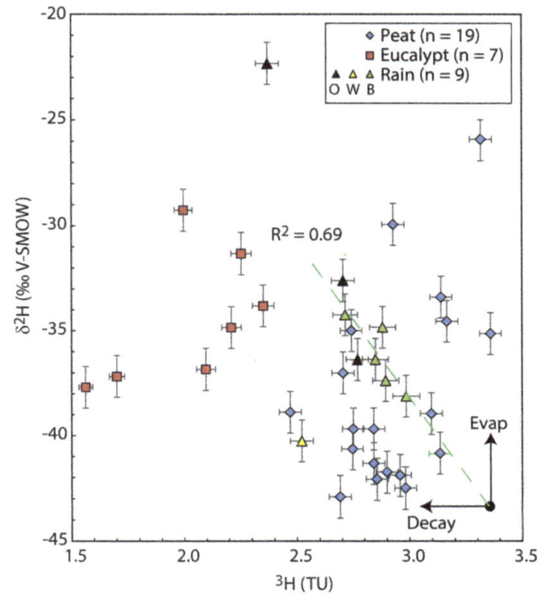

Figure 4. ^3H activities vs. δ^2H values of waters from the Australian Alps (data from Table 1). Error bars are analytical uncertainties (1 ‰ for δ^2H and ^3H from Table 1). Rainfall samples: B indicates Mount Buffalo multi-month samples; W indicates the 2-week winter precipitation sample; O indicates other samples from southeast Australia. Linear regression line is fit to the multi-month Mount Buffalo samples. Arrowed lines show changes in ^3H activities and δ^2H values due to evaporation (evap) and radioactive decay of ^3H.

(Table 1). The sample with the lowest ^3H activity represents mainly snow and low-temperature rainfall collected over a 2-week period in July 2014. The ^3H activities are within the expected range of ^3H activities in rainfall for this region of 2.8 to 3.2 TU (Tadros et al., 2014). Aside from the July 2014 rainfall sample that represents the 2 weeks of winter precipitation, there is an inverse correlation ($R^2 = 0.69$) between the ^3H activities and δ^2H of rainfall (Fig. 4). Since the rainfall with the higher δ^2H values has lower D excess values, there is also a correlation between D excess and ^3H activities. Rainfall from elsewhere in southeast Australia defines broadly similar trends in $\delta^{18}O$ vs. δ^2H values and δ^2H vs. ^3H values (Table 1; Figs. 3, 4).

The peatland waters have ^3H activities that vary from 2.70 to 3.32 TU (median = 2.75 TU) at Mount Buffalo, 2.90 to 3.17 (median = 3.04 TU) at Falls Creek, and 2.47 to 3.36 (median = 2.77 TU) at Dargo (Table 1; Figs. 4, 5). The median ^3H activity of the peatland waters as a whole is 2.88 TU. The ^3H activities of the peatland water from the three sites are closely similar. Additionally, there are no significant differences in ^3H activities of water collected from within the peat and that collected from the streams draining the peat. The ^3H activities of waters collected in February 2015 and those collected in November 2015 also overlap. The ^3H activities of the peatland waters are locally higher than those of the rainfall samples and several of those waters also have lower δ^2H values and higher D excesses than the rainfall (Figs. 3, 4).

The eucalyptus forest streams have ^3H activities of between 1.56 and 2.35 TU (median = 2.10 TU) (Table 1; Figs. 4, 5) that are lower than those of the peatland waters. The ^3H activities of the eucalyptus streams overlap with those of river water from the streams in the upper Ovens catchment at low flow conditions in December 2013 and February 2014, which range from 1.63 to 2.28 TU (median = 2.09 TU) (Cartwright and Morgenstern, 2015; Fig. 5).

3.4 Major ion geochemistry

Cl concentrations of the peatland waters are between 0.5 and 1.1 mg L^{-1}, which are similar to those of rainfall from Mount Buffalo (0.8 to 1.1 mg L^{-1}) (Fig. 5a, Table 1). Peatland water from Falls Creek has lower Cl concentrations (0.5 to 0.9 mg L^{-1}) than that from the other areas; however, there is no significant difference between the Cl concentrations of the water extracted from within the peat and that draining the peatlands. Na concentrations of the peatland water range from 0.8 to 2.1 mg L^{-1}, which are higher than those of rainfall (0.6 to 1.0 mg L^{-1}) (Fig. 5b). The eucalypt for-

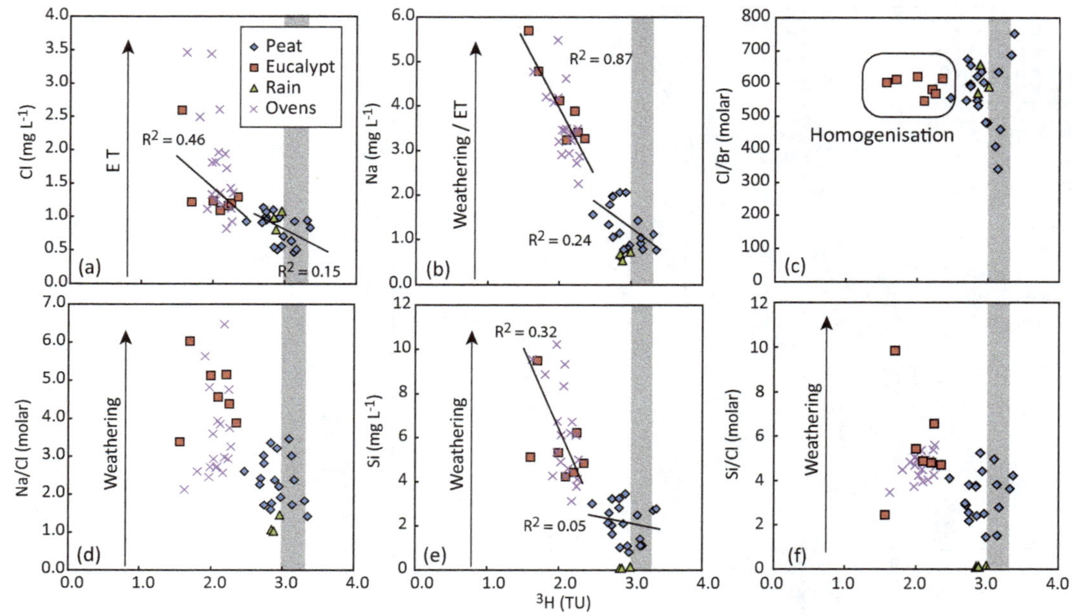

Figure 5. ^{3}H activities vs. Cl **(a)**, Na **(b)**, Si **(e)** concentrations and Cl / Br **(c)**, Na / Cl **(d)**, Si / Cl **(f)** ratios (data from Table 1). Arrowed lines illustrate trends expected from major geochemical processes. Ovens data are from the upper Ovens Valley streams in December 2013 and February 2014 (Cartwright and Morgenstern, 2015); the geochemistry of these larger streams is similar to the eucalypt forest streams in the Victorian Alps.

est streams have higher Cl (1.1 to 2.6 mg L^{-1}) and Na (3.2 to 5.7 mg L^{-1}) concentrations than any of the peatland water (Fig. 5a, b). Molar Na / Cl ratios range from 1.7 to 3.5 in the peatland water and 3.9 to 6.8 in the eucalypt forest streams (Fig. 5d). These Na / Cl ratios are higher than those of Buffalo rainfall, which has Na / Cl ratios of 1.0 to 1.1 that are similar to those generally recorded in southeast Australia (Blackburn and McLeod, 1983; Crosbie et al., 2012).

Molar Cl / Br ratios vary between 340 and 650 in the peatland water and between 550 and 780 in the eucalypt forest streams (Fig. 5c). These ratios are similar to those of Buffalo rainfall (570 to 650) and also similar to the ocean Cl / Br ratio of ~650 (Davis et al., 1998). Other cation / Cl ratios are higher in the eucalypt forest streams compared with the peatland water, and the water from both these sources has higher cation / Cl ratios than rainfall. Si concentrations in Buffalo rainfall are <0.1 mg L^{-1} (Fig. 5e) and molar Si / Cl ratios are ~0.1 (Fig. 5f). By contrast, Si concentrations and Si / Cl ratios in the peatland water and eucalypt forest streams are significantly higher (up to 9.8 mg L^{-1} and 9.5, respectively; Fig. 5e, f).

There are broad inverse correlations between ^{3}H activities and the Na, Cl, and Si concentrations and the Na / Cl and Si / Cl ratios (Fig. 5). In general, the differences in geochemistry between the peatland water and eucalypt forest streams are more marked than the geochemical variations within those groups. For example, the Si vs. ^{3}H trend for the combined peatland water and eucalypt forest streams has an R^2 of 0.63, whereas R^2 values for Si vs. ^{3}H in the peatland water and the eucalyptus forest streams are 0.05 and 0.32, respectively. The correlations between major element concentrations and ^{3}H activities are stronger in the eucalypt forest streams (R^2 values of 0.32 to 0.87) than in the peatland waters (R^2 values of 0.05 to 0.24). The trends in ^{3}H activities vs. major ion concentrations or ratios in the eucalyptus forest streams from the Victorian Alps are similar to those of the upper Ovens catchment as a whole (Fig. 5).

4 Discussion

This section combines the major ion geochemistry, stable isotope data, and ^{3}H activities to determine the mean transit times of the waters from the peatlands and eucalyptus forest streams and assesses the major geochemical processes that impact these waters.

4.1 Modern rainfall ^{3}H activities

Understanding the ^{3}H activities of the water that recharges the catchments is critically important in estimating mean transit times. The three longer-term (9- to 17-month) rainfall samples have ^{3}H activities of between 2.85 and 2.99 TU, which lie within the predicted ^{3}H activity of rainfall in this area of 2.8 to 3.2 TU (Tadros et al., 2014). However, the observation that the peatland water has ^{3}H activities that are locally up to 3.35 TU indicates that it is not possible to use the rainfall ^{3}H activities as the ^{3}H activity of recharge for all of the peatland waters.

There may be spatial variations in rainfall ^{3}H activities, especially between the three areas. The observation, however, that peatland water at Mount Buffalo from <5 km from the rainfall gauge has a ^{3}H activity of up to 3.32 TU suggests that this cannot be the only explanation. The higher ^{3}H activities in the peatland waters potentially reflect preferential recharge by spring rainfall. Maximum ^{3}H activities in Australian rainfall are recorded in early spring (August to September) as this is the time of maximum transport of water vapour with high ^{3}H activities from the stratosphere to the troposphere (Tadros et al., 2014). Stratospheric moisture also has lower δ^2H values and higher D excesses than tropospheric moisture (Bony et al., 2008). Thus, the variation in the relative ratios of stratospheric to tropospheric moisture would also explain the observations that rainfall samples with higher ^{3}H activities have lower δ^2H values (Fig. 4) and higher D excess values (Fig. 3).

Some peatland waters have lower δ^2H values than those of the rainfall. These potentially record recharge by rainfall that has a high proportion of low δ^2H stratospheric moisture. Such rainfall would also be expected to have higher ^{3}H activities than average rainfall. While it is difficult to precisely constrain the ^{3}H of seasonal recharge, extending the rainfall δ^2H vs. ^{3}H trend defined by the Mount Buffalo rainfall samples in Fig. 4 to the lowest δ^2H value of the peatland waters (−43 ‰) yields a ^{3}H activity of ~3.4 TU, which is similar to the highest ^{3}H activities recorded in the peatland waters. This is ~14 % higher than the ^{3}H activities of the rainfall samples, which is well within the seasonal variation of ^{3}H activities of Australian rainfall (Tadros et al., 2014).

The reason that some of the peatland waters with high ^{3}H activities have higher δ^2H values is likely due to subsequent evaporation (Fig. 4). As discussed below, the major ion geochemistry also implies that evaporation has affected these waters. Evaporation at a humidity of 50 to 70 %, which is characteristic of these Alpine areas in summer (Bureau of Meteorology, 2016), produces arrays of δ^{18}O and δ^2H values with slopes of 4 to 5 (Clark and Fritz, 1997). Thus, the δ^{18}O and δ^2H values of the waters (Fig. 3) probably reflects both initial variations in stable isotope ratios and subsequent evaporation. Assuming that the mass-dependant isotope fractionation of ^{3}H / ^{1}H is approximately double that of ^{2}H / ^{1}H, a 10 ‰ increase in δ^2H values would equate to a ~2 % increase in ^{3}H activities, which is approximately the analytical precision.

4.2 Mean transit times

Mean transit times (Table 2; Fig. 6a) were initially calculated using an exponential-piston flow model with an EPM ratio of 0.33 via Eqs. (1) and (2). The soils and shallow aquifers are unconfined and likely to exhibit exponential flow below the water table; however, recharge through the unsaturated

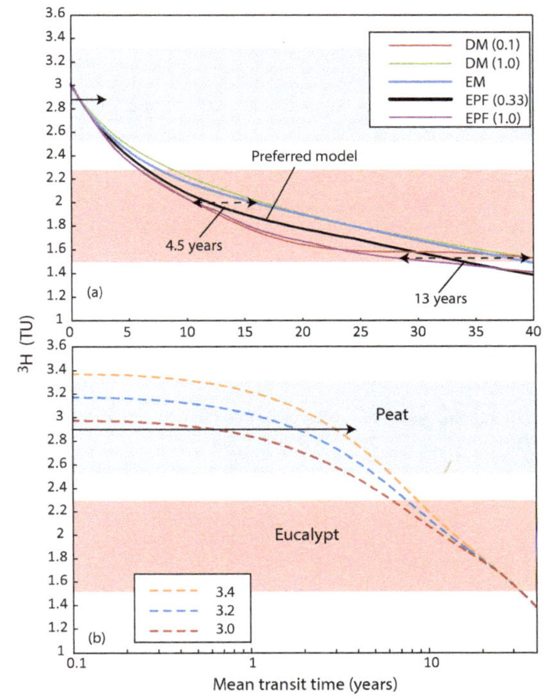

Figure 6. Comparison of mean transit times from different lumped parameter models calculated using Eqs. (1) to (3) assuming a ^{3}H activity of recharge of 3.0 TU (based on the rainfall samples). DM indicates dispersion model with D_P values of 0.1 and 1.0; EM indicates exponential flow model; EPF indicates exponential-piston flow model with EPM ratios of 0.33 and 1.0. **(b)** Mean transit times from the exponential flow model (EPM ratio of 0.33) with ^{3}H activities of modern recharge ranging from 3.0 to 3.4 TU. Shaded fields show ^{3}H activities in the peatland waters and eucalyptus forest streams; arrow is the median ^{3}H activity in the peatland waters (data from Table 1).

zone will most likely resemble piston flow. A similar flow model successfully reproduces time series of ^{3}H activities in upland catchments in New Zealand (Morgenstern et al., 2010). Given that the waters were sampled at baseflow conditions during the late spring or summer, which represent the driest months, it was assumed that there was a single store of water present rather than a mixture of recent rainfall and older water. While the exponential-piston flow model plausibly represents the flow systems, it is not possible to independently test its veracity. Hence, mean transit times were also calculated using the exponential flow model with a higher proportion of piston flow (EPM ratio of 1), the exponential flow model, and the dispersion model with $D_P = 0.1$ and 1.0 which are appropriate values for flow systems of this scale (Małoszewski, 2000) (Table 2; Fig. 6a).

The annual average ^{3}H activities of rainfall in Melbourne, which is located ~200 km to the southwest, from the International Atomic Energy Agency Global Network of Isotopes in Precipitation program as summarised by Tadros et

Table 2. Calculated mean transit times.

Site[a]	Date	Mean transit time (yr)						
LPM[b]		EPF (3.4)[c]	EPF (3.2)	EPF (3.0)	EPF (3.4)	EM (3.4)	DM (3.4)	DM (3.4)
Parameter[d]		0.3	0.3	0.3	1		0.1	1
Peatlands								
Mount Buffalo								
Buffalo Cresta A	26/11/2015	< 1	ND[e]	ND	< 1	< 1	< 1	< 1
Buffalo Cresta A	23/02/2015	3.4	2.2	1.0	3.3	3.5	2.7	4.0
Buffalo Cresta B(P)	23/02/2015	4.1	2.9	1.6	3.9	4.4	3.9	4.9
Buffalo Cresta C	23/02/2015	4.1	2.9	1.6	3.9	4.4	3.9	4.9
Buffalo Cresta D(P)	26/11/2015	4.4	3.2	1.9	4.2	4.8	4.2	5.4
Buffalo Crystal A	23/02/2015	4.2	3.0	1.7	3.9	4.4	3.9	5.3
Buffalo Crystal B	23/02/2015	2.8	1.6	< 1	2.7	2.9	2.8	3.3
Falls Creek								
Falls Cope	24/02/2015	1.4	< 1	ND	1.4	1.5	1.4	1.5
Falls Langford A	24/02/2015	1.7	< 1	ND	1.7	1.7	1.7	1.9
Falls Langford B(P)	24/02/2015	2.4	1.3	< 1	2.4	2.5	2.4	2.9
Falls Rocky A	24/02/2015	2.6	1.4	< 1	2.5	2.7	2.5	3.0
Falls Rocky A	25/11/2015	3.0	1.8	< 1	2.9	3.1	2.9	3.5
Falls Rocky B(P)	25/11/2015	3.3	2.1	< 1	3.2	3.5	3.2	3.9
Falls Rocky C(P)	25/11/2015	1.3	< 1	ND	1.3	1.3	1.3	1.4
Falls Rocky D	25/11/2015	1.4	< 1	ND	1.4	1.5	1.4	1.5
Dargo								
Dargo A	25/11/2015	3.4	2.2	1.0	3.3	3.6	3.2	4.0
Dargo B	25/11/2015	4.5	3.3	2.0	4.3	4.9	4.3	5.6
Dargo C(P)	25/11/2015	6.5	5.4	4.0	6.0	7.4	6.0	8.0
Dargo D	25/11/2015	< 1	ND	ND	< 1	< 1	< 1	< 1
Eucalypt forest								
Falls Forest A	24/02/2015	28.9	28.8	28.6	26.0	39.0	37.2	40.2
Falls Forest B	24/02/2015	11.5	10.8	9.6	9.9	16.4	10.0	17.7
Buffalo Forest A	24/02/2015	9.7	8.8	7.5	8.7	13.0	8.6	14.7
Falls Forest C	25/11/2015	9.1	8.1	6.6	8.0	11.3	7.9	13.2
Falls Forest D	25/11/2015	22.6	22.2	21.6	17.2	30.7	16.7	31.4
Dargo Forest A	25/11/2015	14.0	12.4	11.2	11.3	18.8	11.1	19.6
Buffalo Forest A	26/11/2015	7.8	6.9	5.3	7.0	9.4	7.0	10.8

[a] Sites in Fig. S1 and S1. [b] Lumped parameter model (DM indicates dispersion model; EM indicates exponential flow model; EPF indicates exponential-piston flow model). [c] Assumed ^{3}H activity of modern recharge in brackets. [d] EPM or D_P parameter. [e] Not determined as ^{3}H activity higher than assumed value of recharge.

al. (2014), were used as the ^{3}H input. Rainfall ^{3}H activities reached $\sim$ 62 TU in 1965 and then declined exponentially to present-day values (Tadros et al., 2014). The calculations initially adopted a present-day ^{3}H activity of recharge of 3.4 TU (as discussed above). A ^{3}H activity of 3.4 TU was also used for recharge from prior to the atmospheric nuclear tests.

Mean transit times from the exponential-piston flow model with an EPM ratio of 0.33 are up to 6.4 years with a median of 3.0 years (Table 2). For the range of ^{3}H activities in these peatland waters, similar mean transit times are estimated from the different lumped parameter models (Table 2; Fig. 6a). The exponential-piston flow model with an EPM ratio of 1 yields mean transit times of up to 6.0 years with a median of 2.9 years. The exponential flow model produces mean transit times of up to 7.4 years with a median of 3.1 years. The dispersion model yields mean transit times of up to 6.0 years with a median of 2.8 years ($D_P = 0.1$) and 8.0 years with a median of 3.5 years ($D_P = 1.0$).

For a water with a ^{3}H activity of 3 TU, propagating the analytical uncertainty of ± 2 % produces an uncertainty in mean transit times of ± 0.3 years. Uncertainties in calculated mean transit times for the high ^{3}H samples mainly arise from the assumed ^{3}H activity of the recharging water (Fig. 6b), which, as discussed above, is difficult to constrain precisely. This

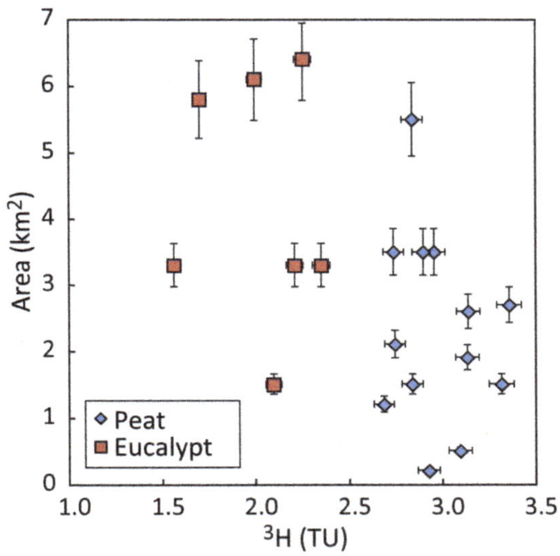

Figure 7. ^{3}H activities vs. catchment areas for the peatlands and eucalyptus forest catchments. Area error bars assume a ±10 % measurement error; ^{3}H error bars are from the analytical uncertainties. Data are from Table 1.

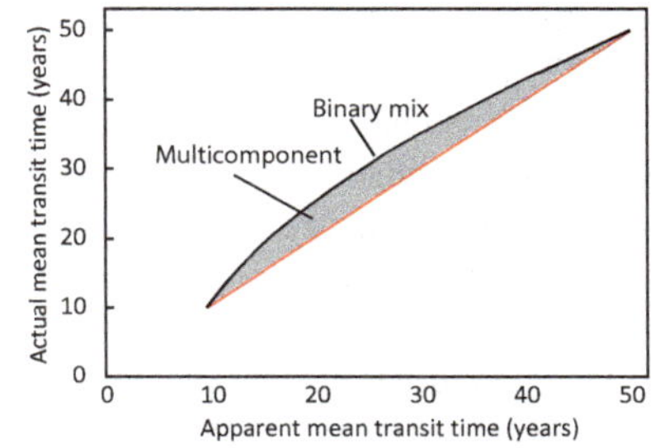

Figure 8. Impact of aggregation on calculated mean transit times. Apparent mean transit times are calculated from the ^{3}H activities of mixtures of waters with actual mean transit times of between 10 and 50 years using the exponential-piston flow model with an EPM ratio of 0.33. Actual mean transit times are calculated from the weighted average mean transit times in the mixed water. The binary curve represents variable mixtures of the two end-members, the shaded field represents mixtures of multiple waters with mean transit times between 10 and 50 years. The red line represents agreement between apparent and actual mean transit times.

will be illustrated for the exponential-piston flow model with an EPM ratio of 0.33, but similar variations would be observed for the other models. Utilising a ^{3}H activity of 3.0 TU for modern and pre-bomb-pulse rainfall based on the ^{3}H activities of the rainfall samples (Table 1) yields mean transit times of up to 4.0 years with a median of < 1 year (Table 2). However, in this case a large number of peatland waters have ^{3}H activities that are higher than the rainfall activities and for those waters a ^{3}H activity of 3.0 TU for the recharging water is not realistic. Adopting a ^{3}H activity of 3.2 TU, which is the upper limit of the estimate for annual rainfall in this area (Tadros et al., 2004) yields mean transit times of up to 5.4 years with a median of 1.8 years. In this case, two peatland waters (Dargo D and Buffalo Cresta A) have ^{3}H activities that are higher than the assumed recharge ^{3}H activities. Given the likelihood that seasonal recharge has occurred, and thus that the ^{3}H activity of the recharging water may be locally variable, it is unrealistic to differentiate mean transit times of < 1 year.

By contrast with the peatland waters, the eucalypt forest streams have much longer mean transit times that are relatively insensitive to the assumed ^{3}H activity of recharge (Fig. 6b). For the exponential-piston flow model with an EPM ratio of 0.33 and a modern and pre-bomb-pulse rainfall activity of 3.0 TU (which assumes recharge has the ^{3}H activity of average rainfall), mean transit times range from 5.3 to 28.6 years with a median of 9.5 years (Table 2). Adopting a ^{3}H activity for recharge of 3.2 TU, the mean transit times range from 6.9 to 28.8 years with a median of 10.8 years and if the ^{3}H activity of recharge was 3.4 TU, the mean transit times are 7.8 to 28.9 years with a median of 11.5 years. These mean transit times are within the range of those of the much larger (6 to 1240 km^2) catchments in the upper Ovens Valley that are also dominated by eucalypt forest. Mean transit times of river water in December 2013 and February 2014 in the upper Ovens Valley calculated using the same exponential-piston flow model and a ^{3}H activity of recharge of 3.0 TU are 7 to 30 years (Cartwright and Morgenstern, 2015).

For the eucalypt forest streams, there are larger uncertainties in mean transit times produced by the different lumped parameter models (Table 2; Fig. 6a). The range in mean transit times for a water with a ^{3}H activity of 2 TU calculated from the exponential flow model, the exponential-piston flow model with EPM ratios of 0.33 and 1, and the dispersion model with $D_P = 0.1$ and $D_P = 1$ is ∼4.5 years. For a water with a ^{3}H activity of 1.5 TU, the range in mean transit times from those models is ∼13 years. Uncertainties as to the most appropriate lumped parameter model thus represent a significant source of uncertainty in these calculations. However, overall, the water in these streams has mean transit times of years to decades and is significantly older than that from the peatlands. Propagating the analytical uncertainty in ^{3}H activities (Table 1) also results in uncertainties in mean transit times of ±0.5 years for a water with a ^{3}H activity of 2 TU.

Aggregation of water with different mean transit times also introduces uncertainty in the calculated mean transit times (Kirchner, 2016a; Stewart et al., 2016). Figure 8 illustrates the impact of mixing different proportions of water with mean transit times of between 10 and 50 years (appropriate for the eucalypt forest waters) on the calculated mean transit times. The actual mean transit time represents the mean transit time that calculated from the weighted average of the transit times of the mixed waters. The apparent mean transit time is calculated from the ^{3}H activity of the mixed water again using the exponential-piston flow model with an EPM ratio of 0.33. The apparent mean transit times are less than the actual mean transit times with the greatest difference (~ 13 %) occurring when there are approximately equal proportions of two end-members with mean transit times of 10 and 50 years. The difference between actual and apparent mean transit times decreases when multiple end-members are aggregated (shaded field). For example, the apparent mean transit time calculated from an equal mixture of nine waters with transit times of 10, 15, 20, 25, 30, 35, 40, 45, and 50 years is 28.9 years, which is within 4 % of the actual mean transit time of 30 years. The relative difference between apparent and actual mean transit times are similar when other lumped parameter models are used. Mixing waters with a range of mean transit times between 1 and 5 years (which would be appropriate for the peatland waters) produces similar relative differences. Thus, aggregation introduces uncertainties that are of similar order of magnitude to those resulting from other uncertainties in the calculations (discussed above).

There are uncertainties in the calculated mean transit times resulting from uncertainties in the ^{3}H activity of recharge, the most appropriate lumped parameter model, and aggregation. However, the conclusion that the mean transit times of water in the eucalypt forest catchments range from several years to decades and are significantly longer than those in the peatlands remains robust. Mean transit times of the peatland waters are unlikely to be more than a few years and may be mainly less than 2 years depending on the ^{3}H activity of recharge. As noted above, the ^{3}H activities of the water from the piezometers in the peatlands are similar to the water that drains the peat. The water that drains the peat is likely derived mainly from the acrotelm (Western et al., 2009; Grover and Baldock, 2013). Presuming that the piezometers which were inserted into the lower levels of the peat sampled water from the catotelm, this observation implies that the catotelm is not a store of older water.

4.3 Major ion tracers

The major ion geochemistry allows the main geochemical processes to be understood and also can provide first-order estimates of mean transit times. Overall, the observation that the peatland water and the eucalypt forest streams have molar Cl / Br ratios that cluster around those of the rainfall implies that, in common with the majority of groundwater and surface water in southeast Australia, the vast majority of Cl is derived from the rainfall and concentrated by evapotranspiration (Cartwright et al., 2006; Herczeg et al., 2001). Other potential sources of Cl (such as dissolution of halite in the unsaturated zone) produce water with high Cl / Br ratios and can thus be discounted.

The higher Cl concentrations in the eucalypt forest streams compared with that from the peatland water reflects higher net evapotranspiration rates in the eucalypt forest catchments. Some of the increase in the concentration of the other major ions (such as Na) in the eucalypt forest streams and peatland water over those in rainfall will also be due to evapotranspiration. The adsorption of Br by organic matter (Gerritse and George, 1988) may locally increase Cl / Br ratios if the pool of organic matter is increasing. By contrast, a net degradation of organic matter releases Br, producing lower Cl / Br ratios. The wider range of Cl / Br ratios in the peatland water may reflect that, locally, the volume of organic matter is increasing, whereas elsewhere it is degrading. The far more homogeneous Cl / Br ratios in the eucalypt forest streams indicate that either the organic matter content of those catchments is relatively stable or that the waters are better mixed and any local variations in Cl / Br ratios have been homogenised. This homogenisation is consistent with longer mean transit times in the eucalypt forest streams.

The increase in cation / Cl and Si / Cl ratios in the peatland and eucalypt forest waters over those in rainfall is interpreted to be due to silicate weathering (cf., Herczeg and Edmunds, 2000). As rainfall contains $<0.1\,\mathrm{mg\,L^{-1}}$ Si, evapotranspiration will not increase Si concentrations appreciable and the vast majority of the Si will be derived from silicate weathering. The eucalypt forest streams have higher cation / Cl and Si / Cl ratios compared with the peatland water and thus records higher degrees of weathering. This is likely due to two factors. Firstly, the eucalypt forest catchments are developed on weathered regolith and the flow of water through the regolith will promote mineral dissolution. Common minerals in the regolith include plagioclase and alkali feldspars (Cartwright and Morgenstern, 2012) and the dissolution of these will provide Si, Na, Ca, and K. By contrast, the peatlands are developed on less weathered rocks, which limits the potential for mineral dissolution; although there clearly has been sufficient weathering to produce the elevated cation / Cl and Si / Cl ratios. There is commonly a thin (<20 cm thick) layer of regolith at the base of the peat and fragments of basement rock within the lower layers of the peat that are likely being weathered by the peatland water. Secondly, the longer mean transit times in the eucalypt forest catchments allow the silicate weathering reactions to progress further than in the peatlands.

4.4 Controls on mean transit times

As with the Ovens catchment as a whole, there is no correlation between ^{3}H activities and catchment areas either within or between the eucalypt forest and peatland catchments (Fig. 7). Elsewhere, inverse correlations between catchment slope and mean transit times have been documented (McGuire et al., 2005). Catchment slopes have not been calculated in this study (and would be difficult do so in the peatlands given the limitations of the DEM). However, the observation that the eucalypt forests occur in gullies that invariably have steeper slopes than the peatlands but contain water with longer mean transit times indicates that, in this case, catchment slopes cannot explain the difference in mean transit times between the two catchment types.

The difference in the mean transit times between the peatland water and the eucalypt forest streams is likely the combination of two factors. Firstly, the high evapotranspiration rates of the eucalypt forests (Allison et al., 1990) results in low recharge rates that in turn increase the transit times. Secondly, differences in the depth of weathering in the catchments and the development of deeper flow paths is also likely to be important. Most of the peatlands are developed on relatively unweathered basement rocks and the majority of the water is stored within the peat itself (Grover and Baldock, 2013; Western et al., 2009). Given that most of the peat is < 2 m thick, this results in a shallow reservoir of water underlain by bedrock with very low hydraulic conductivities. By contrast, groundwater flow in the eucalypt gullies is partially hosted in weathered regolith that is locally several metres thick. The gullies may be developed where weathering is greater due to the presence of joints and/or faults that will also host groundwater flow (Shugg, 1987). The presence of weathered and fractured bedrock on steep slopes conceivably allows long groundwater flow paths to develop that will increase transit times.

The correlation between major ion concentrations and ^{3}H activities may allow a first-order estimate of relative mean transit times to be made, which is useful in extending studies such as this to adjacent catchments. Caution must be exercised, however, in using the data sets as a whole, as the geochemistry may partially reflect differences in the characteristics of the peatland and eucalypt forest catchments (specifically whether the water flows through the regolith or is contained within the peat) rather than solely the mean transit times. Nevertheless, in the catchments studied here there are broad correlations between ^{3}H activities and major ion concentrations within both the peatland water and the eucalypt forest streams, which would permit estimates of relative mean transit times.

Streamflow (Q) is related to the mean transit time (MTT) and the volume of water held in storage (V) via $Q = V$ / MTT (Małoszewski and Zuber, 1982). This relationship is commonly used to estimate V where MTT has been calculated and Q measured (e.g. Morgenstern et al., 2010). However, in ungauged peatland catchments such as these it is possible to estimate Q if the volume of peat can be estimated and MTT is known. The area drained by the stream at Falls Rocky A is 3.5×10^6 m^2 and for a peat thickness of 0.5 to 1 m, which is consistent with observations at this site, $V = 1.75$ to 3.5×10^6 m^3. Adopting a range of mean transit times of 0.5 to 2.5 years (Table 2) yields estimates of Q of 7.0×10^5 to 7.0×10^6 m^3 yr^{-1}, which span the measured average discharge of 5.64×10^6 m^3 yr^{-1}. Presuming that the parameterisation is appropriate, the correspondence between calculated and observed streamflows occurs when mean transit times are < 1 year, which suggests that the ^{3}H activities of rainfall that recharges the peat are not significantly higher than those discussed above. Carrying out these calculations on the Victorian peatland water is difficult as Q tends to ∞ as the MTT approaches 0. In ungauged peatland catchments where mean transit times are longer, however, this would be a viable method of estimating Q.

5 Conclusions

The mean transit times in peatland water in southeast Australia are less than 6.5 years and in many cases substantially shorter. The short mean transit times implies that the peat is susceptible to drying if there are successive years of below-average rainfall. In turn, this makes the peat vulnerable to damage during the periodic bushfires that also occur during the summers in drought periods. The majority of southeast Australia underwent a major drought (the "millennium drought") between 1995 and 2010 (Bureau of Meteorology, 2016) during which time there were major bushfires in the Australian Alps, especially in 2003 and 2009. Although they mainly affected the eucalypt forests, locally these bushfires also impacted the peatlands (Arthur Rylah Institute, 2016).

The observation that the water from within the peat yields similar mean transit times to that draining the peat implies that there is not likely to be older water stored within the deeper layers. This may be because the peat in the Victorian Alps is shallow and consequently stores little water. The peatlands do not act as a long-lived store of water to the river systems. Given their small total area, there would be little impact of peatland drying to the hydrology of the lower reaches of the adjacent catchments. However, drying of the peat could impact the hydrology of the small upland streams that form an integral part of the landscape and ecosystems of the alpine areas. The observation that even in small eucalypt forest catchments the stream water has mean transit times of several years to decades indicates that they are likely to be more resilient to short-term variations in rainfall. As a whole, the river catchments in the Victorian Alps are eucalypt dominated, which implies that the river systems are likely to be sustained by baseflow even during prolonged drought periods, and indeed many of these streams continued to flow through the millennium drought.

More generally, this study illustrates that the combination of ^{3}H and major ion geochemistry allows both the timescales of water movement and the hydrogeochemical processes in these catchments to be understood. As noted by Western et al. (2009), there is a common perception that, due to their high water contents and organic-rich soils, peatlands represent long-term water stores, and this was the premise that this study set out to test. However, this is not the case in this study and possibly not elsewhere. For example, the upland catchments in the United Kingdom that include extensive peat areas have transit times are < 5 years (Kirchner et al., 2012; Soulsby et al., 2006; Hrachowitz et al., 2013; Benettin et al.,

2015) and the Zhurucay area in Ecuador where mean transit times are < 1 year (Mosquera et al., 2016).

The study has determined transit times using individual ^3H measurements, which is possible in the Southern Hemisphere due to the decline of the bomb-pulse ^3H. While this removes the requirement of collecting data over several years and does not require the assumption of steady-state flow systems, it is not possible to test the appropriateness of the lumped parameter models. Differences between the mean transit times estimated from the different lumped parameter models are one of the main uncertainties, especially where the mean transit times are in excess of a few years. Uncertainties in the ^3H activities of recharge and errors caused by aggregation of water with different mean transit times also impact these calculations (as they do the calculations based on time series data). Irrespective of these uncertainties, the fact that the relative difference in mean transit times between waters is readily calculated is valuable in understanding catchments.

Information about the Supplement

Figure S1. Annotated Google Earth® images of Mount Buffalo (Fig. S1a), Falls Creek (Fig. S1b), and Dargo (Fig. S1c) catchments showing extent of the peatlands and the sampled eucalypt forest catchments. Sampling site abbreviations (sites in Table 1). Fig S1a: Cres indicates Buffalo Cresta, Cry indicates Buffalo Crystal, BF indicates Buffalo Forest. Fig. S1b: Cope indicates Falls Cope, Lan indicates Falls Langford, Rky indicates Falls Rocky, FF indicates Falls Forest. Fig. S1c: D indicates Dargo, DF indicates Dargo Forest.

Author contributions. Uwe Morgenstern was responsible for the ^3H analyses. Ian Cartwright undertook the sampling program and oversaw the analysis of the other geochemical parameters. Ian Cartwright and Uwe Morgenstern prepared the manuscript.

Acknowledgements. Funding for this project was provided by Monash University and the National Centre for Groundwater Research and Training. The National Centre for Groundwater Research and Training is an Australian Government initiative supported by the Australian Research Council and the National Water Commission via Special Research Initiative SR0800001. Massimo Raveggi and Rachael Pearson carried out the geochemical analyses at Monash University. E. Timbe, Markus Hrachowitz, and an anonymous reviewer are thanked for their helpful and perceptive comments that improved the paper.

Edited by: M. Hrachowitz

References

Allison, G. B., Cook, P. G., Barnett, S. R., Walker, G. R., Jolly, I. D., and Hughes, M. W.: Land clearance and river salinisation in the western Murray Basin, Australia, J. Hydrol., 119, 1–20, 1990.

Arthur Rylah Institute: Fire Ecology and Recovery, available at: http://www.depi.vic.gov.au/environment-and-wildlife/arthur-rylah-institute/research-themes/fire-ecology-and-recovery, last access: 22 July 2016.

Aravena, R. and Warner, B. G.: Oxygen-18 Composition of Sphagnum, and Microenvironmental Water Relations, Bryologist, 95, 445–448, 1992.

Benettin, P., Kirchner, J. W., Rinaldo, A., and Botter, G.: Modeling chloride transport using travel time distributions at Plynlimon, Wales, Water Resour. Res., 51, 3259–3276, 2015.

Blackburn, G. and McLeod, S.: Salinity of atmospheric precipitation in the Murray Darling Drainage Division, Australia, Aust. J. Soil Res., 21, 400–434, 1983.

Bony, S., Risi, C., and Vimeux, F.: Influence of convective processes on the isotopic composition ($\delta^{18}O$ and δD) of precipitation and water vapor in the tropics: 1. Radiative-convective equilibrium and Tropical Ocean–Global Atmosphere–Coupled Ocean-Atmosphere Response Experiment (TOGA-COARE) simulations, J. Geophys. Res., 113, D19305, doi:19310.11029/12008JD009942, 2008.

Bureau of Meteorology: Commonwealth of Australia Bureau of Meteorology, available at: http://www.bom.gov.au, last access: 22 July 2016.

Cartwright, I. and Morgenstern, U.: Constraining groundwater recharge and the rate of geochemical processes using tritium and major ion geochemistry: Ovens catchment, southeast Australia, J. Hydrol., 475, 137–149, 2012.

Cartwright, I. and Morgenstern, U.: Transit times from rainfall to baseflow in headwater catchments estimated using tritium: the Ovens River, Australia, Hydrol. Earth Syst. Sci., 19, 3771–3785, doi:10.5194/hess-19-3771-2015, 2015.

Cartwright, I., Weaver, T. R., and Fifield, L. K.: Cl / Br ratios and environmental isotopes as indicators of recharge variability and groundwater flow: An example from the southeast Murray Basin, Australia, Chem. Geol., 231, 38–56, 2006.

Cartwright, I., Weaver, T. R., Cendón, D. I., Fifield, L. K., Tweed, S. O., Petrides, B., and Swane, I.: Constraining groundwater flow, residence times, inter-aquifer mixing, and aquifer properties using environmental isotopes in the southeast Murray Basin, Australia, Appl. Geochem., 27, 1698–1709, 2012.

Charman, D. J., Aravena, R., Bryant, C. L., and Harkness, D. D.: Carbon isotopes in peat, DOC, CO_2, and CH_4 in a Holocene peatland on Dartmoor, Southwest England, Geology, 27, 539–542, 1999.

Clark, I. D. and Fritz, P.: Environmental Isotopes in Hydrogeology, Lewis, New York, USA, 1997.

Cook, P. G. and Bohlke, J. K.: Determining timescales for groundwater flow and solute transport, in: Environmental Tracers in Subsurface Hydrology, edited by: Cook, P. G. and Herczeg, A. L., Kluwer, Boston, USA, 1–30, 2000.

Coplen, T. B.: Normalization of oxygen and hydrogen isotope data, Chem. Geol., 72, 293–297, 1988.

Costin, A. B.: Carbon-14 dates from the Snowy Mountains area, southeastern Australia, and their interpretation, Quaternary Res., 2, 579–590, 1972.

Crosbie, R., Morrow, D., Cresswell, R., Leaney, F., Lamontagne, S., and Lefournour, M.: New insights to the chemical and isotopic composition of rainfall across Australia, CSIRO Water for a Healthy Country Flagship, Australia, 2012.

Davis, S. N., Whittemore, D. O., and Fabryka-Martin, J.: Uses of chloride/bromide ratios in studies of potable water, Ground Water, 36, 338–351, 1998.

Department of Environment, Land, Water and Planning: State Government Victoria Department of Environment, Land, Water and Planning Water Measurement Information System, available at: http://data.water.vic.gov.au/monitoring.htm, last access: 22 July 2016.

Dever, L., Hillaire-Marcel, C., and Fontes, J. C.: Isotopic composition, geochemistry and genesis of ice lenses (palsen) in peat bogs, New Quebec, Canada, J. Hydrol., 71, 107–130, 1984.

Dixon, R. K., Brown, S., Houghton, R. A., Solomon, A. M., Trexler, M. C., and Wisniewski, J.: Carbon pools and flux of global forest ecosystems, Science, 263, 185–190, 1994.

Energy and Earth Resources: State Government Victoria Department of Economic Development, Jobs, Transport and Resources, available at: http://www.energyandresources.vic.gov.au/earth-resources/maps-reports-and-data/geovic, last access: 22 July 2016.

Gerritse, R. G. and George, R. J.: The role of soil organic matter in the geochemical cycling of chloride and bromide, J. Hydrol., 101, 83–85, 1988.

Gorham, E.: Northern peatlands: role in the carbon cycle and probable responses to climatic warming, Ecol. Appl., 1, 182–195, 1991.

Grover, S. P. P. and Baldock, J. A.: Carbon decomposition processes in a peat from the Australian Alps, Eur. J. Soil Sci., 61, 217–230, 2010.

Grover, S. P. P. and Baldock, J. A.: The link between peat hydrology and decomposition: Beyond von Post, J. Hydrol., 479, 130–138, 2013.

Grover, S. P. P., McKenzie, B. M., Baldock, J. A., and Papst, W. A.: Chemical characterisation of bog peat and dried peat of the Australian Alps, Aust. J. Soil Res., 43, 963–971, 2005.

Herczeg, A. L. and Edmunds, W. M.: Inorganic ions as tracers, in: Environmental Tracers in Subsurface Hydrology, edited by: Cook, P. and Herczeg, A., Kluwer Academic Publishers, Boston, 31–77, 2000.

Herczeg, A. L., Dogramaci, S. S., and Leaney, F. W.: Origin of dissolved salts in a large, semi-arid groundwater system: Murray Basin, Australia, Mar. Freshwater Res., 52, 41–52, 2001.

Hrachowitz, M., Soulsby, C., Tetzlaff, D., and Speed, M.: Catchment transit times and landscape controls – Does scale matter?, Hydrol. Process., 24, 117–125, 2010.

Hrachowitz, M., Savenije, H., Bogaard, T. A., Tetzlaff, D., and Soulsby, C.: What can flux tracking teach us about water age distribution patterns and their temporal dynamics?, Hydrol. Earth Syst. Sci., 17, 533–564, doi:10.5194/hess-17-533-2013, 2013.

Hughes, C. E. and Crawford, J.: A new precipitation weighted method for determining the meteoric water line for hydrological applications demonstrated using Australian and global GNIP data, J. Hydrol., 464–465, 344–351, 2012.

International Atomic Energy Association: Global Network of Isotopes in Precipitation, available at: http://www.iaea.org/water, last access: 22 July 2016.

Jurgens, B. C., Bohlke, J. K., and Eberts, S. M.: TracerLPM (Version 1): An Excel® workbook for interpreting groundwater age distributions from environmental tracer data, US Geological Survey Techniques and Methods Report 4-F3, 2012.

Kirchner, J. W.: Aggregation in environmental systems – Part 1: Seasonal tracer cycles quantify young water fractions, but not mean transit times, in spatially heterogeneous catchments, Hydrol. Earth Syst. Sci., 20, 279–297, doi:10.5194/hess-20-279-2016, 2016a.

Kirchner, J. W.: Aggregation in environmental systems – Part 2: Catchment mean transit times and young water fractions under hydrologic nonstationarity, Hydrol. Earth Syst. Sci., 20, 299–328, doi:10.5194/hess-20-299-2016, 2016b.

Kirchner, J. W., Tetzlaff, D., and Soulsby, C.: Comparing chloride and water isotopes as hydrological tracers in two Scottish catchments, Hydrol. Process., 24, 1631–1645, 2010.

Leaney, F. and Herczeg, A.: The origin of fresh groundwater in the SW Murray Basin and its potential for salinisation, CSIRO Land and Water Technical Report 7/99, Adelaide, 1999.

Małoszewski, P.: Lumped-parameter models as a tool for determining the hydrological parameters of some groundwater systems based on isotope data, IAHS-AISH Publication, 271–276, 2000.

Małoszewski, P. and Zuber, A.: Determining the turnover time of groundwater systems with the aid of environmental tracers. 1. Models and their applicability, J. Hydrol., 57, 207–231, 1982.

Małoszewski, P., Rauert, W., Stichler, W., and Herrmann, A.: Application of flow models in an alpine catchment area using tritium and deuterium data, J. Hydrol., 66, 319–330, 1983.

Mažeika, J., Guobyte, R., Kibirktis, G., Petrošius, R., Skuratovič, Ž., and Taminskas, J.: The use of carbon-14 and Tritium for peat and water dynamics characterization: Case of Čepkeliai peatland, southeastern Lithuania, Geochronometria, 34, 41–48, 2009.

McDonnell, J. J., McGuire, K., Aggarwal, P., Beven, K. J., Biondi, D., Destouni, G., Dunn, S., James, A., Kirchner, J., Kraft, P., Lyon, S., Małoszewski, P., Newman, B., Pfister, L., Rinaldo, A., Rodhe, A., Sayama, T., Seibert, J., Solomon, K., Soulsby, C., Stewart, M., Tetzlaff, D., Tobin, C., Troch, P., Weiler, M., Western, A., Wörman, A., and Wrede, S.: How old is streamwater? Open questions in catchment transit time conceptualization, modelling and analysis, Hydrol. Process., 24, 1745–1754, 2010.

McDougall, K. L.: The Alpine Vegetation of the Bogong High Plains., Environmental Studies Publication No. 357, Ministry for Conservation, Melbourne, 1982.

McGuire, K. J. and McDonnell, J. J.: A review and evaluation of catchment transit time modeling, J. Hydrol., 330, 534–346, 2006.

McGuire, K. J., McDonnell, J. J., Weiler, M., Kendall, C., McGlynn, B. L., Welker, J. M., and Seibert, J.: The role of topography on catchment-scale water residence time, Water Resour. Res., 41, 1–14, 2005.

Michel, R. L.: Residence times in river basins as determined by analysis of long-term tritium records, J. Hydrol., 130, 367–378, 1992.

Morgenstern, U. and Daughney, C. J.: Groundwater age for identification of baseline groundwater quality and impacts of land-use intensification – The National Groundwater Monitoring Programme of New Zealand, J. Hydrol., 456–457, 79–93, 2012.

Morgenstern, U. and Taylor, C. B.: Ultra low-level tritium measurement using electrolytic enrichment and LSC, Isot. Environ. Healt. S., 45, 96–117, 2009.

Morgenstern, U., Stewart, M. K., and Stenger, R.: Dating of streamwater using tritium in a post nuclear bomb pulse world: continuous variation of mean transit time with streamflow, Hydrol. Earth Syst. Sci., 14, 2289–2301, doi:10.5194/hess-14-2289-2010, 2010.

Morgenstern, U., Daughney, C. J., Leonard, G., Gordon, D., Donath, F. M., and Reeves, R.: Using groundwater age and hydrochemistry to understand sources and dynamics of nutrient contamination through the catchment into Lake Rotorua, New Zealand, Hydrol. Earth Syst. Sci., 19, 803–822, doi:10.5194/hess-19-803-2015, 2015.

Morris, P. J. and Waddington, J. M.: Groundwater residence time distributions in peatlands: Implications for peat decomposition and accumulation, Water Resour. Res., 47, W02511, doi:10.1029/2010WR009492, 2011.

Mosquera, G. M., Segura, C., Vaché, K. B., Windhorst, D., Breuer, L., and Crespo, P.: Insights into the water mean transit time in a high-elevation tropical ecosystem, Hydrol. Earth Syst. Sci., 20, 2987–3004, doi:10.5194/hess-20-2987-2016, 2016.

Page, S. E., Siegert, F., Rieley, J. O., Boehm, H. D. V., Jaya, A., and Limin, S.: The amount of carbon released from peat and forest fires in Indonesia during 1997, Nature, 420, 61–65, 2002.

Shugg, A.: Hydrogeology of the Upper Ovens Valley, Report Victoria Department of Industry, Technology and Resources 5, Melbourne, 1987.

Siegel, D. I., Chanton, J. P., Glaser, P. H., Chasar, L. S., and Rosenberry, D. O.: Estimating methane production rates in bogs and landfills by deuterium enrichment of pore water, Global Biogeochem. Cy., 15, 967–975, 2001.

Soulsby, C., Tetzlaff, D., Rodgers, P., Dunn, S., and Waldron, S.: Runoff processes, stream water residence times and controlling landscape characteristics in a mesoscale catchment: An initial evaluation, J. Hydrol., 325, 197–221, 2006.

Stewart, M. K., Mehlhorn, J., and Elliott, S.: Hydrometric and natural tracer (^{18}O, silica, ^3H and SF_6) evidence for a dominant groundwater contribution to Pukemanga Stream, New Zealand, Hydrol. Process., 21, 3340–3356, 2007.

Stewart, M. K., Morgenstern, U., and McDonnell, J. J.: Truncation of stream residence time: How the use of stable isotopes has skewed our concept of streamwater age and origin, Hydrol. Process., 24, 1646–1659, 2010.

Stewart, M. K., Morgenstern, U., Gusyev, M. A., and Małoszewski, P.: Aggregation effects on tritium-based mean transit times and young water fractions in spatially heterogeneous catchments and groundwater systems, and implications for past and future applications of tritium, Hydrol. Earth Syst. Sci. Discuss., doi:10.5194/hess-2016-532, in review, 2016.

Tadros, C. V., Hughes, C. E., Crawford, J., Hollins, S. E., and Chisari, R.: Tritium in Australian precipitation: A 50 year record, J. Hydrol., 513, 262–273, 2014.

Tetzlaff, D., Birkel, C., Dick, J., Geris, J., Soulsby, C. imbe, E., Windhorst, D., and Celleri, R.: Storage dynamics in hydropedological units control hillslope connectivity, runoff generation, and the evolution of catchment transit time distributions, Water Resour. Res., 50, 969–985, 2014.

Timbe, E., Windhorst, D., Celleri, R., Timbe, L., Crespo, P., Frede, H.-G., Feyen, J., and Breuer, L.: Sampling frequency tradeoffs in the assessment of mean transit times of tropical montane catchment waters under semi-steady-state conditions, Hydrol. Earth Syst. Sci., 19, 1153–1168, doi:10.5194/hess-19-1153-2015, 2015.

van den Berg, A. H. M., Wilman, C. E., Morand, V. J., McHaffie, I. W., Simmons, B. A., Quinn, C., and Westcott, A.: Buffalo 1 : 100 000 map area geological report 124, GeoScience Victoria, Melbourne, 2004.

van der Werf, G. R., Randerson, J. T., Giglio, L., Collatz, G. J., Mu, M., Kasibhatla, P. S., Morton, D. C., DeFries, R. S., Jin, Y., and van Leeuwen, T. T.: Global fire emissions and the contribution of deforestation, savanna, forest, agricultural, and peat fires (1997–2009), Atmos. Chem. Phys., 10, 11707–11735, doi:10.5194/acp-10-11707-2010, 2010.

Western, A., Rutherford, I., Sirawardena, L., Lawrence, R., Ghadirin, P., Coates, F., and White, M.: The Geography and Hydrology of High Country Peatlands in Victoria. Part 2: The Influence of Peatlands on Catchment Hydrology, Arthur Rylah Institute for Environmental Research Technical Report No. 174, Department of Sustainability and Environment, Victoria, Melbourne, 2009.

Zuber, A., Witczak, S., Rozanski, K., Sliwka, I., Opoka, M., Mochalski, P., Kuc, T., Karlikowska, J., Kania, J., Jackowicz-Korczynski, M., and Dulinski, M.: Groundwater dating with ^3H and SF_6 in relation to mixing patterns, transport modelling and hydrochemistry, Hydrol. Process., 19, 2247–2275, 2005.

14

Uncertainty assessment of a dominant-process catchment model of dissolved phosphorus transfer

Rémi Dupas[1], Jordy Salmon-Monviola[1], Keith J. Beven[2], Patrick Durand[1], Philip M. Haygarth[2], Michael J. Hollaway[2], and Chantal Gascuel-Odoux[1]

[1]INRA, Agrocampus Ouest, UMR1069 SAS, 35000 Rennes, France
[2]Lancaster Environment Centre, Lancaster University, Lancaster, LA1 4YQ, UK

Correspondence to: Rémi Dupas (remi.dpas@gmail.com)

Abstract. We developed a parsimonious topography-based hydrologic model coupled with a soil biogeochemistry sub-model in order to improve understanding and prediction of soluble reactive phosphorus (SRP) transfer in agricultural headwater catchments. The model structure aims to capture the dominant hydrological and biogeochemical processes identified from multiscale observations in a research catchment (Kervidy–Naizin, 5 km^2). Groundwater fluctuations, responsible for the connection of soil SRP production zones to the stream, were simulated with a fully distributed hydrologic model at 20 m resolution. The spatial variability of the soil phosphorus content and the temporal variability of soil moisture and temperature, which had previously been identified as key controlling factors of SRP solubilization in soils, were included as part of an empirical soil biogeochemistry sub-model. The modelling approach included an analysis of the information contained in the calibration data and propagation of uncertainty in model predictions using a generalized likelihood uncertainty estimation (GLUE) "limits of acceptability" framework. Overall, the model appeared to perform well given the uncertainty in the observational data, with a Nash–Sutcliffe efficiency on daily SRP loads between 0.1 and 0.8 for acceptable models. The role of hydrological connectivity via groundwater fluctuation and the role of increased SRP solubilization following dry/hot periods were captured well. We conclude that in the absence of near-continuous monitoring, the amount of information contained in the data is limited; hence, parsimonious models are more relevant than highly parameterized models. An analysis of uncertainty in the data is recommended for model calibration in order to provide reliable predictions.

1 Introduction

Excessive phosphorus (P) concentrations in freshwater bodies result in increased eutrophication risk worldwide (Carpenter et al., 1998; Schindler et al., 2008). Eutrophication restricts economic use of water and poses a serious hazard to ecosystems and humans (Serrano et al., 2015). In western countries, reduction of point-source P emissions in the last 2 decades has resulted in a proportionally increasing contribution of diffuse sources, mainly from agricultural origins (Alexander et al., 2008; Grizzetti et al., 2012; Dupas et al., 2015a). Of particular concern are dissolved P forms, often measured as soluble reactive phosphorus (SRP) because they are highly bioavailable and therefore a likely contributor to eutrophication.

To reduce SRP transfer from agricultural soils, it is important to identify the spatial origin of P sources in agricultural landscapes, the biogeochemical mechanisms causing SRP solubilization in soils and the dominant transfer pathways, as well as the potential P resorption during transit. Research catchments provide useful data to investigate SRP transport mechanisms: typically, the temporal variations in water quality parameters at the outlet, together with hydroclimatic variables, are investigated to infer spatial origin and dominant transfer pathways of SRP (Haygarth et al., 2012; Outram et al., 2014; Dupas et al., 2015b; Mellander et al., 2015; Perks et al., 2015). Hypotheses drawn from analysis of water quality time series can be further investigated through hillslope monitoring and/or laboratory experiments (Heathwaite and Dils, 2000; Siwek et al., 2013; Dupas et al., 2015c). When dominant processes are considered reasonably known, it is possible to develop computer models, for two main purposes. First, to validate scientific conceptual models one can test whether model predictions can produce reasonable simulations compared to observations. Of particular interest is the possibility of testing the capability of a computer model to upscale P processes observed at fine spatial resolution (soil column, hillslope) to a whole catchment. Second, if the models survive such validation tests, they can be useful tools to simulate the response of a catchment system to a future perturbation such as changes in agricultural management and climate changes.

However, process-based P models generally perform poorly compared to, for example, nitrogen models (Wade et al., 2002a; Dean et al., 2009; Jackson-Blake et al., 2015). This is of major concern because poor model performance suggests poor knowledge of dominant processes at the catchment scale, and poor reliability of the modelling tools used to support management. The origin of poor model performance might be conceptual misrepresentations, structural imperfection, calibration problems, irrelevant model evaluation criteria and difficulties in properly assessing the information content of the available data when it is subject to epistemic error. All five causes of poor model performance are intertwined; e.g. model calibration strategy depends on model performance evaluation criteria, which depend on the way the information contained in the observation data is assessed (Beven and Smith, 2015).

A key issue in environmental modelling is the level of complexity one should seek to incorporate in a model structure. Several existing P transfer models, such as INCA (Wade et al., 2002a), SWAT (Arnold et al., 1998) and HYPE (Lindstrom et al., 2010) seek to simulate many processes, with the view that complex models are necessary to understand processes and to predict the likely consequences of land use or climate changes. However, these complex models include many parameters that need to be calibrated, while the amount of data available for calibration is often low. An imbalance between calibration requirement and the amount of available observation data can lead to equifinality issues, i.e. when many model structures or parameter sets lead to acceptable simulation results (Beven, 2006). A consequence of equifinality is the risk of unreliable prediction when an "optimal" set of parameters is used (Kirchner, 2006), and large uncertainty intervals when Monte Carlo simulations are performed (Dean et al., 2009). In this situation, it will be worth exploring parsimonious models that aim to capture the dominant hydrological and biogeochemical processes controlling SRP transfer in agricultural catchments. For example, Hahn et al. (2013) used a soil-type-based rainfall–runoff model (Lazzarotto et al., 2006) combined with an empirical model of soil SRP release derived from rainfall simulation experiments over soils with different P content and manure application level/timing (Hahn et al., 2012) to simulate daily SRP load from critical sources areas.

A second key issue, linked to the question of model complexity, concerns model calibration and evaluation. Both calibration and evaluation require assessing the fit of model outputs with observation data. However, observation data are generally not directly comparable with model outputs, because of incommensurability issues and/or because they contain errors (Beven, 2006, 2009). Typically, predicted daily concentrations and/or loads are evaluated against data from grab samples collected on a daily or weekly basis. The information content of these data must be carefully evaluated to propagate uncertainty in the data into model predictions (Krueger et al., 2012). Uncertainty in grab sample data might stem from (i) sampling frequency problems or (ii) measurement problems (Lloyd et al., 2016). Grab sample data represent a specific point in the stream cross section, which can differ from the cross section mean concentration (Rode and Suhr, 2007), and a snapshot of the concentration at a given time of the day, which can differ from the flow-weighted mean daily concentration (McMillan et al., 2012). This difference between observation data and simulation output can be large during storm events in small agricultural catchments, as P concentrations can vary by several orders of magnitude during the same day (Heathwaite and Dils, 2000; Sharpley et al., 2008). Model evaluation can be severely hindered by this difference, because many popular evaluation criteria such as the Nash–Sutcliffe efficiency (NSE) are sensitive to extreme values and errors in timing (Moriasi et al., 2007). During baseflow periods, it is more likely that grab sample data are comparable to flow-weighted mean daily concentrations, as concentrations vary little during the day and they are usually low in the absence of point sources. However, measurement errors are expected to occur at low concentrations, either due to too long storage times or laboratory imprecision when concentrations come close to detection/quantification limits (Jarvie et al., 2002; Moore and Locke, 2013). Uncertainty in the data can also relate to discharge measurement and input data (e.g. maps of soil P content and rainfall data). In this paper we strive to identify and quantify the different sources of uncertainty in the data when the required quality check tests have been performed (on the discharge and SRP concentration data). A generalized likelihood uncertainty estimation (GLUE) "limits of acceptability" approach (Beven, 2006; Beven and Smith, 2015) is used to calibrate/evaluate the model.

This paper presents a dominant-process model that couples a topography-based hydrologic model with a soil biogeochemistry sub-model able to simulate daily discharge and SRP loads. The dominant processes included in the hydrologic and soil biogeochemistry sub-models have been identified in previous analyses of multiscale observational data, which have demonstrated, on the one hand, the control of groundwater fluctuation on connecting soil SRP production zones to the stream (Haygarth et al., 2012; Jordan et al., 2012; Dupas et al., 2015b, d; Mellander et al., 2015), and, on the other hand, the role of antecedent soil moisture and

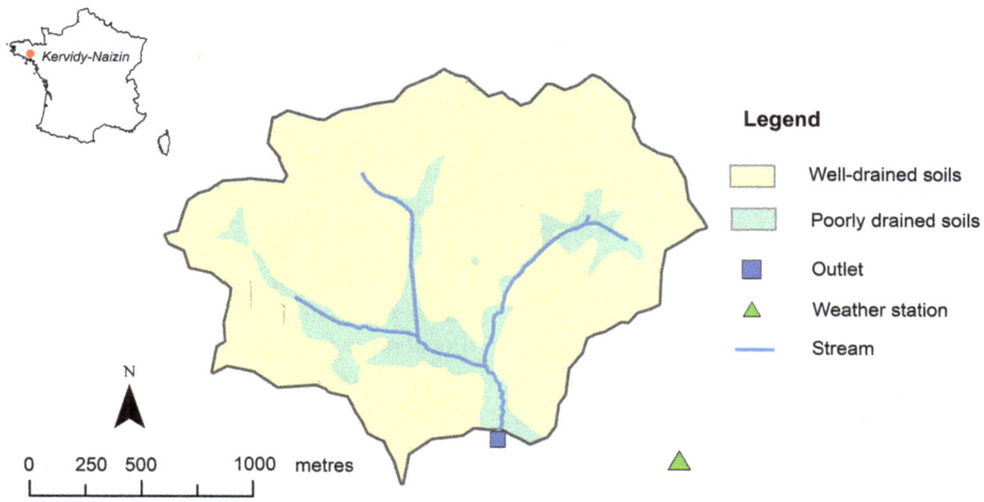

Figure 1. Soil drainage classes in the Kervidy–Naizin catchment (Curmi et al. 1998).

temperature conditions on SRP solubilization in soils (Turner and Haygarth, 2001; Blackwell et al., 2009; Dupas et al., 2015c). Model development and application were performed in the Kervidy–Naizin catchment in western France with the objectives of (i) testing if the model was capable of capturing daily variation of SRP load, thus confirming hypotheses on dominant processes; and (ii) developing a methodology to analyse and propagate uncertainty in the data into model prediction using a "limits of acceptability" approach.

2 Material and methods

2.1 Study catchment

2.1.1 Site description

Kervidy–Naizin is a small (4.94 km^2) agricultural catchment located in central Brittany, western France (48° N, 3° W). It belongs to the AgrHyS environmental research observatory (http://www6.inra.fr/ore_agrhys_eng), which studies the impact of agricultural activities and climate change on water quality (Molenat et al., 2008; Aubert et al., 2013; Salmon-Monviola et al., 2013; Humbert et al., 2015). The catchment (Fig. 1) is drained by a stream of second Strahler order, which generally dries up in August and September. The climate is temperate oceanic, with mean ± standard deviations of annual cumulative precipitation and specific discharge of 854 ± 179 and 290 ± 106 mm respectively, from 2000 to 2014. Mean annual ± standard deviation of temperature is 11.2 ± 0.6 °C. Elevation ranges from 93 to 135 m above sea level. Topography is gentle, with maximum slopes not exceeding 5 %. The bedrock consists of impervious, locally fractured Brioverian schists and is capped by several metres of unconsolidated weathered material and silty, loamy soils. The hydrological behaviour is dominated by the development of a water table that varies seasonally along the hillslope. In the upland domain, consisting of well-drained soils, the water table remains below the soil surface throughout the year, varying in depth from 1 to > 8 m. In the wetland domain, developed near the stream and consisting of hydromorphic soils, the water table is shallower, remaining near the soil surface generally from October to April each year. The land use is mostly agriculture, specifically arable crops and confined animal production (dairy cows and pigs). A farm survey conducted in 2013 led to the following land use subdivisions: 35 % cereal crops, 36 % maize, 16 % grassland and 13 % other crops (rapeseed, vegetables). Animal density was estimated as high as 13 livestock units ha^{-1} in 2010. Estimated soil P surplus was 13.1 kg P ha^{-1} yr^{-1} (Dupas et al., 2015b) and soil extractable P in 2013 (Olsen et al., 1954) was 59 ± 31 mg P kg^{-1} ($n = 89$ samples). A survey targeting riparian areas highlighted the legacy of high soil P content in these currently unfertilized areas (Dupas et al., 2015c). No point-source emissions were recorded, but scattered dwellings with septic tanks were present in the catchment.

2.1.2 Hydroclimatic and chemical monitoring

Kervidy–Naizin was equipped with a weather station (Cimel Enerco 516i) located 1.1 km from the catchment outlet. It recorded hourly precipitation, air and soil temperatures, air humidity, global radiation, wind direction and speed, which are used to estimate Penman evapotranspiration. Stream discharge was estimated at the outlet with a rating curve and stage measurements from a float-operator sensor (Thalimèdes OTT) upstream of a rectangular weir.

To record both seasonal and within storm dynamics in SRP concentration, two monitoring strategies complemented each other from October 2013 to August 2015: a daily manual grab sampling at approximately the same time (between 16:00 and 18:00 local time) and automatic high-frequency sampling during 14 storm events (Teledyne autosampler ISCO 6712 Full-Size Portable Sampler; 24 1 L bottles filled every 30 min). The water samples were filtered on-site, immediately after grab sampling and after 1–2 days in the case of autosampling. They were analysed for SRP (ISO 15681) within a fortnight. To assess uncertainty in daily SRP concentration related to sampling time, storage and measurement errors, a second grab sample was taken at a different time of the day (between 11:00 and 15:00 local time) in 36 instances during the study period. The second sample was analysed within 24 h with the same method; this second data set is referred to as verification data set, as opposed to the reference data set. Among the 36 pairs of comparable daily samples, 12 were taken during storm events and 24 during baseflow periods. To assess uncertainty in high-frequency SRP concentration during storm events due to delayed filtration of autosampler bottles, five grab samples were taken during the course of four distinct storms and were filtered immediately. The same lab procedure was used to analyse SRP.

2.1.3 Identification of dominant processes from multiscale observations

Observations in the Kervidy–Naizin catchment have highlighted that the temporal variability in stream SRP concentrations could not be related to the calendar of agricultural practices but rather to hydrological and biogeochemical processes (Dupas et al., 2015b). The primary control of hydrology on SRP transfer has also been evidenced in several other small agricultural catchments (e.g. Haygarth et al, 2012; Jordan et al., 2012; Mellander et al., 2015). In the Kervidy–Naizin catchment, the groundwater fluctuation in valley bottom areas was identified as the main driving factor of SRP transfer, through the hydrological connectivity it creates when the saturated zone intercepts shallow soil layers (Dupas et al., 2015b).

In situ monitoring of soil pore water at four sites (15 and 50 cm depths) in the Kervidy–Naizin catchment has shown that mean SRP concentration in soils is a linear function of Olsen P (Olsen et al., 1954). This reflects the current knowledge that a soil P test, or alternatively estimation of a degree of P saturation, can be used to assess solubilization in soils (Beauchemin and Simard, 1999; McDowell et al., 2002; Schoumans et al., 2015). This linear relationship derived from the data contrasts however with other studies, where threshold values above which SRP solubilization increases greatly have been identified (Heckrath et al., 1995; Maguire and Sims, 2002).

SRP solubilization in soil varies seasonally according to antecedent conditions of temperature and soil moisture. Dry and/or hot conditions are favourable to the accumulation of mobile P forms in soils, while water-saturated conditions lead to their flushing (Turner and Haygarth, 2001; Blackwell et al., 2009; Dupas et al., 2015c).

2.2 Description of the Topography-based Nutrient Transfer and Transformation – Phosphorus model (TNT2-P)

TNT2 was originally developed as a process-based and spatially explicit model simulating water and nitrogen fluxes at a daily time step (Beaujouan et al., 2002) in meso-scale catchments ($< 50\,\mathrm{km}^2$). TNT2-N has been widely used for operational objectives, to test the effect of mitigation options proposed by local stakeholders or public policymakers (Moreau et al., 2012; Durand et al., 2015), on nitrate fluxes and concentrations in rivers.

TNT2-P uses a modified version of the hydrological sub-model in TNT2-N, to which a P biogeochemistry sub-model was added to simulate SRP solubilization in soils.

2.2.1 Hydrological sub-model

The assumptions in the hydrological sub-model are derived from TOPMODEL, which has previously been applied to the Kervidy–Naizin catchment (Bruneau et al., 1995; Franks et al., 1998). (1) The effective hydraulic gradient of the saturated zone is approximated by the local topographic surface gradient ($\tan\beta$). It is calculated in each cell of a digital elevation model (DEM) at the beginning of the simulation. (2) The effective downslope transmissivity (parameter T) of the soil profile in each cell of the DEM is a function of the soil moisture deficit (S_d). Hydraulic conductivity is assumed to decrease exponentially with depth (parameter m, Fig. 2). Hence, water fluxes (q) are computed as

$$q = T \cdot \tan\beta \cdot \exp(-\frac{S_\mathrm{d}}{m}). \quad (1)$$

Based on these assumptions, TNT2 computes an explicit cell-to-cell routing of fluxes, using a D8 algorithm.

To simulate SRP fluxes, the hydrological sub-model is used to compute water fluxes from each soil layer by integrating Eq. (1) between the maximum depth of the soil layer considered and either

- estimated groundwater level, if the groundwater table is within the soil layer considered,

or

- the minimum depth of the soil layer considered, if the groundwater table above the soil layer considered.

In this application of the TNT2-P model, five soil layers with a thickness of 10 cm are considered. Hence, seven flow components are computed in the model:

- overland flow on any saturated surfaces;
- five sub-surface flow components, one for each soil layer;

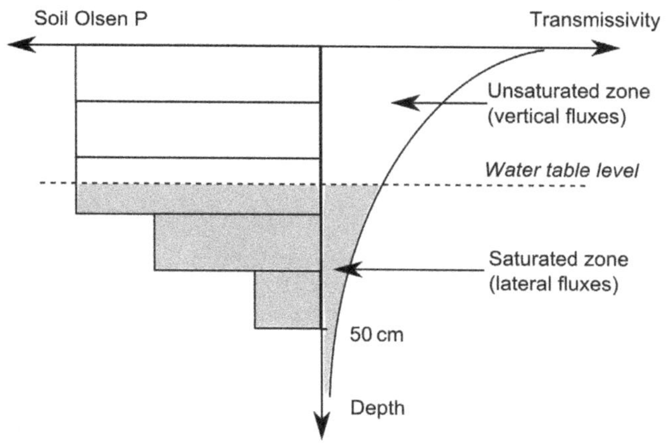

Figure 2. Description of soil hydraulic properties and phosphorus content with depth.

- deep flow, i.e. flow below the five soil layers.

2.2.2 Soil P sub-model

The soil P sub-model is empirically derived from soil pore water monitoring data (Dupas et al., 2015c), specifically assuming that

- the background SRP concentration in the soil pore water of a given layer is proportional to soil Olsen P;
- seasonal increases in P availability compared to background conditions are determined by biogeochemical processes, controlled by antecedent temperature and soil moisture. Data show that SRP availability in the soil pore water increases following periods of dry and hot conditions (Dupas et al., 2015c).

Hence, SRP transfer is modelled with parameters that describe both mobilization and transfer to the stream. A different parameter is used to simulate transfer via overland flow and sub-surface flow.

$$F_{\text{SRP overland}} = \text{Coef}_{\text{SRP overland}} \cdot P_{\text{Olsen}} \cdot q_{\text{overland}}, \quad (2)$$

$$F_{\text{SRP sub-surface}} = \text{Coef}_{\text{SRP sub-surface}} \cdot P_{\text{Olsen}} \cdot q_{\text{sub-surface}}, \quad (3)$$

where $F_{\text{SRP overland}}$ and $F_{\text{SRP sub-surface}}$ are SRP transfer via overland flow and sub-surface flow for a given soil layer respectively, q_{overland} and $q_{\text{sub-surface}}$ are water flows from the same pathways. $\text{Coef}_{\text{SRP overland}}$ and $\text{Coef}_{\text{SRP sub-surface}}$ are coefficients that vary according to antecedent temperature and soil moisture conditions, such as

$$\text{Coef}_{\text{SRP}} = \text{Coef}_{\text{background}} \cdot (1 + F_{\text{T}} \cdot F_{\text{S}}), \quad (4)$$

where Coef_{SRP} is either $\text{Coef}_{\text{SRP overland}}$ or $\text{Coef}_{\text{SRP sub-surface}}$, and F_{T} and F_{S} are temperature and soil moisture factors, respectively. F_{T} and F_{S} are expressed as

$$F_{\text{T}} = \exp\left(\frac{\text{mean (temperature, } i \text{ days)} - T1}{T2}\right), \quad (5)$$

$$F_{\text{S}} = 1 - \left(\frac{\text{mean (water content, } i \text{ days)}}{\text{maximum water content}}\right)^{S1}, \quad (6)$$

where $T1$, $T2$ and $S1$ are parameters to be calibrated. The antecedent condition time length consists in a period of $i = 100$ days. Both soil temperature and soil moisture are estimated by the TNT2 soil module (Moreau et al., 2013). Because soil moisture in the deep soil layers can differ significantly from that of shallow soil layers, two values of F_{S} are calculated for two soil depth ranges: 0–20 and 20–50 cm. The temperature factor F_T was calculated as an average value for the entire 0–50 cm soil profile. Contrary to the water fluxes, SRP fluxes are not routed cell-to-cell because we lack knowledge of the rate of SRP re-adsorption in downslope cells and of the long-term fate of re-adsorbed SRP. Hence, all the SRP emitted from each cell through overland flow and sub-surface flow reaches the stream on the same day. For deep flow, only the immediate riparian flux is used in determining SRP inputs to the river.

No long-term depletion of the different P pools was modelled, because annual P export from the catchment was small compared to the size of soil and sub-soil P pools.

2.2.3 Input data and parameters

Spatial input data required for TNT2-P include

- A DEM in raster format. Here, a 20 m resolution DEM was used; hence, model calculations were made in 12 348 grid cells covering a 4.94 km^2 catchment.
- A map of soil units that could be assumed to have homogeneous hydrological parameter values, in raster format. Here, two soil classes were considered by differentiating well-drained (86 %) and poorly drained soils (14 %) according to Curmi et al. (1998) (Fig. 1). Experimental determination of saturated hydraulic conductivity (29 soil cores) by Curmi et al. (1998) showed significantly different values for soils classified as well-drained and poorly drained in the Kervidy–Naizin catchment. The two units were treated as homogeneous, lacking information about the detailed variability in soil hydraulic characteristics at the model grid scale.
- A map of surface Olsen P in raster format and description of decrease in the Olsen P with depth for five soil layers between 0 and 50 cm. Here, the map of Olsen P in the 0–15 cm soil layer was obtained from statistical modelling with the rule-based regression algorithm CUBIST (Quinlan, 1992) using data from 198 soil samples (2013) in an area of 12 km^2 encompassing the 4.94 km^2 catchment (Matos-Moreira et al., 2015). To describe how Olsen P decreases with depth, land use

information was used. In tilled fields, i.e. all crop rotations including arable crops, Olsen P was assumed to be constant between 0 and 30 cm and to decrease linearly with depth between 30 and 50 cm. In no-till fields, i.e. permanent pasture and woodland, Olsen P was assumed to decrease linearly with depth between 0 and 50 cm. An exponential decrease with depth is more commonly adopted in untilled land (e.g. Haygarth et al., 1998; Page et al., 2005), but a specific sampling in currently untilled areas in the Kervidy–Naizin catchment (Dupas et al., 2015c) has shown that a linear function is more appropriate, probably because of these areas having been ploughed in the past. A previous study has shown that soil Olsen P was the most important factor controlling SRP solubilization in soils of the Kervidy–Naizin catchment (see Sect. 2.1.3); therefore, other parameters in the soil P sub-model (Sect. 2.2.2) were treated as homogeneous in the catchment (the soil classification into well-drained and poorly drained soils only concerned hydrological parameters).

A 20 m resolution was chosen for the DEM and the soil Olsen P raster map to allow for a detailed representation of the interaction of the groundwater table (as simulated by the hydrological model) and the soil Olsen P (as given by the soil Olsen P map). Indeed the soil saturation and soil Olsen P can be very different in a narrow zone close to the stream compared to upslope due to the presence of a 5 to 50 m unfertilized buffer zone with lower Olsen P compared to fertilized fields. The Olsen P value close to the stream has a determining influence on SRP transfer, because this area is the most frequently connected to the stream; therefore, a coarser resolution of the raster maps would degrade representation of the system.

Climate input data include minimum and maximum air temperature, precipitation, potential evapotranspiration and global radiation on a daily basis. The TNT2 model allows for several climate zones to be considered, in which case a raster map of climate zones must be provided to the model. Here, only one climate zone is considered.

In total, the TNT2-P model includes 15 parameters for each soil type, i.e. 30 parameters in total if two soil drainage classes are considered. To reduce the number of model runs necessary to explore the parameter space using Monte Carlo simulations, several parameters were given fixed values, or a constant ratio between the two soil types was set (Table 1). In the hydrological sub-model, the parameters to vary were identified in a previous sensitivity analysis (Moreau et al., 2013). In the soil sub-model, all the parameters were varied.

Finally, only 12 parameters were varied independently (see Table 1). Initial parameter ranges for the hydrological sub-model were based on values from several previous studies in western France (Moreau et al., 2013) and those for the soil sub-model were based on a preliminary manual trial and error procedure. The SRP concentration for deep flow water was based on actual measurement of SRP in the weathered schist (Dupas et al., 2015c). A constant flux value for domestic sources was set at the 1st percentile of the daily flux between 2007 and 2013 (Dupas et al., 2015b).

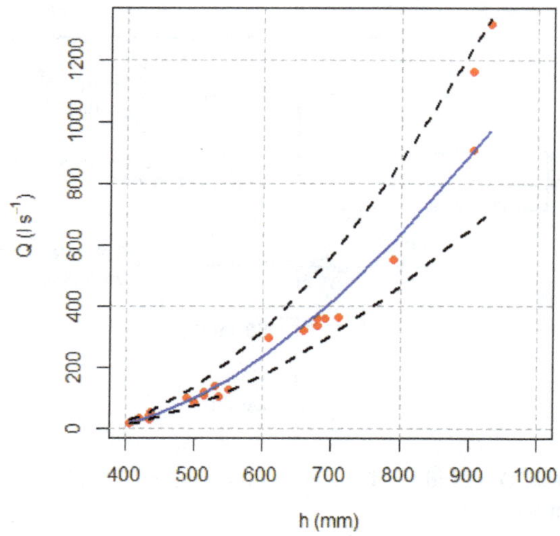

Figure 3. Rating curve in Kervidy–Naizin; acceptability bounds derived from 90 % prediction interval (blue line: fitting regression; black dashes: 90 % prediction interval). Red dots represent the original discharge measurements used to calibrate the stage–discharge rating curve (Carluer, 1998).

2.3 Deriving limits of acceptability from data uncertainty assessment

The Monte Carlo-based GLUE methodology has been widely used in hydrology and is described elsewhere (Beven and Freer, 2001b; Beven, 2006, 2009). Briefly, the rationale of GLUE is that many model structures and parameter sets can give "acceptable" results, according to one or several performance measures. Hence, GLUE considers that all models that give acceptable results should be used for prediction. A key issue in GLUE is to decide on a performance threshold to define acceptable models; typically, modellers set a threshold value of a measure such as the Nash–Sutcliffe efficiency based on their subjective appreciation of data uncertainty or on previously used values. To allow for a more explicit justification of the performance threshold values used, the limits of acceptability approach outlined by Beven (2006) relies on an assessment of uncertainty in the calibration/evaluation data. According to this approach, all model realizations that fall within the limits of acceptability are used for prediction, weighted by a score calculated based on overall performance.

Details on how the limits of acceptability for daily discharge and daily SRP load were derived from uncertainty assessment of the observational data are presented below. Input data, such as weather and soil Olsen P data, also contained uncertainties that were not accounted for explicitly in

Table 1. Initial parameter ranges in the hydrological and soil phosphorus sub-models.

	Abbreviation	Unit	Hydrological (H) or phosphorus (P) model	Range poorly drained soils (min–max)	Range well-drained soils (min–max)
Lateral transmissivity at saturation	*T*	$m^2\,day^{-1}$	H	4–8	$\rightarrow \times 1.5$
Exponential decay rate of hydraulic conductivity with depth	m	$m^2\,day^{-1}$	H	0.02–0.2	0.02–0.2
Soil depth	ho	m	H	0.3–0.8	$\rightarrow \times 1$
Drainage porosity of soil	po	$cm^3\,cm^{-3}$	H	0.1–0.4	$\rightarrow \times 1$
Regolith layer thickness	h1	m	H	5–10	$\rightarrow \times 4$
Exponent for evaporation limit	*A*	–	H	8 (fixed)	$\rightarrow \times 1$
kRC parameter for capillary rise	kRC	–	H	0.001 (fixed)	$\rightarrow \times 1$
n parameter for capillarity rise	N	–	H	2.5 (fixed)	$\rightarrow \times 1$
Drainage porosity of regolith layer	p1	$cm^3\,cm^{-3}$	H	0.01–0.05	$\rightarrow \times 1$
Background P release coefficient for sub-surface flow	$Coef_{SRP\ overland}$	–	P	0–0.015	$\rightarrow \times 1$
Background P release coefficient for overland flow	$Coef_{SRP\ sub\text{-}surface}$	–	P	0–0.25	$\rightarrow \times 1$
Temperature coefficient 1	T1	–	P	5–10	$\rightarrow \times 1$
Temperature coefficient 2	T2	–	P	2–10	$\rightarrow \times 1$
Soil moisture coefficient	*S*1	–	P	0–2	$\rightarrow \times 1$
SRP concentration in deep flow	SRP_deep	$mg\,L^{-1}$	P	0–0.007	$\rightarrow \times 1$

the limits of acceptability due to a lack of data to quantify them.

2.3.1 Discharge

Error in discharge measurement data was assessed from the original discharge measurements used to calibrate the stage–discharge rating curve (Carluer, 1998). The rating curve used in this study was

$$Q = a \cdot (h - h_0)^b, \tag{7}$$

where Q is discharge, h is stage reading, h_0 is stage reading at zero discharge, a and b are calibrated coefficients. Limits of acceptability were defined as the 90 % prediction interval of log–log linear regression (Fig. 3). The acceptability range estimated in this way was ±39 % on average. This uncertainty interval is in the higher range of values found in other studies, e.g. Coxon et al. (2015), who found that mean discharge uncertainty was generally between 20 and 40 % in 500 catchments of the United Kingdom. This relatively large uncertainty interval is due to the fact that it was derived from a prediction interval rather than a confidence interval (the 90 % confidence interval of the log–log linear regression would be 14 % of the mean discharge value during the study period). A prediction interval is an interval in which future observations will likely fall, whereas a confidence interval is an interval in which the mean of repeated observation will likely fall. Because in the TNT2-P model's evaluation we want each observation to fall in the acceptability interval (Sect. 2.3.3), a prediction interval was more appropriate. For daily discharge values below 2 mm day^{-1}, fixed acceptability limits were set at the 90 % prediction interval for a stage measurement corresponding to 2 mm day^{-1}.

2.3.2 SRP load

Uncertainty in "observed" daily load includes uncertainty in discharge (see Sect. 2.3.1) and uncertainty in SRP concentration. The acceptability limit for daily load was estimated by the sum of relative uncertainty assessed for discharge and SRP concentration (in percentage). Uncertainty in SRP concentration stems from sampling frequency problems as one grab sample collected on a specific day is incommensurable with the mean daily concentration or load simulated by the model. Further, measurement errors exist that include the effect of storage time (Haygarth et al., 1995). During baseflow periods, measurement error was expected to be the main source of uncertainty because relative measurement error was large for low concentrations, especially when sample storage time exceeded 48 h (Jarvie et al., 2002), whereas concentrations vary little. During storm events, sampling frequency was expected to be the main source of uncertainty because SRP concentration can vary by 1 order of magnitude within a few hours. Therefore, different acceptability limits were set for both flow conditions. We considered storms as events with > 20 L s^{-1} increase in discharge and the following 24 h.

During baseflow periods, the acceptability limits were derived from the 90 % prediction interval of a linear regression model ($y = a \cdot x + b$) linking pairs of data points sampled on the same day (reference sample between 16:00 and 18:00,

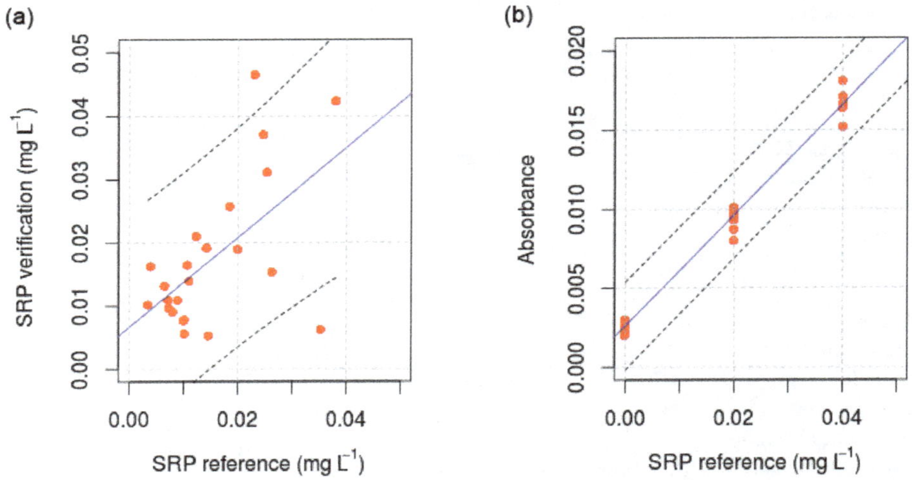

Figure 4. (a) Linear regression model linking the reference data and a verification data set; **(b)** measurement error as estimated from a repeatability test performed by the laboratory in charge of producing reference data (blue line: fitting regression; black dashes: 90 % prediction interval).

verification sample between 11:00 and 15:00) and analysed independently (within a fortnight for the reference sample and within 1–2 days for the verification sample). It was assumed that there was no systematic bias between the two data sets due to different sampling time. The reference SRP concentrations were on average 13 % lower than the verification value but this difference was not statistically significant (Mann–Whitney rank sum test, $p > 0.05$). This method encompasses all various sources of uncertainty, which results in prediction intervals much wider than what would result from a mere repeatability test; at the median concentration ($0.02\,\mathrm{mg\,L^{-1}}$), the estimated prediction interval was 166 % with this method vs. 57 % with a repeatability test (Fig. 4). As for discharge estimates, the high percentage represents a small absolute value ($0.03\,\mathrm{mg\,L^{-1}}$) during baseflow periods.

During storm events, acceptability limits were derived from the 90 % prediction interval of concentration discharge statistical models ($C = a \cdot Q^b$) using high-frequency autosampler data. Two reasons led us to use a statistical model (which also implies the assumption that errors are aleatory and temporally independent): (i) the measurement uncertainty as assessed by the laboratory repetition test was an underestimate of the real uncertainty of autosampler data, because it does not include other major sources of error such as delayed filtration and sample decay during storage; (ii) it was necessary to extrapolate the sub-daily observation to the daily resolution of the model. The limits of this choice will be discussed in Sect. 4.3. An empirical model was used to fit each storm event monitored separately and a delay term was introduced manually in the empirical model when a time lag existed between concentration and discharge peaks. The empirical models were then applied to extrapolate concentration estimation during 2 days at 10 min resolution, for each of the 14 storm events monitored. Finally the 2-day mean

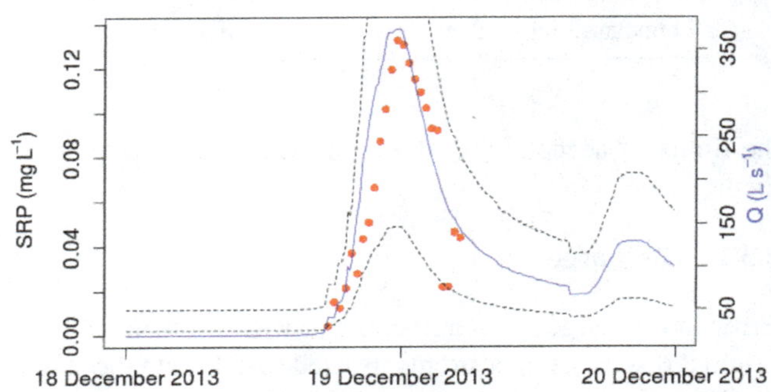

Figure 5. Example of an empirical concentration – discharge model; acceptability bounds derived from 90 % prediction interval. Red circles represent the SRP measurements.

"observed" load was estimated as the mean of 10 min loads and uncertainty limits were derived from the 90 % prediction interval. In model evaluation, the mean of simulated loads during 2 consecutive days was evaluated against the 2-day mean "observed" load for which prediction intervals have been calculated. A 2-day acceptability limit enables all the storm events to be covered (Fig. 5 and Supplement). A 2-day aggregation was necessary here because increased SRP load as a response to each storm event could occur either mainly during the day of the rainfall (if the rainfall occurred early in the morning) or mainly during the day following the rainfall (if the rainfall occurred late in the evening), and with the daily resolution of the input data and model simulation, the information about the timing of the rainfall event was not available to the model.

When comparing autosampler data with data from immediately filtered samples, the ratio obtained had the range 1–1.6 (mean = 1.3); hence, autosampler data were underesti-

Table 2. Starting and ending dates of periods studied.

Name	Starting date	Ending date
Autumn 2013	1 October 2013	31 December 2013
Winter 2014	1 January 2014	31 March 2014
Spring 2014	1 April 2014	31 July 2014
Autumn 2014	1 October 2014	31 December 2014
Winter 2015	1 January 2015	31 March 2015
Spring 2015	1 April 2015	31 July 2015

mates of the true concentration, arguably through adsorption or biological consumption. We used the mean ratio to correct all storm acceptability intervals by 30 % and the range values to extend the upper limit by 60 %. During days with a storm event not monitored at high frequency with an autosampler, we considered that the grab sample data did not contain enough information to derive an acceptability interval for daily SRP load; hence, simulated load was not evaluated for events not monitored at high frequency.

2.3.3 Model runs and selection of acceptable models

To explore the parameter space, 20 000 Monte Carlo realizations were performed to simulate daily discharge and SRP load during the water years 2013–2014 and 2014–2015. The number of Monte Carlo realizations was constrained by the computation time required to run a spatially explicit model in this catchment. A 7-month initialization period was run to reduce the impact of initial conditions on simulated results during the study period, from 1 October 2013 to 31 July 2015.

To be considered acceptable, model runs must fall within the acceptability limits defined in Sect. 2.3.1 and 2.3.2. More specifically, 100 % of simulated daily discharge, 100 % of simulated baseflow SRP load and 100 % of simulated storm SRP load have to fall within the acceptability limits. Thus, 572 acceptability tests were performed for discharge, 378 for baseflow SRP load and 14 for storm SRP loads, i.e. 964 evaluation criteria.

To evaluate the model performance in more detail, normalized scores were calculated during six periods (Table 2). To calculate the scores, a difference was calculated between each of the daily simulated discharge, baseflow SRP load and 2-day storm SRP loads and the corresponding observation. This difference was then normalized by the width of the acceptability limit defined for that day; therefore, the score has a value of 0 in the case of a perfect match with observation, −1 at the lower limit and +1 at the upper limit (Fig. 6a). Finally, the median of this ratio was calculated for each of the six periods to investigate whether the model tended to underestimate or overestimate discharge and loads at different moments of the year and between the two years.

Model runs were successively evaluated for discharge, baseflow SRP load and storm SRP load. To use the models for prediction, each accepted model was given a likelihood weight according to how well it has performed for each of the 964 evaluation criteria. Here the statistical deviation weight was used (truncated to 90 % prediction interval) (Fig. 6b). To "combine" the weights derived from the rating curve and the SRP concentration statistical models, a kernel density estimate (with Gaussian smoothing kernel) was computed to fit 10 000 realizations of the multiplied error models. Calculated weights were then averaged for discharge, baseflow SRP load and storm SRP load respectively, and the final likelihood was calculated as the product of all three averages.

Table 3. Sensitivity analysis of the model to 18 model parameters (insignificant ".", important "*", critical "***"). Parameters significations are detailed in Table 1.

	Discharge	Baseflow SRP load	Storm SRP load
T (poorly drained soils)	.	***	***
m (poorly drained soils)	***	***	***
ho (poorly drained soils)	***	***	.
po (poorly drained soils)	***	***	***
h1 (poorly drained soils)	***	***	.
p1 (poorly drained soils)	***	***	***
T (well-drained soils)	.	***	***
m (well-drained soils)	***	***	***
ho (well-drained soils)	***	***	.
po (well-drained soils)	***	***	***
h1 (well-drained soils)	***	***	.
p1 (well-drained soils)	***	***	***
Coef_sub-surface	.	***	.
Coef_overland	.	***	***
SRP_deep	.	.	.
$S1$	.	***	***
$T1$	.	***	***
$T2$	.	***	***

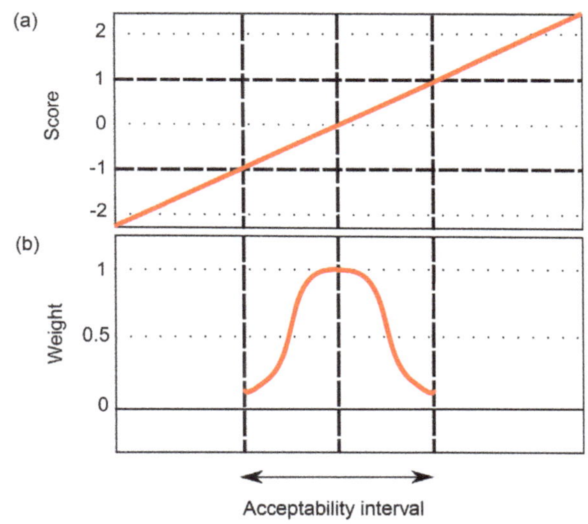

Figure 6. Normalized scores **(a)** and weighting function **(b)**.

The model's sensitivity to each hydrological and soil parameter was performed with a Hornberger–Spear–Young Generalized Sensitivity Analysis (HSY GSA; Whitehead and Young, 1979; Hornberger and Spear, 1981). For each evaluation criteria (daily discharge, daily baseflow SRP load and 2-day storm SRP load), the model runs were split into acceptable and non-acceptable runs according to the above-mentioned acceptability limits. Then a Kolmogorov–Smirnov test was performed to assess whether the distribution of each of the three evaluation criteria differs between acceptable and non-acceptable models for each parameter. Because the Kolmogorov–Smirnov test might suggest that small differences in distribution are very significant when there are larger number of runs, this method is a qualitative guide to relative sensitivity. The p value of the Kolmogorov–Smirnov test is used to discriminate whether the model is critically sensitive ($p<0.01$, "***"), importantly sensitive ($p<0.1$, "*") or insignificantly sensitive ($p>0.1$, ".") to each parameter and for each of the three evaluation criteria.

In addition to the acceptability limit approach, a NSE (Moriasi et al., 2007) was calculated for daily discharge and daily load and concentration to allow for a comparison with other modelling studies where it has been taken as an evaluation criterion.

3 Results

3.1 Presentation of observation data and calculation of acceptability limits

The two water years studied were highly contrasted in terms of hydrology and SRP loads. The water year 2013–2014 was the wettest in the last 10 years, with cumulative rainfall of 1289 mm and cumulative runoff of 716 mm. The water year 2014–2015 was an average year (fifth wettest in the last 10 years), with cumulative rainfall of 677 mm and cumulative runoff of 383 mm. Annual SRP load was $0.35\,\mathrm{kg\,P\,ha^{-1}\,yr^{-1}}$ in 2013–2014 and $0.17\,\mathrm{kg\,P\,ha^{-1}\,yr^{-1}}$ in 2014–2015, i.e. a difference 10 % higher than that of discharge. Observed mean SRP concentration during the study period was $0.024\,\mathrm{mg\,L^{-1}}$.

Figure 7a and b show acceptability limits for daily discharge and daily SRP loads. Note that acceptability limits for discharge were calculated every day, while the acceptability limits for SRP load was calculated on a daily basis during baseflow periods and on a 2-day basis during storm events monitored at high frequency. No SRP load acceptability limit was calculated during storm events when no high-frequency autosampler data were available.

3.2 Model evaluation

First, model runs were evaluated against acceptability limits defined for discharge (Fig. 7c). A total of 5479 out of 20 000 models fulfilled the selection criterion for discharge; i.e. they had 100 % of simulated daily discharge within the acceptability limits. The NSE estimated for these models ranged from 0.75 to 0.93. The normalized scores calculated seasonally (Fig. 8a) show that simulated discharge is often overestimated in autumn and spring, and underestimated in winter.

Then, model runs were evaluated against acceptability limits defined for SRP loads (Fig. 7d). During baseflow periods, 4964 out of 20 000 models fulfilled the selection criterion for SRP loads; i.e. they had 100 % of simulated daily SRP load within the acceptability limits. Among them, 1595 also fulfilled the previous selection criterion for discharge. Normalized scores for baseflow SRP load showed the same trend as for discharge (Fig. 8b), i.e. overestimation in autumn and spring, and underestimation in winter. During storm events, only seven models fulfilled the selection criterion for SRP loads; i.e. they had 14 out of 14 of simulated 2-day storm SRP loads within the acceptability limits, but none of them fulfilled the selection criteria for discharge and baseflow SRP loads. Two storm events were particularly difficult to simulate (number 2 and number 9, Fig. 8c), probably because their acceptability interval was very narrow as a result of only small changes in discharge and concentration. To obtain a reasonable number of acceptable models, we relaxed the selection criterion so that the acceptable models had to simulate 12 out of 14 of storm loads within the acceptability limits, in addition to the selection criteria defined for discharge and baseflow SRP load; 539 models were then accepted. Estimated NSE of these 539 models ranged from 0.09 to 0.81 for daily load and from negative values to 0.53 for daily concentrations (this includes all data from the regular sampling).

3.3 Sensitivity analysis and prediction results

According to the HSY generalized sensitivity analysis, simulated discharge was critically sensitive to 10 out of the 12 hydrological parameters varied (Table 3). Simulated SRP load was critically sensitive to the sub-surface and overland flow parameters during baseflow periods and to the overland flow parameter during storm events. During baseflow periods, SRP load was insignificantly sensitive to the parameter associated with deep flow load. Both baseflow and storm SRP loads were critically sensitive to the parameter related to soil moisture and soil temperature-dependent SRP solubilization ($S1$, $T1$ and $T2$), in addition to 12 and 8 hydrological parameters respectively. This identification of sensitive parameters can be used in future application of the TNT2-P model in the study catchment, as suggested by Whitehead and Hornberger (1984) and Wade et al. (2002b).

Figure 9 shows the daily discharge, SRP load and concentration as simulated by the acceptable models. Simulated SRP load during the water year 2013–2014 ranged from 0.81 to $3.25\,\mathrm{kg\,P\,ha^{-1}\,yr^{-1}}$ (median $=1.68\,\mathrm{kg\,P\,ha^{-1}\,yr^{-1}}$); simulated SRP load during the water year 2014–2015 ranged from 0.14 to $0.73\,\mathrm{kg\,P\,ha^{-1}\,yr^{-1}}$ (me-

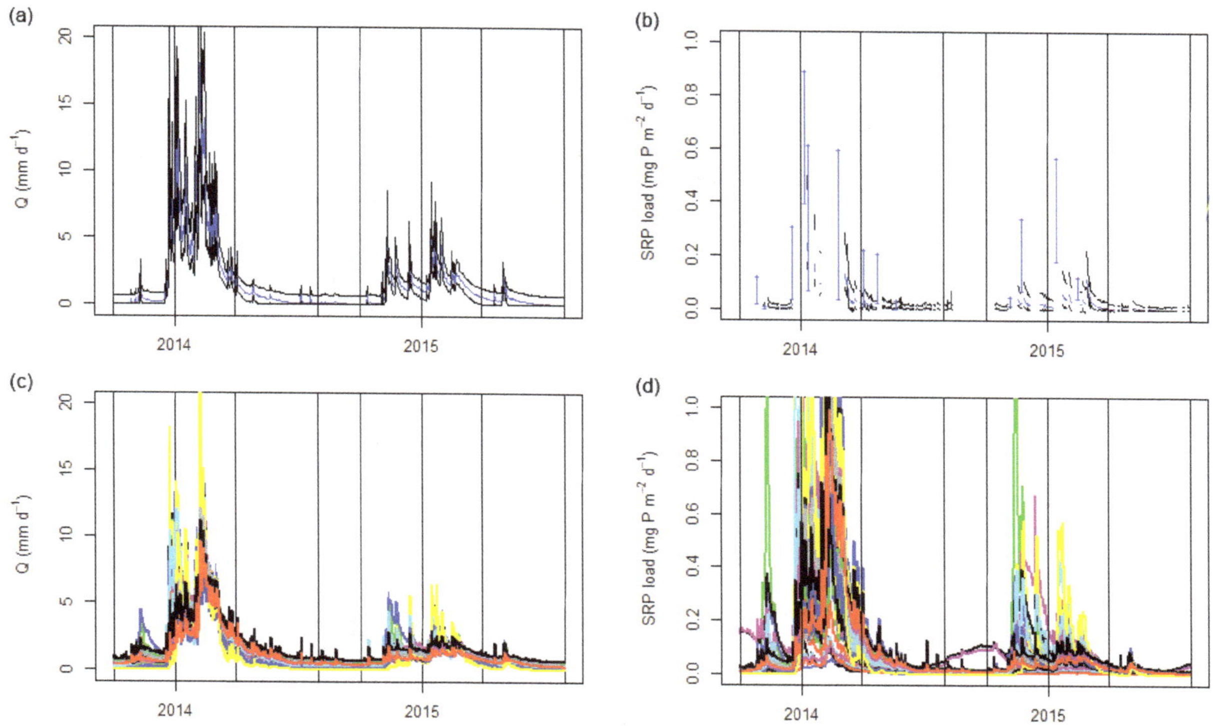

Figure 7. Acceptability limits for daily discharge **(a)** and SRP load **(b)**. Blue lines represent best estimates; black lines represent the acceptability limits. Storm loads acceptability limits are represented by vertical blue lines. An example of 50 model runs simulating discharge **(c)** and daily load **(d)**. Black vertical lines represent the starting and ending dates for each season (Table 2).

dian = $0.34\,\mathrm{kg\,P\,ha^{-1}\,yr^{-1}}$). Best estimate of SRP load according to observation data was $0.35\,\mathrm{kg\,P\,ha^{-1}\,yr^{-1}}$ in 2013–2014 and $0.17\,\mathrm{kg\,P\,ha^{-1}\,yr^{-1}}$ in 2014–2015. According to the model, 49–55 % (median = 52 %) of water discharge and 66–70 % (median = 67 %) of SRP load occurred during storm events. Mean SRP concentrations during the two water years ranged from 0.014 to $0.044\,\mathrm{mg\,L^{-1}}$ (median = $0.029\,\mathrm{mg\,L^{-1}}$), whereas mean observed SRP concentration was $0.024\,\mathrm{mg\,L^{-1}}$.

4 Discussion

4.1 Role of hydrology and biogeochemistry in determining SRP transfer

The fairly good performance of TNT2-P at simulating SRP loads provides further support that the hydrological and biogeochemical processes included into the model are dominant controlling factors in the Kervidy–Naizin catchment (i.e. the modelling hypotheses could not be rejected based on these results, except for two storm events). The primary control of hydrology in controlling connectivity between soils and streams has been highlighted by many studies analysing water quality time series at the outlet of agricultural catchments (Haygarth et al., 2012; Jordan et al., 2012; Dupas et al., 2015c; Mellander et al., 2015). This modelling exercise also provides further support that SRP solubility can be satisfactorily represented by the soil Olsen P content and could vary according to temperature and moisture conditions. The underlying processes have not been identified precisely in the Kervidy–Naizin catchment; independent laboratory experiments have shown that microbial cell lysis resulting from alternating dry and water-saturated periods in the soil could be the cause of increased SRP mobility (Turner and Haygarth, 2001; Blackwell et al., 2009). This could explain the moisture dependence of SRP solubility in the model. Furthermore, net mineralization of soil organic phosphorus could explain the temperature dependence of SRP solubility in the model. These two hypotheses may explain increased SRP solubility in soils in periods of dry and hot conditions and will be further explored by incubation experiment with soils from the Kervidy–Naizin catchments.

4.2 Potential improvements to the model structure according to modelling purpose

The TNT2-P model was designed to test hypotheses about dominant processes and for this purpose, a parsimonious model structure was chosen to include only the processes that were to be tested. This parsimonious model structure might contain some conceptual misrepresentations due to oversimplification, and it might not include all the processes nec-

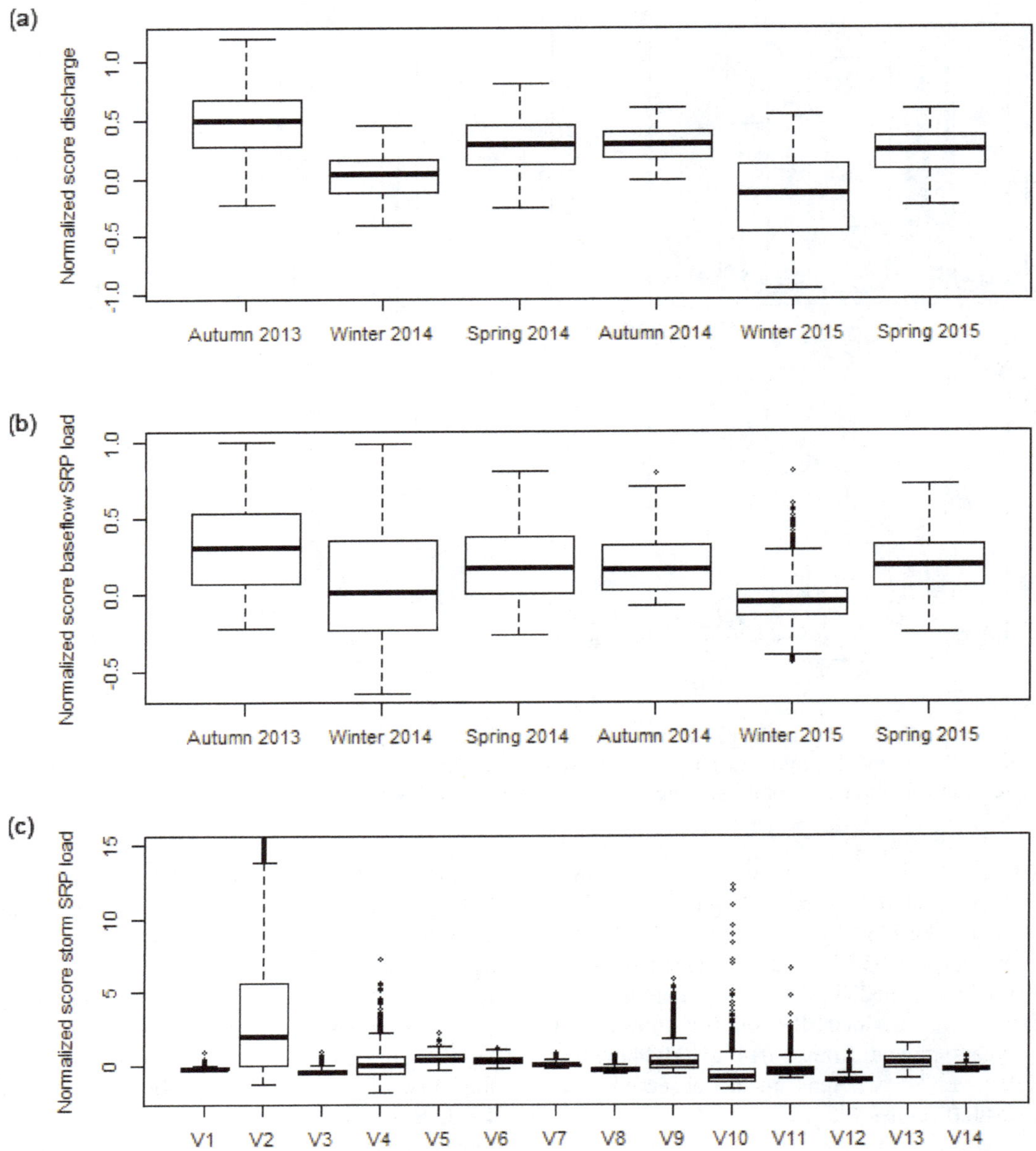

Figure 8. Normalized score for daily discharge **(a)**, baseflow SRP load **(b)** and storm SRP load **(c)**.

essary for the purpose of evaluating management scenarios. This section discusses whether the simplifications made are acceptable in the context of different catchment types, and to which conditions the model could be made more complex by including additional routines for the purpose of evaluating management scenarios.

From a conceptual point of view, the lack of cell-to-cell routing of SRP fluxes might result in erroneous results in some contexts. The fact that all the SRP emitted from each cell through overland flow and sub-surface flow reaches the stream on the same day is generally acceptable for the catchment studied, because groundwater interception of shallow soil layers occurs in the riparian zone only; hence, the signal of SRP mobilization in these soils is generally transmitted to the stream (Dupas et al., 2015c). This simplification, however, does not seem to be acceptable for all the storm events in the study catchment, as the SRP load evaluation criteria had to be relaxed to obtain acceptable model results. It would also not be acceptable in catchments where soil–groundwater interactions are taking place throughout the landscape, e.g. due to topographic depressions or poorly drained soils. In the latter type of catchment, transmission of the SRP mobilization signal to the stream is more complex (Haygarth et

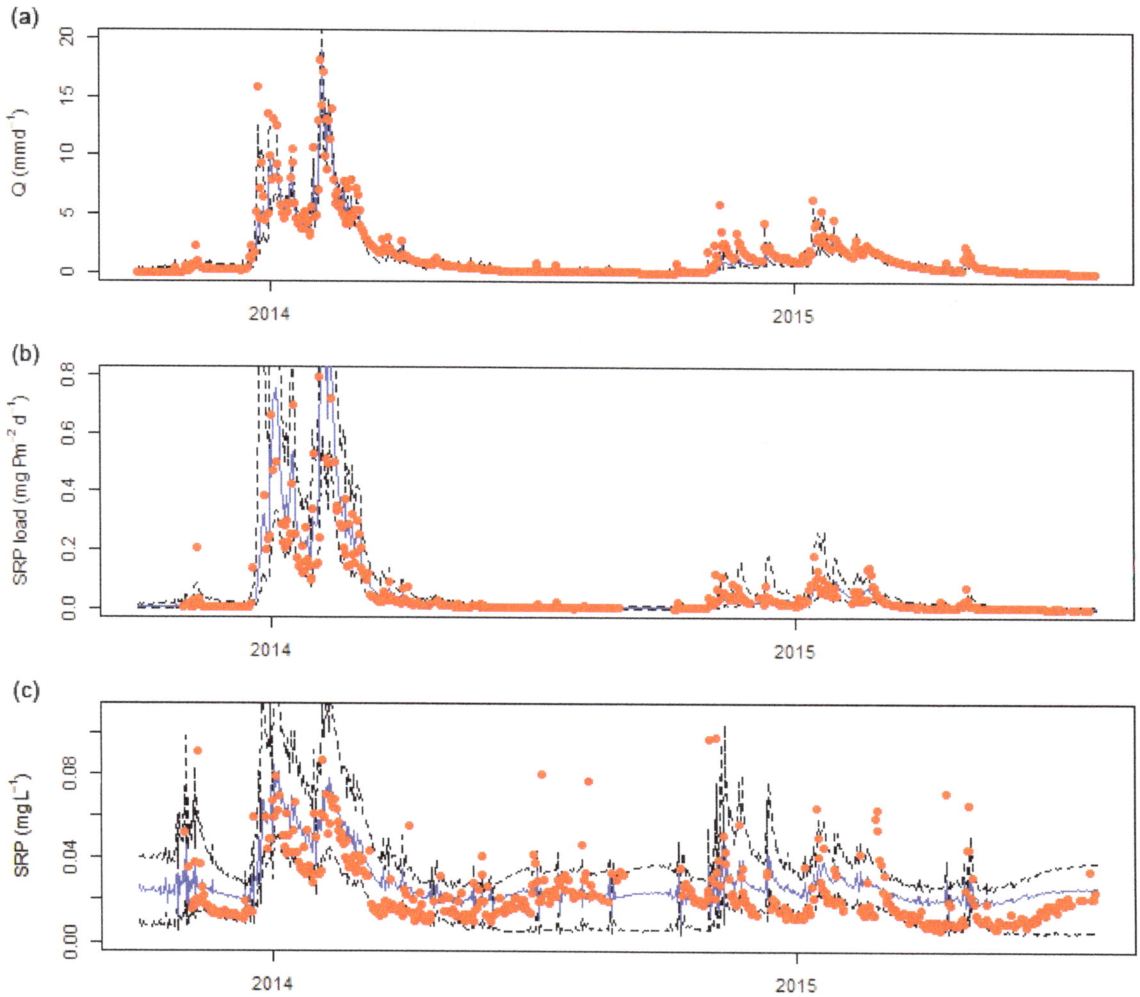

Figure 9. Median and 95 % credibility interval for daily discharge **(a)**, SRP load **(b)** and SRP concentration **(c)**. Red circles represent observational data.

al., 2012); hence, a more complex model structure would be required.

The reason for this simplification was that we lacked knowledge of SRP re-adsorption in downslope cells (or on suspended sediments in the stream network) and on the long-term fate of re-adsorbed SRP. For a more physically realistic representation of processes, it is likely that an explicit representation of flow velocities and pathways would be necessary, along with an explicit representation of several soil P pools. However, such an explicit representation of processes contradicts the idea of a parsimonious model, which was adopted here for the purpose of identifying dominant processes. In this respect, TNT2-P is an aggregative model rather than a fully distributed model although it is based on a fully distributed hydrological model (Beaujouan et al., 2002). The current spatial distribution allows for finer representation of soil–groundwater interactions (i.e. the time-varying extent of the riparian wetland area) than semi-distributed models such as SWAT (Arnold et al., 1998), INCA-P (Wade et al., 2002a) and HYPE (Lindstrom et al., 2010) but at higher computational cost. It would be interesting to test to what extent moving from an aggregative model with fully distributed information to a semi-distributed model would degrade the model performance while reducing computational cost. This could be achieved by grouping cells according to a hydrological similarity criterion like in the Dynamic TOPMODEL (Beven and Freer, 2001a; Metcalfe et al., 2015) and do the same for similarity in soil P content. Reducing computation time is critical in the context of a GLUE analysis because this method requires the parameter space to be sampled adequately to identify those models to be considered acceptable. This is debatable here because 12 parameters were varied and only 20 000 model runs were performed. It is therefore possible that some regions of the parameter space with acceptable models might not have been sampled.

If reducing the number of calculation units proved to reduce computational cost without degrading quality of prediction, it would be possible to include more parameters in the model, for example to simulate SRP re-absorption in downslope cells or include routines to simulate the evolution of soil P content under different management scenarios (Vadas et al., 2012), and still perform a Monte Carlo-based analysis of uncertainty. The question of coupling or not coupling such a soil P routine with the current TNT2-P model will depend on available data and on the length of available time series; studying the evolution of the soil P content requires at least a decade of soil observation data (Ringeval et al., 2014) and probably a longer period of stream data to account for the time delay for a perturbation in the catchment to become visible in the stream (Wall et al., 2013). Thus, the 2 years of daily stream SRP in the Kervidy–Naizin catchment are not enough to build a coupled soil–hydrology model with an elaborate soil P routine. Therefore, as things stand, it is more reasonable to generate new soil Olsen P maps with a separate model such as the APLE model (Vadas et al., 2012; Benskin et al., 2014) or the "soil P decline" model used by Wall et al. (2013), and use these maps as input to TNT2-P.

Because the current model can simulate response to rainfall, soil moisture and temperature, it could be used to test the effect of climate scenarios on SRP transfer. In western France, and more generally in western Europe, the climate for the next few decades is expected to consist of hotter, drier summers and warmer, wetter winters (Jacob et al., 2007; Macleod et al., 2012; Salmon-Monviola et al., 2013) with increased frequency of high-intensity rainfall events (Déqué, 2007). In these conditions, SRP concentrations and load will seemingly increase compared to today's climate as a result of both an increase in SRP solubility in soil due to higher temperatures and more severe drought, and an increase in transfer due to wetter winters and more frequent high-intensity rainfall events. TNT2-P could be used to confirm and quantify the expected increase in SRP transfer from diffuse sources in future climate scenarios, and to determine whether those predicted changes are significant relative to the uncertainty in predictions under current climate variability.

4.3 Improving information content in the data

Despite relatively large uncertainty in the data used in this study, it was possible to build a parsimonious catchment model of SRP transfer for the purpose of testing hypotheses about dominant processes, namely the role of hydrology in controlling connectivity between soils and streams and the role of temperature and moisture conditions in controlling soil SRP solubilization. However, the large uncertainties in the calibration data lead to large prediction uncertainty. For example, the SRP load estimated by the behavioural models from 2013 to 2015 ranged from 0.48 to 1.99 $\text{kg P ha}^{-1}\,\text{yr}^{-1}$; hence, the width of the credibility interval was 150 % of the median (1.0 $\text{kg P ha}^{-1}\,\text{yr}^{-1}$). Similarly, the mean SRP concentration estimated by the behavioural models from 2013 to 2015 ranged from 0.014 to 0.044 mg L^{-1}; hence, the width of the credibility interval was 102 % of the median (0.029 mg L^{-1}). The large uncertainty in the calibration data, along with a lack of long-term information, also prevents including more detailed processes in the soil routine.

To reduce uncertainty in prediction and to build more complex models, several options exist to improve information content in the data. As stated by Jackson-Blake and Starrfelt (2015), "the key to obtaining a realistic model simulation is ensuring that the natural variability in water chemistry is well represented by the monitoring data". The monitoring strategy adopted in the Kervidy–Naizin catchment should theoretically enable one to capture the natural variability in stream SRP concentration, because sampling took place during two contrasting water years, during different seasons and at a high frequency during 14 storm events. The analysis of uncertainty in the data shows that a large part of uncertainty in "observed" SRP concentration originates from sample storage, both unfiltered between the time of autosampling and manual filtration and between filtration and analysis. This is due to SRP being non-conservative. Thus, there is room for improvement in reducing storage time, without further increasing the monitoring frequency. In this respect, the primary interest of investing in high-frequency bankside analysers would lie in their ability to analyse water samples immediately in addition to providing near-continuous data. Because bankside analysers perform measurements in relatively homogeneous conditions, unlike the manual and autosampler data for which storage time of filtered and unfiltered samples vary, a finer quantification of uncertainty in the measurement data would be possible (e.g. Lloyd et al., 2016).

Finally, alternative methods to statistical models could be used to derive acceptability limits (in this study three statistical models are used: the rating curve, the SRP concentration uncertainty during baseflow periods and the storm event interpolation model) because statistical models have at least three shortcomings: (i) they lump the uncertainty linked to the timing of sampling, the immediate or delayed filtration of the samples, the storage time and the analytical error; (ii) the formula chosen adds error to the already existing measurement errors because empirical models are not perfect representations of the system dynamics; (iii) they assume a parametric distribution and temporally independent errors, which are not always verified in practice. As an alternative, nonparametric methods could be used, but these methods generally require a large number of data points and they are not suitable for extrapolation to extreme values.

5 Conclusion

The TNT2-P model was capable of capturing daily variation of SRP loads, thus confirming the dominant processes identified in previous analyses of observation data in the Kervidy–Naizin catchment. The role of hydrology in controlling connectivity between soils and streams, and the role of soil Olsen P, soil moisture and temperature in controlling SRP solubility have been confirmed. The lack of any representation of the short-term effect of management practices did not seem to reduce the model's performance. Their long-term effect on the soil Olsen P could be simulated with an independent model or through an additional sub-model if a longer period of data was available to calibrate it. The modelling approach presented in this paper included an assessment of the information content in the data, and propagation of uncertainty in the model's prediction. The information content of the data was sufficient to explore dominant processes, but the relatively large uncertainty in SRP concentrations would seemingly limit the possibility for including more detailed processes into the model. Data from the near-continuous bankside analyser will probably allow for calibrating more detailed models in the near future.

Acknowledgements. This work was funded by the "Agence de l'Eau Loire Bretagne" via the "Trans-P project". Long-term monitoring in the Kervidy–Naizin catchment is supported by "ORE AgrHyS".

Edited by: J. Freer
Reviewed by: T. Krueger and P. G. Whitehead

References

Alexander, R. B., Smith, R. A., Schwarz, G. E., Boyer, E. W., Nolan, J. V., and Brakebill, J. W.: Differences in phosphorus and nitrogen delivery to the gulf of Mexico from the Mississippi river basin, Environ. Sci. Technol., 42, 822–830, 2008.

Arnold, J. G., Srinivasan, R., Muttiah, R. S., and Williams, J. R.: Large area hydrologic modeling and assessment - Part 1: Model development, J. Am. Water Resour. As., 34, 73–89, 1998.

Aubert, A. H., Gascuel-Odoux, C., Gruau, G., Akkal, N., Faucheux, M., Fauvel, Y., Grimaldi, C., Hamon, Y., Jaffrézic, A., Lecoz-Boutnik, M., Molénat, J., Petitjean, P., Ruiz, L., and Merot, P.: Solute transport dynamics in small, shallow groundwater-dominated agricultural catchments: insights from a high-frequency, multisolute 10 yr-long monitoring study, Hydrol. Earth Syst. Sci., 17, 1379–1391, doi:10.5194/hess-17-1379-2013, 2013.

Beauchemin, S. and Simard, R. R.: Soil phosphorus saturation degree: Review of some indices and their suitability for P management in Quebec, Canada, Can. J. Soil Sci., 79, 615–625, 1999.

Beaujouan, V., Durand, P., Ruiz, L., Aurousseau, P., and Cotteret, G.: A hydrological model dedicated to topography-based simulation of nitrogen transfer and transformation: rationale and application to the geomorphology-denitrification relationship, Hydrol. Process., 16, 493–507, 2002.

Benskin, C. M. H., Roberts, W. M., Wang, Y., and Haygharth, P. M.: Review of the Annual Phosphorus Loss Estimator tool – a new model for estimating phosphorus losses at the field scale, Soil Use and Manage., 30, 337–341, 2014.

Beven, K.: A manifesto for the equifinality thesis, J. Hydrol., 320, 18–36, 2006.

Beven, K.: Environmental Modelling – An Uncertain Future?, Routledge, London, UK, 2009.

Beven, K. and Freer, J.: A dynamic TOPMODEL, Hydrol. Process., 15, 1993–2011, 2001a.

Beven, K. and Freer, J.: Equifinality, data assimilation, and uncertainty estimation in mechanistic modelling of complex environmental systems using the GLUE methodology, J. Hydrol., 249, 11–29, 2001b.

Beven, K. and Smith, P.: Concepts of Information Content and Likelihood in Parameter Calibration for Hydrological Simulation Models, J. Hydrol. Eng., 20, A4014010, 2015.

Blackwell, M. S. A., Brookes, P. C., de la Fuente-Martinez, N., Murray, P. J., Snars, K. E., Williams, J. K., and Haygarth, P. M.: Effects of soil drying and rate of re-wetting on concentrations and forms of phosphorus in leachate, Biol. Fert. Soils, 45, 635–643, 2009.

Bruneau, P., Gascuelodoux, C., Robin, P., Merot, P., and Beven, K.: Sensitivity to space and time resolution of a hydrological model using digital elevation data, Hydrol. Process., 9, 69–81, 1995.

Carluer, N.: Vers une modélisation hydrologique adaptée à l'évaluation des pollutions diffuses: prise en compte du réseau anthropique. Application au bassin versant de Naizin (Morbihan), PhD thesis Université Pierre et Marie Curie, Paris, France, 1998.

Carpenter, S. R., Caraco, N. F., Correll, D. L., Howarth, R. W., Sharpley, A. N., and Smith, V. H.: Nonpoint pollution of surface waters with phosphorus and nitrogen, Ecol. Appl., 8, 559–568, 1998.

Coxon, G., Freer, J., Westerberg, I. K., Wagener, T., Woods, R., and Smith, P. J.: A novel framework for discharge uncertainty quantification applied to 500 UK gauging stations, Water Resour. Res., 51, 5531–5546, 2015.

Curmi, P., Durand, P., Gascuel-Odoux, C., Merot, P., Walter, C., and Taha, A.: Hydromorphic soils, hydrology and water quality: spatial distribution and functional modelling at different scales, Nutr. Cycl. Agroecosys., 50, 127–142, 1998.

Dean, S., Freer, J., Beven, K., Wade, A. J., and Butterfield, D.: Uncertainty assessment of a process-based integrated catchment model of phosphorus, Stoch. Env. Res. Risk A., 23, 991–1010, 2009.

Dequé, M.: Frequency of precipitation and temperature extremes over France in an anthropogenic scenario: Model results and statistical correction according to observed values, Global Planet. Change, 57, 16–26, 2007.

Dupas, R., Delmas, M., Dorioz, J. M., Garnier, J., Moatar, F., and Gascuel-Odoux, C.: Assessing the impact of agricultural pressures on N and P loads and eutrophication risk, Ecol. Indic., 48, 396–407, 2015a.

Dupas, R., Gascuel-Odoux, C., Gilliet, N., Grimaldi, C., and Gruau, G.: Distinct export dynamics for dissolved and particulate phosphorus reveal independent transport mechanisms in an arable headwater catchment, Hydrol. Process., 29, 3162–3178, 2015b.

Dupas, R., Gruau, G., Gu, S., Humbert, G., Jaffrezic, A., and Gascuel-Odoux, C.: Groundwater control of biogeochemical processes causing phosphorus release from riparian wetlands, Water Res., 84, 307—314, 2015c.

Dupas, R., Tavenard, R., Fovet, O., Gilliet, N., Grimaldi, C., and Gascuel-Odoux, C.: Identifying seasonal patterns of phosphorus storm dynamics with dynamic time warping, Water Resour. Res., 51, 8868–8882, 2015d.

Durand, P., Moreau, P., Salmon-Monviola, J., Ruiz, L., Vertes, F., and Gascuel-Odoux, C.: Modelling the interplay between nitrogen cycling processes and mitigation options in farming catchments, J. Agr. Sci., 153, 959–974, 2015.

Franks, S. W., Gineste, P., Beven, K. J., and Merot, P.: On constraining the predictions of a distributed model: The incorporation of fuzzy estimates of saturated areas into the calibration process, Water Resour. Res., 34, 787–797, 1998.

Grizzetti, B., Bouraoui, F., and Aloe, A.: Changes of nitrogen and phosphorus loads to European seas, Glob. Change Biol., 18, 769–782, 2012.

Hahn, C., Prasuhn, V., Stamm, C., and Schulin, R.: Phosphorus losses in runoff from manured grassland of different soil P status at two rainfall intensities, Agr. Ecosyst. Environ., 153, 65–74, 2012.

Hahn, C., Prasuhn, V., Stamm, C., Lazzarotto, P., Evangelou, M. W. H., and Schulin, R.: Prediction of dissolved reactive phosphorus losses from small agricultural catchments: calibration and validation of a parsimonious model, Hydrol. Earth Syst. Sci., 17, 3679–3693, doi:10.5194/hess-17-3679-2013, 2013.

Haygarth, P. M., Ashby, C. D., and Jarvis, S. C.: Short-term changes in the molybdate reactive phosphorus of stored soil waters, J. Environ. Qual., 24, 1133–1140, 1995.

Haygarth, P. M., Hepworth, L., and Jarvis, S. C.: Forms of phosphorus transfer in hydrological pathways from soil under grazed grassland, Eur. J. Soil Sci., 49, 65–72, 1998.

Haygarth, P. M., Page, T. J. C., Beven, K. J., Freer, J., Joynes, A., Butler, P., Wood, G. A., and Owens, P. N.: Scaling up the phosphorus signal from soil hillslopes to headwater catchments, Freshwater Biol., 57, 7–25, 2012.

Heathwaite, A. L. and Dils, R. M.: Characterising phosphorus loss in surface and subsurface hydrological pathways, Sci. Total Environ., 251, 523–538, 2000.

Heckrath, G., Brookes, P. C., Poulton, P. R., and Goulding, K. W. T.: Phosphorus leaching from soils containing different phosphorus concentrations in the Broadbalk experiment, J. Environ. Qual., 24, 904–910, 1995.

Hornberger, G. M. and Spear, R. C.: An approach to the preliminary analysis of environmental systems, J. Environ. Manage., 12, 7–18, 1981.

Humbert, G., Jaffrezic, A., Fovet, O., Gruau, G., and Durand, P.: Dry-season length and runoff control annual variability in stream DOC dynamics in a small, shallow groundwater-dominated agricultural watershed, Water Resour. Res., 51, 7860–7877, 2015.

Jackson-Blake, L. A., Dunn, S. M., Helliwell, R. C., Skeffington, R. A., Stutter, M. I., and Wade, A. J.: How well can we model stream phosphorus concentrations in agricultural catchments?, Environ. Modell. Softw., 64, 31–46, 2015.

Jackson-Blake, L. A. and Starrfelt, J.: Do higher data frequency and Bayesian auto-calibration lead to better model calibration? Insights from an application of INCA-P, a process-based river phosphorus model, J. Hydrol., 527, 641–655, 2015.

Jacob, D., Barring, L., Christensen, O. B., Christensen, J. H., de Castro, M., Dequé, M., Giorgi, F., Hagemann, S., Hirschi, M., Jones, R., Kjellstrom, E., Lenderink, G., Rockel, B., Sanchez, E., Schar, C., Seneviratne, S. I., Somot, S., van Ulden, A., and van den Hurk, B.: An inter-comparison of regional climate models for Europe: model performance in present-day climate, Climatic Change, 81, 31–52, 2007.

Jarvie, H. P., Withers, J. A., and Neal, C.: Review of robust measurement of phosphorus in river water: sampling, storage, fractionation and sensitivity, Hydrol. Earth Syst. Sci., 6, 113–131, doi:10.5194/hess-6-113-2002, 2002.

Jordan, P., Melland, A. R., Mellander, P. E., Shortle, G., and Wall, D.: The seasonality of phosphorus transfers from land to water: Implications for trophic impacts and policy evaluation, Sci. Total Environ., 434, 101–109, 2012.

Kirchner, J. W.: Getting the right answers for the right reasons: Linking measurements, analyses, and models to advance the science of hydrology, Water Resour. Res., 42, W03S04, doi:10.1029/2005WR004362, 2006.

Krueger, T., Quinton, J. N., Freer, J., Macleod, C. J. A., Bilotta, G. S., Brazier, R. E., Hawkins, J. M. B., and Haygarth, P. M.: Comparing empirical models for sediment and phosphorus transfer from soils to water at field and catchment scale under data uncertainty, Eur. J. Soil Sci., 63, 211–223, 2012.

Lazzarotto, P., Stamm, C., Prasuhn, V., and Flühler, H.: A parsimonious soil-type based rainfall-runoff model simultaneously tested in four small agricultural catchments, J. Hydrol., 321, 21–38, 2006.

Lindstrom, G., Pers, C., Rosberg, J., Stromqvist, J., and Arheimer, B.: Development and testing of the HYPE (Hydrological Predictions for the Environment) water quality model for different spatial scales, Hydrol. Res., 41, 295–319, 2010.

Lloyd, C. E. M., Freer, J. E., Johnes, P. J., Coxon, G., and Collins, A. L.: Discharge and nutrient uncertainty: implications for nutrient flux estimation in small streams, Hydrol. Process., 30, 165–152, 2016.

Macleod, C. J. A., Falloon, P. D., Evans, R., and Haygarth, P. M.: The effect of climate change on the mobilization of diffuse substances from agricultural systems, in: Advances in Agronomy, edited by: Sparks, D. L., Advances in Agronomy, 115, 41–77, 2012.

Maguire, R. O. and Sims, J. T.: Soil testing to predict phosphorus leaching, J. Environ. Qual., 31, 1601–1609, 2002.

Matos-Moreira, M., Lemercier, B., Michot, D., Dupas, R., and Gascuel-Odoux, C.: Using agricultural practices information for multiscale environmental assessment of phosphorus risk, Geophysical Research Abstracts, 17, 2015.

McDowell, R., Sharpley, A., and Withers, P.: Indicator to predict the movement of phosphorus from soil to subsurface flow, Environ. Sci. Technol., 36, 1505–1509, 2002.

McMillan, H., Krueger, T., and Freer, J.: Benchmarking observational uncertainties for hydrology: rainfall, river discharge and water quality, Hydrol. Process., 26, 4078–4111, 2012.

Mellander, P. E., Jordan, P., Shore, M., Melland, A. R., and Shortle, G.: Flow paths and phosphorus transfer pathways in two agricultural streams with contrasting flow controls, Hydrol. Process., 3504–3518, doi:10.1002/hyp.10415, 2015.

Metcalfe, P., Beven, K., and Freer, J.: Dynamic TOPMODEL: A new implementation in R and its sensitivity to time and space steps, Environ. Modell. Softw., 72, 155–172, 2015.

Molenat, J., Gascuel-Odoux, C., Ruiz, L., and Gruau, G.: Role of water table dynamics on stream nitrate export and concentration in agricultural headwater catchment (France), J. Hydrol., 348, 363–378, 2008.

Moore, M. T. and Locke, M. A.: Effect of Storage Method and Associated Holding Time on Nitrogen and Phosphorus Concentrations in Surface Water Samples, B. Environ. Contam. Tox., 91, 493–498, 2013.

Moreau, P., Ruiz, L., Mabon, F., Raimbault, T., Durand, P., Delaby, L., Devienne, S., and Vertes, F.: Reconciling technical, economic and environmental efficiency of farming systems in vulnerable areas, Agr. Ecosyst. Environ., 147, 89–99, 2012.

Moreau, P., Viaud, V., Parnaudeau, V., Salmon-Monviola, J., and Durand, P.: An approach for global sensitivity analysis of a complex environmental model to spatial inputs and parameters: A case study of an agro-hydrological model, Environ. Modell. Softw., 47, 74–87, 2013.

Moriasi, D. N., Arnold, J. G., Van Liew, M. W., Bingner, R. L., Harmel, R. D., and Veith, T. L.: Model evaluation guidelines for systematic quantification of accuracy in watershed simulations, T. Asabe, 50, 885–900, 2007.

Olsen, S. R., Cole, C. V., Watanbe, F. S., and Dean, L. A.: Estimation of available phosphorus in soils by extraction with sodium bicarbonate Circ. 939, USDA, Washington, D.C., USA, 1954.

ORE AgrHyS: AgrHyS database, available at: http://www6.inra.fr/ore_agrhys/Donnees, 2015.

Outram, F. N., Lloyd, C. E. M., Jonczyk, J., Benskin, C. McW. H., Grant, F., Perks, M. T., Deasy, C., Burke, S. P., Collins, A. L., Freer, J., Haygarth, P. M., Hiscock, K. M., Johnes, P. J., and Lovett, A. L.: High-frequency monitoring of nitrogen and phosphorus response in three rural catchments to the end of the 2011–2012 drought in England, Hydrol. Earth Syst. Sci., 18, 3429–3448, doi:10.5194/hess-18-3429-2014, 2014.

Page, T., Haygarth, P. M., Beven, K. J., Joynes, A., Butler, T., Keeler, C., Freer, J., Owens, P. N., and Wood, G. A.: Spatial variability of soil phosphorus in relation to the topographic index and critical source areas: Sampling for assessing risk to water quality, J. Environ. Qual., 34, 2263–2277, 2005.

Perks, M. T., Owen, G. J., Benskin, C. M. H., Jonczyk, J., Deasy, C., Burke, S., Reaney, S. M., and Haygarth, P. M.: Dominant mechanisms for the delivery of fine sediment and phosphorus to fluvial networks draining grassland dominated headwater catchments, Sci. Total Environ., 523, 178–190, 2015.

Quinlan, J. R.: Learning with continuous classes, Proceedings of the 5th Australian Joint Conference On Artificial Intelligence, 343–348, 1992.

Ringeval, B., Nowak, B., Nesme, T., Delmas, M., and Pellerin, S.: Contribution of anthropogenic phosphorus to agricultural soil fertility and food production, Global Biogeochem. Cy., 28, 743–756, 2014.

Rode, M. and Suhr, U.: Uncertainties in selected river water quality data, Hydrol. Earth Syst. Sci., 11, 863–874, doi:10.5194/hess-11-863-2007, 2007.

Salmon-Monviola, J., Moreau, P., Benhamou, C., Durand, P., Merot, P., Oehler, F., and Gascuel-Odoux, C.: Effect of climate change and increased atmospheric CO_2 on hydrological and nitrogen cycling in an intensive agricultural headwater catchment in western France, Climatic Change, 120, 433–447, 2013.

Schindler, D. W., Hecky, R. E., Findlay, D. L., Stainton, M. P., Parker, B. R., Paterson, M. J., Beaty, K. G., Lyng, M., and Kasian, S. E. M.: Eutrophication of lakes cannot be controlled by reducing nitrogen input: Results of a 37-year whole-ecosystem experiment, P. Natl. Acad. Sci. USA, 105, 11254–11258, 2008.

Schoumans, O. F. and Chardon, W. J.: Phosphate saturation degree and accumulation of phosphate in various soil types in The Netherlands, Geoderma, 237, 325–335, 2015.

Serrano, T., Dupas, R., Upegui, E., Buscail, C., Grimaldi, C., and Viel, J.-F.: Geographical modeling of exposure risk to cyanobacteria for epidemiological purposes, Environ. Int., 81, 18–25, 2015.

Sharpley, A. N., Kleinman, P. J., Heathwaite, A. L., Gburek, W. J., Folmar, G. J., and Schmidt, J. P.: Phosphorus loss from an agricultural watershed as a function of storm size, J. Environ. Qual., 37, 362–368, 2008.

Siwek, J., Siwek, J. P., and Zelazny, M.: Environmental and land use factors affecting phosphate hysteresis patterns of stream water during flood events (Carpathian Foothills, Poland), Hydrol. Process., 27, 3674–3684, 2013.

Turner, B. L. and Haygarth, P. M.: Biogeochemistry – Phosphorus solubilization in rewetted soils, Nature, 411, 258–258, 2001.

Vadas, P. A., Joern, B. C., and Moore, P. A.: Simulating soil phosphorus dynamics for a phosphorus loss quantification tool, J. Environ. Qual., 41, 1750–1757, 2012.

Wade, A. J., Whitehead, P. G., and Butterfield, D.: The Integrated Catchments model of Phosphorus dynamics (INCA-P), a new approach for multiple source assessment in heterogeneous river systems: model structure and equations, Hydrol. Earth Syst. Sci., 6, 583–606, doi:10.5194/hess-6-583-2002, 2002a.

Wade, A. J., Whitehead, P. G., Hornberger, G. E., and Snook, D.: On Modelling the flow controls on macrophytes and epiphyte dynamics in a lowland permeable catchment: the River Kennet, southern England, Sci. Total Environ., 282–283, 395–417, 2002b.

Wall, D. P., Jordan, P., Melland, A. R., Mellander, P. E., Mechan, S., and Shortle, G.: Forecasting the decline of excess soil phosphorus in agricultural catchments, Soil Use Manage., 29, 147–154, 2013.

Whitehead, P. and Young, P.: Water-quality in river systems – Monte-Carlo analysis, Water Resour. Res., 15, 451–459, 1979.

Whitehead, P. G. and Hornberger, G. M.: Modeling algal behavior in the river thames, Water Res., 18, 945–953, 1984.

15

Technical note: Stage and water width measurement of a mountain stream using a simple time-lapse camera

Pauline Leduc[1], **Peter Ashmore**[1], **and Darren Sjogren**[2]

[1]University of Western Ontario, Department of Geography, London, Ontario, Canada
[2]University of Calgary, Department of Geography, Calgary, Alberta, Canada

Correspondence: Pauline Leduc (pleduc3@uwo.ca)

Abstract. Remote sensing applied to river monitoring adds complementary information useful for understanding the system behaviour. In this paper, we present a method for visual stage gauging and water surface width measurement using a ground-based time-lapse camera and a fully automatic image analysis algorithm for flow monitoring at a river cross section of a steep, bouldery channel. The remote stage measurement was coupled with a water level logger (pressure transducer) on site and shows that the image-based method gives a reliable estimate of the water height variation and daily flow record when validated against the pressure transducer ($R = 0.91$). From the remotely sensed pictures, we also extracted the water width and show that it is possible to correlate water surface width and stage. The images also provide valuable ancillary information for interpreting and understanding flow hydraulics and site weather conditions. This image-based gauging method is a reliable, informative and inexpensive alternative or adjunct to conventional stage measurement especially for remote sites.

1 Introduction

Conventionally, river discharge is gauged using continuous measurement of stage (typically, at temporary sites, using a pressure transducer and data logger) that is converted to continuous discharge data using a stage–discharge curve established for the site. In some cases, installation of a stage recorder is problematic and in complex flows interpretation of stage fluctuations may be uncertain. These conditions may arise, for example, in steep, bouldery or rock bed channels. Image-based measurements may provide equivalent data to the pressure transducer record while giving additional information such as water width, state of flow, water surface configuration and indications of flow hydraulics. For larger rivers, satellite or aerial images may provide useful streamgauging data (e.g. Smith et al., 1996; Gleason and Smith, 2014), but for small streams and very high frequency (minutes) over extended periods, satellite and airborne platforms do not provide sufficient resolution or temporal frequency (Gleason et al., 2015). Ground-based remote sensing provides a wide range of data for many applications for monitoring river flow and morphology especially in smaller channels or where high-frequency data are needed for extended time periods (Bertoldi et al., 2012; Williams et al., 2011; Javernick et al., 2014; Gleason et al., 2015).

Remote sensing based on photogrammetry technology provides an efficient topographic tool and access to topography and hydraulic characteristics (Javernick et al., 2014). However, the large amount of data needed to generate topography makes it difficult to apply on small streams with a high sampling frequency (Gleason et al., 2015).

In relation to flow characteristics, water surface width can be measured from ground-based cameras (Ashmore and Sauks, 2006; Gleason et al., 2015) and correlated with discharge to establish a width–discharge curve in some types of rivers, and local flow velocity has been measured using particle image velocimetry (Creutin et al., 2003; Hauet et al., 2008; Tsubaki et al., 2011; MacVicar et al., 2012; Ran et al., 2016; Stumpf et al., 2016). Direct measurement of stage is less well developed, although Young et al. (2015) obtained a water level and discharge record using manual image processing on a small, steep channel using inexpensive ground-based cameras combined with channel geometry and roughness assumptions. A more automated method that does not require manual image classification and channel geometry and hydraulic assumptions would be useful.

Methods for automated image selection and measurement are also needed in order to process 10^3 or 10^4 images that may come from high-frequency time-lapse red–green–blue (RGB) imagery (Gleason et al., 2015). Here, we test a simple time-lapse camera system for directly measuring stage and water surface width using image classification, and develop automated image selection and classification processes that retain a much larger proportion of the images than the process described by Gleason et al. (2015). We apply the method to monitor flow in a steep, bouldery, glacier-fed mountain stream which presents challenges for any form of flow gauging.

2 Measurement method

2.1 Site

The study site is located on a small, steep, bouldery reach of a stream, approximately 100 m downstream of the outlet from the small pro-glacial lake of the Dome Glacier, in Jasper National Park, Alberta, Canada (Fig. 1).

The site elevation is about 1800 m above sea level and the upstream drainage area is primarily the subglacial drainage of the Dome Glacier which is about 3 km in area. The stream is a left bank tributary of the Sunwapta River, and the primary purpose of the study is to better estimate the total discharge of a braided section of the river downstream of the Sunwapta–Dome Glacier tributary confluence by directly monitoring the Dome Glacier streamflow during the summer meltwater flow season. The Water Survey of Canada gauging station on the Sunwapta River, at the outlet of the pro-glacial Sunwapta Lake, a few hundred metres upstream of the confluence with the Dome Glacier stream (Fig. 1), provides detailed discharge records for the braided reach but does not account for the tributary contribution. Ashmore and Sauks (2006) measured water surface width from oblique time-lapse images on the braided reach of Sunwapta River downstream of this tributary and established a relationship with discharge at the Water Survey of Canada gauging station using a small number of gauging measurements in the braided reach. However, continuous measurement of the discharge of the Dome Glacier stream has not previously been used for monitoring this narrow, steep tributary to directly measure its contribution, daily flow variation and timing of daily peak flow relative to the Sunwapta River discharge. The stream flow is mainly controlled by snow and glacier melt in summer producing a regular diurnal hydrograph with long-period changes due to average air temperature and synoptic weather conditions in the summer. A straight, single thread reach of the channel was chosen for the gauging location.

2.2 Field setting

The main objective of the study was to use ground-based remote sensing to measure the flow characteristics (flow stage and water surface width) and peak flow periods in the daily flow cycle in this pro-glacial stream and assess the flow and timing of peaks relative to the flow of Sunwapta River. Standard pressure transducer measurement of stage is possible at this site (and is used here for comparison with image-based measurements) but we are interested in testing whether reliable image-based measurements are possible to complement or replace stage-only data with water level, water surface width and state of flow information from remote camera monitoring. A Reconyx HyperFire camera was set on the right side of the reach (Fig. 2), clamped to a pole hammered into the rocky ground, facing an almost vertical face of a large boulder on the opposite bank of the river. The entire stream width is visible in the pictures (Fig. 3).

Pictures were taken every 15 min during daylight (typically 06:00–22:00 MST (UTC − 6) at this location in the summer, and the daily peak was usually 16:00–19:00 MST) which corresponds to the sampling interval and timing of the gauging station on the Sunwapta River. During the study period, from 13 June to 22 September 2015, 7284 pictures were taken, saved on an SD card and downloaded at the end of the study period. Two stage boards were installed – one on each side of the stream (Fig. 3). On the left bank, a water level logger was installed in a vertical pipe in the stream bed next to the stage board with stage recorded at 15 min intervals throughout the study period for comparison with the image-based stage data. Level data were compensated for atmospheric pressure. The boulder has an almost vertical surface facing the camera, and it was calibrated for stage measurement in addition to the stage boards.

In the rest of the paper, the phrase "transducer data set" and the notation $H_{\text{transducer}}$ correspond to the stage coming from the pressure transducer, and the phrase "camera data set" and notation H_{camera} correspond to the stage coming from the image analysis.

3 Stage and water width measurement

3.1 Picture quality

A goal of this method is to minimize any manual treatment of the images to select an analysis set of images and to estimate water stage and water width from those images. Consequently, a screening treatment was applied to remove unusable pictures prior to analysis. The initial RGB picture size is 2048 × 1536 pixels, which was saved in .jpg format and converted into grey scale. Dark pictures corresponding to twilight and night were identified using a very low standard deviation of the grey intensity of the picture and automatically deleted from the data set. Over the summer, weather condi-

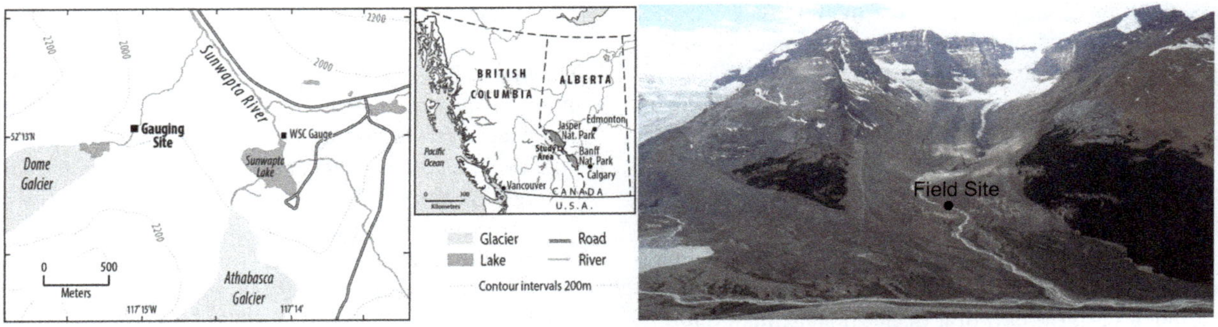

Figure 1. Map of the site location (left) and view of the Dome stream (right); image attribution – J. T. Gardner.

Figure 2. Camera setup. The camera is set on the right bank of the stream, a few metres above the water level. The camera is clamped to a pole hammered into the ground.

Figure 3. Reconyx raw picture showing the installation of stage boards, pressure transducer and boulder used for gauging (centre of picture). Black lines represent the three different tested profiles.

tions also negatively impacted image quality. During rain or snow episodes, the images are blurrier and water drops on the camera block the view. Snow cover on the ground is a lighter shade than the water, the opposite of the normal weather conditions. In the late afternoon and evening, the Sun shines into the camera and induces two kinds of issues. First, sunlight directly hits the camera and the picture is almost entirely saturated. Second, the water surface is saturated by reflections and the boulder facing the camera creates a large dark shadow on the water surface. Each of these issues interferes with the image processing and had to be compensated for.

For all the previous reasons (rain, drops on the lens, snow, Sun effects), and as mentioned in Gleason et al. (2015) and Young et al. (2015), pictures have to be classified. We developed automatic classification processes and set two different output options: (1) the image is removed from the data set or (2) the image is retained but different processes of classification and water surface detection are used for particular conditions (see Sect. 3.2).

3.2 Picture classification

Image quality issues arise throughout the process of water detection and width estimation: poor weather conditions (Fig. 4a), shadows (Fig. 4b), emerged rocks in the stream (Fig. 4c) and light snow cover on the edge on the stream (Fig. 4d).

To deal with those four issues, four different selection tests based on different target zones in the images were applied before or during the stage and water width measurements. All four tests were applied to the grey-scale pictures and in the following descriptions of standard deviations or averages refer to calculations on pixel intensity.

The first test was made to remove pictures taken under bad weather conditions (heavy snow or rain episodes, and sunlight directly into the camera) and it is based on the rocky zone target on the left of the picture (Test 1, area 1 in Fig. 5).

The standard deviation of this surface is high due to the apparent roughness coming from the rocky surface. Under adverse weather conditions, that area is smoother and the standard deviation drops (snow cover with normal weather con-

Figure 4. Images illustrating the main issues affecting image quality and measurements during the stage and width calculation: **(a)** heavy weather conditions, **(b)** boulder shadow and intense water reflection, **(c)** emerged rocks on the main channel at low flows and **(d)** light snow cover on the edge of the stream.

Figure 5. The eight different targets zones used in the picture classification process.

dition is not included in that test because, even with snow, the roughness from the block elevation makes the standard deviation high enough) and the picture was removed from the data sets. Of the 6717 pictures, 12 % were removed after this test (Fig. 6).

During the water surface detection, a significant boulder shadow sometimes interfered with the detection. Figure 5 shows target zones 2 and 3 that were used to detect the boulder shadow (Test 2), combining a height reflectance of the water (on zone 2) and a dark zone on the boulder face (on zone 3). A different water detection threshold was applied for those pictures (see Sect. 3.3).

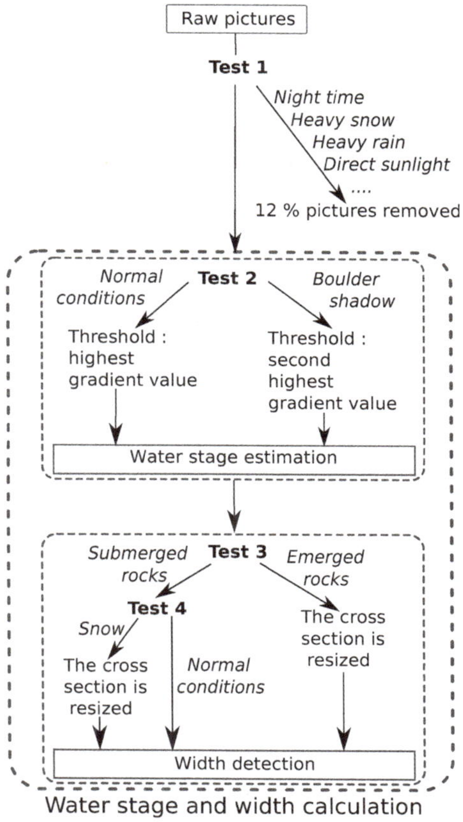

Figure 6. The picture classification process: four different tests are used to remove heavy weather conditions or night pictures from the data set (Test 1), to detect pictures with an important boulder shadow (Test 2), emerged rocks on the main channel at low flow (Test 3) or a light snow cover on the edge of the stream (Test 4).

Two main issues interfering with the water edge detection arise: the rocky bottom of the stream (Test 3) and snow on the river banks (Test 4). In Fig. 5, target zone 4 is used as a water reference, because this part of the stream always had water even at a very low stage. Target zones 5 and 6 are located where rocks emerge at low flows. The mean value of both target zones is compared to the mean value of the reference zone; the target value less than half the reference zone corresponds to submerged rocks. The threshold based on half the value of the reference zone was set empirically after going through a substantial portion of the data set. As described in Sect. 3.6, the width estimate is based on cross sections and the profile with emerged rocks is resized based on the rock emergence/submergence.

The snow cover on the edge of the stream is detected using Test 4 and target zones 7 and 8. The test is based on the lighter colour of the snow, mean values on zones 7 and 8 are compared to the mean value of the picture, and snow cover corresponds to brighter values on the target zone. As for the case of emerged rocks, the cross section used for the width detection is resized.

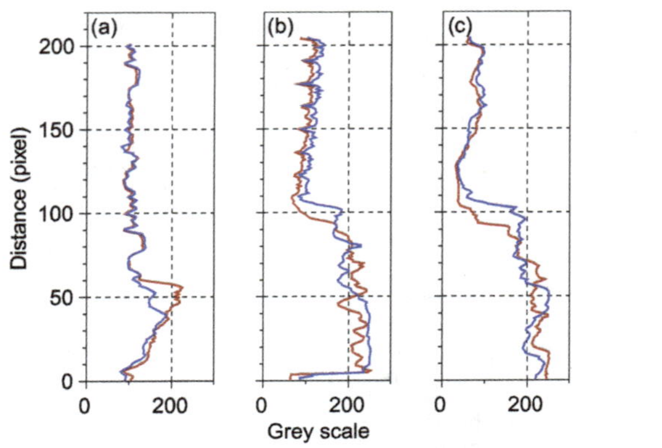

Figure 7. Grey-scale profiles for water surface measurement. **(a)** The first stage board on the right side of the stream, **(b)** the second stage board on the left side of the river and **(c)** on-boulder vertical surface. The red and blue lines show two different water elevations days apart.

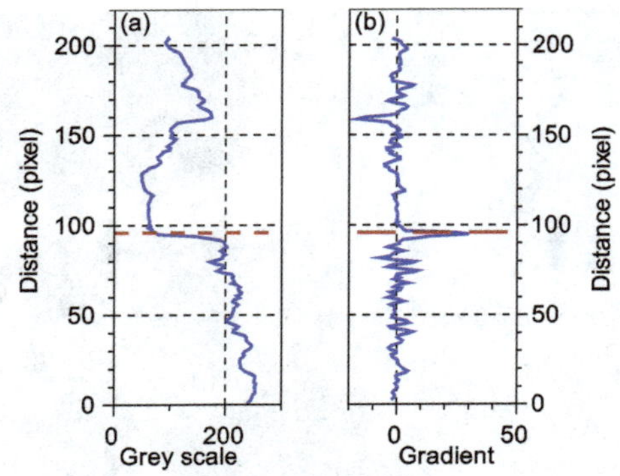

Figure 8. The boulder gauging station profile. The dashed lines represent the water surface. Water is located below those lines. Panel **(a)** indicates the grey-scale profile and **(b)** the gradient of the profile; the calculation step is 2 pixels. The gradient plot makes the inflection point and the water level detection easier.

The four tests are summarized in Table 1.

3.3 Water level

Stage was measured by detecting the water surface line on the images. Instead of a global, manual approach using edge detection (Young et al., 2015), we based the analysis on local site conditions. Black lines in Fig. 3 on both stage boards and the vertical surface of the boulder represent each location where grey-scale profiles from the images were extracted to detect the transition from water to stage board or boulder surface in the image (Fig. 7) and thus locate the water surface in image space.

The water and boulder/stage board transition signal is clear for each water level location. On both stage boards, the image signal is smoother on the water than on the board itself (Fig. 7). However, the transition between rock and water is more obvious on the boulder site (Fig. 7c) than on the left bank stage board (Fig. 7b). On the board located on the right bank of the stream, the transition between the board and the water is not as clear, and the flow stage rise is more difficult to detect. Furthermore, at low stage, the bottom of the scale emerged above the water and the measurement was then impossible. Consequently, this stage board was removed from the analysis. Without the boulder, we would have used the remaining stage board as the water stage estimation but in this particular site the stage board on the far side is used as an independent visual check because the boulder surface gave a clearer signal. We mainly used the boulder station to estimate the flow depth but the boulder is not a requirement for the method and any natural or artificial almost-vertical surface located on the stream could be used.

On the boulder, the water transition corresponds to an obvious inflection point in the image intensity (Fig. 8a) and a local peak in the gradient of the smoothness profile (Fig. 8b).

The inflection point is detected using two conditions. The first, on the gradient profile, was used to pick high gradient values. The second condition was based on a grey shade threshold so that only the lowest values in the grey-scale profile are considered because higher values represent the rock becoming darker when it is wet. Using this combined method, the water line position in pixel coordinates can be automatically detected for each picture. In pictures with the boulder shadow issue (see Sect. 3.2), the second highest gradient point is considered instead of the first.

3.4 Water depth calibration

Given that the boulder surface is almost vertical, and roughly perpendicular to the axis of the camera lens, we assume a linear relationship between the stage and the water surface position in pixel coordinates: Eq. (1), where H_{m} is the stage in metres and d_{pixel} the water surface position on the picture in pixels.

$$H_{\mathrm{m}} = a.d_{\mathrm{pixel}} + b \tag{1}$$

The slope a of Eq. (1) is given by the millimetres–pixel ($\mathrm{mm\,px^{-1}}$) relationship extracted from the pictures. The camera is fixed; therefore, the value is consistent through the entire picture set. Using a measuring device (large ruler) on the boulder surface, we get $a = 5.5 \times 10^{-3}\,\mathrm{m\,px^{-1}}$.

To determine the intercept (b) of Eq. (1), which is the ground reference of the flow stage, we used part of the stage logger data set, taking 100 values randomly, to extract the intercept $b = -1.25\mathrm{m}$. This gives a local datum for the boulder stage. We also calibrated the stage board located on the side of the boulder on the far side of the picture. This water depth calibration is relative to this gauging station and particular to the site.

Table 1. The four different picture classification tests. The standard deviation ranges from 0 to 255. Thresholds are set using both particular and normal condition pictures. The different areas have been chosen to avoid parameter distribution overlap between normal and particular conditions (see the Supplement).

Test number	Target area (Fig. 4)	Picture issue	Conditions (the test is failed if the condition is fulfilled)	Action if test failed
1	1	Nighttime, heavy rain, snow...	Standard deviation lower than 20	Removed from the data set
2	2–3	Boulder shadow	Area 2: mean value higher than 235 – indicating almost direct reflection on the water surface Area 3: mean value lower than 95 – indicating a dark area on the boulder surface	The water detection is based on the second highest gradient point instead of the first
3	4–5–6	Emerged rocks on the main stream	Area 4 is the water colour reference value Areas 5 and 6: the mean value is less than half of the reference value	The width calculation profile is resized
4	7–8	Light snow cover on the edge of the stream	Areas 7 and 8: the mean value is higher than the average of the entire picture and the standard deviation of both area is less than 50 (indicating a smooth area)	The width calculation profile is resized

3.5 Stage validation

Figure 9a shows the comparison between the transducer data set ($H_{\text{transducer}}$) and the camera data set (H_{camera}). The stage prediction from picture analysis is a good estimation of the transducer water level ($R = 0.91$). The water measurement using the stage board located on the side of the boulder with a proper calibration has a lower correlation coefficient ($R = 0.71$). The mean value of the difference of the transducer data set and the camera data set is $\mu = 0.00$ m and the standard deviation is $\sigma = 0.02$ m. Considering a normal distribution (Fig. 9b), the 95 % confidence interval on the error estimation is $[-0.04; 0.04]$ m (the error estimation for the stage board measurement is $[-0.06; 0.06$ m$]$). Pictures corresponding to the cluster of outliers have been manually checked. Those points correspond to pictures where the water surface is correctly detected but corresponds to waves at the gauging station or at the pressure transducer.

At very low discharge, boulder clusters emerge near the left bank creating pools and small channels. This channel configuration creates a pond at the water level logger at very low discharge and probably disconnects the water stage measured using the image analysis from that of the pressure transducer.

Based on this result, the stage measured from image analysis gives a good estimation of the water stage. The high- and low-frequency variations (i.e. daily or monthly variations) on the transducer signal are well reproduced by the camera data set (Fig. 10). The daily snow/ice melt hydrograph which is characteristic of the site, with consistent times of low and high flow each day in the absence of rainstorms, is also shown in Fig. 10.

While the results show that the image-based time-lapse method works well, there are some errors that could be reduced. The hypothesis for our stage measurement and the water depth calibration equation is the constant mm px^{-1} relationship over the boulder surface. The underlying assumptions are that the vertical surface is flat and the camera distortion does not induce a large variation. Realistically, as the boulder is a natural rock, the vertical face is not exactly vertical and we are not able to estimate the distortion variation. The probable inconstant mm px^{-1} relationship may induce the tilt on the scatter plot (Fig. 9a and c). The comparison with the transducer data set shows that at high discharges the camera data set underestimates the transducer data set and at low discharges the camera data set overestimates the transducer data set. The mm px^{-1} variations could induce the slightly curved shape of the scatter plot (Fig. 9), but only a better camera resolution (or image scale) and a more precise geometry between camera and vertical calibration surface would improve the mm px^{-1} relationship.

3.6 Water width measurement

Using the time-lapse images we also estimated the water surface width. As we did for the stage, width was measured by detecting the threshold between the river and the rocks on both banks. During the field work, flow width was also calibrated on one cross section. For picture analysis, considering the rocks masking the view of the water surface and standing waves in the flow, the measurement cross section was moved about 2 m downstream, keeping the same angle across the channel as the calibrated profile (Fig. 11).

On both sides of the stream, some boulders appear at low discharges that are too large to be mobilized by daily high

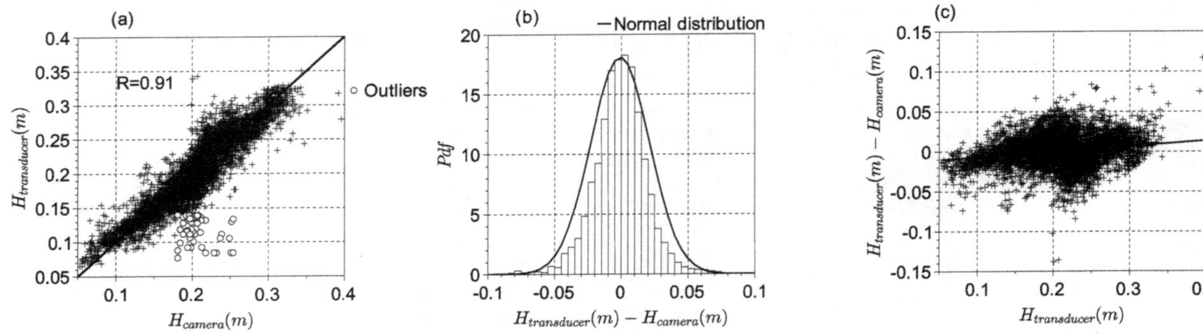

Figure 9. a) Comparison of the transducer data set ($H_{\text{transducer}}$) and the camera data set (H_{camera}) on the scatter plot; the cluster of outliers point represented with open symbols has been manually checked and corresponds to waves at the pressure transducer. **(b)** The error distribution; the normal distribution is set to a mean value of 0.00 m and a standard deviation of 0.02 m. **(c)** The residual plot of $H_{\text{transducer}}-H_{\text{camera}}$ with regard to $H_{\text{transducer}}$; the red line represents the linear regression, showing a tilt on the water depth estimation. At high discharges, the camera data set underestimates the transducer data set and at low discharges the camera data set overestimates the transducer data set.

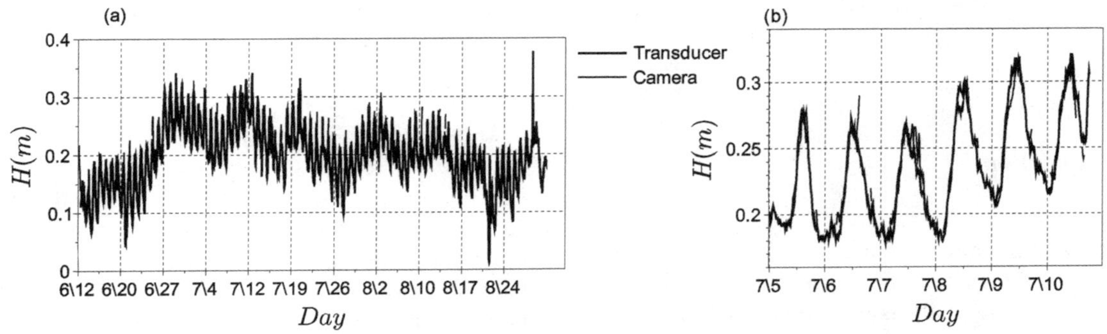

Figure 10. Summer 2015 stage time series. The black line represents the transducer data set ($H_{\text{transducer}}$) and the red line is the camera data set (H_{camera}). The estimated error on the camera stage relative to pressure transducer is around 3 cm. Panel **(a)** shows data from mid-June to the end of July 2015. The camera data set fits the transducer data set. The daily trend as well as the general monthly trend are reproduced **(b)** from 5 to 10 July 2015. At hourly resolution, the trend of the camera stage follows the transducer data closely. However, on the rising stage, the camera data set underestimates the transducer water stage, and on the falling stage, the picture data set slightly overestimates the transducer stage.

flows. On both rocky areas, a test was done to detect if the rocks had emerged (Tests 3 and 4; Fig. 6). If they had, the interrogation area was changed accordingly. On the profile, and as with the water stage, the highest gradient of the grey-scale plot profile was used to detect water edges.

3.7 Water width calibration

A measuring tape was extended across the entire Dome stream. The distance across the channel was measured in 0.5 m intervals and the image distance in pixels was converted to metres (Fig. 12).

The conversion from distance in pixels to distance in metres is done using Eq. (2), where D_{pixel} is the distance in pixels extracted from the picture analysis and D_{m} the distance in metres. The measuring tape was not perfectly straight due to the inherent limitations of field work such as the flow conditions and channel structure; therefore, the conversion into metres may be slightly inconsistent, which could induce the shift in Fig. 10 around $D_{\text{pixel}} = 200$.

$$D_{\text{m}} = 0.05 \times (D_{\text{pixel}} - 10)^{0.95} + 0.45 \quad (2)$$

The width measurement faces two principal issues: the water edge detection and the calibration. The width measurement is tightly linked to the boulders on the side of the stream. As the stream widens, the boulders at the channel edge are submerged and the grey-scale shift at the edge of the water is less sharp. Inaccurate detection cannot be corrected because of a lack of field validation data for the water width.

The calibration is very sensitive to the camera position. Additional information on camera angle and geometry would increase the calibration accuracy and improve the width measurement. Furthermore, moving the cross section a small distance away from the calibrated cross section induced some

Figure 11. The width measurement at low **(a)** and high **(b)** discharges. The left line is the water level at the boulder station. The dashed line is the cross section and the right ticks are the detected flow edges.

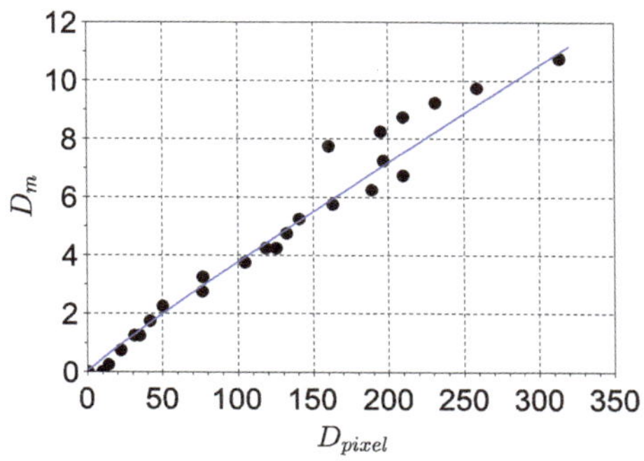

Figure 12. The pixel–metre conversion. D_{pixel} is the flow width in pixels and D_{m} is the flow width in metres. The rating curve was established using a measuring tape across the flow width. The data set has been fitted using Eq. (2).

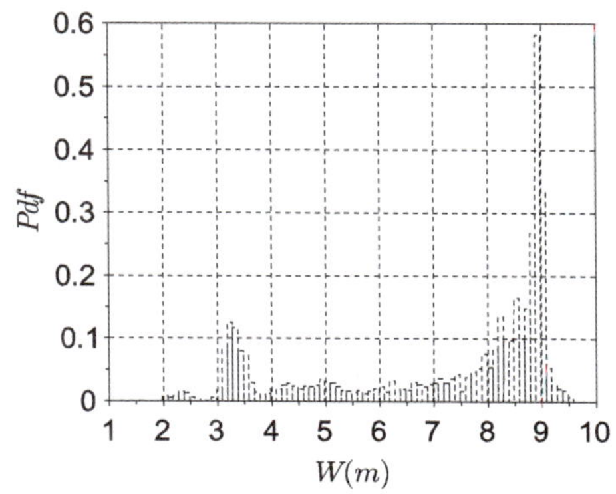

Figure 13. The distribution of measured width. The median class ranging from 4 to 7 m is underrepresented because boulder clusters on each side of the stream make the transition between narrow flows and wide flows fast and nonlinear.

error on the conversion length in metres from length in pixels. Nevertheless, with some refinement, as the distortion of the picture (due to the angle, camera setting which defines the pixel–metre conversion) is a monotone function, there is no major effect on the relative width variations.

3.8 The width observations

At very low discharges, the stream bed is covered by large boulders, clearly seen in the pictures, and rocks are fully submerged at high flows, creating large surface waves. The transition between low and high flows creates secondary channels, and we chose to only consider the main channel and not the side channels at low and medium flows. This induces an underestimation of the width at low discharges. The flow widening is also strongly impacted by those boulder clusters. At low discharge, the main channel is contained in the centre of the bed, and water stage has to be quite high to be over both clusters. The transition between wide and narrow channels is fast, and therefore intermediate widths (between 4 and 7 m wide) are underrepresented in the width data (Fig. 13). The image information reveals these aspects of the hydraulics of the channel (and that affect stage changes) that would not be known with stage data alone.

3.9 The width–stage relationship

In braided channel studies, the wetted surface measurement has been used as a substitute for stage to estimate the discharge (Smith et al., 1996; Ashmore and Sauks, 2006; Gleason et al., 2015). Previous studies have shown the correlation between the wetted surface and the discharge with an exponent from 0.5 (Smith et al., 1996) to 1 (Ashmore and Sauks, 2006). The width response to discharge change is much higher in these braided channels than in many other streams and this gives the potential for using width in addition to, or instead of, stage changes as the primary variable for estimating discharge. In the Dome stream case, although it is not a braided stream, the relatively shallow cross section also gives significant widening of flow with increasing stage and the positive trend is clear with an exponent close to 0.6 (Fig. 14). However, the trend is not linear and data scatter is quite large because of the irregular geometry of

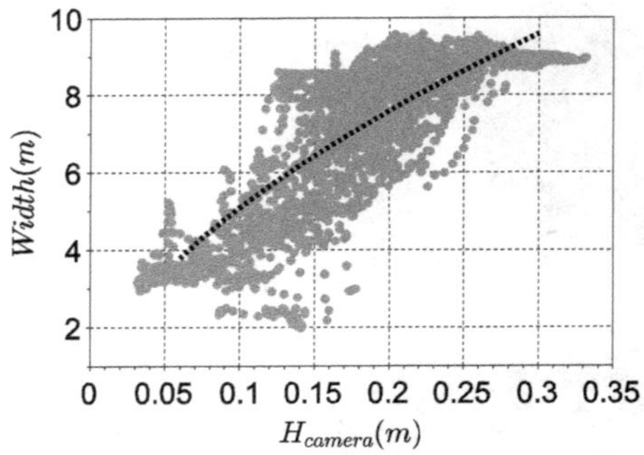

Figure 14. The water surface width as a function of the water height. The stage is taken from the camera data set. The flow width is taken from the width detection and Eq. (2). The trend line equation is $W = 19.19H_{\text{camera}}^{0.58}$.

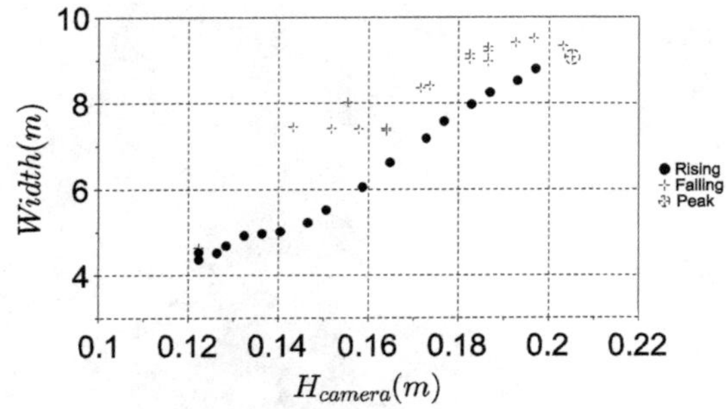

Figure 15. The water surface width as a function of water stage, for falling flows (+), rising flows (•) and the discharge peak (⊕). The graph shows that for a given water level (H_{camera}) the water surface is wider with falling flow than with rising flow conditions. The end of the falling flows usually happens at night; therefore, the data are missing.

the cross section and the bouldery channel edges (especially at low discharge). Nevertheless, the width estimation could be a reliable approximation to the stage measurement. Even if not used directly as a discharge surrogate, the width data give additional information on the hydraulic geometry of the channel that would be difficult to predict theoretically for this type of channel.

The width data also reveal some interesting hysteresis in the flow hydraulics. For 67 % of the 81 daily flow peaks on the Dome stream for which there are data, the mean value of the stream width over the water stage ranges from strictly higher than one-half $H_{\max}$ and $H_{\max}$, $H_{\max}$ being the daily maximum water stage, and is higher on the falling limb of the daily hydrograph than on the rising limb. The width increase ranges from 0.5 to 26 %, with a mean value of 13.9 % and a median value of 9.1 %.

This produces an obvious hysteresis loop in the width–stage plot (Fig. 15), as can often be found in the stage–discharge relationship (Petersen-Øverleir, 2006). The width data derived from image analysis add further information about the channel hydraulics resulting from the complexity of flow associated with the macro-roughness elements in this type of flow.

3.10 Practical aspects of implementing the method

The method we are presenting here requires some field work and some post-processing of the pictures.

The field work, including the site choice, the installation of the equipment and the on-site calibration, can be done in a few hours using two people.

Our method is strongly dependent on the picture classification to retain those that are most reliable for water stage and water width detection. The picture classification is based on interest zones linked to environmental and lighting conditions (night, rain/snow, shadows, water reflection and snow cover), and eight different areas were necessary to classify pictures at our site. Each area has to be chosen related to a particular image issue. For example, to detect nighttime the entire picture can be used, while detection of exposed rocks in the stream channel requires specific small areas of the images. These characteristics are site specific and it takes some trials to identify appropriate criteria and thresholds. However, once established, the actual classification can be done in a few hours and can be used for the same site for extended time periods and repeated installations.

4 Discussion

The method described in this paper is similar to two recent studies proposed by Gleason et al. (2015) and Young et al. (2015), but it differs in general approach, field data and image selection, and processing. Young et al. (2015) assumed a V shape of their studied cross section, so that the edge coordinate is linearly related to the water stage. They estimate the water level from the water edge, without on-site validation data, and use a statistical estimation procedure combined with assumed channel geometry to derive water level changes. In the Dome stream case and with width measurement, the water stage and water width are not linearly related, which gives information on the stream cross section despite the lack of topographic survey and shows that assumptions of the type used by Young et al. (2015) would not be reliable in this case. It also points to the difficulty of reliably predicting flows using a standard resistance assumption in this type of channel. In addition, Young et al. (2015) use manual methods to identify water edges on all images. Gleason et al. (2015) focus on water area detection in a large braided channel and not on direct water stage measurements or on small, steep channels.

The environmental conditions (e.g. Sun position, fog, rain) are the main common difficulties that reduce the picture quality and therefore picture filtering as an important step in the process. Gleason et al. (2015) and Young et al. (2015)

identify similar issues but adopt different approaches. Young et al. (2015) use manual image selection in contrast to our automated selection procedures which make it possible to process a much larger image set and derive much higher-frequency data (15 min vs. 4 h). Gleason et al. (2015) adopt semi-automated image procedures which differ in detail from ours but their procedures result in either retention or rejection of images for measurement, whereas we derive alternative detection criteria (cross-section resizing, peak detection, etc.; see Table 1) for a subset of images rather than eliminating them completely from the data set. Consequently, we are able to retain much higher-frequency monitoring relative to Gleason et al. (2015). Field data on the site characteristics avoid having to make assumptions about the site, such as the cross-section shape (Young et al., 2015), or working without any ground data for validation (Gleason et al., 2015).

Improved image acquisition is the key component for improving remote sensing accuracy and time coverage. The use of inexpensive time-lapse cameras introduces some limitations that can be mitigated. A higher image resolution and a better camera position (reducing sunlight effects or improving the position relative to the calibration surface, for example) would improve the measurement accuracy for both the water stage and the channel width. These refinements are easy to implement and test.

Another obvious limitation is the restriction on daytime images. In the Dome case, night and twilight represent roughly one-third of the day in the summer meltwater period for which data are needed. Using a night-vision camera may extend the effective monitoring times, but we have not tested this. The limitation may be less significant if only certain flow information is needed rather than a 24 h continuous signal. Even without continuous data, useful information on channel hydraulics can also be obtained from this type of monitoring. These procedures and image processing steps may be changed to fit site characteristics or data needs. In this case, the method provided the necessary seasonal stage signal and timing of daily peaks needed for the study objective of comparison between the ungauged tributary and the main channel flow.

5 Conclusions

The results demonstrate the effectiveness of a simple measurement apparatus for flow stage and water surface width: a low-cost time-lapse camera and a few simple field measurements. Fully automatic image processing to select images and to detect the water level and edges makes it possible to process a large number of images to produce a long, high-temporal-resolution data set. It shows that reliable water stage and water width measurement can be measured at small (minutes) time steps over 3 months in this case. The estimated hydraulic parameters reliably reproduce the hourly, daily and monthly variations in flow of this pro-glacial river compared to pressure–transducer stage data. The low cost of the camera (approximately USD 600) and the very easy data collection make the image processing a powerful tool for this type of river monitoring especially on small headwater streams. Image analysis produced a larger variety of data and information than a simple water stage transducer alone can yield. Indeed, pictures provide visible data such as weather conditions (snow cover, freezing conditions, rain), water surface conditions (surface waves, eddies, jumps) and details of the flow hydraulics and geometry over the full range of discharge. Image analysis could also be extended to other hydraulic measurements such as the water slope. The method extends the available methods for inexpensive terrestrial remote sensing of river flow at high-frequency and extended time periods applicable especially to small channels with complex flow.

Competing interests. The authors declare that they have no conflict of interest.

Acknowledgements. This research was supported by a Natural Sciences and Engineering Research Council of Canada Discovery Grant to Peter Ashmore. We thank Sarah Peirce, Lara Middleton and Matilde Welber for field work assistance, and Tobi Gardner for help with the water level logger.

Edited by: Christian Stamm

References

Ashmore, P. and Sauks, E.: Prediction of discharge from water surface width in a braided river with implications for at-a-station hydraulic geometry, Water Resour. Res., 42, W03406, https://doi.org/10.1029/2005WR003993, 2006.

Bertoldi, W., Piégay, H., Buffin-Bélanger, T., Graham, D., and Rice, S.: Applications of Close-Range Imagery in River Research, in: Fluvial Remote Sensing for Science and Management, John Wiley & Sons, Ltd, 341–366, https://doi.org/10.1002/9781119940791.ch15, 2012.

Creutin, J., Muste, M., Bradley, A., Kim, S., and Kruger, A.: River gauging using PIV techniques: a proof of concept experiment on the Iowa River, J. Hydrol., 277, 182–194, https://doi.org/10.1016/S0022-1694(03)00081-7, 2003.

Gleason, C. J. and Smith, L. C.: Toward global mapping of river discharge using satellite images and at-many-stations hydraulic geometry, P. Natl. Acad. Sci. USA, 111, 4788–4791, https://doi.org/10.1073/pnas.1317606111, 2014.

Gleason, C. J., Smith, L. C., Finnegan, D. C., LeWinter, A. L., Pitcher, L. H., and Chu, V. W.: Technical Note: Semi-automated effective width extraction from time-lapse RGB imagery of a remote, braided Greenlandic river, Hydrol. Earth Syst. Sci., 19, 2963–2969, https://doi.org/10.5194/hess-19-2963-2015, 2015.

Hauet, A., Kruger, A., Krajewski, W. F., Bradley, A., Muste, M., Creutin, J.-D., and Wilson, M.: Experimental system for real-time discharge estimation using an image-based method, J. Hydrol. Eng., 13, 105–110, 2008.

Javernick, L., Brasington, J., and Caruso, B.: Modeling the topography of shallow braided rivers using Structure-from-Motion photogrammetry, Geomorphology, 213, 166–182, https://doi.org/10.1016/j.geomorph.2014.01.006, 2014.

MacVicar, B. J., Hauet, A., Bergeron, N., Tougne, L., and Ali, I.: River Monitoring with Ground-Based Videography, in: Fluvial Remote Sensing for Science and Management, John Wiley & Sons, Ltd, 367–383, https://doi.org/10.1002/9781119940791.ch16, 2012.

Petersen-Øverleir, A.: Modelling stage-discharge relationships affected by hysteresis using the Jones formula and nonlinear regression, Hydrolog. Sci. J., 51, 365–388, 2006.

Ran, Q.-H., Li, W., Liao, Q., Tang, H.-L., and Wang, M.-Y.: Application of an automated LSPIV system in a mountainous stream for continuous flood flow measurements, Hydrol. Process., 30, 3014–3029, https://doi.org/10.1002/hyp.10836, 2016.

Smith, L. C., Isacks, B. L., Bloom, A. L., and Murray, A. B.: Estimation of Discharge From Three Braided Rivers Using Synthetic Aperture Radar Satellite Imagery: Potential Application to Ungaged Basins, Water Resour. Res., 32, 2021–2034, https://doi.org/10.1029/96WR00752, 1996.

Stumpf, A., Augereau, E., Delacourt, C., and Bonnier, J.: Photogrammetric discharge monitoring of small tropical mountain rivers: A case study at Rivière des Pluies, Réunion Island, Water Resour. Res., 52, 4550–4570, https://doi.org/10.1002/2015WR018292, 2016.

Tsubaki, R., Fujita, I., and Tsutsumi, S.: Measurement of the flood discharge of a small-sized river using an existing digital video recording system, J. Hydro-Environ. Res., 5, 313–321, https://doi.org/10.1016/j.jher.2010.12.004, 2011.

Williams, R., Brasington, J., Vericat, D., Hicks, M., Labrosse, F., and Neal, M.: Chapter twenty-monitoring braided river change using terrestrial laser scanning and optical bathymetric mapping, Developments in Earth Surface Processes, 15, 507–532, 2011.

Young, D. S., Hart, J. K., and Martinez, K.: Image analysis techniques to estimate river discharge using time-lapse cameras in remote locations, Comput. Geosci., 76, 1–10, https://doi.org/10.1016/j.cageo.2014.11.008, 2015.

PERMISSIONS

All chapters in this book were first published in HESS, by Copernicus Publications; hereby published with permission under the Creative Commons Attribution License or equivalent. Every chapter published in this book has been scrutinized by our experts. Their significance has been extensively debated. The topics covered herein carry significant findings which will fuel the growth of the discipline. They may even be implemented as practical applications or may be referred to as a beginning point for another development.

The contributors of this book come from diverse backgrounds, making this book a truly international effort. This book will bring forth new frontiers with its revolutionizing research information and detailed analysis of the nascent developments around the world.

We would like to thank all the contributing authors for lending their expertise to make the book truly unique. They have played a crucial role in the development of this book. Without their invaluable contributions this book wouldn't have been possible. They have made vital efforts to compile up to date information on the varied aspects of this subject to make this book a valuable addition to the collection of many professionals and students.

This book was conceptualized with the vision of imparting up-to-date information and advanced data in this field. To ensure the same, a matchless editorial board was set up. Every individual on the board went through rigorous rounds of assessment to prove their worth. After which they invested a large part of their time researching and compiling the most relevant data for our readers.

The editorial board has been involved in producing this book since its inception. They have spent rigorous hours researching and exploring the diverse topics which have resulted in the successful publishing of this book. They have passed on their knowledge of decades through this book. To expedite this challenging task, the publisher supported the team at every step. A small team of assistant editors was also appointed to further simplify the editing procedure and attain best results for the readers.

Apart from the editorial board, the designing team has also invested a significant amount of their time in understanding the subject and creating the most relevant covers. They scrutinized every image to scout for the most suitable representation of the subject and create an appropriate cover for the book.

The publishing team has been an ardent support to the editorial, designing and production team. Their endless efforts to recruit the best for this project, has resulted in the accomplishment of this book. They are a veteran in the field of academics and their pool of knowledge is as vast as their experience in printing. Their expertise and guidance has proved useful at every step. Their uncompromising quality standards have made this book an exceptional effort. Their encouragement from time to time has been an inspiration for everyone.

The publisher and the editorial board hope that this book will prove to be a valuable piece of knowledge for researchers, students, practitioners and scholars across the globe.

LIST OF CONTRIBUTORS

Maik Renner and Axel Kleidon
Max-Planck-Institut für Biogeochemie, Jena, Germany

Theresa Blume
GFZ German Research Centre for Geosciences, Section Hydrology, Potsdam, Germany

Markus Weiler
Universität Freiburg, Hydrologie, Freiburg, Germany

Sibylle K. Hassler
GFZ German Research Centre for Geosciences, Section Hydrology, Potsdam, Germany
Karlsruhe Institute of Technology, Institute of Water and River Basin Management, Karlsruhe, Germany

Anke Hildebrandt
Max-Planck-Institut für Biogeochemie, Jena, Germany
Universität Jena, Ecological Modelling Group, Jena, Germany

Stanislaus J. Schymanski
ETH Zürich, Department of Environmental Systems Science, Zurich, Switzerland

Marcus Guderle
Max-Planck-Institut für Biogeochemie, Jena, Germany
Universität Jena, Ecological Modelling Group, Jena, Germany
Technische Universität München, Chair for Terrestrial Ecology, Department of Ecology and Ecosystemmanagement, Munich, Germany

D. J. Peres and A. Cancelliere
Department of Civil Engineering and Architecture, University of Catania, Catania, Italy

R. Hübner, K. Heller and A. Kleber
Institute of Geography, Dresden University of Technology, Helmholtzstr. 10, 01069 Dresden, Germany

T. Günther
Leibniz Institute for Applied Geophysics (LIAG), Stilleweg 2, 30655 Hanover, Germany

Peter A. Troch
University of Arizona, Tucson, AZ, USA

Takeo Yoshida
University of Arizona, Tucson, AZ, USA
National Agricultural and Food Research Organization, Tsukuba, Ibaraki, Japan

A. I. Gevaert
Hydrology and Quantitative Water Management Group, Wageningen University, Wageningen, the Netherlands
Biosphere 2 - College of Science, University of Arizona, Tucson, Arizona, USA
Department of Earth Sciences, VU University, Amsterdam, the Netherlands

A. J. Teuling and R. Uijlenhoet
Hydrology and Quantitative Water Management Group, Wageningen University, Wageningen, the Netherlands

S. B. DeLong
Biosphere 2 - College of Science, University of Arizona, Tucson, Arizona, USA
US Geological Survey, Menlo Park, California, USA

L. A. Pangle
Biosphere 2 - College of Science, University of Arizona, Tucson, Arizona, USA

T. E. Huxman
Biosphere 2 - College of Science, University of Arizona, Tucson, Arizona, USA
Department of Ecology and Evolutionary Biology, University of California, Irvine, California, USA

D. D. Breshears
Biosphere 2 - College of Science, University of Arizona, Tucson, Arizona, USA
School of Natural Resources and the Environment, University of Arizona, Tucson, Arizona, USA

J. Chorover
Biosphere 2 - College of Science, University of Arizona, Tucson, Arizona, USA
Department of Soil, Water and Environmental Science, University of Arizona, Tucson, Arizona, USA

J. D. Pelletier
Biosphere 2 - College of Science, University of Arizona, Tucson, Arizona, USA
Department of Geosciences, University of Arizona, Tucson, Arizona, USA

S. R. Saleska
Biosphere 2 - College of Science, University of Arizona, Tucson, Arizona, USA
Department of Ecology and Evolutionary Biology, University of Arizona, Tucson, Arizona, USA

X. Zeng
Biosphere 2 - College of Science, University of Arizona, Tucson, Arizona, USA
Department of Atmospheric Sciences, University of Arizona, Tucson, Arizona, USA

P. A. Troch
Biosphere 2 - College of Science, University of Arizona, Tucson, Arizona, USA
Department of Hydrology and Water Resources, University of Arizona, Tucson, Arizona, USA

Claudio Paniconi
Institut National de la Recherche Scientifique, Centre Eau Terre Environnement (INRS-ETE), Université du Québec, Quebec City, Canada

Carlotta Scudeler
Institut National de la Recherche Scientifique, Centre Eau Terre Environnement (INRS-ETE), Université du Québec, Quebec City, Canada
Department of Mathematics, University of Padua, Padua, Italy

Mario Putti
Department of Mathematics, University of Padua, Padua, Italy

Luke Pangle
Department of Geosciences, Georgia State University, Atlanta, USA

Damiano Pasetto
Laboratory of Ecohydrology, École Polytechnique Fédérale de Lausanne, Lausanne, Switzerland

Guo-Yue Niu and Peter Troch
Biosphere 2 - Earth Science, University of Arizona, Tucson, USA
Department of Hydrology and Water Resources, University of Arizona, Tucson, USA

Till Volkmann
Department of Hydrology and Water Resources, University of Arizona, Tucson, USA

W. Shao, T. A. Bogaard and M. Bakker
Water Resources Section, Faculty of Civil Engineering and Geosciences, Delft University of Technology, Stevinweg 1, 2600 GA Delft, the Netherlands

R. Greco
Dipartimento di Ingegneria Civile Design Edilizia e Ambiente, Seconda università di Napoli, via Roma 29, 81031 Aversa (CE), Italy

Shabnam Saffarpour, Andrew W. Western and Russell Adams
Department of Infrastructure Engineering, The University of Melbourne, Parkville, 3010, Australia

Jeffrey J. McDonnell
Global Institute for Water Security, University of Saskatchewan, Saskatoon, SK S7N 3H5, Canada
School of Geosciences, University of Aberdeen, Aberdeen, UK

C. P. Ghimire and M. W. Lubczynski
Faculty of Geo-Information Science and Earth Observation (ITC), University of Twente, Enschede, the Netherlands

L. A. Bruijnzeel
Critical Zone Hydrology Group, Faculty of Earth and Life Sciences, VU University, Amsterdam, the Netherlands

M. Bonell
The Centre for Water Law, Policy and Science under the auspices of Unesco, The Peters Building, University of Dundee, Dundee, DD14HN, Scotland, UK

G. Capparelli and P. Versace
Dipartimento di Elettronica, Informatica e Sistemistica, Università della Calabria, Via P. Bucci, 41C, Arcavacata di Rende, CS, Italy

Lukáš Vlček and Kristýna Falátková
Department of Physical Geography and Geoecology, Charles University, Prague, Czech Republic

Philipp Schneider
Department of Geography, University of Zurich, Zurich, Switzerland

D. Penna, M. Borga and P. Tarolli
Department of Land, Environment, Agriculture and Forestry, University of Padova, Agripolis, viale dell'Università 16, 35020 Legnaro (PD), Italy
Dept. of Environmental Systems Science, Swiss Federal Institute of Technology (ETH), Universitätstrasse 16, 8092, Zurich, Switzerland

G. T. Aronica and G. Brigandì
Department of Civil Engineering, University of Messina, Contrada di Dio, Villaggio S. Agata, 98166 Messina, Italy

Index

CPSIA information can be obtained
at www.ICGtesting.com
Printed in the USA
LVHW060158060222
710369LV00002B/55

9 781639 873036